WITHDRAWN
FAIRFIELD UNIVERSITY
LIBRARY

Methods in Enzymology

Volume 215
PLATELETS: RECEPTORS, ADHESION, SECRETION

PART B

METHODS IN ENZYMOLOGY

EDITORS-IN-CHIEF

John N. Abelson   Melvin I. Simon

DIVISION OF BIOLOGY
CALIFORNIA INSTITUTE OF TECHNOLOGY
PASADENA, CALIFORNIA

FOUNDING EDITORS

Sidney P. Colowick and Nathan O. Kaplan

Methods in Enzymology

Volume 215

Platelets: Receptors, Adhesion, Secretion

Part B

EDITED BY

Jacek J. Hawiger

DEPARTMENT OF MICROBIOLOGY AND IMMUNOLOGY
VANDERBILT UNIVERSITY SCHOOL OF MEDICINE
NASHVILLE, TENNESSEE

ACADEMIC PRESS, INC.
Harcourt Brace Jovanovich, Publishers
San Diego   New York   Boston
London   Sydney   Tokyo   Toronto

This book is printed on acid-free paper. ∞

Copyright © 1992 by ACADEMIC PRESS, INC.
All Rights Reserved.
No part of this publication may be reproduced or transmitted in any form or by any means, electronic or mechanical, including photocopy, recording, or any information storage and retrieval system, without permission in writing from the publisher.

Academic Press, Inc.
1250 Sixth Avenue, San Diego, California 92101-4311

United Kingdom Edition published by
Academic Press Limited
24–28 Oval Road, London NW1 7DX

Library of Congress Catalog Number: 54-9110

International Standard Book Number: 0-12-182116-1

PRINTED IN THE UNITED STATES OF AMERICA
92 93 94 95 96 97   EB   9 8 7 6 5 4 3 2 1

Table of Contents

Contributors to Volume 215 ix
Preface xiii
Volumes in Series xv

Section I. Isolation of Platelet Membranes, Subcellular Organelles, and Cytoskeleton

1. Introduction to Platelet Structural and Functional Organization — Jacek Hawiger — 3

A. Platelet Membranes and Granules

2. Isolation and Characterization of Platelet Membranes Prepared by Free Flow Electrophoresis — Neville Crawford, Kalwant S. Authi, and Nashrudeen Hack — 5

3. Homogenization by Nitrogen Cavitation Technique Applied to Platelet Subcellular Fractionation — M. Johan Broekman — 21

4. Isolation of Human Platelet Plasma Membranes by Glycerol Lysis — Joan T. Harmon, Nicholas J. Greco, and G. A. Jamieson — 32

5. Isolation of Dense Granules from Human Platelets — Miriam H. Fukami — 36

B. Platelet Cytoskeleton

6. Studying the Platelet Cytoskeleton in Triton X-100 Lysates — Joan E. B. Fox, Clifford C. Reynolds, and Janet K. Boyles — 42

7. Purification and Characterization of Platelet Actin, Actin-Binding Protein, and α-Actinin — Sharon Rosenberg Schaier — 58

8. Purification and Characterization of Platelet Myosin — James L. Daniel and James R. Sellers — 78

9. Isolation and Characterization of Platelet Gelsolin — Joseph Bryan — 88

10. Purification and Properties of Human Platelet P235: Talin — Nancy C. Collier and Kuan Wang — 99

11. Ultrastructural Analysis of Platelet Contractile Apparatus — JAMES G. WHITE — 109

Section II. Platelet Receptors: Assays and Purification

12. Repertoire of Platelet Receptors — JACEK HAWIGER — 131

A. Receptors for Platelet Agonists

13. 2-Methylthioadenosine [β-^{32}P]Diphosphate: Synthesis and Use as Probe of Platelet ADP Receptors — DONALD E. MACFARLANE — 137

14. Interaction of Nucleotide Affinity Analog 5'-p-Fluorosulfonylbenzoyladenosine with Platelet ADP Receptor: Aggregin — WILLIAM R. FIGURES, L. MARIE SCEARCE, ROBERTA F. COLMAN, AND ROBERT W. COLMAN — 143

15. Thrombin Receptors on Human Platelets — ELIZABETH R. SIMONS, THERESA A. DAVIES, SHERYL M. GREENBERG, AND NANCY E. LARSEN — 155

16. Platelet Membrane Glycoprotein V Purification — DAVID R. PHILLIPS AND MICHAEL C. BERNDT — 176

17. Synthesis of a Yohimbine–Agarose Matrix Useful for Large-Scale and Micropurification of Multiple α_2-Receptor Subtypes — STEVEN E. DOMINO, MARY G. REPASKE, CAROL ANN BONNER, MATTHEW E. KENNEDY, AMY L. WILSON, SUZANNE BRANDON, AND LEE E. LIMBIRD — 181

18. 5-[^{3}H]Hydroxytryptamine and [^{3}H]Lysergic Acid Diethylamide Binding to Human Platelets — JOHN R. PETERS, DAVID P. GEANEY, AND D. G. GRAHAME-SMITH — 201

19. Platelet Serotonin Transporter — GARY RUDNICK AND CYNTHIA J. HUMPHREYS — 213

20. Binding of Platelet-Activating Factor 1-O-Alkyl-2-acetyl-sn-glycero-3-phosphorylcholine to Intact Platelets and Platelet Membranes — FRANK H. VALONE — 224

B. Receptors for Adhesive Proteins

21. Binding of Fibrinogen and von Willebrand Factor to Platelet Glycoprotein IIb–IIIa Complex — JACEK HAWIGER AND SHEILA TIMMONS — 228

22. Platelet Membrane Glycoprotein IIb–IIIa Complex: Purification, Characterization, and Reconstitution into Phospholipid Vesicles — DAVID R. PHILLIPS, LAURENCE FITZGERALD, LESLIE PARISE, AND BEAT STEINER — 244

23. von Willebrand Factor Binding to Platelet Glycoprotein Ib Complex	ZAVERIO M. RUGGERI, THEODORE S. ZIMMERMAN, SUSAN RUSSELL, ROSSELLA BADER, AND LUIGI DEMARCO	263
24. Isolation and Characterization of Glycoprotein Ib	ANDREAS N. WICKI, JEANNINE M. CLEMETSON, BEAT STEINER, WOLFGANG SCHNIPPERING, AND KENNETH J. CLEMETSON	276
25. Platelet Glycocalicin	JOSEPH LOSCALZO AND ROBERT I. HANDIN	289
26. Preparation and Functional Characterization of Monoclonal Antibodies against Glycoprotein Ib	LESLEY E. SCUDDER, EFSTATHIA L. KALOMIRIS, AND BARRY S. COLLER	295
27. Fibronectin Binding to Platelets	JANE FORSYTH, EDWARD F. PLOW, AND MARK H. GINSBERG	311

C. Receptors for Clotting Factors and Other Ligands

28. Mathematical Simulation of Prothrombinase	MICHAEL E. NESHEIM, RUSSELL P. TRACY, PAULA B. TRACY, DANILO S. BOSKOVIC, AND KENNETH G. MANN	316
29. Platelet Factor Xa Receptor	PAULA B. TRACY, MICHAEL E. NESHEIM, AND KENNETH G. MANN	329
30. Binding of Coagulation Factor XIa to Receptor on Human Platelets	DIPALI SINHA AND PETER N. WALSH	361
31. High Molecular Weight Kininogen Receptor	JUDITH S. GREENGARD AND JOHN H. GRIFFIN	369
32. Binding Characteristics of Homologous Plasma Lipoproteins to Human Platelets	ELISABETH KOLLER AND FRANZ KOLLER	383
33. Platelet Insulin Receptor	ANTHONY S. HAJEK AND J. HEINRICH JOIST	398

D. General Approaches to Receptor Analysis

34. Membrane-Impermeant Cross-Linking Reagents for Structural and Functional Analyses of Platelet Membrane Glycoproteins	JAMES V. STAROS, NICOLAS J. KOTITE, AND LEON W. CUNNINGHAM	403

35. Surface Labeling of Platelet Membrane Glycoproteins	DAVID R. PHILLIPS	412
36. Evaluation of Platelet Surface Antigens by Fluorescence Flow Cytometry	BURT ADELMAN, PATRICIA CARLSON, AND ROBERT I. HANDIN	420
37. Identification of Platelet Membrane Target Antigens for Human Antibodies by Immunoblotting	DIANA S. BEARDSLEY	428
38. Crossed Immunoelectrophoresis of Human Platelet Membranes	SIMON KARPATKIN, SABRA SHULMAN, AND LESLIE HOWARD	440
39. Use of Correlative Microscopy with Colloidal Gold Labeling to Demonstrate Platelet Receptor Distribution and Movement	RALPH M. ALBRECHT, OLUFUNKE E. OLORUNDARE, SCOTT R. SIMMONS, JOSEPH C. LOFTUS, AND DEANE F. MOSHER	456

AUTHOR INDEX 481

SUBJECT INDEX 501

Contributors to Volume 215

Article numbers are in parentheses following the names of contributors.
Affiliations listed are current.

BURT ADELMAN (36), *Department of Medical Research, Biogen Inc., Cambridge Massachusetts 02142*

RALPH M. ALBRECHT (39), *Department of Animal Health and Biomedical Sciences, University of Wisconsin, Madison, Wisconsin 53706*

KALWANT S. AUTHI (2), *Platelet Section, Thrombosis Research Institute, Chelsea, London SW3 6LR, England*

ROSSELLA BADER (23), *Hemophilia and Thrombosis Center, Policlinico Hospital, University of Milan, Milan 20122, Italy*

DIANA S. BEARDSLEY (37), *Division of Pediatric Hematology, Yale University School of Medicine, New Haven, Connecticut 06510*

MICHAEL C. BERNDT (16), *Baker Medical Research Institute, Prahran VIC 3181, Australia.*

CAROL ANN BONNER (17), *Department of Pharmacology, Vanderbilt University, Nashville, Tennessee 37232*

DANILO S. BOSKOVIC (28), *Department of Biochemistry, Queen's University, Kingston, Ontario K7L 3N6, Canada*

JANET K. BOYLES (6), *Vanderbilt University School of Medicine, Nashville, Tennessee 37232*

SUZANNE BRANDON (17), *Department of Pharmacology, Vanderbilt University, Nashville, Tennessee 37232*

M. JOHAN BROEKMAN (3), *Divisions of Hematology/Oncology, Departments of Medicine, Department of Veterans Affairs Medical Center, and Cornell University Medical College, New York, New York 10010*

JOSEPH BRYAN (9), *Department of Cell Biology, Baylor College of Medicine, Houston, Texas 77030*

PATRICIA CARLSON (36), *Department of Laboratory Medicine, Hunter Holmes McGuire Veterans Administration Center, Richmond, Virginia 23249*

JEANNINE M. CLEMETSON (24), *Theodor Kocher Institute, University of Berne, CH-3012 Berne, Switzerland*

KENNETH J. CLEMETSON (26), *Theodor Kocher Institute, University of Berne, CH-3012 Berne, Switzerland*

BARRY S. COLLER (24), *Department of Medicine, Division of Hematology, State University of New York at Stony Brook, Stony Brook, New York 11794*

NANCY C. COLLIER (10), *Department of Molecular Microbiology, Washington University School of Medicine, St. Louis, Missouri 63110*

ROBERT W. COLMAN (14), *Sol Sherry Thrombosis Research Center and Department of Medicine, Temple University School of Medicine, Philadelphia, Pennsylvania 19140*

ROBERTA F. COLMAN (14), *Department of Chemistry and Biochemistry, University of Delaware, Newark, Delaware 19716*

NEVILLE CRAWFORD (2), *Department of Biochemistry, Hunterian Institute, Royal College of Surgeons, London WC2A 3PN, England*

LEON W. CUNNINGHAM (34), *Department of Biochemistry, Vanderbilt University School of Medicine, Nashville, Tennessee 37232*

JAMES L. DANIEL (8), *Department of Pharmacology and Thrombosis Research Center, Temple University, Philadelphia, Pennsylvania 19140*

THERESA A. DAVIES (15), *Department of Biochemistry, Boston University School of Medicine, Boston, Massachusetts 02118*

LUIGI DEMARCO (23), *Servisio Immunotrasfusionale e Analisi Cliniche, Centro di Riferimento Oncologico, Aviano, Pordenone, Italy*

STEVEN E. DOMINO (17), *Department of Obstetrics and Gynecology, University of Michigan, Ann Arbor, Michigan 48109*

WILLIAM R. FIGURES (14), *Thrombosis Research Center and Hematology Division, Temple University School of Medicine, Philadelphia, Pennsylvania 19140*

LAURENCE FITZGERALD (22), *University of Utah, Salt Lake City, Utah 84112*

JANE FORSYTH (27), *Committee on Vascular Biology, The Scripps Research Institute, La Jolla, California 92037*

JOAN E. B. FOX (6), *Gladstone Institute of Cardiovascular Disease, University of California, San Fransisco, San Fransisco, California 94141*

MIRIAM H. FUKAMI (5), *Department of Biochemistry, University of Bergen, N-5009 Bergen, Norway*

DAVID P. GEANEY (18), *Department of Clinical Pharmacology, Radcliffe Infirmary, Oxford, OX2 6HE, England*

MARK H. GINSBERG (27), *Committee on Vascular Biology, The Scripps Research Institute, La Jolla, California 92037*

D. G. GRAHAME-SMITH (18), *University Department of Clinical Pharmacology, Radcliffe Infirmary, Oxford OX2 6HE, England*

NICHOLAS J. GRECO (4), *Cell Biology Department, Holland Laboratory, American Red Cross, Rockville, Maryland 20855*

SHERYL M. GREENBERG (15), *Division of Hematology, Brigham and Women's Hospital, Boston, Massachusetts 02115*

JUDITH S. GREENGARD (31), *Department of Molecular and Experimental Medicine, and Committee on Vascular Biology, The Scripps Research Institute, La Jolla, California 92037*

JOHN H. GRIFFIN (31), *Department of Molecular and Experimental Medicine, and Committee on Vascular Biology, The Scripps Research Institute, La Jolla, California 92037*

NASHRUDEEN HACK (2), *Department of Medicine, University of Toronto, Toronto, Ontario M5S 1AB, Canada*

ANTHONY S. HAJEK (33), *Department of Opthamology, Bascom Palmer Eye Institute, University of Miami School of Medicine, Miami, Florida 33101*

ROBERT I. HANDIN (25, 36), *Division of Hematology, Department of Medicine, Brigham and Women's Hospital, Harvard Medical School, Boston, Massachusetts 02115*

JOAN T. HARMON (4), *National Institutes of Diabetes and Digestive and Kidney Diseaes, Division of Diabetes, Endocrinology, and Metabolic Diseases, National Institutes of Health, Bethesda, Maryland 20892*

JACEK HAWIGER (1, 12, 21), *Department of Microbiology and Immunology, A-5321 MCN, Vanderbilt University School of Medicine, Nashville, Tennessee 37232*

LESLIE HOWARD (38), *Hematology/Oncology Section, Veterans Administration, Medical Center, East Orange, New Jersey 07019*

CYNTHIA J. HUMPHREYS (19), *Department of Microbiology, College of Physicians and Surgeons, Columbia University, New York, New York 10032*

G. A. JAMIESON (4), *Cell Biology Department, Holland Laboratory, American Red Cross, Rockville, Maryland 20855*

J. HEINRICH JOIST (33), *Department of Internal Medicine, Division of Bone Marrow Transplantation, Hematology, and Oncology, St. Louis University Medical Center, St. Louis, Missouri 63110*

EFSTATHIA L. KALOMIRIS (26), *Quality Control Assurance, Hoffman–La Roche, Nutley, New Jersey 07110*

SIMON KARPATKIN (38), *Department of Medicine, New York University Medical Center, New York, New York 10016*

MATTHEW E. KENNEDY (17), *Department of Pharmacology, Vanderbilt University, Nashville, Tennessee 37232*

ELISABETH KOLLER (32), *Institut für Medizinischen Physiologie, Universität Wien, Vienna, Austria*

FRANZ KOLLER (32), *Institut für Allegemeine Biochemie, Universität Wien, und Ludwig-Boltemann Forschungs-Stelle für Biochemie, Vienna, Austria*

NICOLAS J. KOTITE (34), *Department of Biochemistry, Vanderbilt University School of Medicine, Nashville, Tennessee 37232*

NANCY E. LARSEN (15), *Biomatrix, Inc., Ridgefield, New Jersey 07657*

LEE E. LIMBIRD (17), *Department of Pharmacology, Vanderbilt University, Nashville, Tennessee 37232*

JOSEPH C. LOFTUS (39), *Committee on Vascular Biology, The Scripps Research Institute, La Jolla, California 92037*

JOSEPH LOSCALZO (25), *Department of Medicine, Brigham and Women's Hospital, Harvard Medical School, Boston, Massachusetts 02115*

DONALD E. MACFARLANE (13), *Department of Medicine, The University of Iowa, Iowa City, Iowa 52242*

KENNETH G. MANN (28, 29), *Department of Biochemistry, University of Vermont College of Medicine, Burlington, Vermont 05405*

DEANE F. MOSHER (39), *Department of Medicine, University of Wisconsin, Madison, Wisconsin 53706*

MICHAEL E. NESHEIM (28, 29), *Departments of Medicine and Biochemistry, Queen's University, Kingston, Ontario K7L 3N6, Canada*

OLUFUNKE E. OLORUNDARE (39), *Department of Pharmacology and Therapeutics, University of Illorin, Illorin, Nigeria*

LESLIE PARISE (22), *Department of Pharmacology, University of North Carolina, Chapel Hill, North Carolina 27599*

JOHN R. PETERS (18), *Department of Medicine, University Hospital of Wales, Cardiff CF4 4XN, United Kingdom*

DAVID R. PHILLIPS (16, 22, 35), *COR Therapeutics, Inc., South San Fransisco, California 94080*

EDWARD F. PLOW (27), *Committee on Vascular Biology, The Scripps Research Institute, La Jolla, California 92037*

MARY G. REPASKE (17), *Department of Pharmacology, Vanderbilt University, Nashville, Tennessee 37232*

CLIFFORD C. REYNOLDS (6), *Gladstone Institute of Cardiovascular Disease, University of California, San Fransisco, San Fransisco, California 94141*

SHARON ROSENBERG SCHAIER (7), *JMR Consulting, Fair Lawn, New Jersey 07410*

GARY RUDNICK (19), *Department of Pharmacology, Yale University School of Medicine, New Haven, Connecticut 06510*

ZAVERIO M. RUGGERI (23), *Roon Research Laboratory for Arteriosclerosis and Thrombosis, Department of Molecular and Experimental Medicine, and Committee on Vascular Biology, The Scripps Research Institute, La Jolla, California, 92037*

SUSAN RUSSELL (23), *Roon Research Laboratory for Arteriosclerosis and Thrombosis, Department of Molecular and Experimental Medicine, The Scripps Research Institute, La Jolla, California 92037*

L. MARIE SCEARCE (14), *Department of Human Genetics, Howard Hughes Medical Institute, University of Pennsylvania School of Medicine, Philadelphia, Pennsylvania 19104*

WOLFGANG SCHNIPPERING (24), *Central Laboratory, Swiss Red Cross Blood Transfusion Service, CH-3000 Berne 22, Switzerland*

LESLEY E. SCUDDER (26), *Department of Medicine, Division of Hematology, State University of New York at Stony Brook, Stony Brook, New York 11794*

JAMES R. SELLERS (8), *Laboratory of Molecular Cardiology, National Heart, Lung and Blood Institute, National Institutes of Health, Bethesda, Maryland 20892*

SABRA SHULMAN (38), *Department of Medicine, New York University Medical Center, New York, New York 10016*

SCOTT R. SIMMONS (39), *Department of Animal Health and Biomedical Sciences, University of Wisconsin, Madison, Wisconsin 53706*

ELIZABETH R. SIMONS (15), *Department of Biochemistry, Boston University School of Medicine, Boston, Massachusetts 02118*

DIPALI SINHA (30), *Department of Biochemistry, Temple University School of Medicine, Philadelphia, Pennsylvania 19140*

JAMES V. STAROS (34), *Department of Molecular Biology, Vanderbilt University, Nashville, Tennessee 37235*

BEAT STEINER (22, 24), *Pharma Division Preclinical Research, Hoffman-La Roche, CH-4002 Basle, Switzerland*

SHEILA TIMMONS (21), *Department of Microbiology and Immunology, Vanderbilt University School of Medicine, Nashville, Tennessee 37232*

PAULA B. TRACY (28, 29), *Department of Biochemistry, University of Vermont College of Medicine, Burlington, Vermont 05405*

RUSSELL P. TRACY (28), *Department of Pathology, University of Vermont, Burlington, Vermont 05405*

FRANK H. VALONE (20), *Department of Medicine, Dartmouth-Hitchcock Medical Center, Lebanon, New Hampshire 03756*

PETER N. WALSH (30), *Thrombosis Research Center, Temple University School of Medicine, Philadelphia, Pennsylvania 19140*

KUAN WANG (10), *Department of Chemistry and Biochemistry, Welch Hall 4.230, University of Texas at Austin, Austin, Texas 78712-1167*

JAMES G. WHITE (11), *Departments of Laboratory Medicine, Pathology, and Pediatrics, University of Minnesota Medical School, Minneapolis, Minnesota 55455*

ANDREAS N. WICKI (24), *ANAWA Laboratories, CH-8602 Wangen, Switzerland*

AMY L. WILSON (17), *Department of Pharmacology, Vanderbilt University, Nashville, Tennesse 37232*

THEODORE S. ZIMMERMAN[1] (23), *Roon Research Laboratory for Arteriosclerosis and Thrombosis, Department of Molecular and Experimental Medicine, The Scripps Research Institute, La Jolla, California 92037*

---

[1] Deceased.

Preface

Blood platelets are tiny corpuscles that continuously survey the inner lining of blood vessels, the vascular endothelium. Any break in the continuity of the vessel wall, leading to hemorrhage, or a break in an atherosclerotic plaque is met with an instant response from platelets. When platelets contact the zone of injury, they spread, aggregate, and form thrombi that seal off the break. The membrane is transformed into batteries of receptors for adenine nucleotides, catecholamines, adhesive molecules, and clotting factors. The platelet interior receives a flux of calcium, flexes its contractile apparatus, and becomes a furnace for enzymatic oxidation of arachidonic acid into endoperoxides which are then transformed into thromboxane A_2, a potent vasoconstrictor and platelet agonist. The platelet storage granules secrete their constituent molecules, adenosine phosphate, serotonin, adhesive proteins, and coagulation factor V, that are targeted toward their own receptors and receptors on other platelets. Clotting enzymes, assembled on the platelet membrane as the prothrombinase complex, generate thrombin from prothrombin. After secretion of serotonin its reuptake involves two membrane mechanisms akin to those in neurons. The platelet, hardly a cell, has gained recognition in human cell biology as a cellular element endowed with a membrane that bears the highest density of receptors per surface unit area among the known blood cells. Mitogenic growth factors, stored in and secreted from platelet α granules, stimulate migration and proliferation of vascular smooth muscle cells and fibroblasts producing the extracellular matrix.

An increasing number of biochemists, cell biologists, hematologists, neuropharmacologists, and pathologists are working with blood platelets as a useful system to study the processes of adhesion, neurotransmitter uptake, secretion, and their regulation. Therefore, Part A (Volume 169) was devoted to the methods of platelet isolation and to the study of their adhesive and secretory functions. This volume (Part B) includes the isolation of subcellular organelles and their characterization, and a spectrum of methods for analyzing platelet receptors.

The idea for these volumes, comprising modern methods of platelet analysis, germinated when the late Editor-in-Chief, Sidney Colowick, encouraged me to embark on this task. Sidney Colowick's untimely death left us not only without one of the founding fathers of this monumental series, but also without a most inspiring friend and advisor. His legacy cannot be measured in words. He was an example of exquisite intellectual elegance in every aspect of his scientific activities. His gentle discourse and unassuming demeanor were matched by his all encompassing mind and heart. These two volumes on platelets are dedicated to him.

It has been a genuine pleasure for me to interact with the staff of Academic Press. They have been helpful and patient in dealing with the difficulties encountered during the preparation of these volumes. In my unit at the New England Deaconess Hospital and Harvard Medical School, Ms. Sheila Timmons and Ms. Marie Bingyou, and later at Vanderbilt University, Ms. Carol Walter, were enormously resourceful in helping me organize and edit these two volumes. Finally, I offer my deepest appreciation to my colleagues, the contributing authors, for their wisdom, expertise, and scholarship.

JACEK HAWIGER

METHODS IN ENZYMOLOGY

VOLUME I. Preparation and Assay of Enzymes
Edited by SIDNEY P. COLOWICK AND NATHAN O. KAPLAN

VOLUME II. Preparation and Assay of Enzymes
Edited by SIDNEY P. COLOWICK AND NATHAN O. KAPLAN

VOLUME III. Preparation and Assay of Substrates
Edited by SIDNEY P. COLOWICK AND NATHAN O. KAPLAN

VOLUME IV. Special Techniques for the Enzymologist
Edited by SIDNEY P. COLOWICK AND NATHAN O. KAPLAN

VOLUME V. Preparation and Assay of Enzymes
Edited by SIDNEY P. COLOWICK AND NATHAN O. KAPLAN

VOLUME VI. Preparation and Assay of Enzymes (*Continued*)
Preparation and Assay of Substrates
Special Techniques
Edited by SIDNEY P. COLOWICK AND NATHAN O. KAPLAN

VOLUME VII. Cumulative Subject Index
Edited by SIDNEY P. COLOWICK AND NATHAN O. KAPLAN

VOLUME VIII. Complex Carbohydrates
Edited by ELIZABETH F. NEUFELD AND VICTOR GINSBURG

VOLUME IX. Carbohydrate Metabolism
Edited by WILLIS A. WOOD

VOLUME X. Oxidation and Phosphorylation
Edited by RONALD W. ESTABROOK AND MAYNARD E. PULLMAN

VOLUME XI. Enzyme Structure
Edited by C. H. W. HIRS

VOLUME XII. Nucleic Acids (Parts A and B)
Edited by LAWRENCE GROSSMAN AND KIVIE MOLDAVE

VOLUME XIII. Citric Acid Cycle
Edited by J. M. LOWENSTEIN

VOLUME XIV. Lipids
Edited by J. M. LOWENSTEIN

VOLUME XV. Steroids and Terpenoids
Edited by RAYMOND B. CLAYTON

VOLUME XVI. Fast Reactions
Edited by KENNETH KUSTIN

VOLUME XVII. Metabolism of Amino Acids and Amines (Parts A and B)
Edited by HERBERT TABOR AND CELIA WHITE TABOR

VOLUME XVIII. Vitamins and Coenzymes (Parts A, B, and C)
Edited by DONALD B. MCCORMICK AND LEMUEL D. WRIGHT

VOLUME XIX. Proteolytic Enzymes
Edited by GERTRUDE E. PERLMANN AND LASZLO LORAND

VOLUME XX. Nucleic Acids and Protein Synthesis (Part C)
Edited by KIVIE MOLDAVE AND LAWRENCE GROSSMAN

VOLUME XXI. Nucleic Acids (Part D)
Edited by LAWRENCE GROSSMAN AND KIVIE MOLDAVE

VOLUME XXII. Enzyme Purification and Related Techniques
Edited by WILLIAM B. JAKOBY

VOLUME XXIII. Photosynthesis (Part A)
Edited by ANTHONY SAN PIETRO

VOLUME XXIV. Photosynthesis and Nitrogen Fixation (Part B)
Edited by ANTHONY SAN PIETRO

VOLUME XXV. Enzyme Structure (Part B)
Edited by C. H. W. HIRS AND SERGE N. TIMASHEFF

VOLUME XXVI. Enzyme Structure (Part C)
Edited by C. H. W. HIRS AND SERGE N. TIMASHEFF

VOLUME XXVII. Enzyme Structure (Part D)
Edited by C. H. W. HIRS AND SERGE N. TIMASHEFF

VOLUME XXVIII. Complex Carbohydrates (Part B)
Edited by VICTOR GINSBURG

VOLUME XXIX. Nucleic Acids and Protein Synthesis (Part E)
Edited by LAWRENCE GROSSMAN AND KIVIE MOLDAVE

VOLUME XXX. Nucleic Acids and Protein Synthesis (Part F)
Edited by KIVIE MOLDAVE AND LAWRENCE GROSSMAN

VOLUME XXXI. Biomembranes (Part A)
Edited by SIDNEY FLEISCHER AND LESTER PACKER

VOLUME XXXII. Biomembranes (Part B)
Edited by SIDNEY FLEISCHER AND LESTER PACKER

VOLUME XXXIII. Cumulative Subject Index Volumes I–XXX
Edited by MARTHA G. DENNIS AND EDWARD A. DENNIS

VOLUME XXXIV. Affinity Techniques (Enzyme Purification: Part B)
Edited by WILLIAM B. JAKOBY AND MEIR WILCHEK

VOLUME XXXV. Lipids (Part B)
Edited by JOHN M. LOWENSTEIN

VOLUME XXXVI. Hormone Action (Part A: Steroid Hormones)
Edited by BERT W. O'MALLEY AND JOEL G. HARDMAN

VOLUME XXXVII. Hormone Action (Part B: Peptide Hormones)
Edited by BERT W. O'MALLEY AND JOEL G. HARDMAN

VOLUME XXXVIII. Hormone Action (Part C: Cyclic Nucleotides)
Edited by JOEL G. HARDMAN AND BERT W. O'MALLEY

VOLUME XXXIX. Hormone Action (Part D: Isolated Cells, Tissues, and Organ Systems)
Edited by JOEL G. HARDMAN AND BERT W. O'MALLEY

VOLUME XL. Hormone Action (Part E: Nuclear Structure and Function)
Edited by BERT W. O'MALLEY AND JOEL G. HARDMAN

VOLUME XLI. Carbohydrate Metabolism (Part B)
Edited by W. A. WOOD

VOLUME XLII. Carbohydrate Metabolism (Part C)
Edited by W. A. WOOD

VOLUME XLIII. Antibiotics
Edited by JOHN H. HASH

VOLUME XLIV. Immobilized Enzymes
Edited by KLAUS MOSBACH

VOLUME XLV. Proteolytic Enzymes (Part B)
Edited by LASZLO LORAND

VOLUME XLVI. Affinity Labeling
Edited by WILLIAM B. JAKOBY AND MEIR WILCHEK

VOLUME XLVII. Enzyme Structure (Part E)
Edited by C. H. W. HIRS AND SERGE N. TIMASHEFF

VOLUME XLVIII. Enzyme Structure (Part F)
Edited by C. H. W. HIRS AND SERGE N. TIMASHEFF

VOLUME XLIX. Enzyme Structure (Part G)
Edited by C. H. W. HIRS AND SERGE N. TIMASHEFF

VOLUME L. Complex Carbohydrates (Part C)
Edited by VICTOR GINSBURG

VOLUME LI. Purine and Pyrimidine Nucleotide Metabolism
Edited by PATRICIA A. HOFFEE AND MARY ELLEN JONES

VOLUME LII. Biomembranes (Part C: Biological Oxidations)
Edited by SIDNEY FLEISCHER AND LESTER PACKER

VOLUME LIII. Biomembranes (Part D: Biological Oxidations)
Edited by SIDNEY FLEISCHER AND LESTER PACKER

VOLUME LIV. Biomembranes (Part E: Biological Oxidations)
Edited by SIDNEY FLEISCHER AND LESTER PACKER

VOLUME LV. Biomembranes (Part F: Bioenergetics)
Edited by SIDNEY FLEISCHER AND LESTER PACKER

VOLUME LVI. Biomembranes (Part G: Bioenergetics)
Edited by SIDNEY FLEISCHER AND LESTER PACKER

VOLUME LVII. Bioluminescence and Chemiluminescence
Edited by MARLENE A. DELUCA

VOLUME LVIII. Cell Culture
Edited by WILLIAM B. JAKOBY AND IRA PASTAN

VOLUME LIX. Nucleic Acids and Protein Synthesis (Part G)
Edited by KIVIE MOLDAVE AND LAWRENCE GROSSMAN

VOLUME LX. Nucleic Acids and Protein Synthesis (Part H)
Edited by KIVIE MOLDAVE AND LAWRENCE GROSSMAN

VOLUME 61. Enzyme Structure (Part H)
Edited by C. H. W. HIRS AND SERGE N. TIMASHEFF

VOLUME 62. Vitamins and Coenzymes (Part D)
Edited by DONALD B. MCCORMICK AND LEMUEL D. WRIGHT

VOLUME 63. Enzyme Kinetics and Mechanism (Part A: Initial Rate and Inhibitor Methods)
Edited by DANIEL L. PURICH

VOLUME 64. Enzyme Kinetics and Mechanism (Part B: Isotopic Probes and Complex Enzyme Systems)
Edited by DANIEL L. PURICH

VOLUME 65. Nucleic Acids (Part I)
Edited by LAWRENCE GROSSMAN AND KIVIE MOLDAVE

VOLUME 66. Vitamins and Coenzymes (Part E)
Edited by DONALD B. MCCORMICK AND LEMUEL D. WRIGHT

VOLUME 67. Vitamins and Coenzymes (Part F)
Edited by DONALD B. MCCORMICK AND LEMUEL D. WRIGHT

VOLUME 68. Recombinant DNA
Edited by RAY WU

VOLUME 69. Photosynthesis and Nitrogen Fixation (Part C)
Edited by ANTHONY SAN PIETRO

VOLUME 70. Immunochemical Techniques (Part A)
Edited by HELEN VAN VUNAKIS AND JOHN J. LANGONE

VOLUME 71. Lipids (Part C)
Edited by JOHN M. LOWENSTEIN

VOLUME 72. Lipids (Part D)
Edited by JOHN M. LOWENSTEIN

VOLUME 73. Immunochemical Techniques (Part B)
Edited by JOHN J. LANGONE AND HELEN VAN VUNAKIS

VOLUME 74. Immunochemical Techniques (Part C)
Edited by JOHN J. LANGONE AND HELEN VAN VUNAKIS

VOLUME 75. Cumulative Subject Index Volumes XXXI, XXXII, XXXIV–LX
Edited by EDWARD A. DENNIS AND MARTHA G. DENNIS

VOLUME 76. Hemoglobins
Edited by ERALDO ANTONINI, LUIGI ROSSI-BERNARDI, AND EMILIA CHIANCONE

VOLUME 77. Detoxication and Drug Metabolism
Edited by WILLIAM B. JAKOBY

VOLUME 78. Interferons (Part A)
Edited by SIDNEY PESTKA

VOLUME 79. Interferons (Part B)
Edited by SIDNEY PESTKA

VOLUME 80. Proteolytic Enzymes (Part C)
Edited by LASZLO LORAND

VOLUME 81. Biomembranes (Part H: Visual Pigments and Purple Membranes, I)
Edited by LESTER PACKER

VOLUME 82. Structural and Contractile Proteins (Part A: Extracellular Matrix)
Edited by LEON W. CUNNINGHAM AND DIXIE W. FREDERIKSEN

VOLUME 83. Complex Carbohydrates (Part D)
Edited by VICTOR GINSBURG

VOLUME 84. Immunochemical Techniques (Part D: Selected Immunoassays)
Edited by JOHN J. LANGONE AND HELEN VAN VUNAKIS

VOLUME 85. Structural and Contractile Proteins (Part B: The Contractile Apparatus and the Cytoskeleton)
Edited by DIXIE W. FREDERIKSEN AND LEON W. CUNNINGHAM

VOLUME 86. Prostaglandins and Arachidonate Metabolites
Edited by WILLIAM E. M. LANDS AND WILLIAM L. SMITH

VOLUME 87. Enzyme Kinetics and Mechanism (Part C: Intermediates, Stereochemistry, and Rate Studies)
Edited by DANIEL L. PURICH

VOLUME 88. Biomembranes (Part I: Visual Pigments and Purple Membranes, II)
Edited by LESTER PACKER

VOLUME 89. Carbohydrate Metabolism (Part D)
Edited by WILLIS A. WOOD

VOLUME 90. Carbohydrate Metabolism (Part E)
Edited by WILLIS A. WOOD

VOLUME 91. Enzyme Structure (Part I)
Edited by C. H. W. HIRS AND SERGE N. TIMASHEFF

VOLUME 92. Immunochemical Techniques (Part E: Monoclonal Antibodies and General Immunoassay Methods)
Edited by JOHN J. LANGONE AND HELEN VAN VUNAKIS

VOLUME 93. Immunochemical Techniques (Part F: Conventional Antibodies, Fc Receptors, and Cytotoxicity)
Edited by JOHN J. LANGONE AND HELEN VAN VUNAKIS

VOLUME 94. Polyamines
Edited by HERBERT TABOR AND CELIA WHITE TABOR

VOLUME 95. Cumulative Subject Index Volumes 61–74, 76–80
Edited by EDWARD A. DENNIS AND MARTHA G. DENNIS

VOLUME 96. Biomembranes [Part J: Membrane Biogenesis: Assembly and Targeting (General Methods; Eukaryotes)]
Edited by SIDNEY FLEISCHER AND BECCA FLEISCHER

VOLUME 97. Biomembranes [Part K: Membrane Biogenesis: Assembly and Targeting (Prokaryotes, Mitochondria, and Chloroplasts)]
Edited by SIDNEY FLEISCHER AND BECCA FLEISCHER

VOLUME 98. Biomembranes (Part L: Membrane Biogenesis: Processing and Recycling)
Edited by SIDNEY FLEISCHER AND BECCA FLEISCHER

VOLUME 99. Hormone Action (Part F: Protein Kinases)
Edited by JACKIE D. CORBIN AND JOEL G. HARDMAN

VOLUME 100. Recombinant DNA (Part B)
Edited by RAY WU, LAWRENCE GROSSMAN, AND KIVIE MOLDAVE

VOLUME 101. Recombinant DNA (Part C)
Edited by RAY WU, LAWRENCE GROSSMAN, AND KIVIE MOLDAVE

VOLUME 102. Hormone Action (Part G: Calmodulin and Calcium-Binding Proteins)
Edited by ANTHONY R. MEANS AND BERT W. O'MALLEY

VOLUME 103. Hormone Action (Part H: Neuroendocrine Peptides)
Edited by P. MICHAEL CONN

VOLUME 104. Enzyme Purification and Related Techniques (Part C)
Edited by WILLIAM B. JAKOBY

VOLUME 105. Oxygen Radicals in Biological Systems
Edited by LESTER PACKER

VOLUME 106. Posttranslational Modifications (Part A)
Edited by FINN WOLD AND KIVIE MOLDAVE

VOLUME 107. Posttranslational Modifications (Part B)
Edited by FINN WOLD AND KIVIE MOLDAVE

VOLUME 108. Immunochemical Techniques (Part G: Separation and Characterization of Lymphoid Cells)
Edited by GIOVANNI DI SABATO, JOHN J. LANGONE, AND HELEN VAN VUNAKIS

VOLUME 109. Hormone Action (Part I: Peptide Hormones)
Edited by LUTZ BIRNBAUMER AND BERT W. O'MALLEY

VOLUME 110. Steroids and Isoprenoids (Part A)
Edited by JOHN H. LAW AND HANS C. RILLING

VOLUME 111. Steroids and Isoprenoids (Part B)
Edited by JOHN H. LAW AND HANS C. RILLING

VOLUME 112. Drug and Enzyme Targeting (Part A)
Edited by KENNETH J. WIDDER AND RALPH GREEN

VOLUME 113. Glutamate, Glutamine, Glutathione, and Related Compounds
Edited by ALTON MEISTER

VOLUME 114. Diffraction Methods for Biological Macromolecules (Part A)
Edited by HAROLD W. WYCKOFF, C. H. W. HIRS, AND SERGE N. TIMASHEFF

VOLUME 115. Diffraction Methods for Biological Macromolecules (Part B)
Edited by HAROLD W. WYCKOFF, C. H. W. HIRS, AND SERGE N. TIMASHEFF

VOLUME 116. Immunochemical Techniques (Part H: Effectors and Mediators of Lymphoid Cell Functions)
Edited by GIOVANNI DI SABATO, JOHN J. LANGONE, AND HELEN VAN VUNAKIS

VOLUME 117. Enzyme Structure (Part J)
Edited by C. H. W. HIRS AND SERGE N. TIMASHEFF

VOLUME 118. Plant Molecular Biology
Edited by ARTHUR WEISSBACH AND HERBERT WEISSBACH

VOLUME 119. Interferons (Part C)
Edited by SIDNEY PESTKA

VOLUME 120. Cumulative Subject Index Volumes 81–94, 96–101

VOLUME 121. Immunochemical Techniques (Part I: Hybridoma Technology and Monoclonal Antibodies)
Edited by JOHN J. LANGONE AND HELEN VAN VUNAKIS

VOLUME 122. Vitamins and Coenzymes (Part G)
Edited by FRANK CHYTIL AND DONALD B. MCCORMICK

VOLUME 123. Vitamins and Coenzymes (Part H)
Edited by FRANK CHYTIL AND DONALD B. MCCORMICK

VOLUME 124. Hormone Action (Part J: Neuroendocrine Peptides)
Edited by P. MICHAEL CONN

VOLUME 125. Biomembranes (Part M: Transport in Bacteria, Mitochondria, and Chloroplasts: General Approaches and Transport Systems)
Edited by SIDNEY FLEISCHER AND BECCA FLEISCHER

VOLUME 126. Biomembranes (Part N: Transport in Bacteria, Mitochondria, and Chloroplasts: Protonmotive Force)
Edited by SIDNEY FLEISCHER AND BECCA FLEISCHER

VOLUME 127. Biomembranes (Part O: Protons and Water: Structure and Translocation)
Edited by LESTER PACKER

VOLUME 128. Plasma Lipoproteins (Part A: Preparation, Structure, and Molecular Biology)
Edited by JERE P. SEGREST AND JOHN J. ALBERS

VOLUME 129. Plasma Lipoproteins (Part B: Characterization, Cell Biology, and Metabolism)
Edited by JOHN J. ALBERS AND JERE P. SEGREST

VOLUME 130. Enzyme Structure (Part K)
Edited by C. H. W. HIRS AND SERGE N. TIMASHEFF

VOLUME 131. Enzyme Structure (Part L)
Edited by C. H. W. HIRS AND SERGE N. TIMASHEFF

VOLUME 132. Immunochemical Techniques (Part J: Phagocytosis and Cell-Mediated Cytotoxicity)
Edited by GIOVANNI DI SABATO AND JOHANNES EVERSE

VOLUME 133. Bioluminescence and Chemiluminescence (Part B)
Edited by MARLENE DELUCA AND WILLIAM D. MCELROY

VOLUME 134. Structural and Contractile Proteins (Part C: The Contractile Apparatus and the Cytoskeleton)
Edited by RICHARD B. VALLEE

VOLUME 135. Immobilized Enzymes and Cells (Part B)
Edited by KLAUS MOSBACH

VOLUME 136. Immobilized Enzymes and Cells (Part C)
Edited by KLAUS MOSBACH

VOLUME 137. Immobilized Enzymes and Cells (Part D)
Edited by KLAUS MOSBACH

VOLUME 138. Complex Carbohydrates (Part E)
Edited by VICTOR GINSBURG

VOLUME 139. Cellular Regulators (Part A: Calcium- and Calmodulin-Binding Proteins)
Edited by ANTHONY R. MEANS AND P. MICHAEL CONN

VOLUME 140. Cumulative Subject Index Volumes 102–119, 121–134

VOLUME 141. Cellular Regulators (Part B: Calcium and Lipids)
Edited by P. MICHAEL CONN AND ANTHONY R. MEANS

VOLUME 142. Metabolism of Aromatic Amino Acids and Amines
Edited by SEYMOUR KAUFMAN

VOLUME 143. Sulfur and Sulfur Amino Acids
Edited by WILLIAM B. JAKOBY AND OWEN GRIFFITH

VOLUME 144. Structural and Contractile Proteins (Part D: Extracellular Matrix)
Edited by LEON W. CUNNINGHAM

VOLUME 145. Structural and Contractile Proteins (Part E: Extracellular Matrix)
Edited by LEON W. CUNNINGHAM

VOLUME 146. Peptide Growth Factors (Part A)
Edited by DAVID BARNES AND DAVID A. SIRBASKU

VOLUME 147. Peptide Growth Factors (Part B)
Edited by DAVID BARNES AND DAVID A. SIRBASKU

VOLUME 148. Plant Cell Membranes
Edited by LESTER PACKER AND ROLAND DOUCE

VOLUME 149. Drug and Enzyme Targeting (Part B)
Edited by RALPH GREEN AND KENNETH J. WIDDER

VOLUME 150. Immunochemical Techniques (Part K: *In Vitro* Models of B and T Cell Functions and Lymphoid Cell Receptors)
Edited by GIOVANNI DI SABATO

VOLUME 151. Molecular Genetics of Mammalian Cells
Edited by MICHAEL M. GOTTESMAN

VOLUME 152. Guide to Molecular Cloning Techniques
Edited by SHELBY L. BERGER AND ALAN R. KIMMEL

VOLUME 153. Recombinant DNA (Part D)
Edited by RAY WU AND LAWRENCE GROSSMAN

VOLUME 154. Recombinant DNA (Part E)
Edited by RAY WU AND LAWRENCE GROSSMAN

VOLUME 155. Recombinant DNA (Part F)
Edited by RAY WU

VOLUME 156. Biomembranes (Part P: ATP-Driven Pumps and Related Transport: The Na,K-Pump)
Edited by SIDNEY FLEISCHER AND BECCA FLEISCHER

VOLUME 157. Biomembranes (Part Q: ATP-Driven Pumps and Related Transport: Calcium, Proton, and Potassium Pumps)
Edited by SIDNEY FLEISCHER AND BECCA FLEISCHER

VOLUME 158. Metalloproteins (Part A)
Edited by JAMES F. RIORDAN AND BERT L. VALLEE

VOLUME 159. Initiation and Termination of Cyclic Nucleotide Action
Edited by JACKIE D. CORBIN AND ROGER A. JOHNSON

VOLUME 160. Biomass (Part A: Cellulose and Hemicellulose)
Edited by WILLIS A. WOOD AND SCOTT T. KELLOGG

VOLUME 161. Biomass (Part B: Lignin, Pectin, and Chitin)
Edited by WILLIS A. WOOD AND SCOTT T. KELLOGG

VOLUME 162. Immunochemical Techniques (Part L: Chemotaxis and Inflammation)
Edited by GIOVANNI DI SABATO

VOLUME 163. Immunochemical Techniques (Part M: Chemotaxis and Inflammation)
Edited by GIOVANNI DI SABATO

VOLUME 164. Ribosomes
Edited by HARRY F. NOLLER, JR., AND KIVIE MOLDAVE

VOLUME 165. Microbial Toxins: Tools for Enzymology
Edited by SIDNEY HARSHMAN

VOLUME 166. Branched-Chain Amino Acids
Edited by ROBERT HARRIS AND JOHN R. SOKATCH

VOLUME 167. Cyanobacteria
Edited by LESTER PACKER AND ALEXANDER N. GLAZER

VOLUME 168. Hormone Action (Part K: Neuroendocrine Peptides)
Edited by P. MICHAEL CONN

VOLUME 169. Platelets: Receptors, Adhesion, Secretion (Part A)
Edited by JACEK HAWIGER

VOLUME 170. Nucleosomes
Edited by PAUL M. WASSARMAN AND ROGER D. KORNBERG

VOLUME 171. Biomembranes (Part R: Transport Theory: Cells and Model Membranes)
Edited by SIDNEY FLEISCHER AND BECCA FLEISCHER

VOLUME 172. Biomembranes (Part S: Transport: Membrane Isolation and Characterization)
Edited by SIDNEY FLEISCHER AND BECCA FLEISCHER

VOLUME 173. Biomembranes [Part T: Cellular and Subcellular Transport: Eukaryotic (Nonepithelial) Cells]
Edited by SIDNEY FLEISCHER AND BECCA FLEISCHER

VOLUME 174. Biomembranes [Part U: Cellular and Subcellular Transport: Eukaryotic (Nonepithelial) Cells]
Edited by SIDNEY FLEISCHER AND BECCA FLEISCHER

VOLUME 175. Cumulative Subject Index Volumes 135–139, 141–167

VOLUME 176. Nuclear Magnetic Resonance (Part A: Spectral Techniques and Dynamics)
Edited by NORMAN J. OPPENHEIMER AND THOMAS L. JAMES

VOLUME 177. Nuclear Magnetic Resonance (Part B: Structure and Mechanism)
Edited by NORMAN J. OPPENHEIMER AND THOMAS L. JAMES

VOLUME 178. Antibodies, Antigens, and Molecular Mimicry
Edited by JOHN J. LANGONE

VOLUME 179. Complex Carbohydrates (Part F)
Edited by VICTOR GINSBURG

VOLUME 180. RNA Processing (Part A: General Methods)
Edited by JAMES E. DAHLBERG AND JOHN N. ABELSON

VOLUME 181. RNA Processing (Part B: Specific Methods)
Edited by JAMES E. DAHLBERG AND JOHN N. ABELSON

VOLUME 182. Guide to Protein Purification
Edited by MURRAY P. DEUTSCHER

VOLUME 183. Molecular Evolution: Computer Analysis of Protein and Nucleic Acid Sequences
Edited by RUSSELL F. DOOLITTLE

VOLUME 184. Avidin–Biotin Technology
Edited by MEIR WILCHEK AND EDWARD A. BAYER

VOLUME 185. Gene Expression Technology
Edited by DAVID V. GOEDDEL

VOLUME 186. Oxygen Radicals in Biological Systems (Part B: Oxygen Radicals and Antioxidants)
Edited by LESTER PACKER AND ALEXANDER N. GLAZER

VOLUME 187. Arachidonate Related Lipid Mediators
Edited by ROBERT C. MURPHY AND FRANK A. FITZPATRICK

VOLUME 188. Hydrocarbons and Methylotrophy
Edited by MARY E. LIDSTROM

VOLUME 189. Retinoids (Part A: Molecular and Metabolic Aspects)
Edited by LESTER PACKER

VOLUME 190. Retinoids (Part B: Cell Differentiation and Clinical Applications)
Edited by LESTER PACKER

VOLUME 191. Biomembranes (Part V: Cellular and Subcellular Transport: Epithelial Cells)
Edited by SIDNEY FLEISCHER AND BECCA FLEISCHER

VOLUME 192. Biomembranes (Part W: Cellular and Subcellular Transport: Epithelial Cells)
Edited by SIDNEY FLEISCHER AND BECCA FLEISCHER

VOLUME 193. Mass Spectrometry
Edited by JAMES A. MCCLOSKEY

VOLUME 194. Guide to Yeast Genetics and Molecular Biology
Edited by CHRISTINE GUTHRIE AND GERALD R. FINK

VOLUME 195. Adenylyl Cyclase, G Proteins, and Guanylyl Cyclase
Edited by ROGER A. JOHNSON AND JACKIE D. CORBIN

VOLUME 196. Molecular Motors and the Cytoskeleton
Edited by RICHARD B. VALLEE

VOLUME 197. Phospholipases
Edited by EDWARD A. DENNIS

VOLUME 198. Peptide Growth Factors (Part C)
Edited by DAVID BARNES, J. P. MATHER, AND GORDON H. SATO

VOLUME 199. Cumulative Subject Index Volumes 168–174, 176–194 (in preparation)

VOLUME 200. Protein Phosphorylation (Part A: Protein Kinases: Assays, Purification, Antibodies, Functional Analysis, Cloning, and Expression)
Edited by TONY HUNTER AND BARTHOLOMEW M. SEFTON

VOLUME 201. Protein Phosphorylation (Part B: Analysis of Protein Phosphorylation, Protein Kinase Inhibitors, and Protein Phosphatases)
Edited by TONY HUNTER AND BARTHOLOMEW M. SEFTON

VOLUME 202. Molecular Design and Modeling: Concepts and Applications (Part A: Proteins, Peptides, and Enzymes)
Edited by JOHN J. LANGONE

VOLUME 203. Molecular Design and Modeling: Concepts and Applications (Part B: Antibodies and Antigens, Nucleic Acids, Polysaccharides, and Drugs)
Edited by JOHN J. LANGONE

VOLUME 204. Bacterial Genetic Systems
Edited by JEFFREY H. MILLER

VOLUME 205. Metallobiochemistry (Part B: Metallothionein and Related Molecules)
Edited by JAMES F. RIORDAN AND BERT L. VALLEE

VOLUME 206. Cytochrome P450
Edited by MICHAEL R. WATERMAN AND ERIC F. JOHNSON

VOLUME 207. Ion Channels
Edited by BERNARDO RUDY AND LINDA E. IVERSON

VOLUME 208. Protein–DNA Interactions
Edited by ROBERT T. SAUER

VOLUME 209. Phospholipid Biosynthesis
Edited by EDWARD A. DENNIS AND DENNIS E. VANCE

VOLUME 210. Numerical Computer Methods
Edited by LUDWIG BRAND AND MICHAEL L. JOHNSON

VOLUME 211. DNA Structures (Part A: Synthesis and Physical Analysis of DNA)
Edited by DAVID M. J. LILLEY AND JAMES E. DAHLBERG

VOLUME 212. DNA Structures (Part B: Chemical and Electrophoretic Analysis of DNA)
Edited by DAVID M. J. LILLEY AND JAMES E. DAHLBERG

VOLUME 213. Carotenoids (Part A: Chemistry, Separation, Quantitation, and Antioxidation) (in preparation)
Edited by LESTER PACKER

VOLUME 214. Carotenoids (Part B: Metabolism, Genetics, and Biosynthesis) (in preparation)
Edited by LESTER PACKER

VOLUME 215. Platelets: Receptors, Adhesion, Secretion (Part B)
Edited by JACEK HAWIGER

VOLUME 216. Recombinant DNA (Part G) (in preparation)
Edited by RAY WU

VOLUME 217. Recombinant DNA (Part H) (in preparation)
Edited by RAY WU

VOLUME 218. Recombinant DNA (Part I) (in preparation)
Edited by RAY WU

VOLUME 219. Reconstitution of Intracellular Transport (in preparation)
Edited by JAMES E. ROTHMAN

Section I

Isolation of Platelet Membranes, Subcellular Organelles, and Cytoskeleton

A. Platelet Membranes and Granules
Articles 2 through 5

B. Platelet Cytoskeleton
Articles 6 through 11

[1] Introduction to Platelet Structural and Functional Organization

By JACEK HAWIGER

The function of platelets evolves from their structure. When one observes platelets at work sealing the cut in a hemorrhaging vessel, their ability to adhere to nonendothelialized surfaces and to aggregate is such an arresting phenomenon that the impression of single-minded functionality cannot be ignored.[1] Platelets circulate in blood as smooth disks of an average volume of 5 to 7.5 μm^3, 14 times smaller than that of erythrocytes. Each day an estimated 3.5×10^{10} platelets per liter of blood are produced by cytoplasmic fragmentation of their mother cell, a megakaryocyte, through two proposed mechanisms.[2,3] Their expected life span is 10 days.[4]

The first encounter of platelets with a damaged vessel wall results in a change of shape from smooth disks into spiny spheres. Pseudopods promote contact with a zone of vascular injury and the subsequent spreading of platelets completes their adhesion.[5] A contractile apparatus harnesses and conveys the energy for this process by rearranging the membrane skeleton and the cytoplasmic actin filaments. In addition to actin and myosin, platelets contain at least nine distinct cytoskeletal proteins involved in assembly, disassembly, and regulation of the contractile apparatus.[6] As in other cells, adhesion is mediated through specific contact sites formed by fibronectin on the outside and extends via integrin receptors to the inside of platelets. Therein, the integrin receptor is clutched by protein P235, analogous to talin.[7] Another adhesive receptor, the GPIb–IX complex, communicating with von Willebrand factor in the extracellular matrix, is linked to short actin filaments that are kept together by an actin-binding protein on the inside of the platelet membrane.[8] An "experiment of nature," in which a deficiency of the GPIb–IX complex in Bernard–Soulier syndrome results in apparently giant platelets, illustrates the role of the

---

[1] J. J. Sixma and J. Wester, *Semin Hematol.* **14**, 265 (1977).
[2] M. Shaklai and M. Tavassoli, *J. Ultrastruct. Res.* **62**, 270 (1978).
[3] J. M. Radley and G. Scurfield, *Blood* **56**, 996 (1980).
[4] L. A. Harker and C. A. Finch, *J. Clin. Invest.* **48**, 963 (1969).
[5] K. S. Sakariassen, R. Mugli, and H. R. Baumgartner, this series, Vol. 169, p. 37.
[6] J. E. B. Fox, *in* "Thrombosis and Hemostasis" (M. Verstate, J. Vermylen, R. Lijnen, and J. Arnout, eds.), p. 175. Leuven Univ. Press, 1987.
[7] T. O'Halloran, M. C. Beckerle, and K. Burridge, *Nature (London)* **317**, 449 (1985).
[8] J. E. B. Fox, *J. Biol. Chem.* **260**, 11977 (1985).

membrane skeleton in maintaining platelet size.[9] The coil of microtubules polymerized circumferentially to the inside of the membrane skeleton provides an additional scaffolding that maintains the discoid shape of platelets.[10]

The surface membrane area increases by an estimated 60% in activated platelets.[11] Pseudopods facilitate the contact of platelets with a vascular surface stripped of endothelium and with each other to form an aggregate. Whereas the predicted radius of pseudopods (0.1 μm) has the "right" dimensions for penetrating the electrostatically repulsive forces on the opposite surface, and for the Ca^{2+}-mediated bonding,[12] The larger protrusions would require the help of complex adhesive molecules.

On close inspection of congeries of platelets heaped together, their paleness due to degranulation and their membranous appearance are particularly striking.[1] Nonactivated platelets carry a load of storage granules. Their contents (nucleotides, serotonin, Ca^{2+} in dense granules, growth factors, adhesive glycoproteins in α granules, and hydrolytic enzymes in lysosomes) are discharged following fusion of the granule membrane with the platelet membrane. Activated platelets acquire glycoprotein constituents from the membrane of secretory granules. For example, platelet activation-dependent granule to external membrane protein (PAGEM) or granule membrane protein 140 (GMP-140) is detected on activated platelets only.[13,14] Inward systems of membrane invaginations, called the surface-connected open canalicular system, provide convoluted channels for transit of secreted constituents. One should be aware that in some species, e.g., oxen, platelets lack the surface-connected open canalicular system involved in membrane-mediated interactions.[15] A dense tubular system, likened to sarcoplasmic reticulum in muscle cells, is prominently involved in Ca^{2+} storage and mobilization.[16]

Wholesale homogenization of platelets by the "brute" force of ultrasound, or by freezing and thawing, will result in a conglomerate of membranes derived from different compartments. Fortunately, gentler techniques such as N_2 cavitation permit isolation of enriched membrane fractions and preserved storage granules for biochemical studies. Such studies are particularly useful in regard to isolation and characterization of membrane receptors and of material stored in granules. Among several

[9] J. G. White, *Hum. Pathol.* **18**, 123 (1987).
[10] J. G. White, this volume [11].
[11] G. V. R. Born, *J. Physiol. (London)* **209**, 487 (1970).
[12] B. A. Pethica, *Exp. Cell Res., Suppl.* **8**, 123 (1961).
[13] C. L. Berman, E. L. Yeo, B. C. Furie, and B. Furie, this series, Vol. 169, p. 311.
[14] R. P. McEver and M. N. Martin, *J. Biol. Chem.* **259**, 9799 (1984).
[15] K. M. Meyers, H. Holmsen, and C. L. Seachord, *Am. J. Physiol.* **243**, R454 (1982).
[16] L. Cutler, G. Rodan, and M. B. Feinstein, *Biochim. Biophys. Acta* **542**, 537 (1978).

classes of membrane receptors reviewed in [12] in this volume, those involved in binding agonists that generate a signal to evoke a response, e.g., secretion, are still not completely understood.

Analysis of platelets, nonactivated and activated, is followed by their dissection to obtain a closer view of their component parts: membranes, cytoskeleton, granules. Another procedure involves studying permeabilized platelets allowing insertion of messenger molecules. Thus, despite their minuscule size, platelets are accessible to more than one method of structural analysis to shed light on their omnipotent functionality in hemostatic and thrombotic processes.

[2] Isolation and Characterization of Platelet Membranes Prepared by Free Flow Electrophoresis

By NEVILLE CRAWFORD, KALWANT S. AUTHI, and NASHRUDEEN HACK

The blood platelet, the smallest of the circulating blood cells ($\sim$2- to 3-μm diameter), has always presented difficulties for biochemical studies involving subcellular fractionation. First, although in the circulation it is a relatively quiescent discoid-shaped cell (Fig. 1a–c), once removed it is readily activated by contact with foreign surfaces or by the generation of excitatory agonists during subsequent handling procedures. Such activation can produce profound morphological changes (from disks to spheres and the formation of filopodia arising from the surface), much reorganization of surface membrane constituents, movement of intracellular granules and their fusion with the surface membrane for release of stored constituents, and, more severely, irreversible aggregation and loss of single cell identity. All these events, which are part of the profile of activation of platelets, present major difficulties for the researcher in the choice of the right anticoagulant and in deciding on the most innocuous isolation and washing procedures, so that changes due to activation are minimized and any analytical or enzymatic data generated can be confidently interpreted in the context of the normal circulating platelet. If this were the extent of the problem it would be formidable, but further difficulties arise in subcellular studies due to the small size of the platelet and its particular resistance to the mechanical shear forces required to rupture the cell, such as one would use in conventional cell homogenization procedures. Moreover, if the physical insult is too severe the integrity of the more sensitive intracellular organelles (lysosomes, mitochondria, and storage granules, etc.) is

FIG. 1. Electron micrographs and diagrammatic representation of quiescent disk-shaped human platelets. (a) Scanning electron micrograph. Magnification: ×5000. (b) Transmission electron micrograph showing intracellular membranes (IM) and granular organelles. Magnification: ×20,000. (c) Diagram of discoid platelet showing internal organelles and cytoskeletal structures. OCS, Open canalicular membrane system (invaginations contiguous with the surface membrane); Mito, mitochondria; DTS, dense tubular membrane system analagous to endoplasmic reticulum; DG, dense granules (5HT storage sites); lys + α gran, lysosomes and α granules (storage sites for procoagulant and mitogenic proteins), indistinguishable morphologically; Mts, actin microfilaments.

destroyed, making it difficult to harvest reasonably discrete subcellular fractions for biochemical studies. Progress has been made, however, in at least partly resolving some of these problems but others still remain about which we have as yet insufficient knowledge to even try to seek the solutions. Two examples are (1) the now well-recognized analytical and functional heterogeneity that exists in the circulating platelet pool: heterogenity in volume, density, surface receptor status, and metabolic compe-

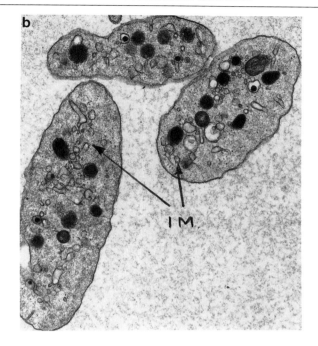

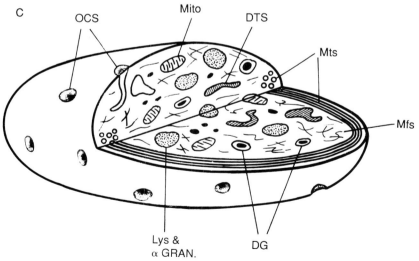

FIG. 1. (continued)

tence, etc., which raises questions about the initial platelet isolation procedures not producing a truly representative cell population, and (2) the certainty that during its sojourn in the circulation the platelet collects and carries around a plasmatic "halo" of loosely associated or adventitiously bound constituents. These reside within the glycocalyx, extrinsic to the components in the plasma membrane bilayer region, and some of these may be of considerable importance in platelet surface interactions and other functions. Again questions are now being asked about the suitability of certain time-honored anticoagulant regimes (and particularly those that complex Ca^{2+}) for the preservation of this loosely adsorbed coating on the circulating cell.

In this chapter we will present some new procedures that we are now routinely using in our laboratory for the isolation and purification of platelet membranes of well-characterized cellular origin. Although they are by no means ideal in the context of some of the aforementioned difficulties, they have led to the definitive localization of some important intracellular processes and analysis of constituents that had hitherto not been possible using more conventional techniques of subcellular fractionation. A cell rupture technique is used that better preserves the integrity of the intracellular granules of the platelets and these can be excluded during the early part of the procedure, so that a mixed membrane fraction can be harvested, essentially free from contamination with the boundary membranes of these storage granules. This mixed membrane fraction can then be separated by free flow electrophoresis into two discrete subfractions representing plasma membrane and intracellular (endoplasmic reticulum-like) membranes and these highly purified membrane fractions have enabled us to determine with more confidence the localization of a number of important membrane constituents and properties.

Materials and Methods

Collection of Blood and Isolation of Platelets

All chemicals and reagents should be of analytical-grade quality wherever possible and all blood- and cell-handling procedures carried out in clear polypropylene or polycarbonate vessels.

Wherever possible platelets are harvested from fresh blood (generally within 2–3 hr of donation) and the anticoagulant regimes and isolation procedures must be customized to the nature of the starting product. An initial isolation procedure, applied to whole blood, has increased the platelet recovery to a level that is now more representative of the full circulating cell pool.

If laboratory donors are used (volumes ~50 ml) the blood is taken into a one-tenth volume of 3.8% (w/v) sodium citrate and a platelet-rich plasma (PRP) is prepared by 200 g centrifugation for 20 min at 22°. With a single centrifugation step the recovery of platelets in the PRP, with respect to original whole-blood platelet counts, rarely exceeds 70%. Variations in yield around this recovery appear to be related to plasma viscosity, side wall effects in the centrifuge tube, and the density heterogeneity of the platelets. To increase the recovery of platelets, the red cell layer after removal of the PRP is diluted and well mixed with a buffer consisting of 36 mM citric acid, 5 mM glucose, 5 mM KCl, 90 mM NaCl, 10 mM EDTA, pH 6.5, and centrifuged at 150 g for 15 min at 22°. Better than 90% recovery of the platelets can be achieved by one such recycling stage. The recycling procedure can be repeated once more to increase the yield (rarely greater than 96%) but the risk of activating the cells is then increased. The original PRP and the recycled supernatant are pooled, centrifuged at 200 g for 5 min to remove any contaminating red cells, and the supernatant acidified to pH 6.4 by the dropwise addition of 0.15 M citric acid (procedure of Lagarde et al.[1]). The platelets are then sedimented at 1200 g for 20 min. This acidification is essential to obtain a subsequent smooth suspension of inactivated cells. The pelleted platelets are resuspended in a buffer consisting of NaCl (150 mM), KCl (4 mM), EDTA (3 mM), and N-2-hydroxyethylpiperazine-N'-2-ethanesulfonic acid (HEPES: 10 mM) at a cell concentration of approximately 10^{10} platelets/ml. The pH of this platelet suspension is adjusted to pH 6.2 if neuraminidase treatment is to be carried out, but otherwise to pH 7.2 for most other purposes.

A similar scaled-up isolation procedure can be used for whole-blood units from the Transfusion Service Laboratories, but if pooled "buffy coat" packs are supplied in which both the residual plasma volume and the cell density are variable and red cell contamination varying from slight to considerable, modifications must be made to the early centrifugation conditions with changes in g force and times appropriate to the material supplied. It is thus unfortunately impossible to offer a standardized procedure for the processing of buffy coat packs, and preliminary trial experiments with cell count monitoring should be carried out to determine the most appropriate g forces and processing times.

---

[1] M. Lagarde, P. A. Bryon, M. Guichardant, and M. Dechavanne, *Thromb. Res.* **17**, 581 (1980).

Neuraminidase Treatment of Whole Platelets

The platelet suspension (up to 10^{10} cells/ml) in the pH 6.2 buffer is first equilibrated to 37° for ~5 min; protease-free neuraminidase (type X; Sigma, St. Louis, MO) is added at a concentration of 0.03–0.05 units (U)/ml and the incubation continued for 20 min. This treatment removes most, if not all, of the neuraminidase-labile sialic acid moieties from the platelet surface membrane, which significantly reduces its electronegativity and improves the resolution of the subsequent free flow electrophoresis fractionation. At the end of the incubation period, the platelet suspension is rapidly diluted with 3 vol of cold washing buffer, which is identical to the aforementioned suspension buffer but with the pH now adjusted to 7.2. The mixture is centrifuged at 1200 g for 15 min, the supernatant removed, and the pelleted cells carefully resuspended in a further volume of washing buffer and recentrifuged under the same conditions. To assist in the later identification of the surface membranes any desired fluorescent or radioactive surface probe such as an antibody or lectin can be applied at this stage or the cells can be iodinated with ^{125}I using the lactoperoxidase procedure. Our method for labeling the platelets with ^{125}I-labeled lectin from *Lens culinaris* by the Greenberg and Jamieson[2] procedure has been earlier reported by Menashi *et al.*[3] The platelet pellet, after washing, is carefully resuspended in cold sonication buffer (0.34 M sorbitol, 10 mM HEPES, pH 7.2) at a concentration approximating 4 ml buffer/g cells (wet weight). This sonication buffer also contains a protease inhibitor cocktail that can be varied according to experimental requirements, but generally includes (at their final concentrations) the following: aprotinin (0.1 U/ml), pepstatin (2.5 μg/ml), and leupeptin (25 μg/ml). Our studies have revealed that the most important components of the cocktail for preserving the integrity of membrane enzymes, and channels involved in ion transport, are pepstatin and aprotinin, and for preparations to be used for polypeptide and glycopeptide characterization leupeptin is an additional required inhibitor. The platelet suspension is then sonicated (Dawes Sonifier, position 6 at maximum tuning) for 10 sec while held at 4°. The suspension is centrifuged (1200 g, 15 min at 4°), the supernatant homogenate removed and saved, and the pellet of unbroken cells and large cell fragments resuspended in a further volume of cold sonication buffer in which the protease inhibitors are again included. A further 10-sec sonication is applied and after centrifugation the two supernatants are pooled and centrifuged at 1200 g for 10 min to deposit any large aggregates. This pooled homogenate is applied to the upper surface of a linear sorbital gradient constructed

---

[2] J. H. Greenberg and G. A. Jamieson, *Biochim. Biophys. Acta* **345**, 231 (1974).
[3] S. Menashi, H. Weintroub, and N. Crawford, *J. Biol. Chem.* **256**, 4095 (1981).

from sorbitol solutions (1.0 and 3.5 M) buffered to pH 7.2 with 10 mM HEPES. Sucrose solutions of similar density characteristics can be used but some minor contaminants in commercial supplies of sucrose leach off certain bound lectins if these are used as surface markers, leading to redistribution artifacts in the separations. Sorbitol appears to be inert in this respect. Approximately 20 ml of homogenate can be applied to a 40-ml sorbitol gradient and the same ratio of "payload" to sorbitol can be used for smaller volume gradients. The density gradient is centrifuged at 42,000 g for 90 min at 4° in a swing-out rotor. Figure 2 shows a flow chart of the platelet isolation and sonication procedures and a diagram of the density gradient before and after centrifugation. The mixed membrane band, which contains both surface and intracellular membrane vesicles, locates discretely in the middle of the gradient and is free from granule contamination as judged by 5-HT (5-hydroxytryptamine) determinations, and lysosome and mitosol mitochondrial marker enzyme assays. This upper membrane zone is removed from the gradient and centrifuged at 100,000 g for 90 min to recover the mixed membranes. These membranes are carefully resuspended in chamber buffer (see below) for the electrophoretic separation.

Free Flow Electrophoresis of the Mixed Membrane Fraction

For free flow electrophoresis we have used either an Elphor VAP 5 or VAP 22 apparatus (Bender & Hobein, Munich, Germany) using chamber and electrode buffers of the following composition:

Chamber buffer: 0.4 M sorbitol, 10 mM triethanolamine, 10 mM acetic acid, adjusted to pH 7.2
Electrode buffer: 100 mM triethanolamine, 100 mM acetic acid, adjusted to pH 7.2

Before the run, the glass plates of the electrophoresis chamber are coated with albumin by filling the chamber with a solution of 3% (w/v) bovine serum albumin (Fraction V bovine albumin; Sigma), allowing it to stand for 1 hr and washing the chamber with 2–3 vol of chamber buffer. The pelleted mixed membranes from the sorbitol density gradient are suspended in chamber buffer to a smooth suspension (~3–4 mg/ml membrane protein) and applied to the separating chamber through a port on the cathodal side of the upper part of the chamber. During sample input, the cell suspension is held at 6° and gently rotated to maintain homogeneity. The following have been found to be the optimum conditions for good separations:

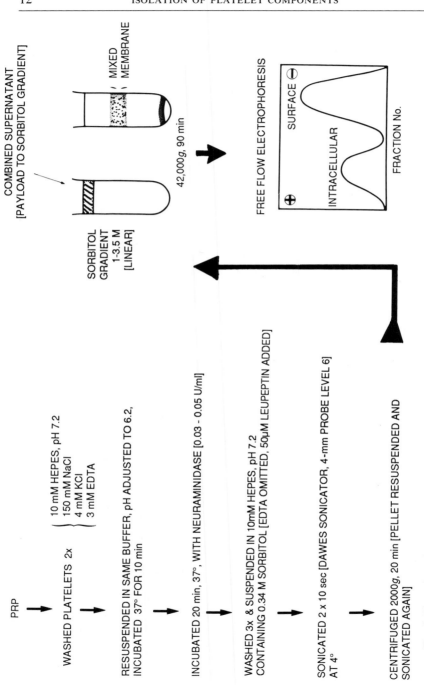

FIG. 2. Flow chart of platelet isolation procedure, neuraminidase treatment, sonication, and sorbitol density gradient fractionation to produce the mixed membrane fraction for further separation by free flow electrophoresis. PRP, Platelet-rich plasma.

Flow Rates

Membrane sample: 1.8–2.0 ml/hr
Chamber buffer: 2.0–2.4 ml/hr/fraction
Chamber temperature: 6.5°
Field strength: 1200 V at 130 mA (~120 V/cm)

Separation can be conducted on a preparative scale and the apparatus is sufficiently temperature and current stable to run continuously for many hours. Under the above operating conditions the complete membrane profile extends between fractions numbers 20–50, which are on the anodal side of the sample port in the electrophoretic chamber. At the end of the electrophoretic run the two membrane peaks (Fig. 3a) can be located in the fraction collector tubes by measuring their absorbance at 280 nm against a chamber buffer blank. Protein analyses of either individual fractions or pooled fractions taken across the peaks are performed by the microtannin procedure of Mejbaum-Katzenellenbogen and Dobryszycka,[4] a method which minimizes interference by sorbitol. A standard curve of bovine albumin (fraction V; Sigma) in the range 0–75 μg is generally used. Individual or pooled fractions taken across the peaks can be concentrated by centrifugation (100,000 g for 90 min) and resuspended for analytical and enzymatic investigations. All these analytical procedures and enzyme assays have been described in our original publications.[3]

Modifications to Procedures for Specific Purposes

1. For studies of Ca^{2+} transport it is necessary to produce a smooth membrane suspension and to keep the subfractions of membrane vesicles well sealed. This is achieved by centrifuging the gradient mixed membrane fraction (100,000 g, 90 min) and the pooled membrane subfractions from the electrophoresis fractions onto a cushion of 3.5 M sorbitol buffered to pH 7.2 with 10 mM HEPES. Studies of other transport systems may also require this modification.

2. Sucrose at 3.0 M can be substituted for sorbitol throughout the procedure but our studies have revealed that sorbitol is less likely to elute membrane-bound lectins used for labeling which would lead to redistribution artifacts.

3. In earlier studies,[3] 1 mM EDTA was included in the sorbitol solutions for preparation of the density gradient. This modification resulted in the surface membrane peak splitting into two distinct components (see Fig. 3b). Both these surface membrane subfractions show bound lectin although the concentration of ^{125}I-labeled lectin from *Lens culinaris* in the

---

[4] W. Mejbaum-Katzenellenbogen and W. M. Dobryszycka, *Clin. Chim. Acta* **4**, 515 (1969).

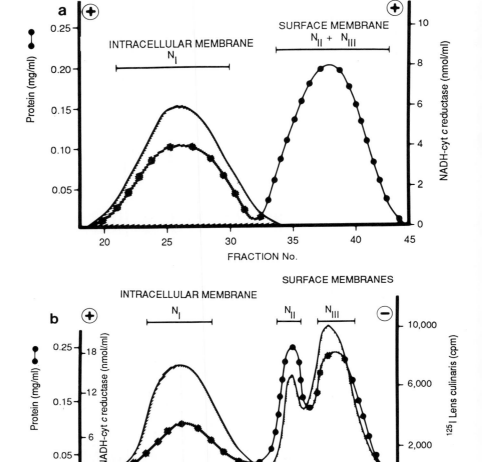

FIG. 3. Free flow electrophoresis profiles of separated mixed membrane fractions from (a) sorbitol gradient not containing EDTA and (b) sorbitol gradient in which 1 mM EDTA was included (showing subfractionation of surface membrane; see text).

least electronegative subfraction is greater than that of the most electronegative surface membrane subfraction. Similarly, specific activities for adenylate cyclase differ in the two fractions, as does the cholesterol and phospholipid content (although the cholesterol–phospholipid ratios are essentially similar). In SDS–polyacrylamide gel separations the most significant difference between these two surface membrane subfractions is, however, in the presence of the heavy chain component (200 kDa) of the contractile protein, myosin. Both surface membrane subfractions show a prominent 43-kDa band that has been identified as actin but the myosin heavy chain is present only in the more electronegative of the two surface membrane subfractions. At present we do not know if this EDTA-dependent phenomenon indicates the separation of two different surface domains of the same platelet or arises through subpopulation heterogeneity of the circulating cell pool, with perhaps the myosin-rich membrane subfraction originating from platelets that have been slightly activated *in vivo* or during their isolation and subsequent *in vitro* handling. For most routine purposes we separate according to the profile shown in Fig. 3a, obtained by omitting EDTA from the sorbitol density gradients.

Analytical and Enzymatic Characterization

Comparative studies with mixed membrane fractions taken directly from the density gradients after fractionating sonicates from neuraminidase-treated and nontreated platelets confirmed that this enzymatic modification to the surface membrane does not affect the lipid and phospholipid composition, the polypeptide profiles, or the specific activities of any of the enzymes measured. However, possible topographical changes in membrane constituents induced by the removal of sialic acid and that might affect membrane properties such as receptor status, nonenzymatic ion transport channels, or transmembrane links with cytoskeletal structures cannot as yet be ruled out, but at present we have no evidence that these are significantly disturbed.

Table I summarizes the analytical data for the human platelet surface and intracellular membranes prepared by the combined density gradient/free flow electrophoresis procedure. Similarly, Table II shows the major enzymes present and their specific activities in the two membrane fractions. Practical details and references to the enzyme assays used in this characterization have been presented earlier.[3,5]

To emphasize the analytical disparity between the surface and intracellular membranes, Figs. 4a and b show two-dimensional SDS–polyacryl-

---

[5] M. Lagarde, M. Guichardant, S. Menashi, and N. Crawford, *J. Biol. Chem.* **257**, 3100 (1982).

TABLE I
Analytical Data for Human Platelet Surface and Intracellular Membranes Isolated by Free Flow Electrophoresis[a]

Constituent	Surface membrane	Intracellular membrane
Cholesterol (μmol/mg)	0.69 ± 0.06	0.27 ± 0.03
Phospholipids (μmol/mg)	0.93 ± 0.06	0.92 ± 0.05
Cholesterol–phospholipid molar ratio	0.74 ± 0.05	0.29 ± 0.01
Sphingomyelin[b]	24.8 ± 1.8	2.6 ± 2.6
Phosphatidylserine[c]	2.8 ± 0.06	3.0 ± 0.07
Phosphatidylethanolamine[c]	41.9 ± 1.5	28.5 ± 1.4
Phosphatidylcholine[c]	50.7 ± 2.5	60.3 ± 2.0
Phosphatidylinositol[c]	4.4 ± 1.5	8.7 ± 0.7
Microviscosity (poise)		
25°	4.7	3.1
37°	2.5	2.0

[a] Values are means ± SD for at least three determinations.
[b] Percentage of total lipids.
[c] Percentage of sum of all glycerophospholipids.

TABLE II
Enzyme Activities for Platelet Surface and Intracellular Membranes Separated by Free Flow Electrophoresis

	Membrane activities	
Enzyme	Surface	Intracellular
Adenylate cyclase[a]	8.0	1.0
NADH–cytochrome c reductase[b]	0	1600
Leucine aminopeptidase[c]	0.2	1.4
5′-Nucleotidase[d]	7.0	220
Ca^{2+},Mg^{2+}-ATPase[b]	ND[e]	60
Phospholipase A_2[d]	1.4	20.9
Diglyceride lipase[d]	2.9	23.1
Cyclooxygenase[a]	50	1200
Thromboxane synthase[a]	50	1600
Lipoxygenase[d]	108	180
Fatty acyl-CoA transferase[b]	11.1	115

[a] Picomoles per minute per milligram.
[b] Nanomoles per minute per milligram.
[c] Micromoles per hour per milligram.
[d] Nanomoles per hour per milligram.
[e] ND, Not detected.

TABLE III
POLYPEPTIDES OF HUMAN PLATELET SURFACE MEMBRANES SEPARATED TWO-DIMENSIONALLY BY POLYACRYLAMIDE GEL ELECTROPHORESIS/ISOELECTRIC FOCUSING[a]

Identity[b]	Approximate molecular weight range ($\times 10^{-3}$)	Approximate pI range
Microtubule-associated polypeptides (MAPs)	300	6.0–5.2
ABP (platelet "filamin")	260	6.0–7.0
P235 (platelet "talin")	235	6.4–7.0
MHC	200	5.7–6.7
GPIa[c]	150–165	5.3–6.0
GPIbα	135–150	5.0–5.8
GPIIa[c]	125–135	5.5–5.7
GPIIbα[c]	115–130	5.3–6.1
GPIIIa[c]	90–95	5.2–5.8
α-Actinin	95	5.9–6.2
GPIIIb (IV)	90–95	5.2–5.4
Gelsolin	85	6.2, 6.3, 6.4
GPV	80–85	6.0–6.5
Actin	43	5.6–5.8
Tropomyosin	29	5.4–5.6
GPIIbβ	25–28	5.1–5.6
GPIbβ	21–23	6.4–7.0
GP17	17–20	6.4–6.7

[a] Under reducing conditions and detected by silver staining.
[b] For glycoprotein nomenclature the Roman numerals and lower-case letters follow the rank order of molecular weight in nonreduced electrophoresis. The α and β notation represents different subunits revealed after reduction.
[c] These glycoproteins have intrachain disulfide bonds and show lower molecular weight values in nonreducing gels.

amide/isoelectric focusing gel preparations of the two membranes run under reducing gel conditions. All the previously known surface membrane glycoproteins (GPI to GPVII), as determined by others from solubilized preparations of whole platelets using SDS–polyacrylamide gel electrophoresis, have been identified in our surface membrane fractions.[6] There are only traces of these glycoproteins in the intracellular membranes.

The importance of the cocktail of proteolytic inhibitors (and particularly the inclusion of leupeptin) in maintaining the integrity of these glycoproteins must be reemphasized, however. Table III lists all the surface

[6] N. Hack and N. Crawford, *Biochem. J.* **222**, 235 (1984).

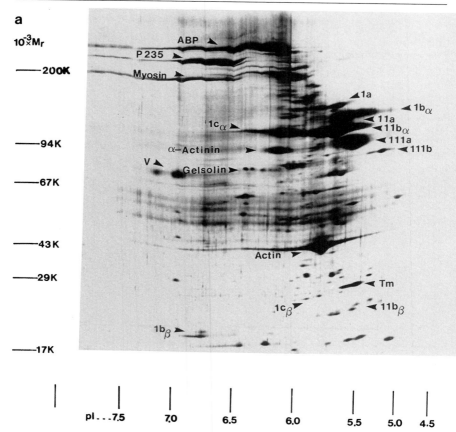

FIG. 4. Two-dimensional sodium dodecyl sulfate (SDS)-polyacrylamide–isoelectric focusing gel separations of (a) human platelet surface membranes and (b) intracellular membranes. Both membrane fractions were prepared from neuraminidase-treated platelets by the combined density gradient/free flow electrophoresis procedures (see text). The gels were run under reducing conditions and silver stained. All components identified with certainty have been labeled. ABP, Actin binding protein; CBP, concanavalin A binding protein not present in surface membranes; V, vinculin; Tm, tropomyosin; for the glycoproteins the conventional nomenclature has been used, with α and β representing the subunits glycoproteins of Ib, Ic, and II.

membrane polypeptides and glycopeptides that are either not detectable or present in trace amounts only in the intracellular membranes. To date only a few of the components of the intracellular membranes have been definitively identified.

Although electron microscopists have coined the term dense tubular membrane system (DTS) for the complex of intracellular membranes

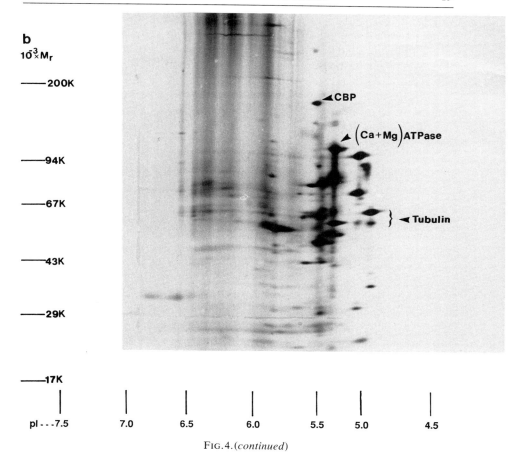

FIG. 4. (continued)

seen coursing through the cytoplasm of platelets, we believe from our analytical, enzymatic, and functional studies that the term "platelet endoplasmic reticulum" is now a more appropriate one for these platelet intracellular membrane elements. These membranes contain the enzymes required for the liberation of arachidonic acid from phospholipids and diacylglycerols, the phospholipase A_2 and diacylglycerol lipase,[7,8] and additionally the enzymes required for the conversion of arachidonic acid to prostanoids, cyclooxygenase and thromboxane synthase.[9] The polypep-

[7] M. Lagarde, K. S. Authi, and N. Crawford, *Biochem. Soc. Trans.* **10**, 241 (1982).
[8] K. S. Authi, M. Lagarde, and N. Crawford, *FEBS Lett.* **180**, 95 (1985).
[9] F. Carey, S. Menashi, and N. Crawford, *Biochem. J.* **204**, 847 (1982).

tide target (a 72-kDa protein) for aspirin acetylation has also been shown to be exclusively localized in the intracellular membranes.[10]

Using ^{45}Ca we have also demonstrated the involvement of these intracellular membrane structures in Ca^{2+} mobilization.[11] They have an active Ca^{2+},Mg^{2+}-ATPase and a protein kinase, both of which appear to be kinetically linked to the sequestration of Ca^{2+}, a property analogous to that of muscle sarcoplasmic reticulum.[12] We have also demonstrated that the intracellular membrane vesicles, preloaded to steady state levels with Ca^{2+}, release ~40–50% of their stored Ca^{2+} 15–30 sec after the addition of inositol 1,4,5-trisphosphate [$Ins(1,4,5)P_3$], a cleavage product of the action of phospholipase C on phosphatidylinositol 4,5-bisphosphate.[13] The half-maximal effect for $Ins(1,4,5)P_3$ is around 0.25 μM, with no further increase in Ca^{2+} release above 0.8 μM. These studies suggest that blood platelets are analogous in these respects to other secreting cells in that the metabolism of phosphatidylinositols is closely linked to Ca^{2+} mobilization from intracellular sequestered sites and that both $Ins(1,4,5)P_3$ and Ca^{2+} may have important second messenger roles in transducing signals generated at the platelet surface membrane by stimulus–receptor interactions.

Clearly, the introduction of free flow electrophoresis into the armory of techniques for studying blood platelets at the subcellular level has many advantages, particularly for the differential separation of surface and intracellular membrane subfractions. It is capable of producing good yields of highly purified membrane vesicles with reasonable functional integrity. The approach could well be extended to the subfractionation and characterization of other platelet intracellular organelles for detailed biochemical studies and may have wider application to other cell entities. For a review of the more general applications of the technique of high-voltage free flow electrophoresis to the study of different cells, to protein fractionation, and for the separation of certain intracellular features the reader is referred to the papers of Hannig and Heidrich.[14–16]

Acknowledgments

We are grateful for the financial support of The Royal Society, the British Heart Foundation, and the Medical Research Council for our studies and our thanks are also due to Heather Watson for typing this manuscript.

[10] N. Hack, F. Carey, and N. Crawford, *Biochem. J.* **223**, 105 (1984).
[11] S. Menashi, K. S. Authi, F. Carey, and N. Crawford, *Biochem. J.* **222**, 413 (1984).
[12] N. Hack, M. Croset, and N. Crawford, *Biochem. J.* **233**, 661 (1981).
[13] K. S. Authi and N. Crawford, *Biochem. J.* **230**, 247 (1985).
[14] K. Hannig and H.-G. Heidrich, this series, Vol. 31, p. 746.
[15] K. Hannig, *Electrophoresis* **3**, 235 (1982).
[16] H. G. Heidrich and K. Hannig, *Methodol. Biochem.* **8**, 91 (1979).

[3] Homogenization by Nitrogen Cavitation Technique Applied to Platelet Subcellular Fractionation

By M. JOHAN BROEKMAN

Background

Since the early 1970s we have utilized the nitrogen cavitation technique to prepare platelet subcellular fractions for biochemical studies. Initially, we distinguished platelet lysosomes (primary lysosomes, organelles that contain acid hydrolases) morphologically and biochemically from platelet α granules.[1] Subsequently, we utilized the technique to establish α granules as the storage site of platelet fibrinogen and platelet factor 4,[2] and to determine differences and similarities in phospholipid and fatty acid composition of various platelet subcellular fractions.[3] Subsequently, we localized additional platelet-specific proteins to α granules: β-thromboglobulin and platelet-derived growth factor,[4] fibronectin,[5] and von Willebrand factor (vWF).[6] More recently, nitrogen cavitation was utilized in a partial characterization of arachidonic acid releasing activity in isolated platelet membranes.[7]

Nitrogen cavitation is a gentle technique for the large-scale preparation of a variety of platelet subcellular fractions. Thus, we have utilized this technique for isolation of α granules, membranes, alluded to above, as well as for the separation of platelet lysosomes and mitochondria.[8]

Platelet Isolation

Reagents and Supplies

Whole blood, collected into ACD (trisodium citrate, 75 mM, glucose, 135 mM, citric acid, 38 mM) anticoagulant (6 vol blood per volume ACD), preferably in a triple pack

---

[1] M. J. Broekman, N. P. Westmoreland, and P. Cohen, *J. Cell Biol.* **60**, 507 (1974).
[2] M. J. Broekman, R. I. Handin, and P. Cohen, *Br. J. Haematol.* **31**, 51 (1975).
[3] M. J. Broekman, R. I. Handin, A. Derksen, and P. Cohen, *Blood* **47**, 963 (1976).
[4] K. L. Kaplan, M. J. Broekman, A. Chernoff, G. R. Lesznik, and M. Drillings, *Blood* **53**, 604 (1979).
[5] M. B. Zucker, M. W. Mosesson, M. J. Broekman, and K. L. Kaplan, *Blood* **54**, 8 (1979).
[6] M. B. Zucker, M. J. Broekman, and K. L. Kaplan, *J. Lab. Clin. Med.* **94**, 675 (1979).
[7] S. T. Silk, K. T. H. Wong, and A. J. Marcus, *Biochemistry* **20**, 391 (1981).
[8] M. J. Broekman, Ph.D. Thesis, p. 65 (1976).

or

Platelet-rich plasma (PRP, purchased from the local blood center, or prepared from whole blood by removal of plasma after centrifugation at 600 g for 10 min)

Citrate solution: 38 mM citric acid, 75 mM trisodium citrate

Tris–citrate buffer: 63 mM Tris (HEPES is a good alternative for many studies), 95 mM NaCl, 5 mM KCl, 12 mM citric acid; adjust pH to 6.5 with concentrated HCl

Washed platelet suspensions are prepared using the plastic bag system at 4°.[9] Initial centrifugation is at 1000 g, 4°, for 10 min). The PRP is expressed into the first satellite bag, and the bag containing erythrocytes and leukocytes is discarded. Platelet cyclooxygenase activity can be checked prior to further platelet processing by monitoring O_2 consumption following collagen stimulation of platelet-rich plasma.[10] Prior to the following centrifugation steps, the satellite bag is blown up with air until taut. This facilitates the separation of liquid phase from pellet. Contaminating erythrocytes and leukocytes remaining in the PRP are removed by a slow spin (225 g, 4°, 10 min). The PRP is expressed into the second satellite bag, and acidified with citrate solution (1 ml citrate solution per 8 g PRP). The bag is filled with air again, and platelets are pelleted (2000 g, 4°, 10 min). The platelet-poor plasma is expressed back into the bag containing erythrocytes and leukocytes pelleted in the 225-g spin, for disposal.

To wash, the platelet pellet is resuspended in 3 ml Tris–citrate buffer by gentle massaging. Tris–citrate buffer (50 ml) is added, and the bag blown taut. The platelets are then pelleted again by centrifugation (2000 g, 4°, 10 min), and resuspended in 3 ml Tris–citrate buffer by gentle massaging. This wash is repeated once to further free the suspension of plasma proteins. Final resuspension is again in Tris–citrate buffer, by gentle massaging of the bag. Platelets so prepared are essentially devoid of polymorphonuclear leukocytes, and only rare erythrocytes or lymphocytes are encountered when examined by phase-contrast microscopy.

To prepare 9 ml or slightly more of final homogenate, the final suspension is transferred to a round-bottom polypropylene tube (see below), and adjusted with Tris–citrate buffer to an appropriate volume: 12 ml was routinely utilized in order to accommodate a sample of the total homoge-

---

[9] M. J. Broekman, *Biochem. Biophys. Res. Commun.* **120**, 226 (1984).
[10] N. M. Bressler, M. J. Broekman, and A. J. Marcus, *Blood* **53**, 167 (1979).

nate for enzyme assays, and to account for losses during nitrogen cavitation homogenization.

We did not use protease inhibitors during most homogenization studies. Nevertheless, platelet subcellular fractions prepared by our technique did not appear to be proteolyzed as judged by Laemmli slab gel electrophoresis (M. J. Broekman, unpublished data, 1977). Addition of EDTA or EGTA to the platelet suspension prior to homogenization to inhibit Ca^{2+}-mediated effects resulted in formation of a gel. The possibility that gel formation can be suppressed by use of the potassium salt of EDTA should be considered. In addition, for critical studies of specific constituents, proteolysis could be inhibited by inclusion of leupeptin, or pretreatment of platelets with diisopropyl fluorophosphate. The latter should of course be handled with extreme care.

Platelet Homogenization

Supplies

Cell disruption bomb, e.g., Parr 4635 (Parr Instrument Co., Moline, IL)
Cylinder of nitrogen ("preparative" grade is sufficiently pure): A large size is preferred
High-pressure two-stage regulator: This is optional, but assists in keeping the pressure in the bomb constant while nitrogen dissolves in its contents
Beaker (250 ml)
Round-bottom and conical polypropylene tubes (50 ml)

The platelet suspension is transferred to the round-bottom polypropylene tube and placed in a beaker, which is filled with sufficient ice-cold water to make the polypropylene tube, when almost empty, float. This ensures that as much as possible of the suspension will be forced out of the bomb into the collection tube. The beaker is placed inside the body of the bomb, which is kept on ice (see Fig. 1).

The lid of the bomb is then assembled on the body of the bomb (to assist in proper sealing, it is useful to wet the O-ring between the lid and body of the bomb). The bomb is connected to the outlet of the nitrogen cylinder. If the high-pressure regulator is not used, the nitrogen filling connection supplied with the bomb is directly connected to the nitrogen cylinder. When using a high-pressure regulator, the CGA fitting of the Parr nitrogen filling connection is removed. The filling connection is then fastened into the outlet of the two-stage high-pressure regulator.

First, all valves are closed. Then, the main valve of the nitrogen cylinder is opened, and the pressure at the outlet of the regulator is adjusted to 1200 psi (80 atm). Subsequently, the outlet valve of the regulator and the inlet valve of the bomb are opened. After ascertaining that the outlet valve of the bomb, as well as the small dump valve on the filling connection, are indeed closed, the big filling valve on the filling connection is slowly opened until a firm but not excessive flow of nitrogen ensues.

The bomb is kept on ice at 1200 psi with the connection between nitrogen cylinder and bomb open, so that pressure inside the bomb remains at 1200 psi, while nitrogen dissolves in the platelet sample. After 15 min the inlet valve of the bomb is closed, as are the big valve on the filling connection and the outlet valve on the two-stage pressure regulator. The nitrogen inside the filling connection and its tubing is vented to the atmosphere by opening the small valve on the filling connection. The filling connection can now be disconnected from the bomb. Subsequently the bomb is transferred (while pressurized) to the bench, care being taken that no ice remains under it (ice may provide lubrication, allowing the bomb to "walk" off the bench).

A conical 50-ml polypropylene collection tube is held under the outlet tube at an angle of 45°, with its opening pointing away from the operator (Fig. 1). The outlet valve is opened, and approximately half of the sample is collected. This collection tube is now put on ice, while the remainder of the sample is "bombed" into a second collection tube, and transferred to the first tube in small aliquots. This serves as a precaution against excessive sample loss, which can occur when the last small amount of sample leaves the bomb with a "sneeze," blowing into and sometimes out of the collection tube.

This first "bombing" is most often not sufficient to disrupt all platelets. Therefore the procedure is repeated two more times, which results in homogenization of more than 98% of the platelets. Separation of homogenate from unbroken cells is not necessary prior to subsequent bombings. Thus samples of platelets, usually derived from 2 units of whole blood, are bombed three times.

Homogenates can be centrifuged (500 g, 4°, 5 min) to remove debris and partially disrupted cells, prior to loading onto sucrose gradients.

Gradient Centrifugation

Supplies

EDTA (100 mM), adjusted to pH 7.4
Sucrose solutions: 10, 30, and 60% sucrose (w/w) solutions are made to contain 5 mM (final concentration) EDTA. It is advisable to prepare

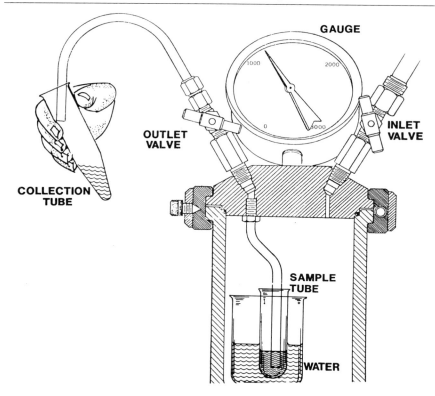

FIG. 1. Nitrogen cavitation bomb set up for homogenization of platelets. See text for details.

a large amount of 66% (w/w) sucrose, since it can be stored almost indefinitely at room temperature without contamination. Tables are available to prepare different concentrations of sucrose solutions from this stock solution.[11]

NaCl (0.154 M) and EDTA (5 mM, pH 7.4)

Ultracentrifugation equipment: e.g., SW 40 swinging bucket rotor (Beckman, Palo Alto, CA)

Linear sucrose gradients are generated from 30 and 60% (w/w) sucrose (containing 5 mM EDTA, pH 7.4). A commercial gradient-forming apparatus (e.g., Beckman #350052) can be used for formation of continuous

[11] O. M. Griffith, in "Techniques of Preparative, Zonal, and Continuous Flow Ultracentrifugation," company brochure. Beckman Instruments, Palo Alto, California, 1976.

gradients. Because of their viscosity, mixing of sucrose solutions of higher concentrations should be performed carefully and thoroughly. Incomplete mixing of 30 and 60% sucrose solutions is clearly visible in the effluent of gradient generators as a schlieren effect. To decrease its viscosity, the 60% sucrose solution should be used at room temperature. Keeping the tubes, into which the gradients are formed, in ice ensures that the gradients are at 4° when ready for loading. Alternatively, roughly linear gradients can be prepared by layering aliquots of sucrose solution of decreasing concentration (60, 55, and 50%, etc.), and leaving this step gradient at 4° overnight. Diffusion will have resulted in an approximately linear gradient the following day.

For a 14-ml ultracentrifuge tube, as utilized in a Beckman SW 40 Ti swinging bucket rotor, 12-ml gradients are layered over a "cushion" of an extra 0.5 ml of 60% (w/w) sucrose. This prevents pelleting of the heaviest fraction (enriched in dense bodies) to the bottom of the gradient tube. Homogenate (1.5 ml) is then carefully layered onto each of the six gradients to be spun. A six-place rotor will also be balanced with three or four tubes, although different maximum speeds may apply under such conditions.

Fraction Collection

Following ultracentrifugation for 90 min at 217,000 g_{max} at 4°, nine bands can be identified as indicated in Fig. 2. The zones are collected by careful pipetting from the top, or by use of commercial fraction collection equipment (e.g. Nyegaard, Oslo, Norway). The collected fractions can be analyzed directly, or sucrose can be removed from particulate fractions by diluting in an appropriate diluent (e.g., 10% sucrose), pelleting by ultracentrifugation, and gentle resuspension [e.g., NaCl/EDTA (154 mM/5 mM, pH 7.4)]. Resuspension can be achieved either with Potter–Elvehjem-type pestles, fitted to the ultracentrifuge tubes utilized for pelleting the isolated fractions, or with disposable 1-ml plastic syringes without needles. Repeated pipetting of the pelleted fractions with the syringes results in gentle resuspension, especially in media containing EDTA.

With the use of these syringes the final volume in which fractions are resuspended can be accurately determined. This is important for calculating total enzyme activity in each fraction, necessary for accurate bookkeeping of the distribution of enzymes over the gradient. This is considered essential for accurate localization of subcellular organelles in the gradient. Figures 3 and 4 show representative distributions of platelet subcellular components achieved by use of our technique. A comparison of Figs.

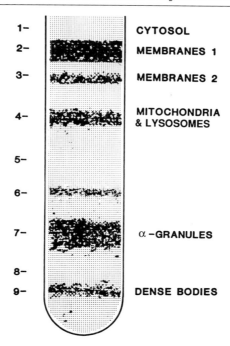

FIG. 2. Schematic representation of ultracentrifugal separation of platelet subcellular organelles following nitrogen cavitation homogenization. Structural identification was by marker enzyme localization and morphological identification. See text and Ref. 1.

3 and 4 shows the clean separation of α granules from lysosomes and mitochondria, as well as from dense bodies. This was also shown by electron micrographs.[1]

Assays

Specific assays can be utilized to determine the presence of various subcellular organelles.[1] The modification[12] by Miller of the Lowry protein determination[13] is conveniently utilized for larger numbers of protein assays. The following assays are used to establish the localization of subcellular organelles:

[12] G. L. Miller, *Anal. Chem.* **31**, 964 (1959).
[13] O. H. Lowry, N. J. Rosebrough, A. L. Farr, and R. J. Randall, *J. Biol. Chem.* **193**, 265 (1951).

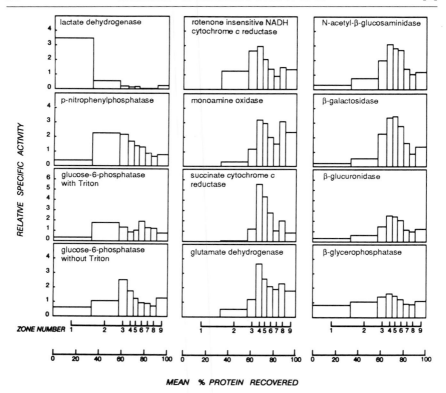

FIG. 3. Marker enzyme distributions over the nine zones depicted in Fig. 2. The data are plotted according to de Duve et al.[20] On the ordinate is the relative specific activity (percentage activity recovered/percentage protein recovered); on the abscissa, the percentage protein recovered per zone. (Reproduced from Broekman et al.,[1] by copyright permission of the Rockefeller University Press.)

1. Lactate dehydrogenase[14] activity is utilized as a marker enzyme for cytosol.

2. Acid p-nitrophenylphosphatase[15] appears to be the easiest marker enzyme for routine determination of plasma membrane localization. We have been unable to find in human platelet preparations sufficient amounts of enzymatic activities, characteristic for plasma membrane in other tissues, such as 5'-nucleotidase or alkaline phosphatase (a good marker for leukocyte plasma membranes). Thus, measurement of the distribution of platelet plasma membrane glycoproteins (e.g., glycoprotein Ib) is probably

[14] F. Wrobleski and J. S. LaDue, *Proc. Soc. Exp. Biol. Med.* **90**, 210 (1955).
[15] H. J. Day, H. Holmsen, and T. Hovig, *Scand. J. Haematol., Suppl.* **7**, 3 (1969).

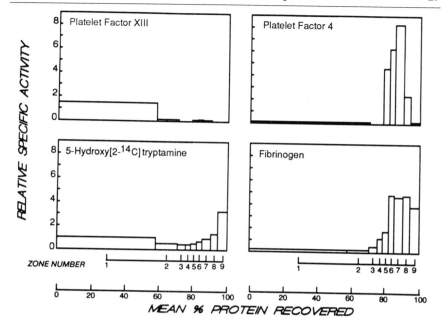

FIG. 4. Distribution of platelet factor XIII, platelet factor 4, fibrinogen, and [^{14}C]serotonin uptake among human platelet subcellular fractions as shown in Fig. 2. The data are plotted[20] as in Fig. 3. (Reproduced from Broekman et al.,[2] by copyright permission of the publisher.)

the most accurate approach available at present to a definitive establishment of the localization of platelet plasma membranes in a given subcellular fractionation technique. If practical, such measurements should be performed on the preparation(s) in which the "new" plasma membrane constituent is analyzed.

3. Rotenone-insensitive NADH–cytochrome c reductase[16] (EC 1.6.99.5) is a marker for endoplasmic reticulum (ER), although it also exhibits minor activity in mitochondria. This is explained by the observation that the mitochondrial outer membrane is contiguous with endoplasmic reticulum.[17] This enzymatic activity can easily be determined analogously to the mitochondrial enzyme succinate–cytochrome c reductase (EC 1.6.99.1): Both depend on the spectrophotometric determination of changes in reduced cytochrome c content (see below). Another ER marker

---

[16] G. L. Sottocasa, B. Kuylenstierna, L. Ernster, and A. Bergstrand, *J. Cell Biol.* **32**, 415 (1967).

[17] P. Siekevitz, *N. Engl. J. Med.* **283**, 1035 (1970).

may be Triton-inhibited glucose-6-phosphatase.[18] We initially suggested this marker in 1974, based on four criteria: High rate of enzymatic hydrolysis, peaking of activity in membrane containing zones, inhibition of activity by Triton X-100 (which should have *increased* lysosomal phosphatases), and comparative inhibition studies where, in contrast to β-glycerophosphate, hydrolysis of glucose 6-phosphate was inhibited less by NaF than by *p*-hydroxymercuribenzoate (Ref. 1).

4. Several enzymes can be utilized for the localization of mitochondria: Monoamine oxidase (outer mitochondrial membranes),[19] succinate–cytochrome *c* reductase (inner mitochondrial membranes),[20] and glutamate dehydrogenase (mitochondrial matrix)[21] were employed previously.[1] In practice, we found succinate–cytochrome *c* reductase the more convenient enzyme. However, enzymatic activity is lost on storage of fractions, and therefore this enzyme was always assayed first, within 24 hr of platelet procurement. Dilution of fractions in distilled water and preincubation of the diluted fraction (30 min, 0°) was necessary to ensure the accessibility of the enzyme to the externally added substrate.[1,20]

5. Enzymes utilized for localization of acid hydrolase-containing organelles include β-glycerophosphatase, β-glucuronidase, β-galactosidase, and *N*-acetyl-β-glucosaminidase. While the first enzyme depends on measurement of free inorganic phosphate in the presence of excess organic phosphate (as does the ER marker glucose-6-phosphatase), the other enzymes utilize *p*-nitrophenol derivatives (as does the membrane marker acid *p*-nitrophenylphosphatase). The latter technique offers greater simplicity and sensitivity and therefore is preferred for routine localization of acid hydrolases in the gradient. If a spectrofluorometer is available, sensitivity can be increased at least one order of magnitude by substituting the 4-methylumbelliferyl derivatives of the appropriate substrates for the *p*-nitrophenol ones without appreciable changes in technique.

6. Assays for human platelet α granules are limited to assays for the presence of specific proteins, rather than enzymes. When such assays are available, this may in fact facilitate identification of α granules. As mentioned above, we originally determined the localization of α granules by electron microscopy,[1] and subsequently identified platelet factor 4 and fibrinogen as α granule constituents.[2] While assays for these proteins may suffice for α granule identification, many other proteins have subsequently

---

[18] B. A. Nichols, P. Y. Setzer, and D. F. Bainton, *J. Histochem. Cytochem.* **32**, 165 (1984).
[19] R. J. Wurtman and J. Axelrod, *Biochem. Pharmacol.* **12**, 1439 (1963).
[20] C. de Duve, B. C. Pressman, R. Gianetto, R. Wattiaux, and F. Appelmans, *Biochem. J.* **60**, 604 (1955).
[21] C. Schnaitman and J. W. Greenawalt, *J. Cell Biol.* **38**, 158 (1968).

been localized to these important platelet secretory organelles. Among the useful proteins for α granule identification are also platelet-derived growth factor,[4,22] thrombospondin,[23-25] histidine-rich glycoprotein,[26,27] and P-selectin (PADGEM, GMP-140, and CD-62).[28-30]

7. We have utilized prelabeling with [^{14}C]serotonin to generate platelets with labeled dense bodies. Nitrogen cavitation homogenization and sucrose density gradient centrifugation yielded fractions in which label, as expected, appeared only in the very densest part of the gradient (Fig. 4). These were exactly the fractions where we had previously observed dense bodies by electron microscopy.[1]

A simplified procedure is utilized to isolate platelet membranes from other organelles, when their separation from each other is deemed unnecessary. Platelet homogenate is loaded onto a suitable volume of 30% (w/w) sucrose (containing 5 mM EDTA, pH 7.4), and spun at 150,000 to 200,000 g_{max} for 1 hr. The membrane band collects just below the interface of cytosol and sucrose, while remaining organelles are pelleted to the bottom of the tube.

We encountered only one disadvantage of our method: Due to retention of part of the suspension within the bomb components, a loss of homogenate occurs. This is minimal (<0.5 ml of homogenate) when utilizing the volumes and quantities of platelets described here. However, homogenization of very small quantities of platelets may lead to relatively large losses (when the volume of homogenate is reduced to <3 ml), or to a very dilute homogenate. Fractions with high activity would then be difficult to prepare.

In summary, platelet homogenization by nitrogen cavitation is a most useful technique for preparation of human platelet subcellular fractions of excellent purity on relatively large scales. The procedure is easy; homogenization is easily performed within 1 hr without need for constant manipulation, and results in an excellent degree of homogenization. As described

---

[22] H. N. Antoniades and P. Pantazis, this series, Vol. 169, p. 210.
[23] I. Hagen, *Biochim. Biophys. Acta* **392**, 243 (1975).
[24] J. M. Gerrard, D. R. Phillips, G. M. R. Rao, E. F. Plow, D. A. Walz, R. Ross, L. A. Harker, and J. G. White, *J. Clin. Invest.* **66**, 102 (1980).
[25] H. S. Slayter, this series, Vol. 169, p. 251.
[26] L. L. K. Leung, P. C. Harpel, R. L. Nachman, and E. M. Rabellino, *Blood* **62**, 1016 (1983).
[27] L. L. K. Leung, P. C. Harpel, and R. L. Nachman, this series, Vol. 169, p. 268.
[28] P. E. Stenberg, R. P. McEver, M. A. Shuman, Y. V. Jacques, and D. F. Bainton, *J. Cell Biol.* **101**, 880 (1985).
[29] C. L. Berman, E. L. Yeo, J. D. Wencel-Drake, M. H. Ginsberg, B. C. Furie, and B. Furie, *J. Clin. Invest.* **78**, 130 (1986).
[30] C. L. Berman, E. L. Yeo, B. C. Furie, and B. Furie, this series, Vol. 169, p. 311.

above and in our original work,[1-7] the purity of identifiable subcellular organelles obtained by sucrose density gradient centrifugation of a nitrogen cavitation homogenate is better than that obtained using other homogenization techniques. A serious drawback of most other methods is the consistent loss of identifiable lysosomal (acid hydrolase-containing) and α granule fractions. This often leads to gross contamination of "plasma membrane" fractions with granule membranes, resulting in misidentification/mislocalization of the cell component of interest. This serious pitfall is essentially averted by the methodology described here.

Acknowledgments

Portions of the work described were supported by the National Institutes of Health (HL 29034, HL 18828, HL 46403, and HL 47073), the New York Heart Association, and the Department of Veterans Affairs. The author was an Established Investigator of the American Heart Association.

Portions of the work were performed under the guidance of Dr. Phin Cohen and Professor Dr. L. L. M. van Deenen in partial fulfillment of requirements for a Ph.D. degree, in collaboration with Drs. R. I. Handin, N. P. Westmoreland, and A. Derksen.

[4] Isolation of Human Platelet Plasma Membranes by Glycerol Lysis

By JOAN T. HARMON, NICHOLAS J. GRECO, and G. A. JAMIESON

Isolation of platelet membranes has presented particular difficulties because of the small volume of platelets and their resistance to mechanical shearing forces and osmotic stress. The glycerol lysis method[1] was developed to provide a reproducible technique for the isolation of well-characterized platelet plasma membranes using equipment generally available in most laboratories. A single band was obtained on density step-gradient centrifugation that could be further resolved to two populations by continuous density gradient centrifugation. Each population was morphologically homogeneous when viewed by electron microscope and consisted of large membranous structures completely devoid of contamination with other subcellular elements. The average diameter of the membranes in the lighter band (d = 1.090) was 1.75 μm and consisted of double concentric membrane structures while the diameter of membranes in the heavier band (d = 1.120) was 0.7 μm and no double structures were visible. (It should

---

[1] A. J. Barber and G. A. Jamieson, *J. Biol. Chem.* **245**, 6357 (1970).

be noted that in the original publication[1] the scales for the electron photomicrographs were incorrectly labeled and should be multiplied by 10.)

Membranes in the two density populations were equally accessible to neuraminidase and trypsin, suggesting that they were not "inside-out, outside-out" ghosts.[2] The membranes themselves showed identical thromboplastic activity but the extracted lipids from the lower density membrane had a higher platelet factor 3 (PF3) activity as measured by the thromboplastin generation test than those from the higher density membrane.[3] Their ability to bind ^{125}I-labeled thrombin in a filtration binding assay was identical[4] and they were devoid of the glycogen synthase activity found in membranes isolated at the stage of density step-gradient centrifugation.[5]

Developments in our understanding of platelet membrane biochemistry have led to an important modification in the procedure. In the original glycerol lysis method, the EDTA concentration was insufficient to inhibit the Ca^{2+}-dependent protease that causes the cleavage of GPIb and the appearance of glycocalicin.[6] The addition of EGTA and a phenylmethylsulfonyl fluoride (PMSF)/leupeptin inhibitor mixture during lysis has completely avoided this problem.[7] More recently we have developed a modified glycerol lysis technique that retains the merits of the initial procedure but has several advantages in terms of yield and simplicity.[8] The principal changes include the use of a hypertonic buffer solution for suspending the platelets prior to glycerol loading and of vertical rotors to shorten centrifugation times.

Modified Glycerol Lysis Procedure

Platelet concentrates are obtained from the blood center less than 24 hr after blood collection. Twenty units of concentrate are satisfactory for a single run but the procedure can be carried out on as little as 1 unit of concentrate. The standard conditions used for the preparation of concentrates in the blood center[9] remove about 95% of the white cells present in the original unit of blood.

[2] A. J. Barber and G. A. Jamieson, *Biochemistry* **10**, 4711 (1971).
[3] A. J. Barber, D. C. Triantaphyllopoulos, and G. A. Jamieson, *Thromb. Diath. Haemorrh.* **28**, 206 (1972).
[4] S. W. Tam and T. C. Detwiler, *Biochim. Biophys. Acta* **543**, 194 (1978).
[5] J. H. Greenberg, A. P. Fletcher, and G. A. Jamieson, *Thromb. Diath. Haemorrh.* **30**, 307 (1973).
[6] D. R. Phillips and M. Jakábová, *J. Biol.Chem.* **252**, 5602 (1977).
[7] J. T. Harmon and G. A. Jamieson, *Biochemistry* **24**, 58 (1985).
[8] J. T. Harmon and G. A. Jamieson, *J. Biol. Chem.* **261**, 13224 (1986).
[9] F. K. Widmann, ed., "Technical Manual for the American Association of Blood Banks," 9th ed. p. 57. American Association of Blood Banks, Arlington, Virginia.

1. The platelet concentrates are poured into 50-ml centrifuge tubes and their pH adjusted to 6.5–6.7 with citric acid (1 M).

2. Platelets are sedimented at 900 g at 22° for 10 min (GLC-1; Sorvall, Newtown, CT). Due to the high concentration of platelets in the concentrates, the supernatant from the first centrifugation is recentrifuged under identical conditions and the second supernatant is discarded.

3. The individual platelet pellets are resuspended in buffer A (see below) and combined as appropriate. The red blood cells and any remaining white cells are removed by short centrifugations (15–30 sec at 900 g, 22°) and careful decantation of the platelet suspension or, alternatively, the platelets may be removed from the red blood cells by teasing them off the top of the red blood cell pellet. The minimum number of centrifugations should be used that ensure the platelets are essentially free of contaminating red cells.

4. After the platelets have been separated from blood cells and plasma, they are resuspended in buffer B (see below) so that the platelets from 4 units of concentrate are contained in 10 ml of buffer B.

5. The platelet suspension is then gently layered on top of 30 ml of a 0–40% linear glycerol gradient made up in a 1 : 10 dilution of buffer B and maintained at 4°: all subsequent operations are conducted at this temperature. At this point, the platelets will not remain at the top of the gradient but will slowly sediment. For this reason, the tubes are placed immediately in a Sorvall HB-4 rotor (Sorvall RC2-B centrifuge) and centrifuged at 1500 g for 30 min before the speed is increased to 5900 g for 10 min.

6. At completion of the centrifugation, the glycerol solution is carefully aspirated off and the final traces removed by wiping the sides of the tube.

7. To effect lysis, a volume of buffer C (see below) equal to four to five times the volume of the platelet pellet and containing leupeptin (50 μg/ml) and PMSF (2 mM) is added rapidly from a syringe fitted with an 18-gauge needle. The platelet pellet is then completely dispersed by rapidly drawing and expelling the platelet–buffer suspension through the needle several times. If desired, this procedure may be followed by vortexing the preparation for several minutes. This is the critical step in the glycerol lysis procedure and occasional calls from investigators regarding unsatisfactory results with the original procedure have invariably been traced to inadequately carrying out this osmotic shock step.

8. The lysed sample is centrifuged at 5900 g for 10 min using the Sorvall HB-4 rotor to remove any unlysed platelets.

9. The supernatant from this low-speed centrifugation is then carefully layered on top of a sucrose step gradient, prepared in 36-ml polyal-

lomer tubes (Sorvall Cat. No. 03141) by layering 10 ml of 33% sucrose in buffer C on top of 5 ml of a saturated solution (66%) of sucrose made up in the same buffer. Since the rotor to be used is a high-speed vertical rotor (Sorvall TV-850) it is necessary to fill the tubes with lysed sample or buffer C to within 0.5 cm from the top of the centrifuge tube and to balance both the tubes and the other accessories necessary for this rotor. Twenty units of platelets generally require 8 tubes with sucrose step gradients. Set the automatic rate controller on the Sorvall ultracentrifuge (OTD-2) to D-SLOW and spin for 30 min at 63,000 g at 4°. The membranes will layer at the top of the 33% sucrose layer. The membranes are removed with an 18-gauge needle attached to a 10-ml syringe with a piece of tubing.

10. The membranes from each gradient tube are diluted to approximately 50 ml with an appropriate buffer selected on the basis of the intended use of the membranes in, for example, binding studies and the membranes are then pelleted by centrifugation (30 min, 40,000 g, 4°). The membrane pellets are resuspended using a 22-gauge needle, pooled, and stored at −70° in small aliquots in the presence of protease inhibitors.

Buffer A
  Glucose, 34.17 g
  Sodium chloride, 75.05 g
  Sodium phosphate monobasic, monohydrate, 5.88 g
  Sodium phosphate dibasic, anhydrous, 10.60 g
  Sodium citrate dihydrate, 14.05 g
  Citric acid anhydrous, 5.04 g
Take up to 1 liter to give a 10-fold concentrated buffer A and freeze in 10- or 20-ml aliquots until needed. When needed, thaw, dilute 10×, and adjust pH to 6.7.

Buffer B: This is a 10-fold concentration of the buffer described by Phillips.[10] Because of its concentration it should be made up in three steps:

1. Combine

   Sodium chloride, 14.1 g
   Glucose, 38.6 g

   Dissolve in 100 ml water

---

[10] D. R. Phillips, *Biochemistry* **11**, 4582 (1972).

2. Combine
   EDTA, 1.02 g
   Tris, 2.6 g
   EGTA, 9.51 g

   Add to 50 ml water and dissolve by the addition of 10 M NaOH
3. Combine solutions 1 and 2, adjust to pH 7.4, and make up to 250 ml

Buffer C: Dilute buffer B 10× and add leupeptin (50 μg/ml) and PMSF (2 mM)

Comments. The yield of membranes by this technique is approximately 3 mg/unit of platelet concentrate, or about 50% increase over the original procedure. Purity is equal to that of the original procedure based on marker enzymes for plasma membranes. The procedure has also been used satisfactorily in our hands for the preparation of plasma membranes following proteolysis of intact platelets with *Serratia marcescens* protease.

[5] Isolation of Dense Granules from Human Platelets

By Miriam H. Fukami

Introduction

Human platelet dense granules are secretory organelles that store most of the calcium in the cell, about 90% of its ADP and lesser amounts of ATP, AMP, and guanine nucleotides, as well as nearly all of the platelet serotonin that is taken up by an active transport mechanism localized on the granule membrane.[1] The high content of stored calcium and nucleotides gives the granules a high buoyant density of about 1.21 in sucrose gradients and makes it possible to separate them readily from the numerous other subcellular organelles present in platelets. The ease with which the dense granules in intact platelets can be labeled with exogenous serotonin provides a convenient method for monitoring granule isolation and purification.

Labeling and Washing of Cells

Platelet-rich plasma is prepared from blood collected into 0.1 vol of an anticoagulant solution such as acid–citrate–dextrose [1.4% (w/v) citric

[1] G. Rudnick, *in* "Platelet Responses and Metabolism" (H. Holmsen, ed.), Vol. 2, p. 119. CRC Press, Boca Raton, Florida, 1987.

acid, 2.5% (w/v) sodium citrate, and 2% (w/v) dextrose, ACD] or 3.8% sodium citrate by centrifugation at 200 g for 15 min at room temperature. The supernatant platelet-rich plasma is carefully removed from the sedimented red cells with a pipette. The platelets can be incubated with [^{3}H]- or [^{14}C]serotonin at this stage if one elects to monitor the isolation of dense granules by isotopic labeling. Incubation for 15 min at 37° with 1 μM serotonin, including the desired amount of isotope, is adequate for labeling. Then 0.1 M sodium EDTA, pH 7.4, is added to the platelet-rich plasma to give a final EDTA concentration of 5 mM, and the platelets are pelleted by centrifugation at 1000 g for 15 min. The platelets are washed once with an ice-cold medium consisting of 0.13 M NaCl, 0.02 M HEPES, and 0.001 mM EDTA, pH 6.5,[2] and resuspended in the same medium. (Platelets obtained from 150 ml of whole blood were resuspended in a volume of 15–20 ml.)

Homogenization

Some special problems exist for the homogenization of platelets that do not occur with tissues such as liver. As with other free cell suspensions, it is difficult to apply a mechanical shear force with a conventional homogenizer such as a Potter–Elvehjem homogenizer, which works very well on minced tissue. The high content of actin and myosin in the platelet ultrastructure seems to confer resilient properties to the plasma membrane, which makes them more resistant to gentle homogenization. Platelets present an additional difficulty since they are activated to release the contents of their secretory organelles by contact with foreign surfaces and vigorous manipulations. (EDTA protects from activation only under moderately gentle conditions, such as low-speed centrifugation and washing.)

Ultrasonication and nitrogen cavitation methods[3] (see also [3] in this volume) are more suitable for application of shear forces necessary for disruption of cell suspensions than mortar-and-pestle techniques. Although ultrasonication has been used successfully by others for the isolation of platelet granules,[4,5] it has not been satisfactory in our hands with respect to reproducibility and yield of intact granules. The method de-

---

[2] Platelet secretion occurs less readily at pH values lower that 7.
[3] P. F. Zurendock, M. E. Tischler, T. P. M. Akerboom, R. Van der Meer, J. R. Williamson, and J. M. Tager, this series, Vol. 56, p. 214.
[4] F. Rendu, M. Lebret, A. T. Nurden, and J. P. Caen, *Br. J. Haematol.* **52**, 241 (1982).
[5] M. Da Prada, J. P. Tranzer, and A. Pletscher, *Experientia* **28**, 1328 (1972).

scribed here utilizes a French pressure cell[6] (and hydraulic press, both from Aminco Corp., Silver Springs, MD) for disruption of platelets that have been pretreated with metabolic inhibitors to block secretion[7] and a protease, nagarse (Enzyme Development Corp., New York, NY), which has been used to digest briefly heart and brain tissue before homogenization.[8-10]

Procedure for Homogenization and Fractionation

The washed platelets obtained from 150 ml of blood and resuspended in 20 ml of the pH 6.5 medium described above are incubated with 20 μM rotenone (0.2 ml of a 2 mM stock solution in ethanol), 10 mM 2-deoxyglucose (0.4 ml of a 0.5 M stock) and 30 mM gluconic acid δ-lactone (108 mg dissolved in 1 ml of washing medium just before addition) at 37° for 10 min. Then 3 mg of nagarse and 10 mg of ATP[11] (both dissolved in 1 ml of medium) are mixed into the suspension, which is kept at room temperature for 5 min. Then 20 mg of soybean trypsin inhibitor in 1 ml of medium is stirred into the suspension to neutralize the nagarse and the platelets are centrifuged at 1000 g for 15 min at 0-4°. All subsequent procedures are carried out at 0-4°. The platelet pellet is resuspended in a cold medium consisting of 0.25 M sucrose, 10 mM HEPES, and 1 mM sodium EDTA, pH 7.4. The suspension is then passed through a precooled French pressure cell[6] at 1000 psi and then centrifuged at 1000 g for 15 min. The supernatant fraction is saved and the pellet is resuspended in 15-20 ml of sucrose medium and subjected to another pass through the pressure cell. After the second homogenate is centrifuged again at 1000 g for 15 min, the combined supernatant fractions are centrifuged at 12,000 g for 20 min to give a pellet (about 5-10 mg protein) that consists of a mixture of platelet organelles, including dense granules, α granules, acid hydrolase-containing vesicles, and mitochondria. Table I shows that the markers for the various organelles are 2 to 3.6 times enriched in this pellet compared to the homogenate. This organelle mixture is suitable for some studies on dense granules and can be used as such after two washes with the sucrose

---

[6] L. Salganicoff and M. H. Fukami, *Arch. Biochem. Biophys.* **153**, 726 (1972). The valve modification described facilities regulation of the flow rate but is not essential.

[7] M. H. Fukami, J. S. Bauer, G. J. Stewart, and L. Salganicoff, *J. Cell Biol.* **77**, 389 (1978).

[8] D. D. Tyler and J. Gonze, this series, Vol. 10, p. 75.

[9] A. L. Smith, this series, Vol. 10, p. 84.

[10] R. E. Basford, this series, Vol. 10, p. 98.

[11] Omission of nagarse gives preparations that are difficult to resuspend and do not separate well on sucrose density gradient centrifugation. ATP seems to prevent the aggregation that occurs occasionally at this step, probably via its action as a potent inhibitor of ADP, small amounts of which may be released from the cells by nagarse.

TABLE I
MARKER ENZYMES IN ORGANELLE MIXTURE

Marker	Relative specific activity[a]	Yield (%)
[^{14}C]Serotonin (dense granules)	2.82 ± 0.5	36.4
α-Glycerophosphate oxidase (mitochondria)	3.04 ± 1.6	49.6
β-N-Acetylglucosaminidase (acid hydrolase vesicles)	1.92 ± 1.3	36.4
Platelet factor 4 (α granules)	2.96 ± 0.5	46.7

[a] The relative specific activity is given as the ratio of the specific activity of the fraction divided by the specific activity of the homogenate, mean ± SD, n = 3–4.

isolation medium if one is certain that the presence of the other organelles does not interfere with the activity to be monitored. The dense granules in this organelle mixture are considerably more stable to leakage[12] than the highly enriched preparation described below.

Subfractionation on a Sucrose Density Gradient

The sucrose stepwise gradient used for subfractionation of the granule mixture consists of seven steps from 0.8 to 2.0 M sucrose in 0.2 M increments. The sucrose steps are buffered with 5 mM HEPES, pH 7.4, and contain 0.1 mM sodium EDTA, pH 7.4. Starting with the 2 M step, 1.5 ml of each step is layered into a tube(s) suitable for use in a Beckman (Palo Alto, CA) SW 40 rotor (or comparable swing-out ultracentrifuge rotor). The gradients are incubated for 1 hr at 37° to soften the interfaces and then cooled on ice to 0°. About 1.5–2.0 ml of the organelle mixture (protein concentration not more than 10 mg/ml) is then layered onto each gradient tube and centrifuged at 100,000 g and 4° for 60 min. At least three bands of organelles and a pellet should be observed.[7] The dense granules are to be found in the pellet, which we have usually resuspended in the homogenization medium and used only for analysis. However, in fact, pig platelet dense granules prepared by the same method and used for nuclear magnetic resonance (NMR) studies at 31° were found to leak and slowly lose their stored contents,[12] and it is reasonable to assume that the human platelet preparation does the same. A modification of the sucrose gradient that resulted in a more stable preparation of pig platelet dense granules[12] and that is probably applicable to the human preparation is described here for those uses that require incubations at temperatures above 0–4°. Bovine

[12] K. Ugurbil, M. H. Fukami, and H. Holmsen, *Biochemistry* **23**, 416 (1980).

TABLE II
MARKER ENZYMES IN SUCROSE GRADIENT-ENRICHED DENSE GRANULES

Marker	Relative specific activity[a]	Yield (%)
[^{14}C]Serotonin (dense granules)	52.3 ± 6.9	19.9
α-Glycerophosphate oxidase (mitochondria)	ND[b]	—
β-N-Acetylglucosaminidase	3.09 ± 1.5	4.0
Platelet factor 4 (α granules)	3.66 ± 0.5	2.1

[a] The relative specific activity is given as the ratio of the specific activity of the fraction divided by the specific activity of the homogenate, mean ± SD, $n = 3-4$.
[b] ND, None detected.

serum albumin (0.6%) was added to all of the gradient steps that were made up with D_2O (the D_2O made it possible to use 1.8 M as the final gradient step). The final resuspension medium was 0.3 M KCl in D_2O, 0.4% bovine serum albumin, and 30 mM phosphate buffer, pH 7.4. Some details of these modifications that were required specifically for proton NMR such as use of D_2O, and a KCl resuspension medium with phosphate buffer, can be changed. The essence of the modification was to use bovine serum albumin and a somewhat hyperosmotic (compared to 0.25 M sucrose) resuspension medium. A metrizamide gradient for human platelet dense granules[4] and a Percoll gradient for pig platelet dense granules[13] have also been described.

Yield and Purity

The pellet consists of about 30–50 μg of protein and contains about 20% of the total serotonin originally present in 150 ml of whole blood starting material. Table II shows that the dense granule marker serotonin is 50-fold enriched compared to the homogenate. The mitochondrial marker was not measurable in this pellet, but the markers for acid hydrolase vesicles and α granules were some threefold enriched, although only 4 and 2% of the starting material was present. This degree of contamination is negligible for most purposes, but the preparation is not suitable for study of unique dense granule proteins except by functional tests. The reason for this limitation is that the other organelles are not only larger and more numerous than dense granules in the platelet, but they also contain primarily proteins. The dense granules, on the other hand, contain other low molecular weight compounds (see below) and very little protein.

[13] S. E. Carty, R. G. Johnson, and A. Scarpa, *J. Biol. Chem.* **256**, 11244 (1981).

TABLE III
LOW-MOLECULAR WEIGHT CONSTITUENTS OF HUMAN PLATELET DENSE GRANULES[a]

Substance	Whole platelets	Secreted amounts	Dense granules
PP_i	6.3 ± 0.05	6.0 ± 0.52	236 ± 53
P_i	16.7 ± 2.5	4.6 ± 1.6	248 ± 65
ATP	23.8 ± 3.5	9.7 ± 1.3	440 ± 96
ADP	14.5 ± 2.4	14.1 ± 1.8	633 ± 142
Ca^{2+}	125 ± 27	88 ± 23	2634 ± 471
Mg^{2+}	—	—	98 ± 33
Serotonin	1.5–2.5	1.5–2.0	90–100

[a] Isolated by the method described in this article (nmol/mg of protein). Values are given as mean ± SD, $n = 4$–6.

Therefore, the relative amounts of protein contributed by the various organelles are disproportionate to their concentration in the preparation.

Properties of Human Platelet Dense Granules

The concentrations of low molecular weight compounds in dense granules are given in Table III. The amounts contained in whole platelets and that released by thrombin are also shown in order to indicate which of the constituents are contained primarily in the dense granules. Most of the platelet content of PP_i,[14] ADP, serotonin, and much of the calcium is secreted and therefore is stored in secondary granules. Significant amounts of ATP and P_i that are required for metabolic activity are present in sites other than in the dense granules and are not secreted. Nuclear magnetic resonance studies on intact human platelets suggest that the nucleotides and calcium in the dense granules form high molecular weight aggregates that are more or less insoluble, which reduce the hyperosmolarity of the granule interior and which bind serotonin.[15,16] Similar studies on pig platelets and isolated pig platelet dense granules in which magnesium is the predominant divalent cation have confirmed the existence of these nucleotide–divalent cation–amine interactions.[12,17,18]

---

[14] M. H. Fukami, C. A. Dangelmaier, J. S. Bauer, and H. Holmsen, *Biochem. J.* **192,** 99 (1980).

[15] J. L. Costa, C. M. Dobson, K. L. Kirk, F. M. Poulsen, C. R. Valeri, and J. J. Vecchione, *FEBS Lett.* **99,** 141 (1979).

[16] K. Ugurbil, H. Holmsen, and R. G. Shulman, *Proc. Natl. Acad. Sci. U.S.A.* **76,** 2227 (1979).

[17] J. L. Costa, C.M. Dobson, K. L. Kirk, F.M. Poulsen, C. R. Valeri, and J. L. Vecchione, *Philos. Trans. R. Soc. London, Ser. B.* **289,** 413 (1980).

[18] M. H. Fukami, H. Holmsen, and K. Ugurbil, *Biochem. Pharmacol.* **33,** 3869 (1984).

Other mechanisms shown to be involved in serotonin uptake and storage in platelet dense granules from species other than human, such as an intragranular pH of 5.5,[19] an ATP-driven proton pump,[20] and a reserpine-sensitive amine transport system on the granule membrane,[20] most probably also exist in human platelet granules. The similarities in serotonin uptake, storage, and secretion by intact platelets of the various species indicate that the granular mechanisms are also similar.[1]

Finally, it should be mentioned that there is a heterogeneous group of inherited bleeding disorders, referred to as storage pool deficiency, both in humans and various animal species in which the platelet dense granule constituents are considerably diminished or absent, as shown both by secretion studies and by electron microscopy.[21]

[19] R. G. Johnson, A. Scarpa, and L. Salganicoff, *J. Biol. Chem.* **253**, 7061 (1978).
[20] G. Rudnick, H. Fishkes, P. J. Nelson, and S. Schuldiner, *J. Biol. Chem.* **255**, 3638 (1980).
[21] K. M. Meyers and M. Ménard, in "The Platelet Amine Storage Granule" (K. M. Meyers and C. D. Barnes, eds.), pp. 149–85. CRC Press, Boca Raton, Florida, 1992.

[6] Studying the Platelet Cytoskeleton in Triton X-100 Lysates

By JOAN E. B. FOX, CLIFFORD C. REYNOLDS, and JANET K. BOYLES

The platelet cytoskeleton is composed primarily of actin filaments. In unstimulated platelets, 40 to 60% of the actin is polymerized into filaments.[1] The majority of these filaments are present in networks throughout the body of the platelet. These are referred to as the cytoplasmic actin filaments. However, a small pool of the filaments appears to be present as part of a two-dimensional latticework that coats the inner surface of the lipid bilayer in much the same way as the red blood cell skeleton coats the membrane of the erythrocyte. This lattice-like structure is referred to as the platelet membrane skeleton. The filaments of the platelet membrane skeleton appear to be cross-linked by actin-binding protein (otherwise known as filamin), spectrin, and other unidentified proteins.[2,3] During platelet activation there is a rapid burst in actin polymerization, and about

[1] J. E. B. Fox, J. K. Boyles, C. C. Reynolds, and D. R. Phillips, *J. Cell Biol.* **98**, 1985 (1984).
[2] J. E. B. Fox, J. K. Boyles, M. C. Berndt, P. K. Steffen, and L. K. Anderson, *J. Cell Biol.* **106**, 1525 (1988).
[3] J. H. Hartwig and M. DeSisto, *J. Cell Biol.* **112**, 407 (1991).

70% of the actin becomes polymerized.[4,5] This polymerization appears to occur primarily in a submembranous location[6] and may result from the addition of actin monomers onto the ends of the filaments in the membrane skeleton.[7] The result is a change in platelet shape and the extension of actin-filled filopodia.[8] An activation-induced reorganization of actin filaments also occurs. The loose networks of actin filaments in unstimulated platelets become highly organized bundles, networks, and rings in activated platelets.[1,5] These changes in actin filament content and organization direct many of the functions of platelets. The changes are regulated by a variety of proteins, such as actin-binding protein, myosin, and α-actinin, each of which can bind to the filaments under certain conditions.[9]

The rapid advances in our understanding of actin filament organization and function can be attributed to the development of biochemical methods for studying actin filaments in Triton X-100-solubilized platelets. These methods originated from the observation that Triton X-100 solubilizes most platelet proteins but does not solubilize actin filaments or many of the proteins that are associated with them.[10]

This chapter will describe ways in which Triton X-100-insoluble cytoskeletons can be isolated from platelets so that they retain much of the composition and organization they had in the intact cell. It will describe how the polypeptide composition of these cytoskeletons can be determined and how the actin filament content can be quantitated.

Isolation of Platelets

Plasma contains factors that depolymerize actin filaments.[11,12] Thus, the study of platelet cytoskeletons requires that platelets be isolated from plasma proteins before being lysed with Triton X-100. During platelet activation, actin polymerization occurs and cytoskeletons reorganize.[1,4,5,9] Consequently, if cytoskeletons are to be studied in unstimulated platelets or if activation-induced changes are to be investigated, care must be taken

[4] L. Carlsson, F. Markey, I. Blikstad, T. Persson, and U. Lindberg, *Proc. Natl. Acad. Sci. U.S.A.* **76,** 6376 (1979).
[5] L. K. Jennings, J. E. B. Fox, H. H. Edwards, and D. R. Phillips, *J. Biol. Chem.* **256,** 6927 (1981).
[6] J. G. White, *Am. J. Pathol.* **117,** 207 (1984).
[7] J. E. B. Fox and M. C. Berndt, *J. Biol. Chem.* **264,** 9520 (1989).
[8] J. F. Casella, M. D. Flanagan, and S. Lin, *Nature (London)* **293,** 302 (1981).
[9] J. E. B. Fox, in "Biochemistry of Platelets" (D. R. Phillips and M. A. Shuman, eds.), p. 115. Academic Press, Orlando, Florida, 1986.
[10] D. R. Phillips, L. K. Jennings, and H. H. Edwards, *J. Cell Biol.* **86,** 77 (1980).
[11] H. E. Harris, J. R. Bamburg, and A. G. Weeds, *FEBS Lett.* **121,** 175 (1980).
[12] R. Thorstensson, U. Göran, and R. Norberg, *Eur. J. Biochem.* **126,** 11 (1982).

to ensure that the platelets are not activated during the isolation procedure. We have found the method described below useful for obtaining unstimulated platelets, as defined by the characteristic discoid shape of platelets in the circulation. However, several other methods have been described.[13,14]

Method 1: Isolation of Unstimulated Platelets by Gel Filtration

Reagents

Anticoagulant: 85 mM sodium citrate, 111 mM dextrose, and 71 mM citric acid

Column buffer: 136 mM NaCl, 2 mM KCl, 10 mM Na$_2$CO$_3$, 0.5 mM NaH$_2$PO$_4$, 0.4 mM MgCl$_2$, 4.4 mM glucose, and 10 mM HEPES, pH 7.4

Prostacyclin (Sigma, St. Louis, MO): Dissolve at 0.5 mg/ml in a buffer containing 100 mM NaCl and 50 mM Tris-HCl, pH 12.0. Store this solution at $-60°$ in 50-μl aliquots (stock I). On the day of use, dilute it to 50 μg/ml in the pH 12.0 buffer (stock II). Prostacyclin has a half-life of minutes at neutral pH. Thus, it should be added to buffers immediately before the buffers are used

Column. Sepharose 2B (packed bed volume, 50 ml; Pharmacia, Uppsala, Sweden) is washed successively on a scintered glass funnel with 250 ml of acetone and 500 ml of 0.9% NaCl.[14,15] If necessary, washing with 0.9% NaCl is repeated until traces of acetone are not detectable. The resin is then equilibrated in column buffer at 37° and poured into a column measuring ~2.5 × 10 cm. If a glass column is used, it should be siliconized first. Alternatively, a plastic column can be used. In this laboratory, we have found it convenient to use a 60-ml plastic syringe. The plunger from such a syringe is removed and a bed support is created using a porous polyethylene disk (2.5-cm diameter; Bio-Rad Laboratories, Richmond, CA) and a 2.5-cm rubber O ring. The column is equilibrated with column buffer containing a 1/100 vol of prostacyclin stock II (final concentration 50 ng/ml). All procedures are performed at 37°. The column resin is preconditioned by passing platelet-poor plasma through it before use, and it is discarded after each platelet isolation procedure.

Procedure. Anticoagulant is equilibrated to 37°, and 5 ml is pipetted into a 35-ml syringe. Blood is drawn from the antecubital vein of a healthy adult donor, using a tourniquet and a 19-gauge butterfly needle. To avoid transferring traces of thrombin resulting from the vascular trauma of venipuncture, 2 to 5 ml of blood is collected into an empty 5-ml syringe. This

---

[13] J. F. Mustard, R. L. Kinlough-Rathbone, and M. A. Packham, this series, Vol. 169, p. 3.
[14] S. Timmons and J. Hawiger, this series, Vol. 169, p. 11.
[15] B. Lages, M. C. Scrutton, and H. Holmsen, *J. Lab. Clin. Med.* **85,** 811 (1975).

syringe and its contents are discarded, and a 35-ml syringe containing the anticoagulant is attached to the needle. The blood and anticoagulant are mixed by inverting the syringe, and 7 ml of anticoagulated blood is transferred to each of five 15-ml plastic conical tubes. All the subsequent steps are performed at 37°.

Blood is centrifuged for 20 min at 160 g to sediment the erythrocytes. Platelets do not sediment at these g forces but remain in the plasma, 5 ml of which is removed and applied to the column of Sepharose 2B. The column outlet is positioned about 10 cm below the top of the resin; in this way, the flow rate is adjusted to ~20 ml/hr. Fractions (1.5 ml) are collected into plastic tubes. Once the plasma has entered the column, the platelets are chromatographed by pumping column buffer onto the top of the column at a flow rate of 20 ml/hr. All tubing and connectors with which the platelets come into contact should be made of plastic or polypropylene. Platelets usually elute between fractions 9 and 14 and can be detected by the turbidity of the suspensions. Plasma proteins elute later in fractions that have a yellow tinge. The concentration of platelets is determined with a Coulter (Hialeah, FL) counter. Peak fractions (generally containing 2 to 4×10^8 platelets/ml) are pooled and incubated in covered tubes at 37° for about 30 min before use.

Evaluation. The major advantage of this method for isolating platelets is that the isolated platelets are totally unstimulated (as judged by electron microscopy, which shows that the platelets retain the discoid shape characteristic of unstimulated platelets). Precautions that allow the isolation of such cells include (1) avoiding the use of surfaces such as glass or metal that can activate platelets, (2) performing all steps at 37°, and (3) including prostacyclin in all buffers. The major disadvantage of this method is that the maximum concentration of platelets obtained is dictated by the concentration present in the platelet-rich plasma applied to the column. It should also be noted that although the half-life of prostacyclin is short, its effect on adenylate cyclase can last up to 180 min.[16] Thus, if platelets are to be used for studies involving their activation with agonists, they should be incubated for at least 3 hr at 37° prior to use. Finally, if it is important to retain the reactivity of the platelets to ADP, apyrase can be included in the column buffer.[13]

Method 2: Isolation of Unstimulated Platelets by Centrifugation

Reagents

CGS: 120 mM NaCl, 13 mM trisodium citrate, and 30 mM dextrose, pH 7.0

---

[16] S. E. Graber and J. Hawiger, *J. Biol. Chem.* **257**, 14606 (1982).

ETS: 154 mM NaCl, 10 mM Tris-HCl, and 1 mM EDTA, pH 7.4

Tyrode's buffer: 138 mM NaCl, 2.9 mM KCl, 12 mM NaHCO$_3$, 0.36 mM NaH$_2$PO$_4$, 5.5 mM glucose, 1.8 mM CaCl$_2$, and 0.4 mM MgCl$_2$, pH 7.4

Prostacyclin, as described in method 1

Procedure. Platelet-rich plasma is prepared from fresh human blood as described in method 1. All subsequent steps are performed at ambient temperatures. The platelet-rich plasma is centrifuged in conical plastic tubes at 730 g for 20 min to sediment the platelets. The original plasma volume of CGS is equilibrated to 37°, mixed with 50 ng/ml prostacyclin, and added to the platelet pellet. The top portion of the pellet is gently resuspended by repeated aspiration into and expulsion from a polyethylene transfer pipette. Care should be taken to avoid resuspension of any contaminating erythrocytes, which, if present, are at the bottom of the pellet and can be detected by their red color. The platelets are sedimented again by centrifugation for 10 min at 730 g, washed one more time in CGS, and then one time in ETS. Before each wash, the buffers are preequilibrated to 37°, and prostacyclin is added. After each wash, the platelets are isolated by centrifugation for 10 min at 730 g. The washed platelets are resuspended to a concentration of 0.1 to 2.0 × 10^9 platelets/ml in a Tyrode's buffer containing 50 ng/ml prostacyclin. A typical yield from 30 ml of blood would be 1 to 2 ml of platelets at a concentration of 1 × 10^9 platelets/ml. Platelet suspensions isolated in this way should be incubated in covered tubes at 37° for at least 60 min before use.

Evaluation. The major advantage of the centrifugation method of isolating platelets is that the washed platelets are isolated as a pellet that can subsequently be resuspended at the concentration of choice and in the buffer of choice. In addition, several published methods are available for radiolabeling the surface glycoproteins of intact platelets.[17,18] These methods include centrifugation steps, which can readily be incorporated into this centrifugation method of isolating platelets.

The major disadvantage of isolating platelets by the centrifugation method is that they tend to lose their discoid shape. This results from centrifugation-induced activation and from performing the procedure at ambient temperatures. Activation is minimized by the inclusion of prostacyclin and by preequilibrating all buffers at 37° before use. We have found that the discoid form is largely restored by incubating the final platelet suspension at 37°. It should be pointed out that some of the steps in the

---

[17] K. J. Clemetson, *in* "Platelet Membrane Glycoproteins" (J. N. George, A. T. Nurden, and D. R. Phillips, eds.), p. 51. Plenum, New York, 1985.

[18] D. R. Phillips, this volume [35].

methods used to radiolabel platelets are performed under conditions that activate platelets (for example, incubation with sodium periodate at 4° in the sodium periodate/sodium boro[^{3}H]hydride method).[17,18] Thus, if radiolabeling procedures are incorporated into this method of isolating platelets, the isolated platelets often not only lose their discoid shape but also extend several filopodia.

An additional disadvantage of the centrifugation method of isolating platelets is that platelets are exposed to EDTA at 37°, a condition that causes dissociation of the GPIIb–IIIa complex in the plasma membrane.[19] Because the GPIIb–IIIa complex is the receptor for the adhesive proteins responsible for platelet aggregation,[20] dissociation of the complex adversely affects platelet function. Although dissociation is reversible on addition of Ca^{2+}, prolonged exposure to EDTA at 37° results in the formation of polymers, which cannot reassociate.[20] Thus, care should be taken to minimize the time of exposure of the platelets to washing buffers. Finally, the use of buffers containing Tris-HCl may adversely influence Ca^{2+} fluxes in stimulated platelets.[21] This problem can be prevented by replacing the Tris-HCl with HEPES.[14]

Method 3: Preparation of Suspensions of Activated Platelets

Reagents

Human thrombin (kindly provided by Dr. John Fenton III of the New York Department of Health, Albany, NY, but also available from Sigma Chemical Co.) is diluted within a few hours of use to a concentration of 10 NIH units/ml in a buffer containing 5% (v/v) polyethylene glycol 6000, 150 mM NaCl, and 10 mM Tris-HCl, pH 7.4. This buffer maintains the stability of thrombin

Procedure. Platelets are isolated by centrifugation (method 2) and resuspended at 37° in a Tyrode's buffer that lacks prostacyclin. Thrombin is added to a final concentration of 0.1 NIH unit/ml, and the suspension is mixed by gentle agitation. Unless one specifically wants to study aggregation-induced events, vigorous vortexing or stirring should be avoided because this causes platelets to aggregate. Clumps of aggregated platelets are resistant to extraction with Triton X-100, leading to sedimentation of many noncytoskeletal proteins with the Triton X-100-insoluble actin

[19] L. A. Fitzgerald and D. R. Phillips, *J. Biol. Chem.* **260**, 11366 (1985).
[20] L. A. Fitzgerald and D. R. Phillips, in "Hemostasis and Thrombosis: Basic Principles and Clinical Practice" (R. W. Coleman, J. Hirsh, V. J. Marder, and E. W. Salzman, eds.), 2nd ed., p. 572. Lippincott, Philadelphia, 1987.
[21] M. A. Packham, M. A. Guccione, M. Nina, R. L. Kinlough-Rathbone, and J. F. Mustard, *Thromb. Haemostasis* **51**, 140 (1984).

filaments. Because aggregation of platelets is a Ca^{2+}-dependent process, an alternative way of preventing aggregation is to use platelets that are suspended in a Tyrode's buffer from which the divalent cations are omitted and in which 1 mM EDTA is included.

Activation-induced changes in the platelet cytoskeleton begin in the initial seconds after the addition of thrombin. Additional stimuli that can be used to activate platelets include collagen (Horm, Munich, Germany), added to a final concentration of 20 μg/ml, and the divalent cation ionophore A23187, which is dissolved in dimethyl sulfoxide and added in a 1/500 vol to give a final concentration of 0.4 μM ionophore and 0.2% (v/v) dimethyl sulfoxide.

Isolation of Triton X-100-Insoluble Cytoskeletons

Reagents

Triton X-100 lysis buffer: 2% (v/v) Triton X-100, 10 mM EGTA, 2 mM phenylmethylsulfonyl fluoride, 100 mM benzamidine, and 100 mM Tris-HCl, pH 7.4. Phenylmethylsulfonyl fluoride rapidly hydrolyzes in aqueous solution and is therefore added immediately before the buffer is used

Procedure

Ice-cold Triton X-100 lysis buffer (750 μl) is pipetted into a microfuge tube (1.5 ml, conical, polypropylene), and an equal volume of platelet suspension is added. The contents of the tube are mixed by inversion. The suspension of platelets clarifies as soon as it is added to the Triton X-100 lysis buffer. The cytoskeletons of unstimulated platelets are not usually visible, but those from activated platelets can normally be seen as insoluble gelatinous material.

The cytoskeletons are immediately isolated by centrifugation at 4° for 2.5 hr at 100,000 g. The supernatant is very carefully removed with a Pasteur pipette, at which point the cytoskeletons can be visualized as a small, translucent pellet.

Evaluation

Two features of the procedure described here are essential for the isolation of cytoskeletons: the use of high g forces to sediment the cytoskeletons and the inclusion of EGTA and protease inhibitors in the lysis buffer.

Actin filaments require high g forces (typically 100,000 g, 2.5 hr) to be sedimented. They can be sedimented at lower g forces only if they are cross-linked into networks or bundles. In the unstimulated platelet, many of the filaments are insufficiently cross-linked (at least in detergent lysates) to be sedimented at lower g forces. In addition, at lower forces the membrane skeleton fragments and separates from the underlying actin filaments. Thus, in the unstimulated platelet, many of the actin filaments require high g forces to be sedimented.

When platelets are activated, the situation is very different. Most filaments, including much of the membrane skeleton, now sediment at low g forces (15,600 g, 4 min). Presumably, activation causes a reorganization of the cytoskeleton, making the structure more cross-linked and more resistant to fragmentation in the lysis buffer and during the centrifugation procedure. This differential recovery of cytoskeletal proteins from unstimulated and activated platelets at low g forces can lead to very misleading results if low g forces are used in studies designed to identify activation-induced changes in the composition of platelet cytoskeletons. (This is discussed in greater detail in the Evaluation subsection under Identification of Actin Filament-Associated Proteins.)

An important feature of the Triton X-100 lysis buffer described here is the presence of the Ca^{2+} chelator, which is necessary to prevent Ca^{2+}-induced depolymerization of cytoplasmic actin filaments by gelsolin.[4,22] The Ca^{2+} chelator is also necessary to prevent the activity of the Ca^{2+}-dependent protease, which is present in platelets[23] and can hydrolyze several cytoskeletal proteins[24] even at micromolar concentrations of Ca^{2+}. Platelets contain considerable amounts of Ca^{2+}; thus, if the platelet suspensions used are at concentrations significantly higher than those suggested here (maximum of 2×10^9 platelets/ml), it may be necessary to increase the concentration of chelator in the Triton X-100 extraction buffer. A second important feature of the Triton X-100 lysis buffer is the presence of serine protease inhibitors. In the absence of these inhibitors, the actin filaments lose much of the three-dimensional organization that they had in the intact cell,[2] presumably because of the hydrolysis of critical cross-linking proteins. This is not a problem if the purpose of the study is to isolate only actin filaments (since these still sediment at 100,000 g). However, it is a major problem if the purpose is to study additional components of the cytoskeleton.

[22] S. E. Lind, H. L. Yin, and T. P. Stossel, *J. Clin. Invest.* **69**, 1384 (1982).
[23] D. R. Phillips and M. Jakábová, *J. Biol. Chem.* **252**, 5602 (1977).
[24] J. E. B. Fox, D. E. Goll, C. C. Reynolds, and D. R. Phillips, *J. Biol. Chem.* **260**, 1060 (1985).

Identification of Actin Filament-Associated Proteins

Procedure

To determine whether a protein is associated with cytoskeletons, platelets are lysed by adding an equal volume of Triton X-100 lysis buffer, and cytoskeletons are isolated by centrifugation as described in Isolation of Triton X-100-Insoluble Cytoskeletons. One way of detecting the polypeptides recovered with cytoskeletons is by electrophoresis on sodium dodecyl sulfate (SDS)–polyacrylamide gels. Cytoskeletons are solubilized in SDS-containing buffer [2% SDS (w/v), 5% 2-mercaptoethanol (v/v), 10% glycerol (v/v), 0.002% bromphenol blue (v/v), and 62.5 mM Tris-HCl, pH 6.8]. Solubilization is facilitated by mechanically agitating the pellet (for example, with a Pasteur pipette). The polypeptide content is analyzed by electrophoresis through slab gels according to the method of Laemmli.[25] Polypeptides can be detected by Coomassie Brilliant blue or silver staining of gels.[26,27] Alternatively, polypeptides can be transferred electrophoretically to nitrocellulose, and the polypeptides can be detected on immunoblots using antibodies against the polypeptides of interest.

Several platelet proteins (in addition to actin filaments and associated proteins) are inherently insoluble in Triton X-100.[26] Thus, cosedimentation of proteins cannot be taken as a criterion for association. The most convincing method of demonstrating that a protein is a component of the cytoskeleton is a morphological one.[2] Platelets are lysed with the Triton X-100 buffer, and 200-μl portions of the lysates are incubated with approximately 10 μg/ml of an affinity-purified antibody against the protein of interest and with 200 μl of protein A coupled to 15-nm-diameter gold beads. Lysates are incubated at 37° for 60 min and then fixed by addition of a 10-fold volume of fixative containing 0.2 M glutaraldehyde and 40 mM lysine in 60 mM sodium cacodylate buffer, pH 7.4.[28] Samples are incubated at ambient temperatures for ~15 min, and the fixed material is then isolated by centrifugation for 15 min at 600 g. The material is washed twice in barbital buffer, pH 7.6, exposed for 15 min at 4° in 0.05 M osmium tetroxide in the same buffer, rinsed three times in ice-cold distilled water, and stained overnight at 4° in 0.05 M uranyl acetate in water. Fixed material is dehydrated in acetone and embedded in Epon 812 (Polysciences, Warrington, PA). Thick (0.2 μm) sections are cut, stained with Reynolds lead citrate and uranyl acetate, and viewed by electron microscopy.[2,28] If the

---

[25] U. K. Laemmli, *Nature (London)* **227**, 680 (1970).
[26] J. E. B. Fox, *J. Clin. Invest.* **76**, 1673 (1985).
[27] J. H. Morrissey, *Anal. Biochem.* **117**, 307 (1981).
[28] J. Boyles, J. E. B. Fox, D. R. Phillips, and P. E. Stenberg, *J. Cell Biol.* **101**, 1463 (1985).

protein of interest is a component of the cytoskeleton, antibodies will label the Triton X-100-insoluble structures, whereas control antibodies will show only minimal labeling.

An important feature of this immunocytochemical labeling procedure is that the labeling is performed before the cytoskeletons are fixed. The reason for this is that the lysine, present in the fixative to preserve actin filaments that are otherwise damaged during the fixation process,[28] often decreases the ability of antigens to bind their antibodies. The protein A-labeled gold is added at the same time as the antibodies because the cytoskeletons do not remain intact during the centrifugation steps that would be required to remove unbound antibodies prior to protein A addition. Because antibody and protein A are present simultaneously, we have found it important to use affinity-purified antibodies for these studies and to titrate carefully the number of platelets and quantity of colloidal gold and antibody.

If the protein of interest is a component of the membrane skeleton, an alternative lysis buffer should be employed for morphological studies. The Tris-containing buffer described under Isolation of Triton X-100-Insoluble Cytoskeletons maintains the membrane skeleton with sufficient organization to allow its sedimentation at 100,000 g, 2.5 hr in biochemical studies but does not retain skeleton in association with the underlying cytoplasmic actin filaments.[4] However, by using low ionic strength buffers such as those utilized for isolating the red blood cell membrane skeleton, the platelet membrane skeleton can be retained at the periphery of the cytoskeleton.[2] Thus, for the morphological detection of components of the membrane skeleton, platelets are lysed by the addition of 9 vol of a buffer containing 1% Triton X-100, 5 mM sodium phosphate, 5 mM EGTA, 1 mM phenylmethylsulfonyl fluoride, and 50 mM benzamidine. Platelet lysates are then incubated with antibodies and protein A and processed for electron microscopy by the methods described in the previous paragraph. It should be noted that this low ionic strength buffer can also be used for biochemical studies in which the cytoskeleton is isolated by centrifugation. However, we have found no advantage to using it for these studies, for although this buffer maintains the membrane skeleton at the periphery of the cytoplasmic actin filaments in the initial lysis of the cell, the membrane skeleton still appears to fragment during centrifugation. In addition, when this buffer is used, the platelet suspension is diluted 10-fold, whereas the Tris buffer (described under Isolation of Triton X-100-Insoluble Cytoskeletons) results in only a 2-fold dilution. Thus, the use of the Tris buffer is often preferable because it results in samples that are less unwieldy and in soluble fractions in which the proteins of interest are less diluted.

A second biochemical method that can be used to determine whether a Triton X-100-insoluble protein is a component of the cytoskeleton is to depolymerize the actin filaments in platelet lysates and examine the effect of this depolymerization on the Triton X-100 insolubility of the component in question. Depolymerization of actin filaments can be accomplished in two ways. In one method, which is useful only for the depolymerization of cytoplasmic actin filaments, platelets are lysed by adding Triton X-100 lysis buffer that does not contain a Ca^{2+} chelator [2% (v/v) Triton X-100, 100 mM Tris-HCl, 100 mM benzamidine, 2 mM phenylmethylsulfonyl fluoride, and 2 mg/ml leupeptin, pH 7.4]. The Ca^{2+} concentration in the lysates is in the millimolar range and therefore permits depolymerization of cytoplasmic actin filaments by gelsolin. It is important to include leupeptin in this lysis buffer to prevent the activity of the Ca^{2+}-dependent protease. Depolymerization of actin by gelsolin is rapid; thus, lysates should be centrifuged in the microfuge immediately after the platelets are lysed.

The second method of depolymerizing actin filaments in Triton X-100 lysates employs DNase I as a depolymerizing agent. This method is useful for depolymerizing actin filaments of the membrane skeleton that, unlike those of the rest of the cytoskeleton, are not depolymerized by gelsolin.[26] In this method, platelets are lysed by adding an equal volume of Triton X-100 lysis buffer that lacks a Ca^{2+} chelator and contains DNase I [2% (v/v) Triton X-100, 100 mM Tris-HCl, 100 mM benzamidine, 2 mM phenylmethylsulfonyl fluoride, 2 mg/ml DNase I, and 2 mg/ml leupeptin, pH 7.4]. It is important to omit the Ca^{2+} chelator, as DNase I is a Ca^{2+}-dependent enzyme. The action of DNase I is slow. Unlike gelsolin, it does not act by severing actin filaments along their length. Rather, it acts by binding actin monomers, thus shifting the equilibrium between filaments and monomers. This results in a time-dependent loss of actin monomers from the ends of the filaments. To allow DNase I to exert its effects, lysates are therefore incubated for at least 30 min at 4°. Any remaining Triton X-100-insoluble actin filaments are subsequently sedimented by centrifugation at 100,000 g for 2.5 hr. Commercial DNase I preparations contain significant protease activity, which must be inhibited. In our laboratory, we have observed that DNase I purchased from some sources retains proteolytic activity even in the presence of the serine protease inhibitors present in the Triton X-100 lysis buffer. However, we have demonstrated that with the DNase I purchased from Boehringer Mannheim (Indianapolis, IN), the protease inhibitors in the lysis buffer are sufficient to inhibit all detectable proteolytic activity. The DNase I purchased from this company is therefore suitable for these experiments.

The criterion for determining actin filament association with a polypeptide of interest is that a reduction of sedimentable actin is accompanied by a reduction in the amount of sedimentable polypeptide, along with a corresponding increase in the amount of the polypeptide in the Triton X-100-soluble fraction. To determine the amount of sedimentable actin that remains, cytoskeletons are solubilized in SDS buffer and electrophoresed through SDS gels containing 7.5% (w/v) acrylamide. The protein is stained with Coomassie Brilliant blue, and actin (the major platelet polypeptide of M_r 43,000) is quantitated by densitometry.[1] The polypeptide content of the Triton X-100-soluble fractions (which are prepared for electrophoresis by adding one-third volume of a four times concentrated SDS solubilization buffer) should also be examined to ensure that no proteolysis was induced by the presence of Ca^{2+} or DNase I.

Evaluation

The study of cytoskeletons in detergent lysates of cells is always complicated by the fact that many protein–protein interactions are dissociated by the detergent because they are of low affinity (i.e., are lost as a result of the dilution that occurs on cell lysis). Thus, the fact that a protein is not associated with the cytoskeleton in Triton X-100 lysates does not mean that it was not associated in the intact platelet. Similarly, many proteins are inherently insoluble in Triton X-100; therefore, it is essential to use criteria other than cosedimentation with the cytoskeleton from Triton X-100-lysed platelets to establish that an association with the cytoskeleton exists.

In studies designed to determine whether a protein is associated with actin filaments, it is very important that platelet lysates are centrifuged at g forces sufficient to sediment all the filaments. This is particularly true when the goal is to detect activation-induced changes in the association of proteins with the cytoskeleton. Studies that demonstrate increased sedimentation of proteins only at low g forces (e.g., 15,600 g, 4 min) are fraught with the possibility of misinterpretation. For example, the glycoprotein Ib complex, a membrane glycoprotein that is known to be associated with the cytoskeleton in both unstimulated and activated platelets, shows increased sedimentation at low g forces from lysates of activated platelets. The reason for this is not that it has undergone an activation-induced association with the cytoskeleton. Rather, it results from an increased cross-linking of the cytoskeleton in activated platelets such that the cytoskeletons (particularly the association between the membrane skeleton and underlying actin filaments) are more resistant to disruption in the lysis buffer and to centrifugation-induced fragmentation. Thus, when studying activation-induced changes in cytoskeletal composition, it is absolutely essential to utilize high g forces (100,000 g, 2.5 hr) to sediment the cytoskeleton.

Methods of Measuring Actin Filament Content

Two methods commonly used to measure the actin filament content of Triton X-100 platelet lysates are the DNase I inhibition assay and the sedimentation assay.

Method 1: DNase I Inhibition Assay

As described elsewhere,[29] monomeric actin binds to DNase I, producing a concentration-dependent inhibition of the ability of DNase I to hydrolyze DNA. In contrast, filamentous actin has no effect on DNase I activity. Filamentous actin depolymerized by guanidine hydrochloride, however, is as active as monomeric actin in inhibiting the catalytic activity of DNase I. The extent of inhibition of DNase I activity in platelet lysates can be used as a measure of monomeric actin concentration; the extent of inhibition of DNase I activity after depolymerization of actin filaments with guanidine hydrochloride measures the total actin concentration; the difference between these measurements is used to determine the actin filament content. Procedures for determining DNase I inhibitory activities are described below.

Reagents

DNA solution: DNA (calf thymus, type 1; Sigma) is cut into fine pieces with scissors and dissolved at 40 μg/ml in 4 mM MgSO$_4$, 1.8 mM CaCl$_2$, and 100 mM Tris-HCl, pH 7.5, by stirring very slowly at room temperature for 24 to 48 hr. High stirring rates should be avoided to prevent shearing of DNA molecules. The solution is filtered, stored at 4°, and brought to 37° before use. The absorbance at 260 nm should be between 0.50 and 0.65

DNase solution: DNase I (beef pancreas, DN 100; Sigma) is dissolved at 0.2 mg/ml in 0.1 mM CaCl$_2$, 0.01 mM phenylmethylsulfonyl fluoride, and 50 mM Tris-HCl, pH 7.5. It is convenient to prepare this solution as a five-times concentrated stock (1 mg/ml) and to store it in aliquots at $-60°$ for extended periods of time

Guanidine hydrochloride buffer: 1.5 M guanidine hydrochloride (ultrapure; Schwarz/Mann, Orangeburg, NY), 1.0 M sodium acetate, 1.0

---

[29] I. Blikstad, F. Markey, L. Carlsson, T. Persson, and U. Lindberg, *Cell* (*Cambridge, Mass.*) **15**, 935 (1978).

mM sodium-ATP (Sigma), 20 mM Tris-HCl, and 6 mM calcium chloride, pH 8.4

Triton X-100 lysis buffer: 2% (w/v) Triton X-100, 10 mM EGTA, and 100 mM Tris-HCl, pH 7.4

Procedure. To measure the activity of DNase I against DNA, DNase I solution is diluted with an equal volume of buffer of the same composition as the Triton X-100 lysates to be assayed (i.e., platelet suspension buffer and Triton X-100 lysis buffer, 1:1). A portion (usually 10 μl) of the diluted DNase I solution is added to 3 ml of DNA solution in a quartz cuvette. The cuvette is covered with Parafilm and inverted. The absorbance at 260 nm is recorded within 15 sec of the addition of the enzyme to the DNA and is monitored continuously thereafter. After an initial lag phase, the absorbance increases linearly with time as DNA is hydrolyzed.[30] The slope of the linear portion of the curve reflects the DNase I activity. The volume of DNase I added should be such that the change in absorbance of the solution is $\sim 10^{-3}$ absorbance units/sec.

To determine the extent of inhibition of DNase I activity in Triton X-100 lysates, platelets suspended at $\sim 2 \times 10^8$ platelets/ml in Tyrode's buffer without divalent cations containing 1 mM EDTA (138 mM NaCl, 2.9 mM KCl, 12 mM NaHCO$_3$, 0.36 mM NaH$_2$PO$_4$, 5.5 mM glucose, and 1 mM EDTA, pH 7.4) are lysed by adding an equal volume of ice-cold Triton X-100 lysis buffer. Lysates are vortexed vigorously, and an equal volume of DNase solution is added. An aliquot (usually 20 μl) of this solution is taken immediately (within 5 sec), added to 3 ml of DNA solution, and the absorbance at 260 nm is continuously monitored. The volume of platelet lysate added should be such that the rate of absorbance change is maintained close to the optimal rate of $\sim 10^{-3}$ absorbance units/sec. The concentration of platelets suggested here is such that lysates from unstimulated platelets produce $\sim 50\%$ inhibition. Thus, a 20-μl aliquot of the platelet–DNase mixture produces about the same rate of absorbance change as a 10-μl aliquot of the buffer–DNase mixture.

The percentage inhibition of the DNase activity is calculated by the formula:

Inhibition (%) produced by platelet aliquot assayed
$$= 100\% - a/b \times 100\% \quad (1)$$

where a is the absorbance units per second produced by a known volume of DNase in the presence of platelet lysate, and b is the absorbance units per second produced by the same volume of DNase in the presence of lysis buffer.

---

[30] J. E. B. Fox, M. E. Dockter, and D. R. Phillips, *Anal. Biochem.* **117**, 170 (1981).

Actin polymerization occurs during platelet activation and decreases the monomeric actin concentration and, therefore, the extent of inhibition of DNase I activity by platelet lysates. The DNase I assay is most accurate when the activity is inhibited between 30 and 70%. Should the extent of inhibition fall below ~30%, the ratio of platelet lysate to DNase I may have to be increased.

The total inhibitory activity in Triton X-100 lysates is assayed in the presence of guanidine hydrochloride buffer. Platelet lysates are vortexed with an equal volume of guanidine hydrochloride buffer. The pH of this buffer is adjusted to 8.4, so that addition of this Ca^{2+}-containing buffer to the EGTA-containing platelet lysate adjusts the final pH to 7.4. If platelet lysates that have different concentrations of chelator and/or divalent cations are used, the pH of the guanidine hydrochloride solution may need to be adjusted accordingly. Similarly, depolymerization of actin by guanidine hydrochloride is dependent on the presence of Ca^{2+}, which also prevents the guanidine hydrochloride-induced inhibition of the ability of monomeric actin to inhibit DNase I activity. The Ca^{2+} concentration of the depolymerizing buffer is such that on addition of this buffer to the EGTA-containing platelet lysates, the final Ca^{2+} concentration is at least 1 mM. Guanidine hydrochloride-treated lysates are incubated at 4° for 5 to 30 min, after which time an equal volume of DNase solution is added. The mixture is vortexed, an aliquot (usually 40 μl) is immediately added to 3 ml of DNA solution, and the absorbance at 260 nm is recorded. The percentage inhibition of DNase activity that is produced by a known volume of platelet lysate is calculated by Eq. (1). The percentage of actin that is filamentous in the original lysate is calculated by the formula:

$$\text{Filamentous actin } (\%) = 100\% - x/y \times 100\% \qquad (2)$$

where x is the percentage inhibition produced by a known volume of the lysate in the absence of guanidine hydrochloride, and y is the percentage inhibition produced by the same volume of the lysate in the presence of guanidine hydrochloride.

Method 2: Sedimentation Assay

Like the DNase inhibition assay, the sedimentation assay is based on the finding that Triton X-100 solubilizes most platelet proteins, including monomeric actin, but does not solubilize actin filaments. In the sedimentation assay, the filamentous actin is separated from monomeric actin by centrifugation, and the amount of sedimented actin is quantitated after SDS–polyacrylamide gel electrophoresis.

Reagents

Triton X-100 lysis buffer: 2% (v/v) Triton X-100, 10 mM EGTA, 100 mM Tris-HCl, 100 mM benzamidine, and 1 mM phenylmethylsulfonyl fluoride, pH 7.4

Procedure. Ice-cold Triton X-100 lysis buffer is pipetted into a centrifuge tube, and an equal volume of platelets is added. The contents of the tube are mixed by inversion, and the filaments are then separated from unpolymerized actin by centrifugation at 4° at 100,000 g for 3 hr. The supernatant is carefully removed, and the translucent pellet is solubilized in SDS-containing buffer as described in Identification of Actin Filament-Associated Proteins. Solubilized proteins are electrophoresed through slab gels containing 7.5% acrylamide in the resolving gel and 3% acrylamide in the stacking gel. Protein is stained with Coomassie Brilliant blue, and the actin is quantitated by densitometry of wet gels. The amounts of sample electrophoresed should be such that the density of actin falls within the linear region of the densitometer. This range should be established beforehand by scanning gels containing varying amounts of sample. The amount of filamentous actin is determined by expressing the amount of actin sedimented from a known volume of platelet lysate as a percentage of the total actin in an equal volume of total platelet lysate (which is solubilized by adding one-third volume of a four-times concentrated solubilization buffer and is then electrophoresed and quantitated in the same way as the sedimented material).

A modification of this assay frequently used to measure the actin filament content of platelets is to isolate the actin filaments from Triton X-100 lysates by centrifugation at low g forces (for example, 15,600 g for 4 min). Although some of the actin filaments in Triton X-100 lysates do indeed sediment at these g forces, as noted, many of the filaments, particularly those in lysates from unstimulated platelets, are not sufficiently cross-linked to allow their sedimentation at these g forces. Thus, this method is not suitable for quantitative determinations.

Evaluation. One of the major concerns in methods used to evaluate the actin filament content of platelets is to ensure that the actin monomer and polymer pools are stable after lysis. Because factors present in plasma can depolymerize actin filaments, these assays must be performed on washed platelet suspensions. Similarly, platelets are lysed with a buffer that contains a Ca^{2+} chelator to prevent gelsolin-induced depolymerization of actin filaments in the lysates. Some of the solutions that have been used for lysing platelets can cause extensive changes in the state of actin between the time of lysis and the time of the initiation of the assay. This not only introduces problems in relating the results obtained with cell lysates to the actin filament content of intact cells, but also introduces errors in comparing the actin filament content of samples assayed at different times after lysis. The conditions described here minimize these problems.[30]

Findings from the DNase I inhibition assay and the sedimentation assay have been compared, and the two methods have been found to give similar results.[5] Of the two assays, the most commonly used is the DNase I inhibition assay, mainly because it is much more rapid and allows assays of more samples than is possible by the method that requires sedimentation of actin filaments in an ultracentrifuge. The major problem with the DNase I inhibition assay is in calculating the rates of hydrolysis of DNA by DNase I. The increase in absorbance of the DNA is sigmoidal, with the linear portion of the curve being proportional to the concentration of DNase I. Calculation of the rates of hydrolysis can be time consuming and can involve subjective decisions, so that small changes in the amount of unpolymerized actin in platelet lysates may not be accurately quantitated. By using the rate of change of absorbance suggested here and by ensuring that the extent of inhibition is between 30 and 70%, these drawbacks can be minimized. In addition, a computer program has been developed and used for direct data acquisition and for analysis of the linear portion of the reaction curve.[30] The use of this program has allowed the reproducible detection of small changes in the actin filament content of platelets. This program is available on request.[30]

Acknowledgments

We thank Pamela Steffen for technical assistance; Michele Prator and Kate Sholly for manuscript preparation; and Russell Levine, Barbara Allen, Al Auerbach, and Sally Gullatt Seehafer for editorial assistance.

[7] Purification and Characterization of Platelet Actin, Actin-Binding Protein, and α-Actinin

By SHARON ROSENBERG SCHAIER

Human blood platelets, highly contractile cells essential for the process of hemostasis, contain large amounts of actin and other contractile proteins. Under certain pathological conditions, e.g., a break in a blood vessel, platelets are stimulated to undergo dramatic morphological changes. They change in shape from discoid to spheroid, send out long, slender filopodia, centralize their organelles, secrete the contents of their granules, aggregate with each other to form a hemostatic plug, and retract these aggregates.[1] Contractile proteins have been proposed to be directly or indirectly involved in most of these processes.[2]

Crude actomyosin was originally isolated from platelets by Bettex-Galland and Lüscher in 1959.[3,4] It was shown to be similar to actomyosins purified from skeletal muscle sources in its viscosity, solubility, sensitivity toward ATP, and ability to be dissociated into actin and myosin. Although it is difficult to identify and localize actin and myosin filaments in thin sections of intact resting platelets due to the dense fibrogranular material filling the cytoplasmic matrix and to fixation techniques that destroy microfilaments, submembrane microfilaments have been seen peripheral to the circumferential band of microtubules.[1] In addition, platelet membrane fractions were shown to contain a large amount of F-actin capable of enhancing the Mg-ATPase of muscle myosin by two- to three-fold[5] and crude actomyosin preparations were found to be contaminated with membranes,[6] suggesting their close association.

In 1976, Lucas *et al.* reported on the direct isolation of platelet cytoskeletons, which contained most of the actin and a high molecular weight (subunit M_r 260,000) actin-binding protein (ABP).[7] Platelet ABP was originally named based on its similarities to the high molecular weight protein isolated from rabbit alveolar macrophages by Hartwig and Stossel.[8,9] They found this protein to be composed of two identical subunits capable of cross-linking actin filaments into a gel or into tight bundles depending on the ABP-to-actin ratio.[10]

α-Actinin, which was originally isolated from rabbit skeletal muscle,[11] has also been identified in platelets, where it has been seen in close

---

[1] J. G. White, in "The Circulating Platelet" (S. A. Johnson, ed.), p. 45. Academic Press, New York, 1971.
[2] R. S. Adelstein and T. D. Pollard, *Prog. Hemostasis Thromb.* **4**, 37 (1978).
[3] M. Bettex-Galland and E. F. Lüscher, *Nature (London)* **184**, 276 (1959).
[4] M. Bettex-Galland, H. Portzehl, and E. F. Lüscher, *Nature (London)* **193**, 777 (1962).
[5] N. Crawford, in "Platelets in Biology and Pathology" (J. L. Gordon, ed.), p. 121. Elsevier/North-Holland Biomedical Press, Amsterdam, 1976.
[6] R. Niederman and T. D. Pollard, *J. Cell Biol.* **67**, 72 (1975).
[7] R. C. Lucas, T. C. Detwiler, and A. Stracher, *J. Cell Biol.* **70**, (2, pt. 2), 259a (abstr.) (1976).
[8] J. H. Hartwig and T. P. Stossel, *J. Biol. Chem.* **250**, 5696 (1975).
[9] T. P. Stossel and J. H. Hartwig, *J. Biol. Chem.* **250**, 5706 (1975).
[10] J. H. Hartwig nd T. P. Stossel, *J. Mol. Biol.* **134**, 539 (1979).
[11] S. Ebashi, F. Ebashi, and K. Maruyama, *Nature (London)* **203**, 645 (1964).

association with the plasma membrane.[12,13] The muscle protein is a dimer with an apparent subunit molecular weight of 100,000 by sodium dodecyl sulfate (SDS)–polyacrylamide gel analysis and has the ability to cross-link and cause the gelation of actin filaments *in vitro*.[14-16]

In addition to myosin, ABP, and α-actinin, which promote the cross-linking of actin filaments, there are other accessory actin-binding proteins in platelets that function to prevent the polymerization and/or cross-linking of actin. Profilin is a 16,000-Da protein that binds to G-actin in a 1:1 molar ratio and renders it polymerization resistant.[17-19] A 91,000-Da protein found in platelet extracts that binds to G-actin in the presence of Ca^{2+} [20,21] has been identified as gelsolin.[22] Gelsolin has the ability to break actin filaments in the presence of Ca^{2+}, thereby increasing the amount of ABP required for the onset of incipient gelation of the actin.[23]

Other proteins found in platelets that probably have a role in contraction and platelet motility analogous to their counterparts in muscle include tropomyosin,[24] calmodulin,[25] and myosin light chain kinase.[26]

Experimental Procedures

Preparation of Platelets

All operations are carried out at 4° using plastic or siliconized glass utensils. Platelet concentrates from freshly drawn human blood are centrifuged at 350 *g* (20 min) to remove contaminating cells and then respun at 1000 *g* (20 min) to pellet the platelets. The platelets are gently resuspended and washed twice by centrifugation (1000 *g*, 20 min) in a solution con-

---

[12] J. M. Gerrard, J. V. Schollmeyer, D. R. Phillips, and J. G. White, *Am. J. Pathol.* **94**, 509 (1979).
[13] E. G. Puszkin and S. Puszkin, *Protides Biol. Fluids* **26**, 451 (1979).
[14] G. R. Holmes, D. E. Goll, and A. Suzuki, *Biochim. Biophys. Acta* **253**, 240 (1971).
[15] Z. A. Podlubnaya, L. A. Tskhovrebova, M. M. Zaalishvili, and G. A. Stefanenko, *J. Mol. Biol.* **92**, 357 (1975).
[16] J. V. Schollmeyer, G. H. R. Rao, and J. G. White, *Am. J. Pathol.* **93**, 433 (1978).
[17] L. Carlsson, L. E. Nyström, I. Sundkvist, F. Markey, and U. Lindberg, *J. Mol. Biol.* **115**, 465 (1977).
[18] H. E. Harris and A. G. Weeds, *FEBS Lett.* **90**, 84 (1978).
[19] F. Markey, U. Lindberg, and L. Eriksson, *FEBS Lett.* **88**, 75 (1978).
[20] F. Markey, T. Persson, and U. Lindberg, *Cell (Cambridge, Mass.)* **23**, 145 (1981).
[21] L. L. Wang and J. Bryan, *Cell (Cambridge, Mass.)* **25**, 637 (1981).
[22] S. E. Lind, H. L. Yin, and T. P. Stossel, *J. Clin. Invest.* **69**, 1384 (1982).
[23] H. L. Yin, K. S. Zaner, and T. P. Stossel, *J. Biol. Chem.* **255**, 9494 (1980).
[24] I. Cohen and C. Cohen, *J. Mol. Biol.* **68**, 383 (1972).
[25] G. C. White, S. N. Levine, and A. N. Steiner, *Am. J. Hematol.* **10**, 359 (1981).
[26] J. L. Daniel and R. S. Adelstein, *Biochemistry* **15**, 2370 (1976).

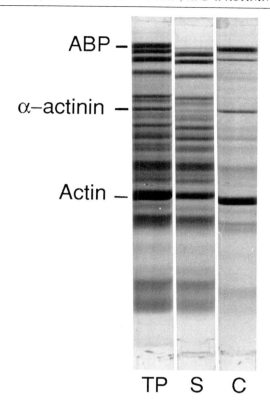

Fig. 1. Isolation of the platelet cytoskeleton. Sodium dodecyl sulfate-polyacrylamide gels (5.5%) of total platelet proteins (TP), Triton-soluble platelet supernatant (S), and Triton-insoluble cytoskeleton (C) after 3000 g centrifugation. Reproduced from Rosenberg et al.[39] by copyright permission of the American Society of Biological Chemists.

taining 126 mM NaCl, 5 mM KCl, 0.3 mM EDTA, 10 mM NaPO$_4$, pH 7.4 (platelet washing solution). Sodium dodecyl sulfate-polyacrylamide gel electrophoresis is used to visualize the protein composition of platelets (Fig. 1, gel TP) and quantitative densitometry has shown that actin and the proteins that bind to it constitute half of the total protein of platelets (actin, 25–30%; myosin, 12%; ABP, 8%; α-actinin, 3%).[27,28]

[27] R. C. Lucas, M. Gallagher, and A. Stracher, in "Contractile Systems in Non-Muscle Tissues" (S. V. Perry, A. Margreth, and R. S. Adelstein, eds), p. 133. North-Holland Publ., New York, 1976.
[28] S. Rosenberg, J. Lawrence, and A. Stracher, Cell Motil. 4, 317 (1982).

Isolation of Platelet Cytoskeleton

The platelet cytoskeleton is an excellent starting material for the purification of actin, ABP, and α-actinin since it contains little else. The microtubule system is disrupted by the isolation buffer and no intermediate filament system has yet been found in platelets. The cytoskeleton is isolated by adding washed platelets (diluted 1 : 1 in platelet washing solution) to 10 vol of ice-cold 1% (v/v) Triton X-100, 40 mM KCl, 10 mM imidazole chloride, 10 mM EGTA, 2 mM NaN$_3$, pH 7.0 (Triton solubilization buffer). A flocculent precipitate appears immediately and is collected at 3000 g (2 min) after settling for 12 min on ice. The Triton-insoluble cytoskeleton generally contains 60–80% of the platelet actin, 70–95% of the total ABP, 90% of the α-actinin, and small, variable amounts of myosin (Fig. 1, gel C). When the platelets are treated with the Triton solubilization buffer directly on a carbon-coated grid, negatively stained with uranyl acetate, and examined with an electron microscope, the cytoskeleton reveals itself as branched bundles of actin filaments.[29] More recently, Fox et al.[30] also used electron microscopic techniques to show the Triton cytoskeleton as a platelet-shaped structure consisting of a network of long actin filaments with a more amorphous F-actin- and ABP-containing layer outlining the periphery.

All the components of the Triton solubilization buffer are important. If any components are changed in concentration, added, or omitted, the amount or composition of the cytoskeletal precipitate may be altered.

Addition of EDTA. The inclusion of 10 mM EDTA or the substitution of EDTA for EGTA in the Triton solubilization buffer leads to a doubling of the amount of myosin included in the cytoskeletal pellet (Table I, conditions 2 and 3). This is probably due to the formation of rigor bonds between myosin and the actin filaments of the cytoskeleton caused by the chelation of magnesium ions. This myosin-containing cytoskeleton has been named the "contractile apparatus" and was found to be capable of hydrolyzing MgATP at a rate of 10–50 nmol P_i/min/mg of myosin.[31]

Addition of MgATP. The addition of 2.5 mM MgATP to the Triton solubilization buffer causes an effect opposite to that of EDTA. Myosin filaments dissociate from the actin cytoskeleton, so less myosin precipitates with the cytoskeletal pellet (Table I, condition 4).

Ionic Strength. When the concentration of KCl in the Triton solubiliza-

---

[29] S. Rosenberg, A. Stracher, and R. C. Lucas, *J. Cell Biol.* **91**, 201 (1981).
[30] J. E. B. Fox, J. K. Boyles, M. C. Berndt, P. K. Steffen, and L. K. Anderson, *J. Cell Biol.* **106**, 1525 (1988).
[31] R. C. Lucas, S. Rosenberg, S. Shafiq, A. Stracher, and J. Lawrence, *Protides Biol. Fluids* **26**, 465 (1979).

TABLE I
PROTEIN COMPOSITION OF PLATELET CYTOSKELETONS
ISOLATED UNDER VARIOUS CONDITIONS

	Total protein (%) in cytoskeleton[a]		
Additions	ABP	Myosin	Actin
1. None	83.8[b]	24.9	57.2
2. 10 mM EDTA	51.8	47.9	48.0
3. 10 mM EDTA (no EGTA)	53.0	48.0	45.0
4. 2.5 mM MgATP	73.9	10.8	58.5
5. 150 mM KCl (not 40 mM)	26.8	1.3	41.7

[a] Twenty-five microliters of washed platelets (resuspended 1:1 in platelet washing buffer) was added to 250 μl of Triton solubilization buffer (1% Triton X-100, 40 mM KCl, 10 mM imidazole chloride, 10 mM EGTA, 2 mM NaN$_3$, pH 7.0) with the various changes listed above. After 12 min on ice, the cytoskeletons were collected at 3000 g, washed once in their respective Triton solutions, and twice in 40 mM KCl, 1 mM EGTA, 10 mM PIPES, and 2 mM NaN$_3$, pH 6.8. Triton-soluble proteins were collected at 1000 g after their precipitation with 9 vol of cold acetone. Equivalent loads of the Triton pellets and supernatants were electrophoresed on 5.5% SDS–polyacrylamide gels, which were subsequently scanned in a Helena Quick Scan densitometer and the area under the peaks determined.
[b] All percentages given represent the average of four experiments. Reproduced from Rosenberg et al.[28] by copyright permission of Alan R. Liss, Inc.

tion buffer is increased from 40 to 150 mM, the recovery of ABP is reduced from 84 to 27% of the total content of ABP (Table I, condition 5) concomitant with a 27% decrease in the actin recovery. This is probably due to the decreased interaction between ABP and actin at higher salt concentrations,[29] enabling the freed proteins to be solubilized by the Triton buffer.

EGTA versus CaCl$_2$. The requirement for the high concentration of EGTA (10 mM) in the Triton solubilization buffer is due to two conditions:

1. Platelets contain a Ca^{2+}-activated, neutral protease[32,33] that attacks ABP and the 230,000-Da protein.[29,33] If EGTA is omitted from the Triton

---

[32] D. R. Phillips and M. Jakábová, *J. Biol. Chem.* **252**, 5602 (1977).
[33] J. A. Truglia and A. Stracher, *Biochem. Biophys. Res. Commun.* **100**, 814 (1981).

TABLE II
Effect of Thrombin on Platelet Cytoskeleton

Time after thrombin addition (sec)	Total protein (%) in cytoskeleton[a]		
	ABP	Myosin	Actin
0	68.0[b]	23.5	59.7
5	72.6	44.3	64.6
30	81.2	51.5	69.0
120	86.2	71.3	70.4

[a] Fifty microliters of washed platelets (resuspended 1:1 in platelet washing buffer) were stimulated with 5 μl of 2×10^{-7} M thrombin for the times listed above, at which point 500 μl of cold Triton solubilization buffer plus 5 nM hirudin (thrombin inhibitor) were added and the samples were placed on ice. Triton-insoluble cytoskeletons were sedimented at 3000 g (2 min) after settling for 12 min on ice. Triton-soluble proteins were collected at 1000 g (5 min) after their precipitation with 9 vol of cold acetone. Equivalent loads of the Triton-insoluble pellets and soluble supernatants were electrophoresed on 5.5% SDS–polyacrylamide gels, which were subsequently scanned and the area under the peaks determined.

[b] The data represent the average of two experiments. Reproduced from Rosenberg et al.[28] by copyright permission of Alan R. Liss, Inc.

solubilization buffer, the release of membrane-limited Ca^{2+} by Triton leads to the complete disappearance of these two proteins within 10 min at 4°.

2. When EGTA is omitted from the Triton solubilization buffer, allowing the free Ca^{2+} level to rise, but leupeptin is included to prevent proteolysis,[32] no cytoskeletal complex can be pelleted while the ABP remains intact in the Triton-soluble supernatant.[29] Therefore, there appears to be a Ca^{2+}-directed event, aside from proteolysis, that promotes the *in vivo* disassembly of the cytoskeleton.

Effect of Platelet Activation. The platelet cytoskeleton undergoes a marked change in composition when isolated from platelets that were previously stimulated with 20 mM thrombin (Table II). The most profound effect is on myosin association, which increases from 23% (resting platelet) to 71% of the platelet total myosin content recovered with the cytoskeleton when it is isolated 2 min after thrombin activation. The yield of ABP increases from 68 to 86% of the total ABP present in the cytoskeleton. Actin association increases as well, from 60 to 70%.

Purification of Actin-Binding Protein

The cytoskeleton isolated with the Triton solubilization buffer is washed once in this buffer and then twice in 40 mM KCl, 10 mM PIPES,

1 mM EGTA, 2 mM NaN$_3$, pH 6.8 (buffer A) by resuspending and recentrifuging at 3000 g for 2 min. Since these low-speed pellets are very loose and slippery, the supernatants should be poured off through two layers of cheesecloth.

The washed pellet is then resuspended in a small volume (one to two times that of the original packed platelets) of 0.6 M KCl, 10 mM PIPES, 2 mM NaN$_3$, pH 6.8 (high-salt buffer), homogenized for 2 min on ice, and clarified of the high-salt-insoluble proteins at 25,000 g (15 min). The insoluble protein pellet is reextracted with high-salt buffer either immediately or overnight at 4° to increase the yield of solubilized ABP and actin. The residual high-salt-insoluble protein is stored in 50% glycerol at $-20°$ and is used for subsequent extractions in the purification of α-actinin.

To remove the actin from this high-salt-soluble supernatant, 5 mM EGTA–ATP is added and the sample is stored at 4° for 16 hr. Kane originally described the phenomenon in sea urchin eggs, in which actin forms birefringent bundles of filaments when ATP is added to a gelled extract solubilized by high salt.[34] The same is found to be true with human platelet actin. When the high-salt EGTA–ATP solution is centrifuged at 25,000 g (15 min) the pellet is seen to consist mostly of actin, leaving ABP and any contaminating proteins behind in the supernatant. This "bundling" step thereby provides a mechanism for easily separating actin from ABP once they are dissociated from each other in high salt. The actin bundle pellet is used for the subsequent purification of actin, while the bundle supernatant is used in the preparation of ABP (see Fig. 2).

Polyethylene glycol (PEG) 6000 (50%, w/v) is added to this ABP-rich supernatant until a final concentration of 5% is reached. After sitting for 10 min on ice, the solution is centrifuged at 10,000 g (12 min), sedimenting the ABP and separating it from most of its major contaminants. Because PEG fractionates essentially on the basis of molecular weight,[35] the use of gel-filtration chromatography (which is more time consuming, dilutes the protein, and reduces its yield) is avoided.

The 5% PEG pellet is resuspended in 10 mM PIPES, 1 mM EGTA, 2 mM NaN$_3$, pH 6.8 (column buffer) and its subsequent clarification (25,000 g, 30 min) usually removes the small myosin contamination. This PEG-purified ABP is generally found to be 90% pure when analyzed by quantitative densitometry. It is critical that any F-actin be removed from the sample to be PEG fractionated because its presence causes a clumping of the actin with ABP in the PEG precipitate, making it almost impossible to resuspend and recover. F-Actin could be present here if the prior bundling step was not complete. Incomplete bundling occurs if the actin

---

[34] R. E. Kane, *J. Cell Biol.* **66**, 305 (1975).
[35] K. C. Ingham, *Arch. Biochem. Biophys.* **186**, 106 (1978).

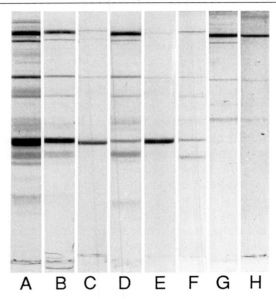

FIG. 2. Purification of ABP. Sodium dodecyl sulfate-polyacrylamide gels (5.5%) of the starting cytoskeleton (A), 25,000 g supernatant (B), and pellet (C) after solubilization of the cytoskeleton in high-salt buffer. A 25,000 g supernatant (D) and pellet (E) after the bundling of actin in the high-salt soluble fraction with EGTA-ATP. A 10,000 g supernatant (F) and pellet (G) after the precipitation of the bundle supernatant with 5% PEG, and the 5% PEG pellet resuspended and clarified at 25,000 g (H). Reproduced from Rosenberg et al.[29] by copyright permission of The Rockefeller University Press.

concentration in the high-salt sample is too low, as will be seen in the next section on actin purification.

Clarified, PEG-purified ABP in column buffer is applied to a DEAE-Sephacel column that is then washed with column buffer. The adherent proteins are eluted with a linear 0–0.4 M KCl gradient (Fig. 3). Actin-binding protein is seen to elute free of its contaminating protein at approximately 0.2 M KCl. The ABP peak is pooled, concentrated by dialysis against glycerol, and dialyzed into buffer A. Purified ABP can be stored for weeks in the refrigerator without evidence of proteolysis or loss of its ability to bind to actin.

Purification of Actin

Two methods can be used to purify platelet actin:

1. Actin-rich bundles, generated during the ABP purification (Fig. 2, gel E), are resuspended in high-salt buffer and rebundled overnight with

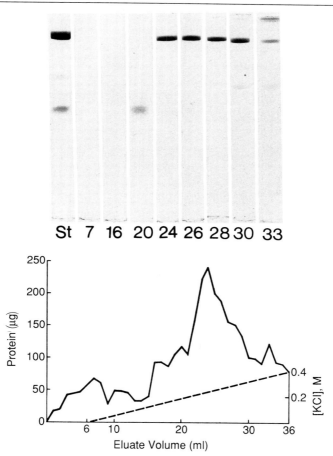

FIG. 3. Column purification of ABP. Three milligrams of PEG-purified ABP is applied to a 0.7 × 5.0 cm DEAE-Sephacel column equilibrated with column buffer. The column is washed with 2–3 vol of this buffer and then eluted in 1-ml fractions with 30 ml of a 0–0.4 M KCl linear concentration gradient (prepared in the column buffer). Protein concentrations are determined from aliquots of every other fraction using the Bio-Rad dye-binding assay. The dotted line represents the salt concentration across the gradient. The inset shows 5.5% SDS–polyacrylamide gels of selected column fractions and the starting material applied to the column (St). Reproduced from Rosenberg et al.[29] by copyright permission of The Rockefeller University Press.

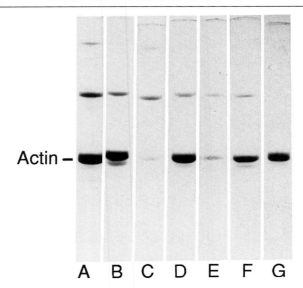

FIG. 4. Actin purification by bundling. Sodium dodecyl sulfate-polyacrylamide gels (5.5%) of actin bundles (A), a by-product of the ABP purification (Fig. 2, gel E), which are rebundled in 5 mM EGTA–ATP to yield a cleaner actin preparation (B) and its resulting supernatant (C). These cleaner bundles are again cycled to yield another pellet (D) and supernatant (E). The third bundle pellet is resuspended in buffer A, homogenized, and centrifuged at 15,000 g to yield a pellet (F) and a supernatant (G) of purified actin. Reproduced from Rosenberg et al.[29] by copyright permission of The Rockefeller University Press.

5 mM EGTA–ATP (Fig. 4). This process can be repeated several times, each time resulting in a further purification of the actin. Every bundling step excludes more of the α-actinin, which does not bind well to the actin in 0.6 M KCl. Finally, when only a trace of α-actinin remains, the actin bundles are resuspended in buffer A, homogenized on ice, and centrifuged at 15,000 g (15 min) to pellet any actin–α-actinin complexes. The resulting supernatant of uncomplexed actin is free of contamination. The pelleted actin complex can be resolubilized in buffer A, homogenized, and recentrifuged to increase the yield of soluble, purified actin.

2. The actin-rich, high-salt-insoluble material generated during the ABP purification (Fig. 2, gel C) can be used to prepare actin. The pellet is dissolved in and dialyzed against actin depolymerization buffer (2 mM Tris-HCl, 0.2 mM CaATP, 0.5 mM 2-mercaptoethanol, pH 8.0). Actin is then purified through repeated cycles of polymerization and depolymerization according to Spudich and Watt.[36]

[36] J. A. Spudich and S. Watt, *J. Biol. Chem.* **246**, 4866 (1971).

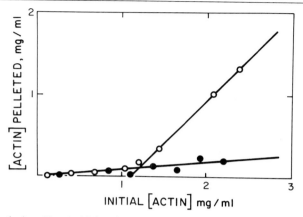

FIG. 5. Actin bundling in high salt. Human platelet (open circles) and rabbit skeletal muscle actin (closed circles) are diluted to various starting concentrations in equal volumes of 0.6 M KCl. Then 5 mM EGTA–ATP is added and after 16 hr at 4° the samples are centrifuged at 25,000 g (15 min) to pellet any newly formed bundles. Bundle pellets are resuspended in equal volumes of 0.6 M KCl and Lowry protein determinations [O. H. Lowry, N. J. Rosebrough, A. L. Farr, and R. J. Randall, *J. Biol. Chem.* **193**, 265 (1951)] are performed on the bundle pellets and supernatants as well as on the original actin dilutions to determine the extent of bundle formation. Reproduced from Rosenberg *et al.*[29] by copyright permission of The Rockefeller University Press.

The highly purified platelet actin retains its ability to self-associate into 25,000 g sedimentable bundles in 0.6 M KCl, 5 mM EGTA–ATP (Fig. 5). Its critical concentration for bundling under these conditions is 1.1 mg/ml, very similar to that of sea urchin egg actin, which bundles at concentrations greater than 1.2–1.3 mg/ml.[37] Thus far, actin bundling appears to be a property of nonmuscle actins only, since skeletal muscle actin will not self-associate under identical conditions.[29,37] The ability of the purified actin to bundle shows that it is an intrinsic property of the actin and is not due to any other proteins in the high-salt extract of the platelet cytoskeleton (e.g., ABP, α-actinin). It is the critical actin bundling concentration that makes it necessary, during the ABP purification, to resuspend the washed cytoskeleton in only a small volume of high-salt buffer. More high-salt buffer will dilute and solubilize more of the ABP and actin from the cytoskeleton, thus increasing their eventual yields. However, incomplete bundling due to too low an actin concentration will cause the unbundled, soluble actin filaments to PEG precipitate with the ABP in a tight clump that cannot be resolubilized. To increase the yield of ABP and actin, it is

[37] J. Bryan and R. E. Kane, *J. Mol. Biol.* **125**, 207 (1978).

better to perform a second high salt extraction of the cytoskeleton and combine the PEG pellets of these two preparations.

Platelet actin is made up of the β and γ isoelectric varieties, which are present in a 5:1 ratio.[18,31] Gordon et al.[38] found, however, after studying various highly purified nonmuscle actins, that although there were quantitative differences in their critical concentrations, there were no qualitative differences in their polymerization properties as compared to muscle actin, which is all of the α variety. Platelet actin polymerizes to form 7-nm filaments indistinguishable from those of muscle sources although greater concentrations of platelet actin are necessary to see the onset of polymerization.

Purification of α-Actinin

Platelet cytoskeletons previously extracted once or twice with high-salt buffer for the ABP and actin purifications are stored in 50% high-salt buffer and 50% glycerol at $-20°$. Samples from many stored preparations are pooled to use for the α-actinin purification (Fig. 6) since the prior ABP extractions removed much of the α-actinin as well.

The pooled high-salt sample is homogenized on ice for several minutes and then clarified at 25,000 g (15 min). The insoluble pellet is reextracted with high-salt buffer and the two soluble supernatants are combined. These extractions remove most of the α-actinin from the insoluble pellet along with some actin, ABP, and myosin.

Actin is removed from this preparation by precipitation with 30% ammonium sulfate at 10,000 g (10 min). The rest of the protein is precipitated by increasing the ammonium sulfate concentration to 50% and centrifuging at 12,000 g (10 min).

The 30–50% ammonium sulfate fraction is resuspended in and dialyzed against column buffer, clarified at 25,000 g (20 min), and applied to a DEAE-Sephacel column equilibrated with the same buffer (Fig. 7). After washing with 2–3 vol of this buffer, the proteins are eluted with a linear 0–0.4 M KCl gradient. Actin-binding protein elutes at approximately 0.2 M KCl as described earlier, while the α-actinin elutes later, at approximately 0.225–0.25 M KCl. Fractions of the greatest purity (monitored by gel electrophoresis) are pooled, dialyzed into column buffer, and rechromatographed on the DEAE-Sephacel column. The α-actinin obtained from this procedure is greater that 95% pure as demonstrated by quantitative densitometry and is stable for months in the refrigerator without evidence of proteolysis.

[38] D. J. Gordon, J. L. Boyer, and E. D. Korn, *J. Biol. Chem.* **252**, 8300 (1977).

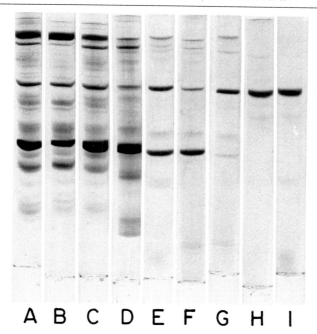

FIG. 6. Purification of α-actinin. Sodium dodecyl sulfate-polyacrylamide gels (5.5%) of the starting Triton-insoluble cytoskeleton (A), 25,000 g supernatant (B) and pellet (C) of the first high-salt extraction, high-salt insoluble pellet (D) and high-salt soluble supernatant pool (E) of two subsequent extractions, 0–30% ammonium sulfate fraction (F), 30–50% ammonium sulfate fraction (G), pool of the first DEAE-Sephacel column (H), and pool of the second DEAE-Sephacel column (I). All gels are heavily loaded to highlight the impurities of these preparations and gels (C) and (D) may be unrepresentative of their samples since these insoluble pellets are difficult to keep resuspended. Reproduced from Rosenberg *et al.*[39] by copyright permission of the American Society of Biological Chemists.

Platelet α-actinin has a molecular weight of 105,000, slightly heavier than that of skeletal muscle α-actinin (M_r 100,000), and runs as a compact but discrete doublet on 7.5 and 10% SDS–polyacrylamide gels.[39] Human platelet α-actinin cross-reacts with antibody prepared against rabbit cardiac α-actinin and is seen to be very similar but not identical to rabbit skeletal muscle α-actinin on one-dimensional, partial proteolysis peptide maps.[39]

[39] S. Rosenberg, A. Stracher, and K. Burridge, *J. Biol. Chem.* **256**, 12986 (1981).

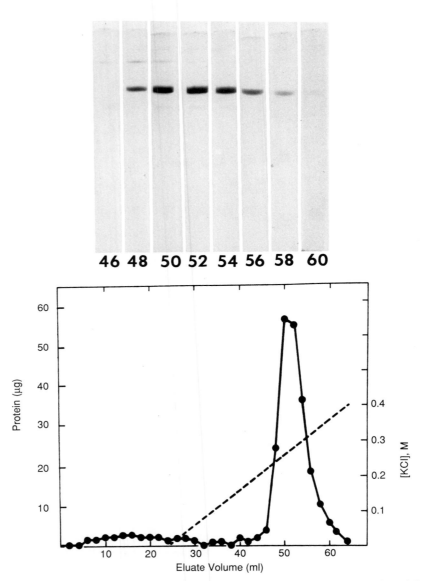

FIG. 7. Column purification of α-actinin. The 30–50% ammonium sulfate fraction of the α-actinin purification (Fig. 6, gel G) is applied to a 1.5 × 7.0 cm DEAE-Sephacel column equilibrated with column buffer. The column is washed with 2–3 vol of this buffer and then eluted in 1-ml fractions with a 40-ml linear 0–0.4 M KCl concentration gradient (prepared in column buffer). Protein concentrations are determined from aliquots of every other fraction using the Bio-Rad dye-binding assay. The dotted line represents the salt gradient across the column. The inset shows 5.5% SDS–polyacrylamide gels of selected column fractions. Reproduced from Rosenberg et al.[39] by copyright permission of the American Society of Biological Chemists.

Recombination of Actin-Binding Protein and α-Actinin with Actin

After purification, both ABP and α-actinin retain their ability to bind to purified actin. This interaction is influenced by several factors.

Effect of Ionic Strength. As the ionic strength of the incubation solution used increases, the amount of the cross-linker bound to actin decreases.[29,39]

Effect of State of Actin. When ABP is added to preformed F-actin filaments, it binds until a maximum ratio of 1 : 9 (ABP to actin, mole to mole) is reached. However, if G-actin is polymerized in the presence of ABP, saturation does not occur until 3.4 actin monomers are bound by an ABP dimer. This greater ABP-to-actin ratio will permit a more highly branched structure to form. Actin-binding protein cannot induce the polymerization of G-actin under nonpolymerizing conditions but it does increase the sedimentability (148,000 g) of actin undergoing salt-induced polymerization by reducing both the time and concentration of actin required to see the onset of this sedimentation.[40] Actin-binding protein can be accelerating the salt-induced polymerization of actin either by acting as a nucleation site on which new actin filaments can grow or by cross-linking newly formed actin oligomers into larger aggregates that are more easily sedimented.

Effect of Ca^{2+}. In whole-platelet extracts both ABP and α-actinin exhibit Ca^{2+}-sensitive binding to actin but after purification only the α-actinin retains this property (Table III). Since purified ABP is shown to bind equally well to actin in the presence or absence of 2.5 mM $CaCl_2$, its *in vivo* Ca^{2+}-sensitive interaction must be mediated by a third platelet protein not present in the purified preparations. Purified platelet α-actinin, however, continues to show a decreased binding to actin in the presence of $CaCl_2$ although this is found not to be true for skeletal muscle or smooth muscle α-actinin. That Ca^{2+}-sensitive binding of α-actinin to actin may only be a property of nonmuscle α-actinins has been confirmed by others.[41]

Actin-Binding Protein versus α-Actinin. Actin-binding protein and α-actinin are found to compete with each other for binding sites on actin. This is probably due to steric restraints since these proteins are large compared with the actin monomer and can block each other's binding sites. When platelet α-actinin and ABP are coincubated with actin in the absence of $CaCl_2$, the levels of binding of both of these proteins drop significantly from the controls. (The amount of platelet α-actinin bound decreases from 22.8 to 15.2 μg and the amount of ABP bound to actin decreases from 16.6 to 5 μg.) In the presence of Ca^{2+}, however, platelet

---

[40] S. Rosenberg and A. Stracher, *J. Cell Biol.* **94**, 51 (1982).
[41] K. Burridge and J. R. Feramisco, *Nature (London)* **294**, 565 (1981).

TABLE III
Binding of Actin-Binding Protein and α-Actinin to Actin

	Protein (μg) in pellet[a]				
Condition	Ca²⁺	Actin	Platelet α-actinin	Skeletal muscle α-actinin	ABP
+ ABP	−	39.0			16.6
	+	35.7			14.4
+ Platelet α-actinin	−	32.0	22.8		
	+	33.0	12.0		
+ Skeletal muscle α-actinin	−	38.6		18.7	
	+	37.2		21.0	
+ ABP + platelet α-actinin	−	37.4	15.2		5.0
	+	36.2	6.6		7.9
+ ABP + skeletal muscle α-actinin	−	38.9		13.5	10.2
	+	38.2		15.1	8.6

[a] Forty micrograms of skeletal muscle F-actin were incubated with various combinations of the actin-binding proteins indicated above (48 μg of platelet α-actinin, 48 μg of skeletal muscle α-actinin, or 18 μg of platelet ABP) in 40 mM KCl, 10 mM PIPES. 0.25 mM EGTA, pH 6.8, ±2.5 mM CaCl$_2$. After 15 min at room temperature, the samples were centrifuged at 148,000 g (17 min) in a Beckman airfuge (forces able to sediment F-actin). The resulting pellets and supernatants were processed for SDS–polyacrylamide gel electrophoresis and the gels were scanned for their protein content as described earlier. Reproduced from Rosenberg et al.[39] by copyright permission of the American Society of Biological Chemists.

α-actinin binds weakly (50% decreased from the minus Ca²⁺ condition), and consequently 50% more ABP is able to bind to the actin. Skeletal muscle α-actinin also competes with ABP for binding sites on actin although it does not exhibit Ca²⁺-sensitive binding.

Effect of Actin-Binding Protein and Myosin on Actin Depolymerization

Due to the controversy over the state of actin in the resting platelet and the increased amount of actin associated with the cytoskeletons of thrombin-activated platelets,[20,42–44] it was of interest to determine the effect of ABP and myosin binding on the polymerization state of platelet actin. Can the large amount of actin in the cytoskeletons of resting platelets

---

[42] L. Carlsson, F. Markey, I. Blikstad, T. Persson, and U. Lindberg, *Proc. Natl. Acad. Sci. U.S.A.* **76**, 6376 (1979).

[43] D. R. Phillips, L. K. Jennings, and H. H. Edwards, *J. Cell Biol.* **86**, 77 (1980).

[44] L. K. Jennings, J. E. Fox, H. H. Edwards, and D. R. Phillips, *J. Biol. Chem.* **256**, 6927 (1981).

TABLE IV
EFFECT OF ACTIN-BINDING PROTEIN AND
MYOSIN ON DEPOLYMERIZABILITY OF PLATELET
ACTIN

Condition	Actin (%) in pellet[a]
Actin alone	36.6
Actin + ABP[b]	60.9
Actin + myosin[b]	74.7

[a] Samples of platelet actin (2 mg/ml), platelet ABP (1 mg/ml) and chicken gizzard myosin (1 mg/ml) were dialyzed versus 125 mM KCl, 10 mM Tris-HCl, pH 7.4. Aliquots (100 μl) of platelet actin were then combined with either 50 μl of myosin, 50 μl of ABP, or 50 μl of the dialysis buffer, and the mixtures were dialyzed overnight at 4° against the Triton solubilization buffer. The samples were then transferred from the dialysis bags to airfuge tubes and were centrifuged at 148,000 g for 20 min in a Beckman airfuge (forces sufficient to pellet free F-actin but not free ABP or myosin). Supernatants and pellets were separated and processed for SDS–polyacrylamide gel electrophoresis. Equivalently loaded gels were scanned and the area under the peaks determined.

[b] All of the ABP and myosin were found in the high-speed pellet. Reproduced from Rosenberg et al.[28] by copyright permission of Alan R. Liss, Inc.

isolated with the Triton solubilization buffer (as compared to other solutions) be due to conditions that preserve the binding of ABP, myosin, and α-actinin (low-salt, high EGTA)? Is the increased amount of actin associated with the cytoskeleton of activated platelets due to new polymerization or to a prevention of its depolymerization during isolation by the incorporation of new ABP and myosin into the cytoskeleton? The experiment described in Table IV was designed to answer these questions. Purified platelet F-actin (in 125 mM KCl), at a final concentration of 1.35 mg/ml (estimated to be its concentration after a 1 : 20 dilution of packed platelets into the washing and Triton solubilization buffers[45]), is dialyzed ± one-fourth its weight of ABP or myosin (again, mimicking their *in vivo* ratios)

---

[45] Packed platelets (1000 g, 20 min) are determined to contain 88 mg/ml protein. If actin constitutes 30% of the total protein of platelets, its concentration in the packed platelets is 26.4 mg/ml. After being diluted 1 : 1 with platelet washing buffer and 1 : 10 with Triton solubilization buffer, the final actin concentration is 1.35 mg/ml.

against Triton solubilization buffer. The results show that in the absence of any accessory proteins only 37% of the actin could be detected as filamentous by sedimentation. When ABP or myosin are present during the dialysis, however, 61 and 75%, respectively, of the actin remains as filamentous. Therefore, it is seen that both ABP and myosin have the ability to protect the actin from depolymerizing when it is added to the cold, chelated, Triton solubilization buffer. If these proteins exhibit the same behavior *in vivo*, then isolation conditions that prevent the disassembly of the cytoskeleton (low ionic strength, low $CaCl_2$, high EGTA, thrombin activation) will protect the actin from depolymerizing by virtue of its being bound up by ABP, α-actinin, and myosin. Other isolation conditions that may result in the removal of ABP and other accessory proteins from the actin leave it free to depolymerize when diluted into these solutions.

Concluding Remarks

For the reader interested solely in the purification of actin, ABP, and α-actinin, the methods described above provide a procedure that was perfectly satisfactory in the author's hands. There are experimenters who claim, however, that this cytoskeleton is not representative of that in the resting platelet. This is not critical to the biochemist who purifies these proteins but is important to the physiologist studying platelet function. Some of the discrepancies can be explained as follows.

1. *Effect of local anesthetics:* Other workers have used local anesthetics to keep platelets morphologically resting[46,47] although these have been shown to cause the release of Ca^{2+} from membranes,[48,49] cause the disruption of the cytoskeletal organization in tissue-cultured cells,[50,51] and result in the degradation of ABP from platelets.[52] Allowing the Ca^{2+} level in platelets to rise to preproteolysis levels has been found to disassemble and thereby prevent the isolation of the cytoskeleton,[29] which may be why Nachmias[53] describes the cytoskeleton as composed of amorphous aggregates devoid of any filaments.

2. *Different isolation solutions:* The higher ionic strength and lower

[46] V. T. Nachmias, J. S. Sullender, and A. Asch, *Blood* **50**, 39 (1977).
[47] V. T. Nachmias, J. S. Sullender, and J. R. Fallon, *Blood* **53**, 63 (1979).
[48] H. Hauser and R. M. C. Dawson, *Biochem. J.* **109**, 909 (1968).
[49] P. S. Low, D. H. Lloyd, T. M. Stein, and J. A. Rogers, III, *J. Biol. Chem.* **254**, 4119 (1979).
[50] G. L. Nicholson and G. Poste, *J. Supramol. Struct.* **5**, 65 (1976).
[51] G. L. Nicholson, J. R. Smith, and G. Poste, *J. Cell Biol.* **68**, 395 (1976).
[52] E. I. B. Peerschke, *Blood* **68**, 463 (1986).
[53] V. T. Nachmias, *J. Cell Biol.* **86**, 795 (1980).

EGTA concentrations used in the isolation solutions of Carlsson et al.[42] and Phillips et al.[43] could be responsible for removing ABP and α-actinin from actin, thereby allowing the actin to depolymerize in the lysis buffer (as seen in Table IV). They report that only 30–40% of the platelet actin is filamentous after Triton lysis but their equally low yields of ABP support the depolymerization hypothesis.

3. *Actin dilution effect:* In the work described in this chapter, the platelets are diluted to a final concentration of $1.0–2.5 \times 10^9$/ml in the Triton solubilization buffer and the final actin concentration is approximately 1.4 mg/ml.[45] Since Gordon et al.[38] found that the critical concentration required for the polymerization of actin in 0.1 M KCl at 5° is 0.51 mg/ml, the mere dilution of platelets into the chelated Triton solubilization buffer at 5° can result in a depolymerization of 36% of the actin. Carlsson et al.[42] and Phillips and co-workers,[43,54] who work with $0.5–5.0 \times 10^8$ platelets/ml, will therefore have an even greater problem controlling the dilution-induced depolymerization of actin. This factor alone can account for their lower yields of F-actin.

4. *Effect of cold:* Fox et al.[54] suggested that the high yield of F-actin achieved with the Triton solubilization buffer at 4° was due to a cold-induced activation of the platelets. Using the method outlined in this chapter, they were able to recover 58% of the cell actin (and 86% of the ABP) with the cytoskeleton. However, when using a platelet suspension (2.6×10^8 platelets/ml) prepared by centrifugation at 37°, they isolated only 25–29% of the actin after a 2× dilution into their isolation solution or a 10× dilution into Triton solubilization buffer. The lower actin yield that they saw may not be due to the ambient temperature but to the fact that they were trying to isolate actin filaments from platelets that were 100× more dilute than the platelets used at 4°. (see point 3 above.)

5. *Effect of centrifugation:* Fox et al.[54] presented evidence that most filaments present in centrifuged platelets already exist and do not arise from centrifugation-induced activation as suggested by Nachmias.[53] They found that the filament content of resting platelets (measured by DNase and centrifugation assays) prepared by gel filtration or by centrifugation ± cytochalasin D (to prevent new actin polymerization) was very similar (40–48%). Electron micrographs, which showed that centrifuged platelets had changed shape and extended many filopodia although they had a filament content comparable to that of preparations of discoid platelets, provide evidence that filopodial formation did not require new polymerization of G-actin but only a rearrangement of preexisting actin filaments.

[54] J. E. B. Fox, J. K. Boyles, C. C. Reynolds, and D. R. Phillips, *J. Cell Biol.* **98**, 1985 (1984).

[8] Purification and Characterization of Platelet Myosin

By JAMES L. DANIEL and JAMES R. SELLERS

The platelet cytoplasm is dominated by platelet myosin, platelet actin, and a group of proteins that regulate both the polymerization and interaction of myosin and actin.[1] Morphologic studies on platelets suggest that contractile proteins play an essential part in platelet activation.[1] These facts make the platelet a particularly good system in which to study the function of nonmuscle contractile proteins.

By analogy with muscle systems, the force produced by platelet actomyosin, which produces a physiologic response, is thought to result from an ATP-dependent interaction between platelet actin and myosin. In striated muscle, this interaction is regulated by Ca^{2+} binding to the troponin–tropomyosin complex located on the actin filament. In platelets, the best evidence indicates that this interaction is regulated by phosphorylation of the 20,000-D light chain of platelet myosin.[2] In turn the phosphorylation state of myosin is regulated by Ca^{2+} by a calmodulin-dependent kinase.[3,4] The study of the phosphorylation state of myosin in intact platelets has provided a useful tool for the study of the role of contractile events in platelet activation.[1,5–7]

We will describe a method to obtain platelet myosin in large quantity. Protease inhibitors are essential to this method because platelet myosin is easily cleaved into several fragments. We will also describe methods to characterize platelet myosin with respect to ATPase activity and phosphorylation state. We will present methods to convert myosin from one phosphorylation state to another and a method to characterize the phosphorylation state of myosin in intact platelets. Finally, a method will be described to characterize which amino acids are phosphorylated using one-dimensional peptide mapping. Other methods for characterization of

---

[1] J. L. Daniel and G. P. Tuszynski, *in* "Hemostasis and Thrombosis" (R. Colman, E. Salzman, V. Marder, and A. Marcus, eds.), p. 664. Lippincott, New York, 1987.
[2] R. S. Adelstein and M. A. Conti, *Nature (London)* **256,** 597 (1975).
[3] R. Dabrowska and H. Hartshorne, *Biochem. Biophys. Res. Commun.* **85,** 1352 (1978).
[4] D. R. Hathaway and R. S. Adelstein, *Proc. Natl. Acad. Sci. U.S.A.* **76,** 1653 (1979).
[5] J. L. Daniel, I. R. Molish, H. Holmsen, and L. Salganicoff, *Cold Spring Harbor Conf. Cell Proliferation* **8,** 913 (1981).
[6] J. L. Daniel, I. R. Molish, M. Rigmaiden, and G. Stewart, *J. Biol. Chem.* **259,** 9826 (1984).
[7] M. E. Bromberg, R. W. Sevy, J.L. Daniel, and L. Salganicoff, *Am. J. Physiol.* **249,** C297 (1985).

platelet and other myosins are found in Vol. 85 of "Methods in Enzymology."

Preparation of Platelet Myosin

Reagents

Platelet washing buffer: 0.154 M NaCl, 5 mM EDTA, 10 mM Tris–HCl, (pH 7.4)

Extraction buffer: 0.5% Triton X-100, 0.5 M NaCl, 50 mM 3-(N-morpholino)propane sulfonate (MOPS) (pH 7.0), 5 mM EDTA, 10 mM EGTA, 5 mM ATP, 2 mM dithiothreitol, 0.1 mM phenylmethylsulfonyl fluoride (PMSF), 10 mg/liter pepstatin (Vega Biochemicals, Tucson, AZ), 10 mg/liter aprotinin (Boehringer Mannheim, Indianapolis, IN), 1.0 mg/liter N-tosyl-1-phenylalanine chloromethyl ketone (TCPK), 10 mg/liter leupeptin (Vega), 10 mg/liter antipain (Vega), 5 mg/liter chymostatin (Vega), 0–1 mM diisopropyl fluorophosphate (DFP), 10 mg/liter tosyllysine chloromethyl ketone (TLCK)

Buffer A: 25 mM NaCl, 10 mM MOPS (pH 7.0), 10 mM MgCl$_2$, 0.1 mM EGTA, 1 mM dithiothreitol, 0.1 mM PMSF, 5 mg/liter leupeptin

Buffer B: 0.5 M NaCl, 10 mM MOPS (pH 7.0), 0.1 mM EGTA, 1 mM dithiothreitol, 3 mM NaN$_3$

Magnesium sulfate, 0.5 M

ATP, 0.1 M (pH 7.0)

Saturated (NH$_4$)$_2$SO$_4$ solution containing 1 mM EDTA. The pH is adjusted to 7.0 with NH$_4$OH

Procedure

Platelet concentrates, either fresh or outdated, can be obtained from a blood bank. Any anticoagulant may be used. Alternatively, animal blood may be obtained fresh from a slaughterhouse. In this case, platelet-rich plasma must be prepared first. The platelet concentrates are centrifuged at 200 g_{max} at 20° for 5 min to remove most of the red blood cells. The suspension of platelets is centrifuged at 3000 g_{max} for 5 min at 20° and the platelet pellet gently resuspended (avoiding the red cells) in the platelet washing buffer (20°). This washing procedure is repeated twice and the final suspension of platelets is centrifuged at 16,300 g_{max} for 20 min to obtain a well-packed pellet. At this point, the pellet can be frozen in liquid N$_2$ and stored at −70° or used immediately.

All subsequent steps are done at 4°. The platelets are disrupted by homogenization in a Teflon–glass tissue grinder in 4–5 vol of extraction buffer. The homogenate is centrifuged for 30 min at 200,000 g_{max}. The supernatant is made 30 mM in MgSO$_4$ and an additional 5 mM ATP is added. These additions are made while maintaining the pH between 6.8 and 7.0 with the addition of Tris base. Saturated ammonium sulfate is slowly added with stirring to a final saturation of 20%. The solution is centrifuged for 10 min at 12,000 g_{max}. This step results in the precipitation of excess Triton X-100. The supernatant is slowly brought to 40% saturation with ammonium sulfate and centrifuged for 10 min at 12,000 g_{max}. This supernatant is brought to 60% of saturation with ammonium sulfate and is again centrifuged for 20 min at 13,000 g_{max}. The resulting pellet is dissolved in buffer A and is dialyzed overnight against the same buffer. The contents of the dialysis sack are sedimented at 40,000 g for 20 min and the pellet is resuspended in a small amount of dialysis solution. If the crude myosin had less than 20% actin contamination as determined by SDS–polyacrylamide gel electrophoresis, it is made 0.5 M in NaCl and 10 mM in MgATP (by addition of stock solutions of 5 M and 0.1 M, respectively) and applied directly to a Sepharose 4B column (2.5 × 90 cm) which is equilibrated with buffer B. Prior to application of the crude myosin, 50 ml of the column buffer containing 1 mM MgATP is applied to the column. If the crude myosin showed larger actin contamination, a second ammonium sulfate fractionation is performed in the presence of 0.5 M NaCl, 10 mM MgCl$_2$, and 5 mM ATP as described above. The fraction sedimenting between 45 and 60% of NH$_4$SO$_4$ saturation was used in this case. The resulting pellet is solubilized in column buffer and chromatographed as described above.

Comments

This procedure has been modified from the one originally used by Adelstein et al.[8] The main advantage of the present procedure is that it yields myosin that is fully suitable for studies of the actin-activated ATPase activity. Approximately 1 mg of pure myosin is usually obtained from each unit of platelets. For an example of the elution profile from the Sepharose 4B column see the original article.[8] It is possible that not all the protease inhibitors are needed, but the preparation has not been tried without the complete mixture. Leupeptin and EGTA are probably the most important to prevent activation of Ca^{2+}-dependent protease (calpain). This method has only be used with fresh platelets and the amounts of head and rod fragments present may increase if outdated platelets are used.

[8] R. S. Adelstein, T. D. Pollard, and W. M. Kuehl, *Proc. Natl. Acad. Sci. U.S.A.* **68**, 2703 (1971).

Assay of ATPase Activities of Platelet Myosin

Reagents

Actin: Prepared from rabbit skeletal muscle by the method of Spudich and Watt.[10] Note that a method to prepare platelet actin is given in [7] in this volume and can be used. We use rabbit skeletal muscle actin since a greater quantity of rabbit muscle actin can be easily attained

Assay buffer A: 6.25 mM MgCl$_2$, 0.125 mM EGTA, 18 mM Tris–HCl (pH 7.5).

Assay buffer K: 0.625 M KCl, 2.5 mM EDTA, 18 mM Tris–HCl (pH 7.5) 10 mM ATP

Isobutanol and benzene (1:1, v/v)

H$_2$SO$_4$ (3 N) containing 4% (w/v) silicotungstic acid

10% (w/v) Ammonium molybdate,

H$_2$SO$_4$ (0.73 N) in 96% ethanol

Stannous chloride (10% w/v) in 12 N HCl

Procedure

Actin-Activated Mg-ATPase. This assay is done at 25°. Actin and myosin (5 and 2 mg/ml final concentrations, respectively) are mixed in 0.5 M KCl to assure proper association of the two proteins. This mixture (0.1 ml) is added to 0.8 ml of assay buffer A and the solution gently stirred. ATP (0.1 ml) is added to start the reaction [final ionic conditions: 50 mM KCl, 5 mM MgCl$_2$, 0.1 mM EGTA, 15 mM Tris–HCl (pH 7.5), 1.0 mM ATP]. Aliquots (0.1 ml) are taken at 1, 2, 3, 4, and 5 min after ATP addition into glass test tubes containing 2 ml of isobutanol–benzene and 0.5 ml of 3 N H$_2$SO$_4$–silicotungstic acid.[11] This mixture is vortexed for 1 sec to assure that all reactions are quenched. When all assays are completed, 0.2 ml of NH$_4$MbO$_4$ is added and the mixture is vortexed for 10–15 sec. The test tubes are put aside until the two phases separate. One milliliter of the organic upper phase is carefully removed and added to 2 ml of the solution of 0.73 N H$_2$SO$_4$ in ethanol. One milliliter of the 10% stannous chloride solution is diluted to 25 ml with 1 N H$_2$SO$_4$ and 0.1 ml of this dilution is added to each assay mixture. A stable blue color develops that can be read immediately at 720 nm.

As a more sensitive alternative, this assay can be done using

---

[9] Deleted in proof.
[10] J. A. Spudich and S. Watt, *J. Biol. Chem.* **246**, 4866 (1971).
[11] T. D. Pollard and E. D. Korn, *J. Biol. Chem.* **284**, 4682 (1973).

[γ-^{32}P]ATP. The radioactive phosphate liberated is counted after the phosphomolybdate complex is separated into the isobutanol–benzene phase. This avoids the need to develop the color.

K$^+$-EDTA ATPase. This assay is used for a rapid indication of the location of myosin on chromatography columns. The assay is similar to the one above except that 0.05 ml of the column fraction is mixed with 0.45 ml of buffer K. A 0.2-ml aliquot is removed after 1 min and phosphate liberated is measured.

Comment

The ATPase activity is calculated from the linear portion of the time course. The phosphate assay is appropriate for the concentrations of phosphate obtained from the given quantities of proteins. Another suitable assay for phosphate can be substituted for the one described but the concentrations of proteins or times of incubation may have to be adjusted.

Phosphorylation and Dephosphorylation of Platelet Myosin

Reagents

Turkey gizzard smooth muscle myosin light chain kinase prepared by the method of Adelstein and Klee.[12]

Calmodulin prepared by the method of Klee[13] or obtained from commercial sources.

The catalytic subunit of smooth muscle phosphatase I prepared by the method of Pato and Adelstein.[14]

Phosphorylation buffer: 4 mM magnesium acetate, 50 mM Tris–HCl (pH 7.4), 0.25 mM CaCl$_2$, 0.5 mM ATP

Procedure

Phosphorylation of Myosin. Myosin (2–4 mg/ml) is added to the phosphorylation buffer. Myosin kinase (4 × 10^{-8} M) and calmodulin (10^{-7} M) are incubated with the myosin for 10 min at 25°. The reaction can be stopped by the addition of 2 mM EGTA.

Dephosphorylation of Myosin. Myosin is dephosphorylated by the addition of 1 mM EGTA to inactivate myosin kinase. The myosin is incubated with the catalytic subunit of smooth muscle phosphatase I (1.5 μg/ml) for 10 min at 25°.

---

[12] R. S. Adelstein and C. B. Klee, this series, Vol. 85, p. 298.
[13] C. B. Klee, *Biochemistry* **16**, 1017 (1989).
[14] M. D. Pato and R. S. Adelstein, this series, Vol. 85, p. 308.

Comment

The myosin phosphorylated under these conditions remains fully phosphorylated for several weeks at 4°. The phosphorylation state of myosin can be assessed using alkaline–urea polyacrylamide gel electrophoresis (below) or, for more rapid determination, by counting the incorporation of $^{32}PO_4$ from [γ-^{32}P]ATP. This is measured as described by Corbin and Reeman.[15]

Measurement of Myosin Light Chain Phosphorylation

Materials. It is important that good quality reagents be used for preparation of the gels. We have used acrylamide gel reagents from Bio-Rad (Richmond, CA), dithiothreitol from Calbiochem (La Jolla, CA), or ICN (Costa Mesa, CA), and urea, an ultrapure grade, from ICN.

Reagents

Acrylamide stock solution: A 40% (w/v) solution of acrylamide containing 1.08% (w/v) bisacrylamide

Concentrated (12×) buffer stock: 0.24 M Tris–glycine (pH 8.6). *Note:* This buffer is very important. We have obtained good results using 29.2 g/liter of Tris base and 40 g/liter of glycine

Running buffer:
1. Top chamber: 72 g ultrapure urea and 25 ml stock buffer diluted to 300 ml with water
2. Bottom chamber: 100 ml stock buffer diluted to 1200 ml with water.

Sample buffer: 0.5 ml buffer stock, 3.6 g ultrapure urea, 160 mg dithiothreitol, 2.35 ml H_2O, one or two drops 0.5% (w/v) bromphenol blue

Note: Add dithiothreitol to buffer after all the urea is dissolved. This buffer should not be heated above 37° in an effort to dissolve the urea

Top gel (3.5% acrylamide; makes about 12 ml): 6 g ultrapure urea, 1.04 ml Tris–glycine buffer stock, 1.1 ml acrylamide stock, 5.1 ml H_2O, 5 μl N,N,N',N'-tetramethylethylenediamine (TEMED), 100 μl 10% (w/v) ammonium persulfate (freshly made)

Bottom gel (10.0% acrylamide; makes about 30 ml): 14.4 g ultrapure urea, 2.5 ml Tris–glycine buffer stock, 7.5 ml acrylamide stock, 7.5 ml H_2O, 15 μl TEMED, 180 μl 10% (w/v) ammonium persulfate (freshly made)

---

[15] J. D. Corbin and E. M. Reeman, this series, Vol. 38, p. 287.

Procedure

This procedure is adapted[16] from the original of Perrie and Perry[17] as modified by Siemankowski and Dreizen.[18] The samples are assumed to be perchloric acid precipitates either from whole platelets or from isolated proteins. The pellet is rinsed once with an appropriate amount of H_2O (usually 1 ml) to remove most of the perchloric acid. Samples can be frozen overnight at this point. Sample buffer is then added to each pellet and the pellets are resuspended with a small stirring rod. The pellets are sonicated in a sonication bath for 30 min to assure that the sample is completely dissolved. Take care not to overheat samples. All the reagents for the bottom gel except the ammonium persulfate are mixed in a vacuum flask and deaerated for 10 min. The ammonium persulfate is added to the mixture and the solution is immediately poured into the gel apparatus, leaving room at the top for the sample wells (about 3 cm). The solution is overlaid with a small volume of isobutanol or water and allowed to polymerize (about 30 min). The top gel is prepared and poured in a similar manner using a comb to form sample wells. The samples are applied to the gels and electrophoresed at 8–9 mA for each gel plate being used (i.e., for a 15 mm × 10 cm gel). The electrophoresis is stopped 45–60 min after the bromphenol blue marker dye has come off the bottom of the gel. The gels are stained overnight or for 1 hr in 0.05% (w/v) Coomassie Brilliant blue R-250, 10% (v/v) acetic acid, 30% (v/v) methanol, 60% (v/v) H_2O. The gels are destained in the same solution except that the dye is omitted. The percentage myosin phosphorylation is calculated by scanning the wet gel and determining the relative amounts of phosphorylated and dephosphorylated light chain. The gel may be dried after soaking in water for long-term storage.

Comments

Several factors are important to obtain good results: (1) The pH of the buffer, if the pH is too high, the 20,000-D light chain is poorly resolved. We have found that the safest way to obtain reproducible results is to add a weighed amount of glycine to the stock buffer rather than to titrate the buffer on a pH meter. (2) The urea used, lesser grades of urea contain appreciable amounts of cyanate that can react with lysine residues and cause the appearance of ghost bands.[17] All urea-containing buffers are made fresh, since urea spontaneously decomposes to produce cyanate.

---

[16] J. L. Daniel, I. R. Molish, and H. Holmsen, *J. Biol. Chem.* **256**, 7510 (1980).
[17] W. T. Perrie and S. V. Perry, *Biochem. J.* **119**, 31 (1970).
[18] R. F. Siemankowski and P. Driezen, *J. Biol. Chem.* **253**, 8648 (1979).

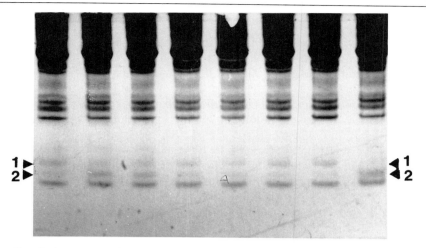

FIG. 1. A representative urea polyacrylamide gel of acid pellets from intact platelets. Band 1 represents the dephosphorylated 20-kDa light chain and band 2 represents the phosphorylated light chain. The most phosphorylated samples are in the second, third, and seventh lanes from the left. Other examples of this system can be seen in Refs. 6 and 16.

We also avoid using excessive heat with these buffers. (3) Adequate reduction, we have used a high concentration of dithiothreitol to assure that all proteins are completely reduced. The urea in the sample buffer precludes boiling the samples. (4) Complete solubilization, we have used sonication to assure that the samples are completely dissolved. This step also allows time for disulfide bonds to be reduced by dithiothreitol. (5) The stacking gel should be given sufficient time to fully polymerize. In our experience this requires at least 90 min. An example of a urea gel is shown in Fig. 1.

One-Dimensional Phosphopeptide Mapping of Myosin Light Chain Phosphorylation Sites

Preparation of Samples

Reagents

TPCK trypsin (Worthington solution), 1 mg/ml in 1 mM HCl
Isopropanol 25% (v/v)
Methanol 10% (v/v)
NH_4HCO_3, 100 mM, pH 8.0

Procedure

The following procedure has been developed as a quick and powerful tool for examining the sites on the myosin light chain that are phosphorylated by protein kinase C or myosin light chain kinase. An example of its use is Nakabayashi *et al.*[19] The procedure is considerably faster and is more economical, sensitive, and reliable than the previously employed two-dimensional phosphopeptide mapping procedures.[20]

The analysis begins with the radioactive phosphorylated light chain samples excised from the Coomassie blue-stained gels. Alternatively, the light chain may be detected using radioautography of unstained gels. The gel slice is placed in a 15-ml polypropylene conical tube and washed three times for 15 min with 5 ml 25% isopropanol (v/v) followed by three washes for 15 min with 10% methanol. The slice is minced and dried under a hot lamp.

The dried gel slice is suspended in 950 μl of a freshly made of 100 mM NH_4CO_3, pH 8.0, and 50 μl of TPCK trypsin (1 mg/ml) in 1 mM HCl is added. The sample is digested for 24 hr at 37°. Another 50 μl of trypsin solution is added followed by an additional 4–6 hr of digestion. The solution surrounding the gel particles is removed and lyophilized in a Savant speed vac. The dried samples are redissolved in 50 μl of NH_4CO_3 solution and used in the following procedure.

One-Dimensional Mapping

Reagents

Solution of acrylamide, 29.1% (w/v)
Solution of bisacrylamide, 0.9% (w/v)
Ampholytes (Pharmacia-LKB) pH 2.5–4.0, pH 3.5–5.0, pH 5–8
Ammonium persulfate, 10% (w/v)

The mold for casting the gel consists of two glass plates with a 0.5-mm rubber gasket spacer. A piece of PAGE gel bond film is applied to one plate using glycerol as an adherent according to the manufacturer's instructions.

The gel mixture is composed of 7.2 ml of acrylamide solution, 7.2 ml bisacrylamide solution, 0.6 ml of the three ampholines, 10.8 g urea, and enough H_2O to bring the volume to 30 ml. The solution is degassed and polymerization initiated by addition of 20 μl TEMED and 100 μl of 10%

---

[19] H. Nakabayashi, J. R. Sellers, and P-K. Huang, *FEBS Lett.* **294**, 144 (1991).
[20] S. Kawamoto, A. R. Bengur, J. R. Sellers, and R. S. Adelstein, *J. Biol. Chem.* **264**, 2258 (1989).

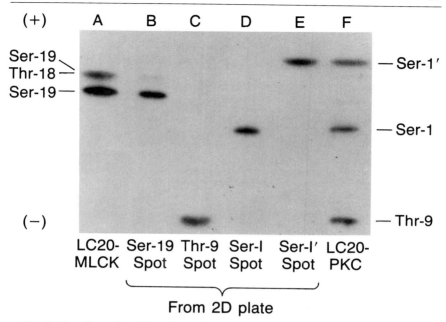

FIG. 2. One-dimensional phosphopeptide mapping of myosin light chain phosphorylation sites. Lane A shows a tryptic digest of light chains phosphorylated to about 1.2 mol PO$_4$/mol with myosin light chain kinase (MLCK). MLCK is known to preferentially phosphorylate Ser-19, but will also phosphorylate Thr-18. The latter phosphorylation results in a diphosphorylated peptide. Lane F shows the peptide map of light chains phosphorylated with protein kinase C. Protein kinase C phosphorylates Thr-9 and Ser-1 and/or Ser-2. The amino terminal sequence of the light chain is Ac-Ser-Ser-Lys-Arg. Ser-1 and Ser-1' most likely correspond to Ac-Ser-Ser-Lys-Arg and Ac-Ser-Ser-Lys, respectively. Lanes B–E represent spots scraped from plates of two-dimensional peptide maps using the system described by Kamamoto et al.[20] showing the correspondence between the two systems. The isoelectric points of the Ser-19 band is 4.5; Ser-1', 3.6; Ser-1, 5.0; and Thr-9, 7.5.

ammonium persulfate. The mixture is poured into the apparatus and allowed to polymerize at room temperature.

Following polymerization, the gel, which is tightly adhered to the PAGE gel bond film, is removed from the glass plates and transferred to the Multiphor II electrophoresis apparatus that is cooled to 4°. Wicks soaked in cathode buffer (1 M NaOH) and anode buffer (1 M H$_3$PO$_4$) are applied to opposite sides of the gel. The gel is preelectrophoresed for 10 min at 30 W, 1000 V.

Peptide samples are applied on the cathode side using small filter papers as described in the manufacturer's instructions and the gel is electrophoresed for 1 hr at a constant voltage of 1000 V with an initial setting of

25 W. (The power drops to about 7 W during the run.) At the end of run, the gel is removed and dried using a vacuum gel drying apparatus for 1 hr at 55°. The dried gel is exposed to X-ray film.

Figure 2 shows an autoradiogram of a representative gel using turkey gizzard smooth muscle myosin light chains as standard. Platelet myosin light chains show identical migration patterns.

[9] Isolation and Characterization of Platelet Gelsolin

By Joseph Bryan

Gelsolin (G) is an M_r 90,000 Ca^{2+}-binding protein[1,2] that can form high-affinity complexes with actin (A).[3-5] The molecule was originally isolated from macrophages[6] and has two Ca^{2+}-binding sites[1,2] and two actin-binding sites.[4] The binary complex, GA_1Ca_1, has been isolated from activated platelets[1,7] and is stable in EGTA. The formation of GA_2Ca_2 from GA_1Ca_1 and actin is Ca^{2+} sensitive with a K_D in EGTA of approximately 30 nM and <0.1 nM in 100 μM Ca^{2+}.[3] *In vitro,* gelsolin will sever actin filaments.[2,7,8] When gelsolin is added to G-actin under polymerizing conditions, GA_2Ca_2 and higher oligomers form that serve as nuclei for filament elongation.[1,4,5,9,10] Gelsolin remains tightly bound to the preferred end[5,7] of these nucleated filaments in either Ca^{2+} or EGTA and elongation occurs at the nonpreferred end.[1,5,7] This chapter describes two methods for the purification of the gelsolin–actin complex, one utilizing DNase I–agarose, the other using more conventional biochemical methods. In addition, a method to isolate gelsolin free of actin is detailed. We have used a variety of methods to detect gelsolin during the purification steps, including a fluorescent actin nucleation assay[1] and high shear viscometry.[7] Sedimentation assays have also been reported.[5] Using the methods described, gel-

[1] M. C. Kurth, L.-L. Wang, J. Dingus, and J. Bryan, *J. Biol. Chem.* **258**, 10895 (1983).
[2] H. L. Yin and T. P. Stossel, *J. Biol. Chem.* **255**, 9490 (1980).
[3] J. Bryan and M. C. Kurth, *J. Biol. Chem.* **259**, 7480 (1984).
[4] M. C. Kurth and J. Bryan, *J. Biol. Chem.* **259**, 7473 (1984).
[5] H. L. Yin, J. H. Hartwig, K. Maruyama, and T. P. Stossel, *J. Biol. Chem.* **256**, 9693 (1981).
[6] H. L. Yin and T. P. Stossel, *Nature (London)* **281**, 583 (1979).
[7] L.-L. Wang and J. Bryan, *Cell (Cambridge, Mass.)* **25**, 637 (1981).
[8] H. L. Yin, K. S. Zaner, and T. P. Stossel, *J. Biol. Chem.* **19**, 9494 (1980).
[9] F. Markey, T. Persson, and U. Lindberg, *Biochim. Biophys. Acta* **709**, 122 (1982).
[10] R. Tellam and C. Frieden, *Biochemistry* **21**, 3207 (1982).

solin can be detected using sodium dodecyl sulfate (SDS)–polyacrylamide gel electrophoresis and purified from platelets even without using functional assays. Two purification strategies employing immunoaffinity and ion-exchange chromatography are discussed in the latter part of this chapter.

Isolation of Platelet Gelsolin and Gelsolin–Actin Complexes

Reagents

Preparation of Platelet Extracts

EDTA–saline: 125 mM NaCl, 5 mM KCl, 1 mg/ml glucose, 0.1 mM EDTA, 20 mM sodium phosphate, pH 6.5
Lysis buffer: 0.34 M sucrose, 1 mM ATP, 0.5 mM $MgCl_2$, 0.2 mM dithioerythritol, 5 mM EGTA, 10 mM piperazine-N,N'-bis(2-ethanesulfonic acid) (PIPES), pH 6.9
Dialysis buffer: 50 mM NaCl, 0.1 mM EGTA, 1.0 mM NaN_3, 10 mM imidazole, pH 7.4

DNase I–Agarose Chromatography

Column of DNase I–agarose (10 to 25 ml) equilibrated with buffer 1
Buffer 1: 50 mM NaCl, 0.1 mM $CaCl_2$, 10 mM imidazole, pH 7.4
Buffer 2: 200 mM NaCl, 0.1 mM $CaCl_2$, 10 mM imidazole, pH 7.4
Buffer 3: 200 mM NaCl, 5 mM EGTA, 10 mM imidazole, pH 7.4
Buffer 4: 3 M guanidine hydrochloride, 5 mM $CaCl_2$, 10 mM imidazole, pH 7.4

DEAE-Sephacel Chromatography

Column of DEAE-Sephacel (2.5 × 50 cm; Pharmacia Fine Chemicals, Piscataway, NJ) equilibrated with buffer 5
Buffer 5: 1 mM EGTA, 0.1 mM NaN_3, 10 mM imidazole, pH 7.4
Buffer 6: 500 mM NaCl, 1 mM EGTA, 0.1 mM NaN_3, 10 mM imidazole, pH 7.4

Sephadex G-200 Chromatography

Column of Sephadex G-200 (2.5 × 90 cm; Pharmacia Fine Chemicals) equilibrated with buffer 7
Buffer 7: 0.8 M KCl, 1 mM EGTA, 0.1 mM NaN_3, 10 mM imidazole, pH 7.4

Hydroxylapatite Chromatography

Column of hydroxylapatite (1.5 × 30 cm, HT grade; Bio-Rad, Richmond, CA) equilibrated with buffer 8
Buffer 8: 5 mM sodium phosphate, pH 7.0
Buffer 9: 200 mM sodium phosphate, pH 7.0

DEAE-Urea Chromatography

Column of DEAE-Sephacel (1 × 5 cm; Pharmacia Fine Chemicals) equilibrated with buffer 10
Buffer 10: 1 mM EGTA, 20 mM Tris–HCl, pH 7.4
Buffer 11: 6 M urea, 1 mM EGTA, Tris–HCl, pH 7.4
Buffer 12: 6 M urea, 0.3 M NaCl, 1 mM EGTA, 20 mM Tris–HCl, pH 7.4

Miscellaneous

Ammonium sulfate
Reagents for SDS–polyacrylamide gel electrophoresis[11]

Platelet Extract Preparation

Platelet concentrates are used within 5–6 days after drawing from blood donors. White and red blood cells are removed by centrifugation at 200 g for 5 min in a clinical centrifuge. Then the platelets are pelleted at 2000 g for 15 min at room temperature. The platelets are then washed three times in EDTA–saline. For large numbers of platelets, the HG-4L rotor is used in a Sorvall RC3 centrifuge (Du Pont Instruments-Sorvall, Newton, CT). The final platelet pellets are resuspended in 3 ml of lysis buffer per unit of platelet concentrate, then lysed by sonication with a sonifier (Branson Sonic Power, Danbury, CT). Reasonably complete lysis is obtained using 20 sonication bursts of 30 sec each interspersed with 30-sec cooling periods with the container in ice water. The volume of the sonicate is measured and solid KCl is added to a final concentration of 0.6 M, then stirred for 2 hr at 4°. This crude extract is centrifuged at 100,000 g for 60 min and the supernatant is recovered. This high-salt supernatant can be frozen rapidly in liquid nitrogen and stored at −70° indefinitely or used immediately.

Formation of Contractile Gel. To deplete the extract partially of actin and other proteins the high-salt extract is dialyzed against approximately 100 vol of dialysis buffer. A dense precipitate forms that consists of actin, myosin, α-actinin, and other actin-associated proteins. This can be removed by centrifugation at 5000 g for 15 min and used for the isolation of

---

[11] U. K. Laemmli, *Nature (London)* **227**, 680 (1970).

actin and various actin-associated proteins.[12-15] (see [7] in this volume). The supernatant is clarified further by centrifugation at 100,000 g for 60 min.

First Ammonium Sulfate Precipitation. A gelsolin-enriched fraction is obtained by addition of ammonium sulfate to 35% (0.2 g/ml of the 100,000 g supernatant) followed by gentle stirring at 4° for 30 min. The 35% pellet can be collected by centrifugation at 20,000 g for 20 min and saved for isolation of platelet vinculin.[13] The 35% supernatant is brought up to 50–55% saturation by addition of 0.14 g of ammonium sulfate per milliliter of supernatant followed by 30 min of slow stirring at 4°. The 55% pellet is collected by centrifugation at 20,000 g for 20 min. This pellet contains the bulk of the gelsolin–actin complexes and can also be used to isolate the platelet profilin–actin complex.[16] For use in the DNase I–agarose procedure, the pellet is taken up in 0.5 ml of buffer 1 per milliliter of platelet concentrate and dialyzed for 12 to 18 hr against 100 vol of this buffer. Alternatively, for input into the DEAE-Sephacel method, the 55% pellets are resuspended in 0.5 ml of buffer 5 per milliliter of platelet concentrate and dialyzed for 12 to 18 hr against 100 vol of this buffer.

Method Ia: DNase I–Agarose Method for Isolation of Gelsolin–Actin Complexes[7]

This method depends on the high-affinity binding between DNase I and actin and the fact that gelsolin forms complexes with multiple actin monomers, GA_2Ca_2 and higher, that are stronger in the presence of Ca^{2+}. The 55% fraction, which contains GA_2Ca_2, excess actin, and actin–profilin complexes, is applied to a column of DNase I–agarose at a flow rate of approximately 10–15 ml/hr. Fractions (2.5 ml) are collected. We have used a column bed volume of approximately 25 ml, 2.5 × 5 cm, with an actin-binding capacity of 15–20 mg of actin determined using rabbit skeletal muscle G-actin, then applied 10–12 ml of the 55% fraction equilibrated with buffer 1. The column is washed with buffer 2 until the protein concentration falls below about 5 μg/ml as determined by the Bradford assay.[17] The gelsolin–actin complex is then eluted by applying buffer 3 to the column. The DNase I–agarose is regenerated by washing with four column volumes of buffer 4 followed by five column volumes of buffer 1. The first

---

[12] D. C. Hou, J. D. Dingus, G. C. Rosenfeld, and J. Bryan, *J. Cell Biol.* **97**, 276a (1983).
[13] G. C. Rosenfeld, D. C. Hou, J. Dingus, I. Meza, and J. Bryan, *J. Cell Biol.* **100**, 669 (1985).
[14] S. Rosenberg, A. Stracher, and R. C. Lucas, *J. Cell Biol.* **91**, 201 (1981).
[15] S. Rosenberg, A. Stracher, and K. Burridge, *J. Biol. Chem.* **25**, 12986 (1981).
[16] F. Markey, T. Persson, and U. Lindberg, *FEBS Lett.* **88**, 75 (1978).
[17] M. M. Bradford, *Anal. Biochem.* **72**, 248 (1976).

fractions of the 3 M guanidine hydrochloride wash, buffer 4, contain actin, profilin, some residual gelsolin, and an M_r 55,000–58,000 protein of unknown function.[7] The EGTA wash, buffer 3, contains mainly actin and gelsolin at peak concentrations of 200–300 µg/ml. The yield is 1–2 mg of total proteins per 20 ml of the 55% fraction. If required, the contaminants can be removed effectively by concentration of the protein using 55% ammonium sulfate, resuspension in a small volume, approximately 0.5 ml, of EGTA–saline or buffer 5 followed by chromatography on a small Sephadex G-150 column (1.5 × 50 cm). Alternatively, the EGTA fractions from multiple runs can be pooled, concentrated, and gel filtered. The gelsolin elutes as an M_r 130,000 species GA_1Ca_1 complexed with one actin.[1]

The DNase I–agarose is stable for more than 1 year if regenerated quickly and stored in buffer 1 plus 0.1 mM NaN_3 at 4°.

Method Ib: Isolation of Gelsolin–Actin Complexes by Ion-Exchange, Gel-filtration, and Hydroxylapatite Chromatography[1]

Platelet extracts or the 55% ammonium sulfate fraction can be fractionated by conventional biochemical procedures to yield the gelsolin–actin complex, GA_1Ca_1.

DEAE-Sephacel Chromatography. The 55% ammonium sulfate fraction from 50 to 100 units of platelet concentrate is applied to a 2.5 × 50 cm column of DEAE-Sephacel equilibrated with buffer 5. The sample is loaded and the column is then washed with 300 ml of buffer 5. The adsorbed protein is eluted with a 1-liter linear gradient of 0–0.5 M NaCl (buffer 6). The gelsolin–actin complex can be detected using high-shear viscosity,[7] by its effect on actin filament sedimentation,[5] by assaying for nucleation activity,[1] or by SDS–gel electrophoresis and immunoblotting with antigelsolin.[1] The complex elutes at approximately 0.23 M NaCl, slightly ahead of the actin–profilin complex assayed by its DNase I inhibition activity.[16] The gelsolin-rich fractions, judged either by nucleation activity or Coomassie Blue-stained SDS–polyacrylamide gels, are pooled and brought to 55% saturation with ammonium sulfate. After slow stirring for 30 min at 4° the precipitate is collected by centrifugation at 20,000 g for 20 min and resuspended in 5 ml of buffer 7. This material is dialyzed for 12–15 hr against 100 vol of buffer 7, then clarified by centrifugation at 100,000 g for 60 min.

Gel Filtration on Sephadex G-200. The supernatant is applied to a 2.5 × 90 cm column of Sephadex G-200 equilibrated with buffer 7 and eluted at a flow rate of 8–10 ml/hr. Fractions (6.5 ml) are collected. Gelsolin, complexed with one actin monomer, elutes with an apparent molecular weight of 130,000 at approximately fraction numbers 35–40.

The nucleation activity peak is reasonably symmetric and the protein is easily detectable by SDS–polyacrylamide gel electrophoresis. The gelsolin-rich fractions are again collected and concentrated by precipitation with 55% ammonium sulfate. The pellets are taken up in 10 ml of 5 mM sodium phosphate at pH 7 (buffer 8) and dialyzed versus 100 vol of buffer 8 at 4°.

Hydroxylapatite Chromatography. The dialyzed fraction is clarified by centrifugation at 100,000 g for 60 min, then applied to a 1.5 × 30 cm column of hydroxylapatite. The protein is eluted with a 400-ml gradient of phosphate from 5 to 200 mM (buffer 9) at a flow rate of approximately 10 ml/hr. Fractions (6.5 ml) are collected. The gelsolin–actin complex elutes at approximately 0.1 M phosphate. The nucleation activity elutes as a single symmetric peak coincident with the gelsolin and actin bands visualized by SDS–polyacrylamide gel electrophoresis. The major contaminants at this stage are two proteolytic products of gelsolin, one at M_r 47,000 and the other at M_r 40,000. These contaminate the lead edge of the peak and can be eliminated by judicious selection of fractions. Alternatively, the complex can be concentrated and chromatographed on Sephadex G-150 as indicated in the DNase I purification section. These peptides do not appear to be tightly bound to actin and are separated by gel filtration.[1] The complex can be dialyzed into any desired solution at this stage and stored at 4°. We do detect breakdown of the complex into gelsolin and inactive actin when the complex is stored at 4° in dilute salt solutions with 0.1–1.0 mM EGTA and without Mg^{2+} or ATP, but the half-life appears to be > 4–6 weeks under those conditions. Our attempts to freeze the complex at −20° or rapidly in liquid nitrogen for long-term storage have led to significant inactivation. For isolation of gelsolin we have dialyzed the pooled gelsolin-rich fractions against 100 vol of buffer 10.

Method II: DEAE-Urea Chromatography for Isolation of Gelsolin

The material obtained from the DNase I–agarose procedure or the hydroxylapatite step can be used to obtain gelsolin free of actin. Approximately 5 mg of the gelsolin–actin complex, equilibrated with buffer 10, is applied to a small 1 × 5 cm column of DEAE-Sephacel. Essentially all of the protein is adsorbed and the column is washed with 10 vol of 6 M urea (buffer 12) to dissociate the gelsolin–actin complex. The protein is eluted with a 0–0.3 M NaCl gradient in 6 M urea (buffer 12). Gelsolin elutes at approximately 0.11–0.12 M NaCl. The flow rate should be approximately 4 ml/hr with 1-ml fractions collected. The peak fractions contain gelsolin at approximately 200 μg/ml. We see minor contamination, <2–5%, at this stage from gelsolin breakdown products identified by immunoblots with anti-gelsolin. Gelsolin can also be stored for extended periods in buffer 5 at 4°, but is inactivated by freezing.

Yields

The yields from these procedures are variable. The principal source of variability that is difficult to control is the quality of the platelet concentrates themselves. We try to obtain platelets from local blood banks as soon as they are outdated, usually 3 to 4 days after drawing, and continue to agitate these by slow rotation at room temperature. Nevertheless, large quantities of platelets at a single age are difficult to obtain and, using immunoblots, we have noted varying degrees of breakdown of gelsolin and other actin-associated proteins. A second source of variability is the degree of activation. In the absence of activation inhibitors such as aspirin and tetracaine, simple manipulation and washing causes activation and formation of the gelsolin–actin complex,[4] but we have not determined the completeness of this activation. Finally, a third source of variability is the amount of gelsolin removed by formation of the contractile gel. We estimate very roughly, from immunoblots of the various fractions, that this may amount to as much as 20–30% of the total gelsolin originally present. This loss may be avoided partially by not adding EGTA to the initial lysis and high-salt extraction steps and by increasing the volume of lysis buffer used to 10 ml/unit of platelet concentrate. This leads to rather massive destruction of the high molecular weight proteins,[7,18,19] including actin-binding protein, a 235K protein, and myosin. Little precipitate is formed at the first dialysis step and all of the gelsolin is presumably complexed with actin. The trade-off is the loss of material that can be used for isolation of additional actin-associated proteins.[12–15] Using the procedures described we can routinely isolate 10–12 mg of GA_1Ca_1 or 5–6 mg of gelsolin from 100 units of expired platelet concentrate. In addition, the plasma recovered can be used for the isolation of brevin.[20,21]

Activity Assays

Actin–gelsolin interactions have been studied using a variety of methods. Low-shear viscosity measurements have been used to show that gelsolin, purified on DNase I-Sepharose or DEAE-cellulose,[7,8] reduces the length of F-actin when added to preformed filaments. The mechanism of this restriction is not completely clear and could be accounted for by either

---

[18] D. A. Harris and J. H. Schwartz, *Proc. Natl. Acad. Sci. U.S.A.* **78**, 6798 (1981).
[19] H. E. Harris and A. G. Weeds, *Biochemistry* **22**, 2728 (1983).
[20] N. C. Collier and K. Wang, *J. Biol. Chem.* **257**, 6937 (1982).
[21] D. R. Phillips and M. Jakábová, *J. Biol. Chem.* **252**, 5602 (1977).

severing or fragmentation/capping mechanisms. Electron microscopy of negatively stained specimens has been used to demonstrate progressively shorter filament distributions when G-actin is polymerized with increasing concentrations of gelsolin,[7,8] or added to filaments,[8] and to show that growth occurs only from the nonpreferred end of gelsolin capped filaments.[5,7] Flow birefringence has also been used to show a length restriction.[5,8] The ability of gelsolin–actin complexes to nucleate growth of capped filaments has been demonstrated using high-shear viscometry[7] and by monitoring the assembly of fluorescently labeled actin monomers. Using 4-chloro-7-nitrobenz-2-oxa-1,3-diazole (NBD)-Cl-labeled actin[1] we have shown that GA_1Ca_1 purified using the procedure described here will nucleate the polymerization of filaments in either the presence of Ca^{2+} or EGTA and that these filaments are capped at the preferred end. Similar results have been obtained using pyrene-labeled actin[10] and the platelet gelsolin–actin complex purified using DNase I–Sepharose. These fluorescent probes have been used to show that the critical concentration or steady state concentration of actin is elevated from approximately 0.2 to 1.0 μM in the presence of gelsolin.[1,10] This shift is presumably due to capping of the kinetically active barbed ends of the filaments (at low critical concentration). The interactions between gelsolin, GA_1Ca_1, and actin have also been studied using fluorescently labeled actin.[3] This work indicates that gelsolin has two high-affinity actin-binding sites that are Ca^{2+} regulated. Binding of NBD–actin to the first site to form GA_1Ca_1 does not produce a change in fluorescence, but binding to the second site does. A somewhat similar pattern has been reported for brevin using pyrene-labeled actin.[19]

Mixture experiments have been reported that show that macrophage gelsolin will disrupt gels of actin, myosin, and actin-binding protein if the Ca^{2+} concentration is elevated and allow contraction to occur.[22] We have used pyrene-labeled actin to study the effects of gelsolin and GA_1Ca_1 on dilution-induced F-actin depolymerization.[22a] The results are consistent with the idea that gelsolin will sever filaments and increase the number of depolymerizing ends, but will not cap the preferred ends. GA_1Ca_1 at nanomolar concentrations, on the other hand, does not appear to sever, but does strongly cap the preferred ends in either Ca^{2+} or EGTA.

Additional Purification Strategies

After the earlier development of gelsolin purification methods described above, two additional purification strategies, immunoaffinity chromatography and ion-exchange chromatography in the presence and

---

[22] O. I. Stendahl and T. P. Stossel, *Biochem. Biophys. Res. Commun.* **92**, 675 (1980).
[22a] J. Bryan and L. M. Coluccio, *J. Cell. Biol.* **101**, 1236 (1985).

absence of Ca^{2+}, have been employed to purify plasma[23,24] and cytoplasmic[24,25] gelsolins. In addition, attention has been directed at the localization of the various binding domains,[23,26-29] gelsolin cDNAs from several species have been cloned and sequenced and shown to be highly homologous,[29,30-32] and specific fragments have been expressed using different systems.[33,34]

Large-Scale Preparation of Human Plasma Gelsolin

Outdated platelet concentrate is activated by addition of Tris-HCl, pH 7.5 and $CaCl_2$ to 25 mM followed by warming to 37° for 45 min. The clotted serum is cooled at room temperature for 1-2 hr to allow retraction, then centrifuged at 10,000 g for 30 min to remove the clot. The serum is dialyzed against three changes, 4 vol/change, of 25 mM Tris-HCl, 0.5 mM $CaCl_2$, pH 7.5. After dialysis NaN_3 is added to 1 mM. The dialyzed serum is centrifuged at 25,000 g for 30 min to remove aggregate material, adjusted to 35 mM NaCl, then loaded at a flow rate of 60 ml/min onto an 11.5 × 30 cm column of DEAE-Sephacel equilibrated at room temperature with 25 mM Tris-HCl, 50 mM NaCl, 0.5 mM $CaCl_2$, and 1 mM NaN_3, pH 7.5. Typically the dialyzed serum is stored frozen at −20°, then thawed and centrifuged before use. Typically 1 liter of centrifuged, dialyzed serum is loaded. The first 500 to 700 ml of eluate is discarded; the next 1500 ml is collected and cooled to 4°. The location of gelsolin can easily be followed by running SDS gels on small aliquots during the elution. EGTA is added to 10 mM, the pH is checked and adjusted to pH 7.5 with 1 M Tris if necessary, then this material is adsorbed to a second DEAE-Sephacel column (11.5 × 8 cm) equilibrated at 4° with 25 mM Tris-HCl, 50 mM NaCl, 0.1 mM EGTA, 1 mM NaN_3, pH 7.5. Plasma gelsolin passes through the first column under these conditions, but is adsorbed to the second and can be recovered approximately 95-98% pure by elution with

[23] J. Bryan, *J. Cell Biol.* **106**, 1553 (1988).
[24] S. Hwo and J. Bryan, *J. Cell Biol.* **102**, 227 (1986).
[25] C. Chaponnier, H. L. Yin, and T. P. Stossel, *J. Exp. Med.* **165**, 97 (1987).
[26] J. Bryan and S. Hwo, *J. Cell Biol.* **102**, 1439 (1986).
[27] C. Chaponnier, P. A. Janmey, and H. L. Yin, *J. Cell Biol.* **103**, 1473 (1986).
[28] D. J. Kwiatkowski, P. A. Janmey, J. E. Mole, and H. L. Yin, *J. Biol. Chem.* **260**, 15232 (1986).
[29] H. L. Yin, K. Iida, and P. A. Janmey, *J. Cell Biol.* **106**, 805 (1987).
[30] D. J. Kwiatkowski, T. P. Stossel, S. H. Orkin, J. E. Mole, H. R. Colten, and H. L. Yin, *Nature (London)* **323**, 455 (1986).
[31] M. Way and A. G. Weeds, *J. Mol. Biol.* **203**, 1127 (1988).
[32] B. Mulac-Jericevic and J. Bryan, *J. Cell Biol.* **108**, abst (1989).
[33] D. J. Kwiatkowski, P. A. Janmey, and H. L. Yin, *J. Cell Biol.* **108**, 1717 (1989).
[34] M. Way, J. Gooch, B. Pope, and A. G. Weeds, *J. Cell Biol.* **109**, 593 (1989).

a 4-liter linear gradient from 0.05 to 0.3 M NaCl. We usually pool the flow-through fractions from three runs of the first DEAE column, i.e., 3 liters of serum, and adsorb this to the second DEAE column. The peak of gelsolin from the gradient is pooled and concentrated by ammonium sulfate precipitation (3.8 g/10 ml of eluate). The essential elements of this purification scheme are the use of sufficient DEAE resin to adsorb the bulk of the serum proteins, particularly albumin, at the first step. We have found 1.5 vol of resin per volume of serum to be adequate, but routinely use approximately twice this amount. We have also found that having some salt present, 25–50 mM, improves reproducibility and minimizes gelsolin adsorption to the first DEAE column. The recovery is 80–100 mg of plasma gelsolin after the ammonium sulfate precipitation step, dialysis against the desired buffer, and centrifugation for 10 min in a microfuge to remove minor amounts of aggregated material. Typically we store gelsolin on ice in a final dialysis buffer, 10 mM Tris-HCl, 0.1 mM EGTA, pH 7.5 or freeze it rapidly in liquid nitrogen and hold at $-80°$ for extended storage.

Immunoaffinity Chromatography

This method requires specific antibodies. Two procedures have been described that employ murine monoclonal antibodies.[24,25] The approach employing monoclonal antibodies that recognize a Ca^{2+}-induced conformation of gelsolin[24] is described here.

Growth in Serum-Free Medium

The IgG- and IgA-secreting hybridomas, designated 8G5 and 4F8, respectively, are obtained as described in Hwo and Bryan.[24] Hybridomas are grown in CEM serum-free medium supplemented with transferrin, insulin, and selenium (Scott Laboratories, Inc., Fiskeville, RI) at an initial density of 2.5×10^5 cells/cm^2 in T-175 flasks (Falcon labware; Becton Dickinson, Oxnard, CA). After 3 days, the medium is removed and separated from the cells by low-speed centrifugation at 200 g for 5 min. The cells are gently resuspended in fresh medium and reseeded into T-175 flasks. Approximately 5 liters of serum-free medium is collected in this fashion and processed to obtain purified antibody.

Concentration of Serum-Free Medium and Purification of Antibodies

Serum-free harvest fluid is concentrated ~15-fold using a Millipore Minitan system (Millipore/Continental Water Systems, Bedford, MA) following the manufacturer's instructions. The IgG in the concentrate from 8G5 is purified on protein A–agarose (Sigma Chemical Co., St. Louis,

MO) by adsorbing in 120 mM NaCl, 25 mM Tris-HCl, pH 8.0 and eluting with 0.1 M sodium citrate, pH 4.5. The recovery is ~110 mg of 8G5 IgG per 5 liters of serum-free medium determined using an extinction coefficient of 1.4 at OD$_{280}$ for a concentration of 1 mg/ml.

The IgA in the concentrate from 5 liters of 4F8 harvest fluid is purified by chromatography on a 1.7 × 30 cm column of DEAE-Sephacel and elution with a 0–0.3 M NaCl gradient. The equilibrating buffer is 0.01 M Tris-HCl, pH 8.0. The fractions containing antibody and transferrin, determined by SDS–gel electrophoresis, are pooled and concentrated by precipitation with 50% (NH$_4$)$_2$SO$_4$. The precipitate is resuspended in 120 mM NaCl, 0.01 M Tris-HCl, pH 8.0 and chromatographed on a 2.5 × 100 cm column of Sephadex G-150 superfine to remove transferrin. The IgA antibodies elute in the void volume and can be concentrated by (NH$_4$)$_2$SO$_4$ precipitation.

Coupling of Antibody to Cyanogen Bromide-Activated Agarose

Purified 8G5 IgG or 4F8 IgA is dialyzed exhaustively against 0.5 M NaCl, 0.1 M NaHCO$_3$, pH 8.8, then coupled to CNBr-activated Sepharose 4B (Sigma Chemical Co.) following the procedures outlined by Pharmacia (Pharmacia Fine Chemicals). Briefly, the CNBr–Sepharose is hydrated by washing five times in 1 mM HCl. The final gel pellet is washed rapidly with 0.5 M NaCl, 0.1 M NaHCO$_3$, pH 8.8, and then resuspended in approximately an equal volume of antibody solution at 9 mg/ml in 0.5 M NaCl, 0.1 M NaHCO$_3$, and allowed to react overnight at 4° with gentle agitation. The supernatant is removed by centrifugation at 2500 g for 2 min and the gel pellet is resuspended in 0.2 M glycine, pH 8.5, and reacted overnight at 4°. The centrifugation is repeated and the gel is washed once with 0.5 M NaCl, 0.1 M NaHCO$_3$, pH 8.8, once with 0.5 M NaCl, 0.1 M sodium acetate, pH 4.0, and once with 0.5 M NaCl, 0.1 M NaHCO$_3$, pH 8.8. Finally, the antibody–Sepharose is packed in a column measuring ~2.0 × 3.5 cm and equilibrated with Tris–saline at pH 8.0. We estimate the coupling efficiency at ~90%.

Purification of Gelsolin Using Immunoreagents

The immunoreagents described above have been used to purify human plasma gelsolin and both free and actin-bound human platelet gelsolin. The method for the plasma form is illustrated here.

Serum, 100 ml prepared as described above, is loaded on an 8G5 IgG–Sepharose column. The column is washed with 150 ml of 0.3 M KCl, 25 mM Tris-HCl, pH 7.5 containing 0.1 mM CaCl$_2$ or until the OD$_{280}$ is below 0.05. Bound gelsolin is eluted with 0.3 M KCl, 25 mM Tris-HCl, pH

7.5 containing 5 mM EGTA. The elution can be followed using SDS–gel electrophoresis.

The use of these particular antibodies to purify actin-bound gelsolin is described in more detail in Hwo and Bryan.[24] Chaponnier *et al.*[25] have described the use of murine monoclonal antibodies to study changes in the concentration of gelsolin–actin complexes in macrophages.

Acknowledgments

The author wishes to thank Shuying Hwo, Robyn Lee, Wen-Gen Lin, David Hou, and Perry Sedlar for their excellent technical assistance during the years this work was done. Special thanks also to Brenda Cipriano, who provided help and careful preparation of the manuscript. This work was supported by NIH Grants HL26973 and GM26091 to J.B.

[10] Purification and Properties of Human Platelet P235: Talin

By NANCY C. COLLIER and KUAN WANG

Introduction

When a resting platelet is activated, a dramatic reorganization of the cytoskeleton occurs. Although few microfilaments are observed in the resting platelet, activation results in a burst of microfilament formation.[1] These changes in microfilament morphology are temporally and spatially controlled by a large number of actin-modulating proteins that have been identified in the platelet. These proteins serve to sequester monomeric actin, restrict length of actin filaments, modulate rate and polarity of actin polymerization, sever polymeric actin, or cross-link actin filaments.[2] One of these proteins, P235, is a major cytoplasmic component (3–5% of total protein).[3] P235 is implicated in the contractile activities of platelet, because the specific *in situ* degradation of P235 and filamin accompanies the loss of the ability of the platelet to extend pseudopodia.[4] Furthermore, both proteins are degraded by endogenous calcium-dependent proteases during

---

[1] J. C. Lewis, *Cell Muscle Motil.* **5**, 341 (1984).
[2] J. E. B. Fox, in "Biochemistry of Platelets" (D. R. Philips and M. A. Shuman, eds.), p. 115, Academic Press, Orlando, Florida, 1986.
[3] N. C. Collier and K. Wang, *J. Biol. Chem.* **257**, 6937 (1982).
[4] V. T. Nachmias, J. S. Sullender, and J. R. Fallon, *Blood* **53**, 63 (1979).

thrombin-induced platelet aggregation.[5] In solution, purified P235 affects actin polymerization. When present with G-actin during salt-induced polymerization, P235 reduces the rate of actin polymerization in a calcium- and calmodulin-sensitive manner.[6-9] The resultant actin filaments are shorter but the total amount of actin filaments remains unaffected.[6-9] Thus, P235 may play an important role in the actin-mediated changes in the platelet and its activity may be regulated by its phophorylation state, proteolytic processing, accessory proteins, and calcium level.

P235 has been shown to bear similarity to talin, a vinculin- and integrin-binding phosphoprotein found in other cell types at the adhesion plaques where actin filaments terminate as the plasma membrane.[10-13] Its role in transmembrane linkage of extracellular matrix to actin cytoskeleton in platelet is suggested by its redistribution to submembranous location in activated platelets.[13]

Purification of Human Platelet P235

Platelet Concentrates

Our starting material is human platelet concentrates. The use of concentrates has the advantage over whole blood since it eliminates several centrifugal purification steps. A small amount of contaminating erythrocytes and leukocytes is easily removed by differential centrifugation. Concentrates, which quickly become outdated (within 3 days) for transfusion purposes, are usually available at minimal or no cost from blood banks and hospitals. Since we have found that P235 and filamin remain intact in freshly outdated concentrates (within 3 to 7 days of venipuncture), we routinely use these for purification of contractile proteins. Fresh platelet concentrates, needed to confirm the results of some experiments, are available at cost.

[5] J. E. B. Fox, D. E. Goll, C. C. Reynolds, and D. R. Phillips, *J. Biol. Chem.* **260**, 1060 (1985), and references listed therein.
[6] N. C. Collier and K. Wang, *FEBS Lett.* **143**, 205 (1982).
[7] N. C. Collier and K. Wang, *Biophys. J.* **41**, 86a (1983).
[8] N. C. Collier and K. Wang, *J. Cell Biol.* **97**, 289a (1983).
[9] N. C. Collier, Ph.D. dissertation, University of Texas at Austin (1982).
[10] T. O'Halloran, M. C. Beckerle, and K. Burridge, *Nature (London)* **317**, 449 (1985).
[11] L. Molony, D. McCaslin, J. Abernethy, B. Paschal, and K. Burridge, *J. Biol. Chem.* **262**, 7790 (1987).
[12] A. Horwitz, K. Duggan, C. Buck, M. C. Beckerle, and K. Burridge, *Nature (London)* **320**, 531 (1986).
[13] M. C. Beckerle, D. E. Miller, M. E. Bertagonolli, and S. J. Locke, *J. Cell Biol.* **109**, 3333 (1989).

Proteolytic Degradation

P235 is highly susceptible to proteolytic degradation, in particular to the endogenous Ca^{2+}-dependent proteases.[3-5] Several measures are therefore taken to minimize proteolysis. First, platelets are washed at room temperature in the presence of EDTA in plasticware to avoid activation that leads to P235 degradation.[14] Second, EDTA is included in all steps during and after the extraction of platelet proteins to prevent proteolysis induced by Ca^{2+}-activated proteases. We have found that when the EDTA concentration in the extraction buffer was reduced below 0.3 mM, extensive degradation of P235 occurred. The inclusion of leupeptin (at 20 μg/ml), an inhibitor of calcium-dependent protease, did not further improve P235 stability. Covalent inhibitors such as iodoacetamide or N-ethylmaleimide were not used to avoid potential alkylation of P235. Even with these precautions, purified P235 is slowly degraded into large fragments of 220 and 200 kDa on storage at 4° for several days.

Assay

The purity and yield of P235 are assayed by sodium dodecyl sulfate (SDS)–gel electrophoresis. Polyclonal antiserum directed against P235 is used, when needed, to monitor P235 degradation by immunoblot techniques. The discontinuous SDS–gel method of Laemmli[15] is modified as follows: Proteins are analyzed on a 3 to 12% linear gradient polyacrylamide separating gel without the stacking gel. This modification results in a sharp protein banding pattern while resolving both high and low molecular weight proteins.

Protein in the presence of Triton X-100 is determined by the method of Bensadoun and Weinstein.[16] Protein in other buffers is determined by the Hartree method.[17] Crystalline bovine serum albumin (BSA, type V; Sigma, St. Louis, MO) is used as the standard. The concentration of BSA was determined by absorbance ($E_{280}^{1\%} = 6.6$).

Buffers

Platelet Washing Buffers

Buffer 1: 0.12 M NaCl, 0.129 M sodium citrate, 0.03 M glucose, pH 7.4
Buffer 2: 0.154 M NaCl, 0.005 M EDTA, 0.01 M Tris-HCl, pH 7.4

---

[14] Other investigators have used aspirin or prostacyclin in the wash medium to prevent platelet activation.
[15] U. K. Laemmli, *Nature* (*London*) **227**, 680 (1970).
[16] A. Bensadoun and D. Weinstein, *Anal. Biochem.* **70**, 241 (1976).
[17] E. F. Hartree, *Anal. Biochem.* **48**, 422 (1972).

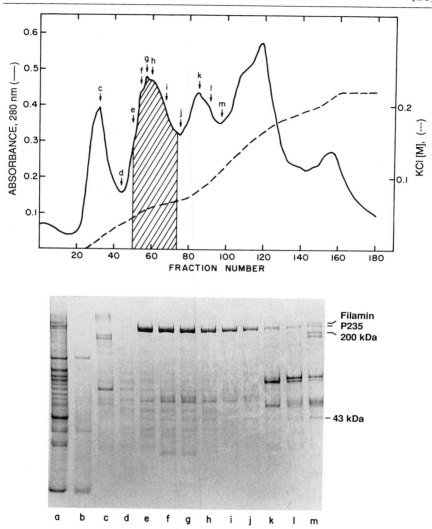

FIG. 1. *Top:* DEAE-cellulose column chromatography of crude platelet extract. Extract from the washed platelets of 25 units of platelet concentrate was applied to a Whatman (Clifton, NJ) DE-52 column and eluted with a salt gradient as described in text. Fractions enriched in P235 but exclusive of filamin and 200 kDa polypeptide (shaded area) were pooled (pool DEAE). *Bottom:* SDS–polyacrylamide gel patterns of selected fractions from the elution profile: (a) whole extract; (b) flow-through; (c) fraction 32; (d) fraction 45; (e) fraction 50; (f) fraction 52; (g) fraction 55; (h) fraction 58; (i) fraction 66; (j) fraction 75; (k) fraction 85; (l) fraction 90; (m) fraction 95.

Extraction and Purification Buffers

Buffer 3: 50 mM Tris-HCl, 3 mM EDTA, 0.5 mM dithiothreitol (DTT), pH 9.0

Buffer 4: 50 mM Tris-HCl, 3 mM EDTA, 0.1 mM DTT, pH 8.0

Extraction[18]

Platelet concentrates are first freed of residual erythrocytes and leukocytes by a low-speed centrifugation in polycarbonate tubes (235 g_{max} for 5 min at room temperature). The supernatant containing the platelets is removed, either by careful decanting or by aspiration. This step is repeated if the supernatant still appears to contain erythrocytes. Pellet the platelets by centrifugation at 8000 g_{max} for 30 min at 20° in preweighed polycarbonate tubes. Wash twice by resuspension and centrifugation in buffer 1 (10 ml/g wet weight) at 8000 g_{max} for 30 min, then once at 23,500 g_{max} for 30 min in buffer 2.[19] After chilling on ice for 20 min, the washed platelets are resuspended in ice-cold buffer 3 (5 ml/g wet weight). With swirling, add an equal volume of ice-cold buffer 3 containing 1% (w/v) Triton X-100. A flocculent precipitate forms at this point. Centrifuge the mixture at 8000 g_{max} for 10 min at 4°. This low-speed centrifugation pellets the Triton–EDTA-insoluble cytoskeleton, which can be used for subsequent purification of filamin, α-actinin, and other cytoskeletal proteins.[20,21] The supernatant is then spun at 75,000 g_{max} for 1 hr at 4° in a Beckman (Palo Alto, CA) type 30 rotor. This extraction procedure preferentially extracts P235 while maintaining the majority of the 200-kDa polypeptide (containing myosin heavy chain and degradation products of filamin and P235) and some filamin in the insoluble residue. The clear, light yellow supernatant (whole extract) contains all the P235 and about 50% of the total platelet proteins (Fig. 1, bottom, lane a).

DEAE-Cellulose Ion-Exchange Chromatography

This is the key step in the purification of P235. The high speed supernatant is directly applied to a Whatman (Clifton, NJ) DE-52 column equili-

---

[18] As a precautionary measure to avoid transmittable pathogens, wear disposable gloves and disinfect all labware by soaking in Wescodyne (AMSCE, Medical Products Division) according to the manufacturer's suggestions.

[19] The packed platelets can be stored by flash freezing in liquid nitrogen and then kept at $-70°$ (or at $-20°$ in a non-frost-free freezer) for at least several months. To proceed with extraction, ice-cold buffer 3 is added directly to the frozen pellet.

[20] S. Rosenberg, A. Stracher, and R. C. Lucas, *J. Cell Biol.* **91**, 201 (1981).

[21] S. Rosenberg, A. Stracher, and K. Burridge, *J. Biol. Chem.* **256**, 12986 (1981).

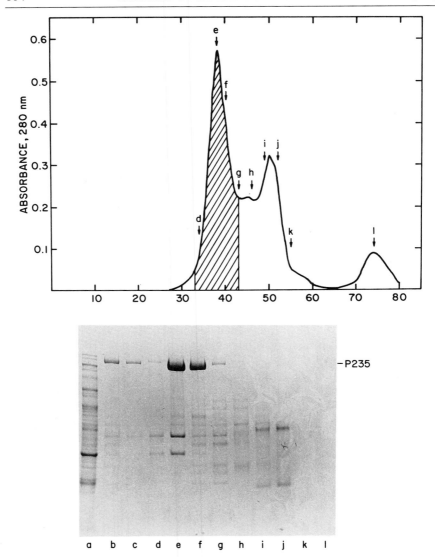

FIG. 2. *Top:* Gel-filtration chromatography of concentrated pool DEAE. Pool DEAE was concentrated by dry dialysis against Aquacide IIA and applied to a Sephacryl S-300 column as described in text for alternative procedure B. Fractions enriched in P235 (shaded area) were pooled (pool S-300). *Bottom:* SDS–polyacrylamide gel patterns of selected fractions from the elution profile: (a) whole extract; (b) pool DEAE; (c) concentrated pool DEAE; (d) fraction 34; (e) fraction 38; (f) fraction 41; (g) fraction 43; (h) fraction 46; (i) fraction 49; (j) fraction 52; (k) fraction 56; (l) fraction 75.

brated in buffer 4 without pH adjustment or dialysis.[22] The dimensions of the column are determined by the amount of protein applied. For 25 units of platelet concentrates, a 1.5 × 57 cm column is optimal. The extract is applied at a fast flow rate (approximately 50 to 75 ml/hr) to avoid gelation on the column. Approximately 80% of the total protein in the high-speed supernatant, including all filamin, P235, and any 200-kDa polypeptide, adheres to the column. Wash the column with at least 10 column volumes of buffer 4. Elute proteins with a linear salt gradient consisting of 5 column volumes, each of buffer 4, and buffer 4 containing 0.2 M KCl (at 20 ml/hr for a 1.5 × 57 cm column). P235 generally elutes with a sharp leading edge between 0.05 and 0.07 M KCl and a long trailing edge, up to 0.14 M KCl. If the optimal conditions are followed as outlined above, P235 will elute as a separate peak (Fig. 1). Pool fractions enriched in P235 but exclusive of filamin and 200-kDa polypeptide (pool DEAE) [Fig. 2, top and bottom (lane b)]. It is important that fractions be pooled according to gel analysis rather than by the expected salt concentration since inclusion of fractions containing filamin (eluting between 0.09 and 0.14 M KCl) and 200-kDa polypeptide (eluting between 0.09 and 0.12 M KCl) interferes with subsequent purification steps and lowers the yield. P235 in pool DEAE is 40 to 60% pure and can be stored at least 4 days at 4° without detectable degradation.

Alternative A: Phosphocellulose Followed by Gel Filtration

This alternative is our original procedure, which uses phosphocellulose chromatography to remove the majority of low molecular weight contaminants in pool DEAE ($M_r \sim$ 83,000, 69,000, 67,000, 65,000, 62,000, 42,000, and 26,000). Pool DEAE is applied without dialysis to a Whatman P-11 column equilibrated in buffer 4. Again the size of the column depends on the amount of protein in pool DEAE. Usually 0.5 to 2 mg of protein is applied per milliliter of P-11 (and for 25 units of concentrates, a 1.5 × 12.5 cm column is optimal). Wash the column with 7.5 column volumes of buffer 4 and develop with a linear salt gradient consisting of four column volumes of buffer 4 and buffer 4 with 0.5 M NaCl. A slower flow rate (e.g., 10 ml/hr for the 1.5 × 12.5 cm column) is dictated here. P235 elutes as a separate and major peak between 0.11 and 0.2 M NaCl. A trailing edge containing minor amounts of P235 is eluted after 0.2 M NaCl. P235 in pool

---

[22] If the extract is inadvertently left standing for more than several hours, gelation occurs, which can be clarified by centrifuging at 18,000 rpm for 30 min at 4° in a SS-34 rotor. The clear supernatant, containing most of the P235, can then be applied to the DEAE ion-exchange column.

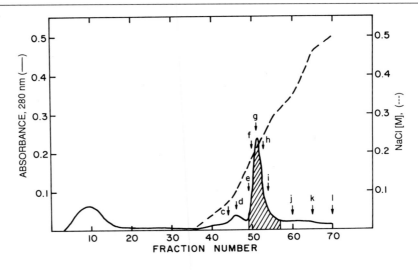
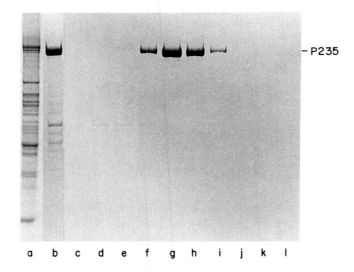

FIG. 3. *Top:* Phosphocellulose column chromatography of pool S-300. Pool S-300 was applied to a Whatman P-11 column and eluted with a salt gradient as described in text. Fractions containing purified P235 (shaded area) were pooled (pool PC). *Bottom:* SDS–polyacrylamide gel patterns of selected fractions from the elution profile: (a) whole extract; (b) pool S-300; (c) fraction 44; (d) fraction 46; (e) fraction 49; (f) fraction 50; (g) fraction 51; (h) fraction 52; (i) fraction 53; (j) fraction 60; (k) fraction 65; (l) fraction 70.

PC is at least 80% pure and can be stored for about 3 days at 4° without formation of degradation fragments of 220 and 200 kDa. Concentrate pool PC by dry dialysis with Aquacide IIA (Calbiochem, San Diego, CA).[23] Apply concentrated pool PC to a gel-filtration column (BioGel A5M, Bio-Rad, Richmond, CA) equilibrated in buffer 4. (For 25 units, we apply 2 ml of concentrated pool PC to a 1.6 × 75 cm column with a flow rate of 6 ml/hr.) P235 is resolved into two peaks. The first peak (peak I) appears in the void volume with a Stokes radius greater than 85 Å. The second one (peak II) elutes with a K_{av} between 0.33 and 0.44 with a Stokes radius of 67 Å. The relative proportion of P235 in these two peaks varied in separate experiments. In general, peak I contained 25 to 40% of the total P235. Peak II was absent only in 1 out of more than 20 preparations. Peak I was slightly contaminated (90% pure) and peak II was at least 95% pure. Both peaks of P235 have identical electrophoretic mobility on SDS gels and both react with antibodies raised against P235. Neither peak contains interchain disulfides because their electrophoretic mobilities are not altered by reduction with thiols. When peak I is reapplied to the column, both peak I and peak II appear, suggesting a slow conversion between perhaps different conformations or aggregation states. It is also noted that the 220-kDa fragment is eluted slightly ahead of peak II P235 on the gel-filtration column, indicating a larger Stokes radius for the 220-kDa fragment.

Alternative B: Gel Filtration Followed by Phosphocellulose

Because P235 purified by the above procedure gradually degrades into 220- and 200-kDa fragments either on storage at 4° or during prolonged incubation at higher temperatures required for actin interaction studies, we have developed an alternate method that yields a final product that is much more resistant to degradation. This was achieved by reversing the order of column chromatography, employing gel filtration prior to phosphocellulose chromatography. After fractionation on DEAE-cellulose, concentrate pool DEAE by dry dialysis against Aquacide IIA. Dialyze the concentrated pool against buffer 4 and apply to the gel-filtration column, either BioGel A5M or Sephacryl S-300 as described above (Fig. 2). P235 (Pharmacia, Piscataway, NJ) elutes solely in the included volume of the column, with a K_{av} equal to that of peak II P235 of the original procedure.

---

[23] Other methods of concentration, such as Amicon (Danvers, MA) filtration and stepwise elution from small DEAE columns, can also be used, but in our hands resulted in greater loss of P235.

TABLE I
MOLECULAR PROPERTIES OF HUMAN PLATELET P235

Property	Value/Description
Polypeptide chain mass[3]	235 kDa
Subunit composition[3]	Dimer
Molecular weight[3]	470,000
Subunit isoelectric point[2]	6.6
Stokes radius[3]	67 Å
Sedimentation coefficient[3]	9.8S
Frictional coefficient[3]	1.6
Amino acid composition[3,11]	Similar to chicken gizzard talin
Protein interaction[6-13]	Actin, vinculin, integrin
Immunological property[3,10]	Distinct from filamin, spectrin, fibronectin, myosin; similar to talin
Cellular distribution[9,13]	Cytoplasm, membrane

The pool from the gel-filtration column (pool S-300, Fig. 3, right) is variably contaminated with a number of lower molecular weight proteins ($M_r \sim$ 104,000, 88,000, 73,000, 60,000, 53,000, 49,000, and 30,000). In order to remove these lower molecular weight contaminants, apply the peak fractions of P235 to a phosphocellulose column and develop as described above (Fig. 3). P235 elutes under salt conditions from the phosphocelluse column identical to those described for alternative procedure A. The yield and purity of P235 purified by this alternate procedure are about the same as those obtained by the original procedure. P235 purified by this method is substantially more resistant to degradation and has been stored up to 2 weeks at 4° without signs of degradation.

Comments

Reproducible and fairly simple procedures have been developed to purify P235 from human platelets in the absence of denaturants. The purification procedures described here yield consistently 2 to 5 mg of P235 from 25 units of platelet concentrate. The alternate procedure was developed to remove proteolytic contamination of P235 purified by the original method. The alternate procedure yields P235 that is substantially free of protease activities. However, the conformation states of P235 purified in this manner are uncharacterized. This procedure is useful to avoid proteolytic degradation of P235 whenever studies dictate prolonged incubation at room or higher temperature. The molecular properties of P235 (peak II of alternative procedure A) are summarized in Table I.

Acknowledgments

This work is supported in part by grants from the National Institutes of Health (HL31491 and CA09192) and the American Heart Association, Texas Affiliate, Inc., and an Established Investigatorship of the American Heart Association.

[11] Ultrastructural Analysis of Platelet Contractile Apparatus

By JAMES G. WHITE

Introduction

Blood platelets are anucleate cells whose primary purpose is to help preserve the integrity of the cardiovascular system.[1,2] Due to their small size and relative transparency platelets were the last of the blood cells to be recognized as a separate and distinct type with a unique function.[3] Yet the characteristics of small size and clear cytoplasm that delayed discovery of the platelet proved advantageous for ultrastructural study.[4] Platelets were the first cells to be evaluated in the electron microscope, and ultrastructural techniques have been employed for several decades in platelet investigations.[5,6] These studies were among the first to suggest that platelets were a form of muscle cell.[7,8] Elements of platelet anatomy now known to be constituents of the cytoskeleton and cytocontractile apparatus were recognized in early experiments before the terms became popularized.[9-11]

---

[1] C. Achard and M. Aynaud, *C. R. Seances Soc. Biol. Ses Fil.* **63**, 593 (1907).
[2] J. Roskam, *Arch. Int. Physiol.* **20**, 241 (1923).
[3] A. H. T. Robb-Smith, *Br. J. Haematol.* **13**, 618 (1967).
[4] C. Wolpers and H. Ruska, *Klin. Wochenschr.* **23**, 1077 (1939).
[5] J. F. David-Ferreira, *Int. Rev. Cytol.* **17**, 99 (1964).
[6] T. Hovig, *Ser. Hematol.* **1**, 3 (1968).
[7] M. Bettex-Galland and E. F. Luscher, *Nature (London)* **184**, 276 (1959).
[8] M. Bettex-Galland and E. F. Luscher, *Thromb. Diath. Haemorrh.* **4**, 178 (1963).
[9] J. G. White, *in* "The Platelet" (F. K. Mostafi and K. M. Brinkhous, eds.), p. 83. Williams & Wilkins, Baltimore, Maryland, 1971.
[10] J. G. White, *in* "The Circulating Platelet" (S. A. Johnson, ed.), p. 45. Academic Press, New York, 1971.

Although biochemical approaches have been critical to the development of our knowledge, electron microscopy continues to provide basic concepts of the organization and function of the platelet contractile system and cytoskeleton. The purpose of this chapter is to discuss some of the methods useful for analyzing the cytocontractile apparatus at the ultrastructural level.

Overview

Microscopic techniques of various kinds have always been important in investigations of blood platelets. Improved lenses for light microscopes, introduction of phase-contrast systems, and Nomarski optics all had an impact, but it was the electron microscope that made possible the evaluation of specific subcellular constituents of the platelet contractile cytoskeleton.[4-8] Still, refinements in microscopes for studying platelet morphology would have been useless without comparable improvements in methods used to prepare platelets for *in vitro* study and for fixing and embedding the cells for evaluation in the electron microscope. Changes in procedures necessary to isolate platelets from whole blood in an undamaged, resting state, stimulate the cells under conditions resembling those found *in vivo*, and other changes preserving resting and activated platelets in a near pristine condition were both essential before advances in optical systems could be used effectively.[5,9-11]

Detailed descriptions of the anticoagulants used, procedures for collection of blood and separation of platelets, functional evaluation of responses in the platelet aggregometer or in unstirred suspensions, and step-by-step instructions for the fixation, dehydration, and embedding of platelets for study in the electron microscope are reviewed elsewhere. The reader is referred to Ref. 12 for a discussion of essential techniques beyond the purview of this chapter.

Platelet Anatomy and Structural Physiology

Before describing specific techniques for analyzing the contractile cytoskeleton, it will be useful to consider some of the general features of platelet anatomy and functional response. Platelets in circulating blood and in well-prepared samples of platelet-rich plasma (PRP) have a characteristic discoid form. Thin sections of the discoid cells reveal a variety of

---

[11] J. G. White, *in* "Platelet Aggregation" (J. Caen, ed.), p. 15. Masson, Paris, 1971.
[12] J. G. White, *in* "Measurements of Platelet Function" (L. A. Harker and T. S. Zimmerman, eds.), Vol. 8, p. 1. Churchill-Livingstone, Edinburgh, 1983.

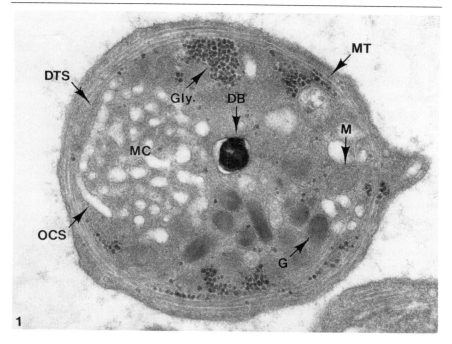

Fig. 1. Thin section of discoid human platelet sectioned in the equatorial plane. The cell is from a sample of citrate platelet-rich plasma (C-PRP) fixed first in glutaraldehyde and then osmic acid as described in the text and in Ref. 12. A circumferential microtubule (MT) supports the discoid form. Numerous granules (G), a few mitochondria (M), and occasional dense bodies (DB) making up the organelle zone are randomly dispersed in the cytoplasmic matrix or sol–gel zone. Glycogen (Gly) is concentrated in masses or in single particles. Clear channels of the open canalicular system (OCS) follow tortuous courses from the surface membrane deep into the cytoplasm. The surface membrane and OCS constitute the peripheral zone. Elements of the dense tubular system (DTS) are often associated with the circumferential MT. Interactions between elements of the two membrane systems result in the formation of an interwoven membrane complex (MC). Magnification: ×36,000.

formed elements embedded in a dense matrix (Fig. 1). In order to simplify structural features and relate them to functional and biochemical activities, the anatomy of the platelet has been divided into four major regions.[10,11] The peripheral zone consists of the membrane and closely associated structures providing the surface of the platelet and walls of the tortuous channels making up the surface-connected open canalicular system. An exterior coat, or glycocalyx, rich in glycoproteins provides the outermost covering of the peripheral zone. Its chemical constituents provide the receptors for stimuli triggering platelet activation and for binding fibrinogen and other adhesive proteins in adhesion–aggregation reactions. The middle layer of the peripheral zone is a typical unit membrane. It is rich in asymmetrically distributed phospholipids that provide an essential surface following activation for interaction with coagulant proteins. Stimulation also results in cleavage from membrane phospholipids of arachidonic acid, the substrate for prostaglandin synthesis, and of inositol trisphosphate, important for calcium transport. The area just underneath the surface membrane represents the third component of the peripheral zone. It is closely linked to the surface through transmembrane proteins that translate signals received on the outside into chemical messages and physical alterations characteristic of the platelet hemostatic response. The sol–gel zone is the matrix of the platelet cytoplasm. Formed elements include microtubules supporting the discoid shape of unaltered platelets and actin filaments involved in shape change, pseudopod extension, internal contraction, and secretion. Fiber systems of this zone are major constituents of the platelet cytoskeleton. The organelle zone consists of α granules, electron-dense bodies, peroxisomes, lysosomes, mitochondria, and masses, as well as discrete particles of glycogen, randomly dispersed in the cytoplasm. It serves in metabolic processes and for storage of enzymes, a nonmetabolic pool of adenine nucleotides, serotonin, a variety of adhesive, coagulant, and platelet-specific proteins, and calcium destined for secretion. Membrane systems constitute a fourth zone. The dense tubular system has been shown to be the site where calcium important for triggering contractile events is sequestered. Also, it is the site where enzymes important in endoperoxide and thromboxane synthesis are localized. The surface-connected open canalicular system provides access to the interior for plasma-borne substances and an egress route for products of the release reaction. Together with elements of the dense tubular system, channels of the open canalicular system form specialized membrane complexes closely resembling the relationships of transverse tubules and sarcotubules regulating calcium flux in embryonic muscle cells.

Platelets respond to a wide variety of chemical, physical, and other stimuli in a characteristic manner.[10–13] The cells quickly lose their discoid shape, become relatively spherical in form and extend long, spiky pseudopods. Changes in surface contour are accompanied by transport of randomly dispersed internal organelles to cell centers. The internal transformation is associated with constriction of the circumferential microtubule into tight coils around the centrally concentrated organelles (Fig. 2). Assembly of the actin microfilament system is responsible for both shape

---

[13] J. G. White and J. M. Gerrard, in "Handbook of Inflammation" (G. Weissman, ed.)., Elsevier, Amsterdam, p. 83. 1980.

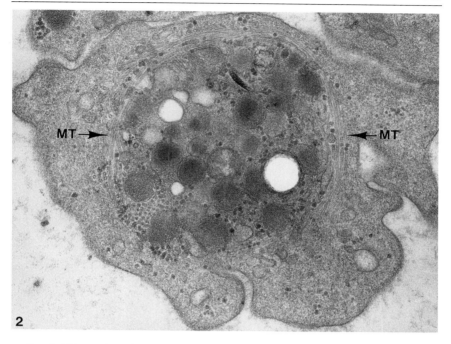

FIG. 2. Thin section of activated platelet from sample of C-PRP treated with taxol for 30 min and then mixed once with thrombin at 0.2 U/ml. The sample was not stirred before fixation 45 sec later. Despite the precaution, small aggregates formed in this sample. Cytoplasmic organelles in the cell shown are crushed together in the platelet center and encircled by a tight web of microtubules (MT). The cell has lost its lentiform appearance, has become irregular, and appears attached to two adjacent platelets. Magnification: ×36,000.

change and internal contraction. As contents of secretory organelles are extruded, the web of tubules and microfilaments continues to contract until the microtubule coils fracture, leaving a mass of contractile gel in the central zone (Fig. 3). Microtubule fragments are dispersed and appear more frequently in pseudopods as contraction of aggregates or clots progresses.

The platelet physical responses described above are typical of cells suspended in a fluid phase. Interaction with damaged vessels, foreign surfaces, and in clots under isometric tension is believed to produce a different sequence of alterations.[14] However, the similarities are far greater than the differences, and suggest that the platelet contractile system adapts

[14] J. M. Gerrard and J. G. White, in "Pathobiology Annual" (H. L. Iochim, Ed.), Vol. 6, p. 31. Appleton-Century-Crofts, New York, 1979.

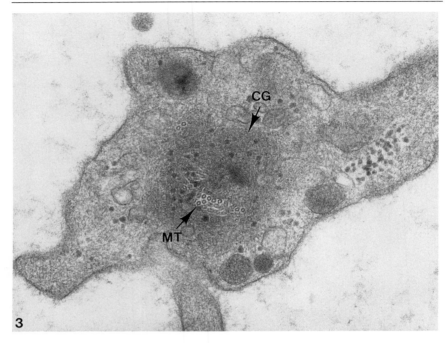

FIG. 3. Platelet from C-PRP exposed to thrombin (0.2 U/ml) without stirring for 3 min before fixation. The cell is irregular in form. Contents of most storage organelles have been secreted. The coils of microtubules (MT) have been crushed and are buried in a mass of contractile gel (CG) in the platelet center. Magnification: ×50,000.

in a unique manner to the conditions demanded of it in order to exercise hemostatic function.

Ultrastructural Methods for Evaluation of the Platelet Contraction Cytoskeleton

Thin-Section Techniques

The development of methods of fixation capable of preserving platelet discoid shape made it possible to follow typical feature of platelet transformation in the electron microscope. Although a description of these methods has been reported,[12] certain critical aspects of the procedure are worth emphasizing. Care must be taken in selection of the anticoagulant, aspirating blood, and separating platelet-rich plasma from leukocytes and red blood cells in order to preserve the discoid shape of unstimulated

TABLE I
STOCK SOLUTIONS FOR WHITE'S SALINE

Component	Weight (g)
A. NaCl	14.0
KCl	0.75
$MgSO_4$	0.55
$Ca(NO_3)_2 \cdot 4H_2O$	1.5
Add distilled H_2O to 100 ml. Refrigerate	
B. $NaHCO_3$	1.1
$Na_2PHO_4 \cdot 7H_2O$	0.22
KH_2PO_4, anhydrous	0.052
Phenol red	0.01
Add distilled H_2O to 100 ml. When dissolved at room temperature, bubble CO_2 in for about 15 sec, if necessary. Mix well and adjust pH to 7.4. Refrigerate	

platelets. Some of the steps recommended in the literature, such as using large-bore needles and allowing blood to flow directly from the needle into tubes, represent excessive caution, but reasonable care is important. Resting platelets, stimulated cells, and aggregates should be fixed in suspension at room temperature or 37°. To accomplish this without also precipitating plasma proteins, it is necessary to use a low concentration of fixative. We routinely use 0.1% (v/v) glutaraldehyde in White's saline for the initial fixation (Tables I and II), but other concentrations of aldehyde and different buffers can be used. After initial fixation the samples are sedimented to buttons and refixed in a higher concentration of glutaral-

TABLE II
ZETTERQUIST'S VERONAL BUFFER

Component	Weight (g)
1. Stock buffer	
Sodium barbital	14.7
Sodium acetate $\cdot 3H_2O$	9.7
Dilute to 500 ml with distilled H_2O	
2. Stock salt solution	
NaCl	20
KCl	1
$CaCl_2$	0.5
Dilute to 250 ml with distilled H_2O	

TABLE III
FIXATIVES

1. Glutaraldehyde, 0.1% in White's saline
   Distilled H$_2$O, 8.9 ml
   White's A, 0.5 ml
   White's B, 0.5 ml
   Glutaraldehyde (10%), 0.1 ml
2. Glutaraldehyde, 3% in White's saline
   Distilled H$_2$O, 6 ml
   White's A, 0.5 ml
   White's B, 0.5 ml
   Glutaraldehyde (10%), 3 ml
3. OsO$_4$, 1%. Prepare and keep in a dark room in a glass-stoppered bottle:

   OsO$_4$ aqueous (4%), 2.5 ml
   Stock buffer (Zetterquist's), 2.0 ml
   HCl (0.1 N), 0.68 ml
   HCl (0.1 N), 2.0 ml
   Distilled H$_2$O, 2.82 ml

   Keeps 1 week in refrigerator

dehyde in White's saline followed by osmic acid in Zetterquist's buffer (Table III). Selection of the aldehyde and osmic acid concentrations as well as the buffers in which they are dissolved are subject to considerable variation because no single method is perfect. For example, one factor of importance in preserving the cytocontractile architecture is the concentration of osmic acid. The fixative is known to destroy actin filaments[15] and low concentrations combined with brief exposure may offer some protection during subsequent steps of dehydration and embedding.

Another procedure designed to preserve the actin filaments for study in thin sections of platelets has been introduced. Boyles suggested adding 50 mM lysine with or without 1% (v/v) Triton X-100, 5 mM EGTA, and 2% (v/v) glutaraldehyde to 50 mM sodium cacodylate at pH 7.4.[16] Membranes and soluble proteins are extracted with the detergent, but the surface membrane-associated short actin filaments along with microtubules are reasonably well stabilized by this procedure and mark the contour of the extracted membrane cytoskeleton (Fig. 4). Application of the technique together with phalloidin to activated platelets reveals considerably more detergent-resistant actin than in resting cells.[17] Much of the

[15] P. Maupain-Szamier and T. Pollard, *J. Cell Biol.* **77**, 837 (1978).
[16] J. E. B. Fox, J. K. Boyles, C. C. Reynolds, and D. R. Phillips, *J. Cell Biol.* **98**, 1985 (1984).
[17] G. Escolar, M. Krumwiede, and J. G. White, *Am. J. Pathol.* **123**, 86 (1986).

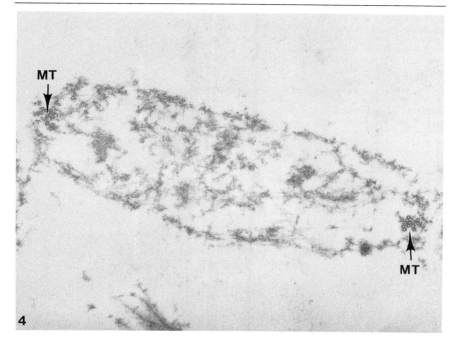

FIG. 4. Actin shell of human platelet. The cell has been extracted with detergent in the presence of lysine and phalloidin before fixation. All membranes and organelles have been removed, but the circumferential band of microtubules (MT) is still intact. A thin film of actin outlines the site of the surface membrane and actin filaments are evident in the cytoplasm. Magnification: ×32,400.

residual actin is in a collar around centrally concentrated organelles (Fig. 5). The increased amount of detergent-resistant actin stabilized by lysine and phalloidin (Fig. 5) in activated platelets is in harmony with the concept that most of the actin in resting platelets is in the globular state (G-actin) and assembles into filaments (F-actin) only after stimulation. Localization around centrally concentrated organelles is in accord with observations made on thin sections of activated platelets[9-12] and in whole-mount preparations.[18] Unfortunately, the disruption of other platelet structures by this treatment makes it difficult to recognize the fundamental organization of the cytocontractile apparatus.

Several agents have been used to protect actin filaments during preparation of cells for study in thin sections at the ultrastructural level. Phalloi-

[18] J. G. White, *Am. J. Pathol.* **117**, 207 (1984).

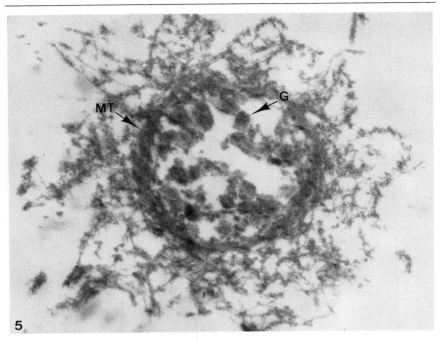

FIG. 5. Actin shell of thrombin-activated human platelet. The cell is from a suspension of washed platelets exposed to thrombin (0.3 U/ml) for 2 min before extraction with detergent in the presence of phalloidin and lysine. Actin outlines the site of the extracted surface membrane. Filaments are randomly dispersed in the cytoplasm. The greatest concentration of actin is associated with the constricted ring of MT exclosing partially extracted granules (G). Magnification: ×32,400.

din, although extremely toxic, has been found in several studies to protect actin from damage by osmic acid.[19] Thiocarbohydrizide is a bipolar compound that binds to osmium. By sandwiching thiocarbohydrizide between two phases of osmium treatment, cell exposure to osmium can be minimized without destroying actin filaments.[20] Glutaraldehyde can also damage actin filaments by extensive cross-linking on prolonged exposure. Tropomyosin bound to F-actin filaments protects the assembled protein from damage by the aldehyde fixative, as well as from the toxic effects of osmic acid.[21] Although these several approaches are reputed to have worked well in other cells and tissues, they have not been evaluated as

[19] T. Wieland, *Adv. Enzyme Regul.* **15**, 285 (1977).
[20] M. Aoki and M. Tavassoli, *J. Histochem. Cytochem.* **29**, 682 (1981).
[21] S. S. Lehrer, *J. Cell Biol.* **90**, 459 (1981).

yet for their influence on the platelet cytocontractile system. Therefore, we would recommend the procedure introduced by Boyles[16] or our modification of it[17] for use at the present time.

Chemical agents, other than those used to protect actin filaments from fixation damage, have also contributed significantly to the value of thin-section techniques for evaluation of the platelet contractile system. Taxol extracted from the stem bark of the western yew has been helpful in confirming the contractile nature of platelet internal transformation.[22] Studies employing a colchicine-binding assay had suggested that the circumferential microtubule supporting the discoid shape of resting platelets dissolves within 15 sec after exposure to aggregating agents and reassembles 1–4 min later in a new location.[23,24] The finding conflicted sharply with results obtained by indirect immunofluorescence and thin-section electron microscopy, which had shown that the circumferential microtubule undergoes constriction following platelet activation, but does not dissolve.[9-12] Thus, the disassembly–reassembly hypothesis contradicted the concept that internal reorganization in activated platelets is a contractile process. Taxol is a useful agent because it inhibits disassembly of microtubules by conditions, such as cold,[25] and chemical agents like vincristine.[26] Treatment of platelets with taxol prior to stimulation with aggregating agents had no effect on shape change, pseudopod formation, internal transformation, circumferential microtubule constriction, secretion, irreversible aggregation, or clot retraction.[22] The findings demonstrated that a force-generating process, rather than dissolution, was responsible for central displacement of microtubule rings (Fig. 2).

Thin-section techniques have also been useful for evaluation of the platelet contractile system during isometric clot retraction.[27] Under these conditions fibrin strands and platelets become oriented in the long axis of force generation (Fig. 6). Microtubules and actin microfilaments are similarly arranged. The organization of spindle-shaped platelets is strikingly similar to that of skeletal muscle cells. Results of these studies demonstrate that platelets can adapt their contractile system to meet varying demands.

Cytochalasin B (CB) has also proven useful in efforts to define the nature and function of the cytocontractile apparatus by thin-section techniques.[11] The agent has been shown in previous work to inhibit the assem-

---

[22] J. G. White and G. H. R. Rao, *Am. J. Pathol.* **112**, 207 (1983).
[23] M. Steiner and Y. Ikeda, *J. Clin. Invest.* **63**, 443 (1979).
[24] D. M. Kenney and F. C. Chao, *J. Cell. Physiol.* **103**, 289 (1980).
[25] J. G. White, *Am. J. Pathol.* **108**, 184 (1982).
[26] J. G. White and G. H. R. Rao, *Blood* **60**, 474 (1982).
[27] I. Cohen, J. M. Gerrard, and J. G. White, *J. Cell Biol.* **93**, 775 (1982).

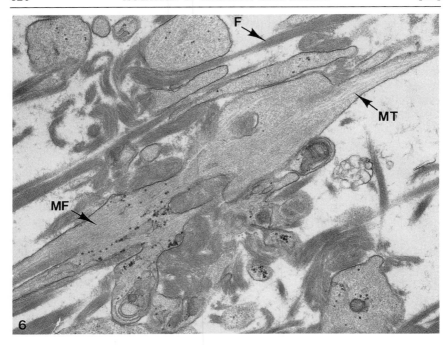

FIG. 6. Clot retraction. The clot shown here was fixed at maximum tension during contraction under isometric conditions. Platelets and platelet pseudopods are oriented together with fibrin (F) strands in the axis of tension. Microtubules (MT) and microfilaments (MF) inside the platelet body are also aligned. The organization strongly resembles a syncytium of smooth muscle cells. Magnification: ×18,900.

bly of platelet actin molecules into filaments.[28] In addition, CB causes dissociation of the freshly formed actin filaments when added after platelet stimulation. However, it does not appear to reduce the 40–50% of actin filaments already present in resting platelets before activation. The platelet response to stimulation was presented above. Following exposure to aggregating agents, the cells lose their resting discoid form, extend pseudopods, and undergo internal contraction. Randomly dispersed cytoplasmic organelles are concentrated in cell centers enclosed within a web of closefitting microtubule coils and microfilaments. Treatment with CB alters the sequential transformation.[11] It makes discoid platelets more spherical in appearance, possibly by influencing the relationship between the actin filament lattice of resting cells and actin-binding protein. By blocking

[28] J. E. B. Fox and D. R. Phillips, *Nature (London)* **292**, 650 (1981).

formation of new actin filaments, CB inhibits formation of pseudopods when treated platelets are exposed to aggregating agents. Addition of the agent to activated platelets that have already undergone shape change causes dissociation of newly formed actin filaments and disappearance of pseudopods. Cytochalasin B does not prevent contraction of the 40–50% assembled actin filaments when resting platelets are stimulated. A dense spot formed by contracted actomyosin is present in stimulated cells. However, CB appears to amputate the connection between the actin filament system and the platelet membrane, microtubule coil, and α granules. As a result, the microtubule coil remains peripherally located and α granules randomly oriented, despite contraction of the actomyosin gel.

Exposure of platelets to CB before chilling added another facet of interest.[29] Although the microtubules dissolved after chilling in CB-treated platelets, the discoid form was preserved in many. Cold causes assembly of actin filaments, a phenomenon prevented by CB. Thus, assembly of actin appears to be as important as disappearance of microtubule coils for the shape change induced by low temperature.

It was pointed out earlier that tropomyosin can protect actin from the damage caused by exposure to osmium.[21] Another protein fragment, heavy meromyosin (HMM), can act as a specific marker of actin in thin-section and whole-mount studies. In addition to selectively labeling actin, HMM also indicates the direction in which the filaments are oriented.[30]

Thus, thin-section techniques have been very helpful for characterizing the cytocontractile system of blood platelets. However, the tendency of osmium to destroy actin filaments makes interpretation difficult. The use of chemical agents that selectively stain actin, influence its assembly into filaments, or protect them from destruction by osmic acid have helped overcome this problem.

Whole-Mount Method

Prior to the introduction of thin-section techniques, the whole-mount method was the only way to examine platelets in the electron microscope. Early studies were helpful in recognizing the transformation of discoid platelets into dendritic and spread forms as a result of interaction with the grid surface.[5,31] The hyaline nature of the cytoplasm, however, prevented resolution of filamentous elements. Shadowing the grids with carbon and platinum or other metals at a fixed angle demonstrated the presence of fibrous elements in the cytoplasm of spread platelets. The application of

---

[29] J. G. White and M. Krumwiede, *Blood* **41**, 823 (1973).
[30] V. Nachmias, J. Sullander, and A. Asch, *Blood* **50**, 39 (1977).
[31] M. Bessis and M. Burstein, *C. R. Soc. Biol. Ses Fil.* **142**, 27 (1948).

negative staining improved the resolution of these elements, and permitted description of the organization of filaments in surface-activated cells.[31] In recent years critical point drying[32] has largely replaced negative staining procedures in an attempt to avoid the damage expected from drying whole-mounted cells in air.[33-35] A wide variety of procedures have been introduced to improve preservation and visualization of the cytoskeleton in platelets prepared by the whole-mount method before negative staining or critical point drying. Fixation is commonly employed while the platelets are in suspension[33] or at intervals after contact with the grid surface.[33-35] This step is usually followed by detergent extraction with Nonidet or Triton X-100, but may be preceded by it. Hoglund et al.[36] introduced the technique of simultaneous fixation and detergent extraction, which has produced excellent results. The use of stereo pair image reconstruction combined with high-voltage or intermediate high-voltage electron microscopy enhances results obtained by the whole-mount method.[37]

The variety of cell preparation procedures, fixation and detergent extraction protocols, staining and critical point drying methods, and image processing techniques can yield quite different, sometimes conflicting results. Detergent extraction, for example, is very helpful for removing membrane and detergent-soluble proteins that obscure filament organization. However, detergent alone, or the combination of Triton X-100 and glutaraldehyde, can perturb the relationships between actin filaments in bundles by removing membrane constituents.[18] The parallel associations between filaments are better preserved if membrane can be left in place.

Another major difference resulting from procedural variations stems from the drying techniques. Samples that are critical point dried[32-35,37] have a cytoplasmic organization very different from those that are negatively stained and dried in air.[18,31,38] The cytoplasm of critical point-dried platelets resembles a trabecular lattice produced by a diversity of filament sizes[39] (Fig. 7). Platelets prepared by simultaneous treatment with glutaraldehyde and Triton X-100, negatively stained, and air dried reveal filaments of a single range of diameters, 5-6 nm. Their organization in parallel bundles

[32] T. F. Anderson, *Trans. N.Y. Acad. Sci.* [2] **13**, 120 (1951).
[33] V. T. Nachmias, *J. Cell Biol.* **86**, 795 (1980).
[34] J. C. Mattson and C. A. Zuichers, *Ann. N.Y. Acad. Sci.* **370**, 11 (1981).
[35] J. C. Lewis, M. S. White, T. Prater, L. G. Taylor, and K. S. Davis, *Exp. Mol. Pathol.* **37**, 370 (1982).
[36] A. S. Hoglund, R. Karlson, E. Arro, B. E. Frederiksson, and U. Lundberg, *J. Muscle Res. Cell Motil.* **1**, 127 (1980).
[37] J. C. Loftus, J. Choate, and R. M. Albrecht, *J. Cell Biol.* **98**, 2019 (1984).
[38] J. G. White and J. J. Sauk, *Blood* **64**, 470 (1984).
[39] J. C. Lewis, M. S. White, T. Prater, K. R. Porter, and R. J. Steele, *Cell Motil.* **3**, 589 (1983).

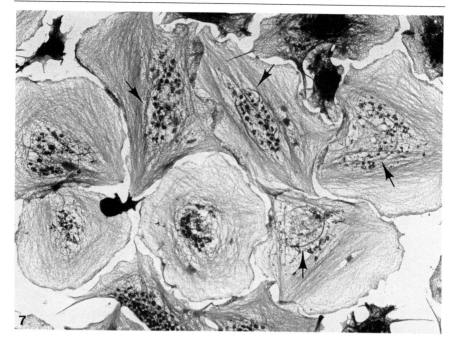

FIG. 7. Critical point drying of whole-mounted platelets. Ultrastructure of spread platelets fixed and simultaneously extracted in detergent 1 hr after surface activation and prepared by critical point drying. Most of the cells in this example are late dendritic or spread forms. Residual coils of microtubules are apparent in three cells (arrows), and substantial remains of dissociated coils or fragments of microtubule rings can be identified in most of the other cells. Actin filaments fill the cytoplasm of the spread cells and are organized in a variety of arrangements. Magnification: ×4500.

and a peripheral weave, as well as other arrangements, are quite different from the trabecular lattice produced by the critical point method. Possible reasons for the differences have been discussed by Small[40] and, more recently, by Bridgman and Reese.[41]

The method we have found most useful combines the simultaneous extraction and fixation introduced by Hoglund et al.[36] with the modifications suggested by Small.[40] The Triton X-100–glutaraldehyde mixture consisted of 0.5% (v/v) Triton X-100 and 0.25% (v/v) glutaraldehyde. Detergent extraction and fixation are carried out at room temperature, 37°, or in the cold. Grids carrying spread platelets are washed briefly in Tris-

[40] J. V. Small, J. Cell Biol. **91**, 695 (1981).
[41] P. C. Bridgman and T. S. Reese, J. Cell Biol. **99**, 1655 (1984).

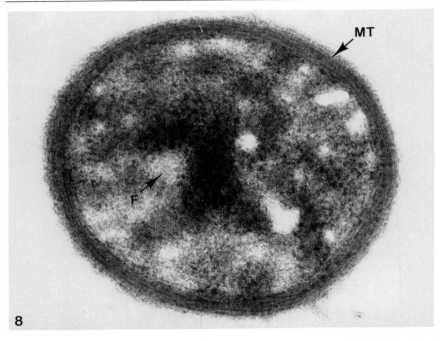

Fig. 8. Negative stain whole mount. Discoid platelet from a drop of C-PRP fixed and detergent extracted 1 min after contact with a grid and then negatively stained. The circumferential band of microtubules (MT) forms the margin of the cell. Only a few assembled actin filaments are apparent in the cytoplasm. Magnification: ×36,000.

buffered saline followed by a cytoskeleton buffer [NaCl, 127 mM; KCl, 5 mM; Na$_2$HPO$_4$, 1.1 mM; KH$_2$PO$_4$, 0.4 mM; NaHCO$_3$, 4 mM; glucose, 5.5 mM; MgCl$_2$, 2 mM; EGTA, 2 mM; piperazine-N,N'-bis(2-ethanesulfonic acid) (PIPES), 5 mM; pH 6.0–6.1). After washing in the cytoskeletal buffer the cells are transferred to the Triton X-100–glutaraldehyde mixture for 1 min. After a brief wash in cytoskeleton buffer, the grids are stored on coverslips in the same buffer containing 2.5% glutaraldehyde for 2 hr before negative staining for electron microscopy. Staining in sodium silicotungstate is carried out at room temperature. Grids are rinsed two times in distilled water and transferred sequentially through 4 drops of bacitracin (40 mg/ml in water; Sigma Chemical Co., St. Louis, MO) in a plastic petri dish and drained briefly on the edge with filter paper. They are then passed through four drops of 3% (w/v) sodium silicotungstate, finally drained of excess stain, and allowed to air dry.

This resulted in preservation of microtubule coils in discoid and den-

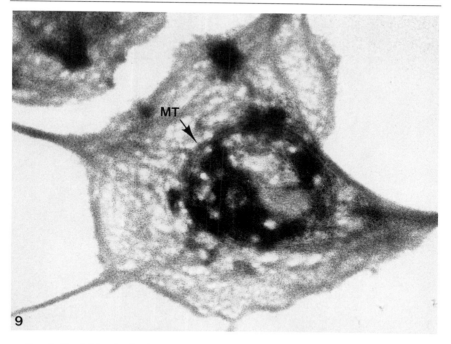

FIG. 9. Dendritic platelet from sample prepared in the same manner as the cell in Fig. 8 after contact with the grid for 30 min. The margin of the cell, pseudopods, and cytoplasm contain numerous filaments. Microtubule rings (MT) are constricted in the cell center. Magnification: ×16,800.

dritic and most spread platelets, which other procedures had failed to do (Figs. 8–10). Negative staining revealed several different arrangements of actin, but only one average diameter for the filaments. Some perturbation in microtubule substructure was noted in this procedure, and removal of membranes, air drying, and glutaraldehyde fixation all influence actin filament structure and organization. Further refinements will undoubtedly improve these results.

Immunocytochemistry

Identification of specific protein constituents associated with the contractile cytoskeleton of platelets has depended largely on biochemical, immunochemical, and immunofluorescence techniques. Immunofluorescence microscopy has been most useful for subcellular localization of

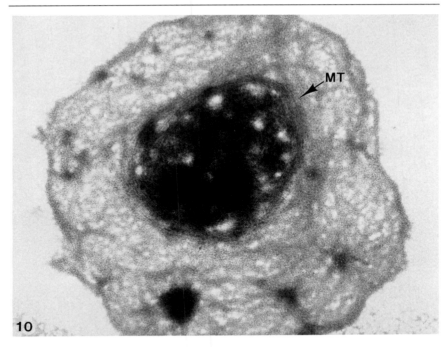

FIG. 10. Negatively stained whole mount of spread platelet prepared as described in Figs. 8 and 9. The cytoplasm and peripheral margin contain numerous assembled actin filaments. Microtubule rings (MT) are constricted in the central region of the cell. Magnification: ×16,800.

platelet contractile proteins.[42,43] Actin, myosin, tropomyosin, α-actinin, filamin, and vinculin have been identified in resting and surface activated platelets by this approach. Attempts to characterize the contractile cytoskeleton at the ultrastructural level with the aid of immunocytochemistry have been limited. Puszkin et al.[44] used ferritin-labeled antibodies to localize α-actinin in platelets treated with saponin. The label appeared to be concentrated just under the cell membrane. Loftus and Albrecht[45] have followed redistributin of the fibrinogen receptor on the outside surface of platelets after surface activation. Receptor redistribution was accompa-

[42] E. Bebus, K. Weber, and M. Osborn, Eur. J. Cell Biol. 24, 45 (1981).
[43] H. H. Billett, C. S. P. Jenkins, J. Maimon, T. H. Spaet, and E. G. Puszkin, J. Lab. Clin. Med. 103, 534 (1984).
[44] E. G. Puszkin, C. S. P. Jenkins, C. Ores-Carton, and M. B. Zucker, J. Lab. Clin. Med. 105, 52 (1985).
[45] J. C. Loftus and R. M. Albrecht, J. Cell Biol. 99, 822 (1984).

nied by cytoskeletal reorganization. The association of fibrinogen receptors with the contractile cytoskeleton provides an indirect way of following rearrangement after platelet stimulation. It will be of interest to see if other surface antigens undergo a similar redistribution in activated platelets.

Summary

Methods for evaluating the organization, subcellular distribution, and rearrangements in specific constituents of the platelet contractile cytoskeleton are rapidly being developed. The procedures presented and discussed here have yielded useful information. However, a great deal more is required before the role of contractile elements of the cytoskeleton in platelet physiology will be fully understood. The new approaches employing better preservation of platelet structure and immunocytochemistry will undoubtedly provide this knowledge in the near future.

Section II

Platelet Receptors: Assays and Purification

A. Receptors for Platelet Agonists
Articles 13 through 20

B. Receptors for Adhesive Proteins
Articles 21 through 27

C. Receptors for Clotting Factors and Other Ligands
Articles 28 through 33

D. General Approaches to Receptor Analysis
Articles 34 through 39

[12] Repertoire of Platelet Receptors

By JACEK HAWIGER

The main function of platelets is to plug holes in leaky blood vessels. This physiologically useful function has been pathologically subverted when, instead of cuts in blood vessels, platelets encounter breaks in cholesterol-rich atherosclerotic deposits. Then, they adhere to the ruptured atherosclerotic plaque, form aggregates, and effectively block blood flow through the narrowed vessel. Both physiologic and pathologic responses of platelets involve a number of agonists, adhesive proteins, and clotting factors. All of them interact with their respective receptors on the platelet membrane. The repertoire of receptors on platelets is astounding when one considers their tiny size, 14 times smaller than that of an erythrocyte.

Agonists that are generated at the zone of vascular injury include ADP, thrombin, epinephrine, thromboxane A_2, and platelet-activating factor (PAF) in the fluid phase and collagen and other components of the extracellular matrix in the subendothelium.[1] Thrombin and thromboxane A_2 are the most potent agonists for human platelets. The detectable activation of platelets separated from plasma proteins can be observed at the concentration of thrombin as low as 0.05 NIH unit (U)/ml.[2] Thrombin interacts with a specific receptor belonging to the superfamily of the receptors that are coupled to G proteins and phospholipase C and are distinguished by seven transmembrane segments. Only proteolytically active thrombin can generate a signal on binding to the receptor because its extracellular aminoterminal piece must be cleaved before the receptor is sterically rearranged to evoke the signal transduced by G protein(s) and phospholipase C.[3] This unique mode of thrombin receptor "activation" distinguishes it from other G protein-coupled receptors with seven transmembrane domains: α_2-adrenergic receptor, serotonin receptors, and platelet-activating factor receptor. Although platelets reportedly possess only about 200 copies of the α_2-adrenergic receptor,[4] its role in the regulation of the adhesive receptor function of the GPIIb–IIIa complex and the Na^+/H^+ exchange is

---

[1] J. Hawiger, this series, Vol. 169, p. 191.
[2] T. Fujimoto, S. Ohara, and J. Hawiger, *J. Clin. Invest.* **69**, 1212 (1982).
[3] T. K. Vu, D. T. Hung, V. I. Wheaton, and S. R. Coughlin, *Cell (Cambridge, Mass.)* **64**, 1057 (1991).
[4] R. Kerry and M. C. Scrutton, in "The Platelets: Physiology and Pharmacology" (G. L. Longercher ed.), p. 113. Academic Press, Orlando, Florida, 1985.

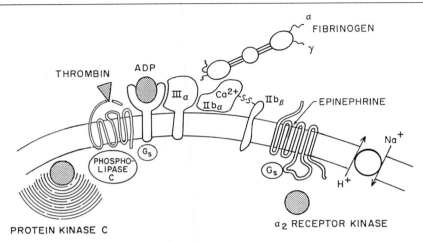

FIG. 1. Diagrammatic presentation of three receptors for agonists on human platelets involved in activation of the fourth receptor, the integrin GPIIb–IIIa, from the nonbinding to the binding mode. The thrombin receptor and the α_2-adrenergic receptor for epinephrine belong to the superfamily of receptors with seven putative transmembrane domains. They are coupled to G_s protein and phospholipase C. The cytoplasmic loops of the epinephrine receptor can be phosphorylated by α_2-adrenergic receptor kinase or possibly by protein kinase C. The Na^+/H^+ pump is involved in epinephrine-evoked effects. The ADP receptor, classified pharmacologically as purinergic P_{2y} receptor, awaits molecular identification. It is distinct from the fibrinogen receptor, the integrin GPIIb–IIIa, although it is intimately involved in its regulation (see text for more details).

prominent.[5-9] Like the thrombin receptor, the α_2-adrenergic receptor spans the cell membrane seven times, threading α-helix segments composed of 26 predominantly hydrophobic amino acids.[10] There is a functional link between α_2-adrenergic receptor and the GPIIb–IIIa complex, which is occupied by fibrinogen in order to evoke a signal related to an increase in the Ca^{2+} level and alkalinization of the cytoplasm (Na^+/H^+ antiporter).[7,8] The serine- and threonine-rich third cytoplasmic domain of the α_2-adrenergic receptor is a potential substrate for receptor specific kinase, or the ubiquitous protein kinase C (Fig. 1).

[5] J. S. Bennett and G. Vilaire, *J. Clin. Invest.* **64**, 1393 (1979).
[6] E. I. Peerschke, *Blood* **60**, 71 (1982).
[7] J. D. Sweatt, I. A. Blair, E. J. Cragoe, and L. E. Limbird, *J. Biol. Chem.* **261**, 8660 (1986).
[8] H. S. Banga and E. R. Simons, *Proc. Natl. Acad. Sci. U.S.A.* **83**, 9197 (1986).
[9] S. J. Shattil, A. Budzynski, and M. C. Scrutton, *Blood* **73**, 150 (1989).
[10] B. K. Kobilka, H. Matsui, T. S. Kobilka, T. L. Yang-Fang, U. Francke, M. G. Caron, R. J. Lefkowitz, and J. W. Regan, *Science* **238**, 650 (1987).

The platelet receptor for serotonin [5-hydroxytryptamine (5-HT)] belongs to the distinct 5-HT-2 subtype of serotonin receptors present also in the cerebral cortex.[11] It is linked to activation of phospholipase C and increased phosphoinositol turnover. Stimulation of 5-HT-2 receptor evokes shape change in human platelets and irreversible aggregation of cat platelets. It is blocked by nanomolar concentrations of ketanserin. Molecular cloning and deduced amino acid sequence of 5-HT-2 receptor led to the realization that together with the 5-HT-1c subtype, they form a family distinct from the 5-HT-1a subtype. All of them, however, are members of the superfamily of G protein-coupled neurotransmitter and hormone receptors. They have their amino-terminal part located outside of the cell and seven hydrophobic membrane-spanning α helices linked together by three extracellular and three intracellular loops.[12] Related receptors in this superfamily are the visual opsins, muscarinic cholinergic receptor, tachykin receptors, dopamine D_2 receptors, yeast pheromone receptors, and adrenergic receptors, described above.

Serotonin bound to human platelets is transported to the dense granules through two distinct uptake mechanisms, very similar to those in neurons.[13,14] The first uptake mechanism carries serotonin through the plasma membrane. It is coupled to a ouabain-sensitive Na^+, K^+-ATPase and it is inhibited by antidepressants such as imipramine. The second uptake mechanism translocates serotonin to dense granules. It requires a carrier and an electrochemical proton gradient operated by a H^+-translocating ATPase. This mechanism is blocked by reserpine. Its putative carrier protein was affinity labeled with azido[^{125}I]iodoketanserin in isolated rabbit platelet dense granules.[15]

Platelet-activating factor (PAF) is a unique phospholipid-like agonist with potent platelet aggregatory, proinflammatory, and smooth muscle contractile activities.[16] Complementary DNA for PAF receptor from guinea pig lung was isolated and functionally expressed.[17] The deduced sequence indicates that the receptor is made of 342 amino acids (M_r 38,982)

---

[11] P. L. Bonate, *Clin. Neuropharmacol.* **14**, 1 (1991).
[12] D. Julius, *Proc. Natl. Acad. Sci. U.S.A.* **87**, 928 (1990).
[13] A. Pletscher, in "Essays in Neurochemistry and Neuropharmacology" (M. B. H. Yondim, W. Lovenberg, D. F. Sharman, and J. R. Lagnado, eds.), Vol. 3, p. 49. Wiley, Chichester, 1977.
[14] G. Rudnick, H. Fishkes, P. J. Nelson, and S. Schuldiner, *J. Biol. Chem.* **255**, 3638 (1980).
[15] A. M. Cesura, A. Bertocci, and M. Da Parda, *Eur. J. Pharmacol.* **186**, 95 (1990).
[16] R. E. Whatley, G. A. Zimmerman, T. M. McIntyre, and S. M. Prescott, *Prog. Lipid Res.* **29**, 45 (1990).
[17] Z. Honda, M. Nakamura, I. Miki, M. Minami, T. Watanabe, Y. Seyama, H. Okado, H. Toh, K. Ito, T. Miyamoto, and T. Shimizu, *Nature (London)* **349**, 342 (1991).

forming 7 hydrophobic putative transmembrane segments characteristic of G protein-coupled receptors.

The ADP receptor on platelets is assigned to the P_{2y} class of purinergic receptors, while the P_1 class recognizes adenosine.[18,19] The signal generated by ADP induces rapid shape change in platelets, exposure of binding sites for fibrinogen and von Willebrand factor (vWF) on GPIIb–IIIa, and platelet aggregation. In addition, the ADP receptor mediates the inhibition of adenylate cyclase in platelets stimulated by PGE_1.[20] Among the nucleoside di- and triphosphates, GDP is a much less potent agonist than ADP; GTP and UTP are much less potent antagonists than ATP.[18] In spite of its important role in activation of platelets, there is a surprising lack of firm data on the biochemical characterization of the ADP receptor in platelets. The reported number of bound ligand molecules ranged from less than 1000 to 160,000 per platelet, and the affinity of binding ranged from nanomolar to micromolar.[21] To overcome the difficulties due to degradation of ligands (ATP and ADP) by membrane ectoenzymes,[21,22] paraformaldehyde-fixed platelets were employed to study the binding of ADP and its analogs.[21] However, such "fixed," metabolically inert platelets are not suitable for correlation of receptor occupancy with stimulus–response coupling. They displayed a very high and low number of high-affinity sites for [^{3}H]ADP (160,000 and 400,000, respectively).[21] The putative platelet receptor of M_r 100,000 was labeled by the covalent probe, 5'-p-fluorosulfonylbenzoyladenosine (FSBA), which blocked platelet shape change, aggregation, and binding of fibrinogen. The relationship of FSBA-labeled protein ("aggregin") to GPIIIa has been analyzed.[23] The ADP receptor can generate a signal transduced by G proteins and phospholipase C (Fig. 1).

Platelets are programmed to recognize and seal any breaks in continuity of the endothelial cell carpet lining the vascular wall. This function implies the existence of a molecular recognition mechanism ("sensors") signaling to platelets that the subendothelial extracellular matrix is exposed as a result of endothelial cell injury or detachment. This role of "molecular sensors" is performed by platelet receptors for adhesion molecules in the

---

[18] L. Needham and N. J. Cusack, *Eur. J. Pharmacol.* **134**, 9 (1987).
[19] G. Burnstock, in "Cell Membrane Receptors for Drugs and Hormones: A Multi-disciplinary Approach" (R. W. Strub and L. Balis, eds.), p. 107. Raven Press, New York, 1978.
[20] J. Hawiger, M. L. Steer, and E. W. Salzman, in "Hemostasis and Thrombosis: Basic Principles and Clinical Practice" (R. W. Colman, J. Hirsh, V. J. Marder, and E. W. Salzman, eds.), Chapter 40, p. 710. Lippincott, Philadelphia, 1987.
[21] J. R. Jefferson, J. T. Harmon, and G. A. Jamieson, *Blood* **71**, 110 (1988).
[22] J. D. Pearson, L. L. Slakey, and J. L. Gordon, *Biochem J.* **214**, 273 (1983).
[23] R. W. Colman, W. R. Figures, Q. X. Wu, S. Y. Chung, T. A. Morinelli, G. P. Tuszynski, R. F. Colman, and S. Niewiarowski, *Arch. Biochem. Biophys.* **262**, 298 (1988).

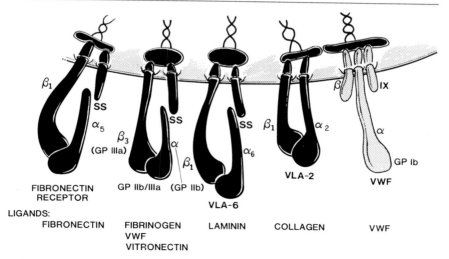

FIG. 2. Integrin and nonintegrin receptors on human platelets. This diagram depicts a fragment of a platelet with its membrane studded with receptors for adhesive proteins. The integrin receptors shown in black are Ca^{2+}-held heterodimers made of two nonidentical subunits, α and β. The external amino-terminal parts of the receptors interact with adhesive proteins in blood and/or extracellular matrix. The intracellular, carboxyl-terminal parts interact with the cytoskeleton. Among the seven distinct integrin in β subunit families, at least two families are represented in platelets: β_1 and β_3. The β_1 family is represented by $\alpha_2\beta_1$ (collagen receptor), $\alpha_5\beta_1$ (fibronectin receptor), and $\alpha_6\beta_1$ (laminin receptor). The β_3 integrin family is represented by $\alpha_{IIb}\beta_3$ (GPIIb–IIIa receptor for fibrinogen, vWF, and vitronectin). The nonintegrin receptor is shaded. It is composed of a complex of GPIb and GPIX. The GPIb, made of α and β subunits, has the extracellular amino-terminal part interacting with vWF and the intracellular carboxyl-terminal part interacting with the cytoskeleton.

extracellular matrix. These receptors belong to integrins and nonintegrins (Fig. 2). The nonintegrin receptor GPIb–IX complex interacts with von Willebrand factor (vWF) deposited in the extracellular matrix. We are still baffled with the apparent inability of this receptor to bind vWF in plasma, while avidly interacting with vWF in the exposed subendothelial extracellular matrix. Thus, this nonintegrin receptor, along with members of the β_1 and β_3 families of integrins, is kept in check by intact endothelial cells barring access to the highly reactive extracellular matrix. The collagen receptor, made of glycoproteins Ia and IIa ($\alpha_2\beta_1$ in integrin nomenclature), is unusual.[24] It recognizes collagen deposited in the extracellular matrix

[24] W. D. Staatz, J. J. Walsh, T. Pexton, and S. A. Santoro, *J. Biol. Chem.* **265**, 4778 (1990).

and it generates a signal that is transduced in the platelet through a poorly understood mechanism to induce secretion.

The procoagulant function of platelets has been recognized for a long time. A putative platelet "procoagulant" was coined "platelet factor 3" on the basis of accelerated clotting of platelet-rich plasma in the presence of the surface-active agent, kaolin.[25,26] Elegant advances in the protein chemistry and enzymology of clotting factors have led to the realization that the assembly of the "prothrombinase" complex, generating thrombin from prothrombin, and assembly of the "tenase" complex, generating factor Xa from its zymogen, are preferentially taking place on the membranes of activated platelets with a resultant 300,000 times increase in the rate of thrombin generation.[27,28] Platelet factor V expressed on the membrane of activated platelets plays a key role in the assembly of the prothrombinase complex. Among the known platelet agonists, thrombin, collagen, and calcium ionophore A23187 were able to induce the assembly of the prothrombinase complex on the platelet membrane undergoing vesiculation.[29] Among other receptors on platelets we will address in this section are those involved in binding of insulin and low-density lipoproteins.

The methodology of platelet receptors follows general approaches to their characterization, identification, purification, reconstitution, cDNA cloning, and sequencing. Then, structure–function analysis leads to establishing the structural basis for molecular recognition of ligands, receptor links to signal transducers, and regulation of receptors from the nonbinding to the binding mode. The negative regulation of receptors is so far attributed to adenylate cyclase and guanylate cyclase and cyclic nucleotide-activated kinases.[20]

Although platelet receptors dominate the pages of this section of the volume, several methods and experimental approaches transcend the boundaries of platelets, so richly endowed in representatives of many receptor families. These methods can be applied to other cells such as megakaryocytes, endothelial cells, and neurons because they share with platelets many common features of receptors involved in their nonadhesive and adhesive functions.

---

[25] T. H. Spaet and J. Cintron, *Br. J. Haematol.* **11**, 269 (1965).
[26] P. Gautheron, E. Dumont, and S. Renaud, *Thromb. Diath. Haemorrh.* **32**, 382 (1974).
[27] S. Singh, D. G. Lowe, D. S. Thorpe, H. Rodriguez, W. J. Kuang, L. J. Dangott, M. Chinkers, D. V. Goeddel, and D. L. Garbers, *Nature (London)* **334**, 708 (1988).
[28] E. L. Gordon, J. D. Pearson, and L. L. Slakey, *J. Biol. Chem.* **261**, 15496 (1986).
[29] P. J. Sims and T. Wiedmer, *J. Biol. Chem.* **264**, 17049 (1989).

[13] 2-Methylthioadenosine [β-^{32}P]Diphosphate: Synthesis and Use as Probe of Platelet ADP Receptors

By Donald E. Macfarlane

Introduction

Adenosine diphosphate has two distinguishable actions on platelets.[1,2] The first is the initiation of a chain of events leading to platelet aggregation, and the second is the blockade of the activation of adenylate cyclase by such agents as prostacyclin.[3] There is some evidence that these two effects are mediated by the different receptors (see Conclusion, below).

In order to determine if these effects are mediated by classical pharmacological receptors, we have synthesized several radioactive analogs of ADP that have substantially greater agonist potency than natural ADP. One of these analogs was the photolabile molecule, 2-azidoadenosine diphosphate. This reagent exists as a mixture of two or more slowly interconverting forms in aqueous solution, which complicates its use as a radioligand.[4]

2-Methylthioadenosine diphosphate (2-MeSADP) was first synthesized by Gough *et al.* and shown to be an aggregating agent of sheep platelets with a potency five times that of ADP.[5] Several synthetic routes have been reported.[5,6] We found the following to be technically straightforward, and it uses the freely available adenosine N^1-oxide as its starting material (Scheme 1).[7] The overall yield is adequate for the purpose at hand.

Adenosine N^1-oxide (**I**) (1 g, 3.4 mmol; Sigma Chemical Co., St. Louis, MO) is refluxed with 5 N NaOH (5 ml) for 15 min, cooled, and diluted with 30 ml H$_2$O. The mixture is passed through a Dowex 50 column (NH$_4^+$ form, 50-ml bed volume) and eluted further with 100 ml H$_2$O. The pooled eluate is evaporated at 40° and coevaporated with ethanol in a rotary evaporator to yield an amber-colored gum (**II**).[8] This is dissolved

---

[1] D. E. Macfarlane, *in* "Platelet Function and Metabolism" (H. Holmsen, ed.), CRC Press, Boca Raton, Florida, 1987.
[2] R. J. Haslam and N. J. Cusack, *in* "Purinergic Receptors" (G. Burnstock, ed.), p. 221. Chapman & Hall, London, 1987.
[3] D. E. Macfarlane and D. C. B. Mills, *J. Cyclic Nucleotide Res.* **7**, 1 (1981).
[4] D. E. Macfarlane, D. C. B. Mills, and P. C. Srivastava, *Biochemistry* **21**, 544 (1982).
[5] G. Gough, M. H. Maguire, and F. Penglis, *Mol. Pharmacol.* **8**, 170 (1972).
[6] N. J. Cusack and S. M. O. Hourani, *Br. J. Pharmacol.* **77**, 329 (1982).
[7] D. E. Macfarlane, P. C. Srivastava, and D. C. B. Mills, *J. Clin. Invest.* **71**, 420 (1981).
[8] K. Kikugawa, H. Suehiro, R. Yanase, and A. Aoke, *Chem. Pharm. Bull.* **25**, 1959 (1977).

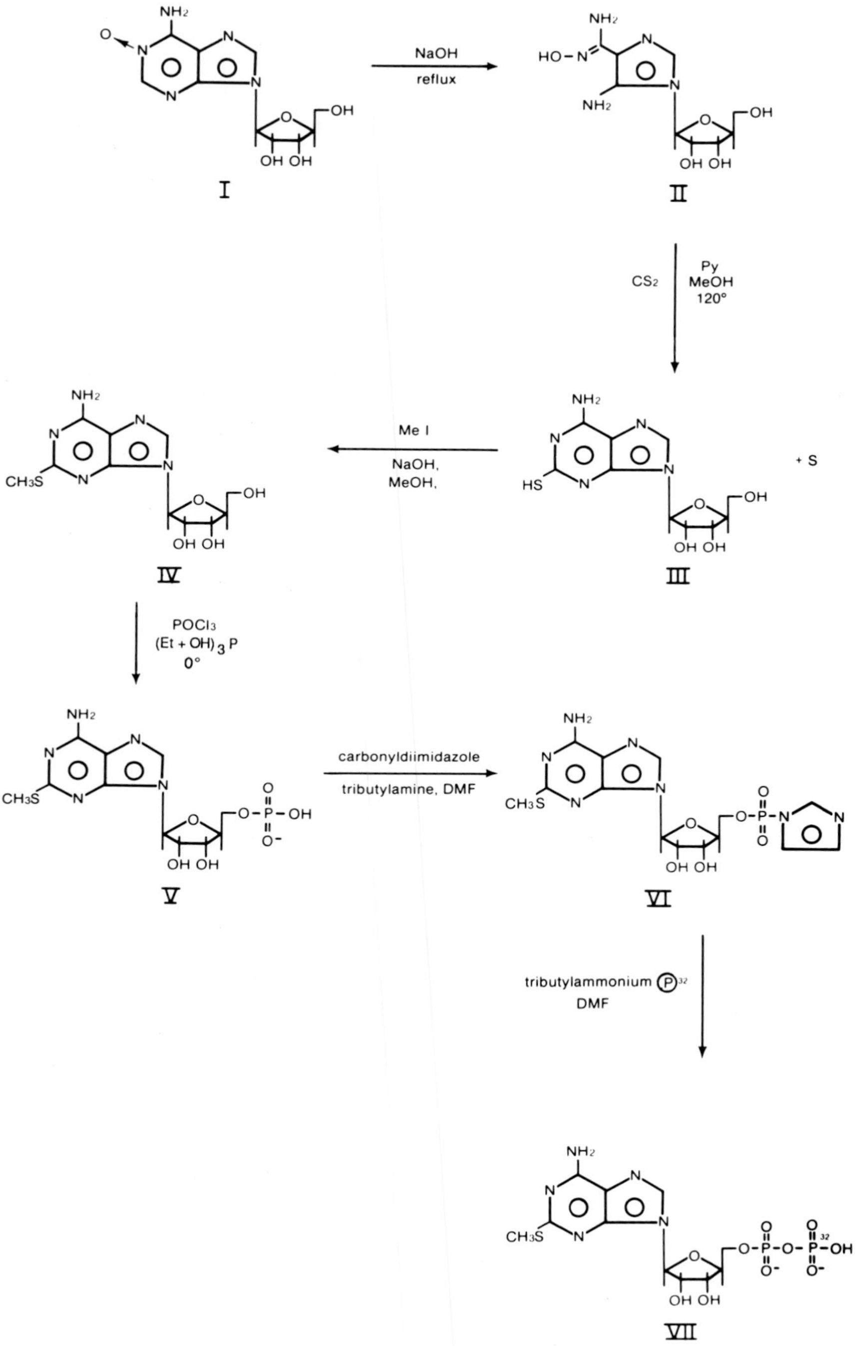

SCHEME 1. Synthesis of 2-methylthioadenosine [β-^{32}P]diphosphate.

in a mixture of 10 ml methanol, 10 ml pyridine, and 5 ml carbon disulfide, and heated in a Parr reaction bomb (80-ml volume) at 120° for 6 hr. The reaction mixture is evaporated and triturated with 30 ml acetone. The insoluble residue is collected by centrifugation, and is extracted with 8 ml 2.5 N NH$_4$OH. The remaining insoluble material (sulfur) is removed by filtration, and the extract is added to 24 ml n-butanol–acetic acid (2:1). The mixture is allowed to stand at 4°, whereupon golden yellow needles of 2-thioadenosine (**III**) separate, and are collected and dried on filter paper (yield about 35%).[8]

2-Thioadenosine [**III**, 317 mg (1 mmol)] is dissolved in 2 ml H$_2$O, 2 ml methanol, and 1 ml 1 N NaOH, and cooled on ice. Methyl iodide [0.4 ml (4 mmol)] is then added, and the reaction mixture is cooled on ice for 3 hr. The separated product, 2-methylthioadenosine (**IV**), is collected by filtration, washed with ice-cold water, dried (yield 90%), recrystallized from warm methanol, and dried over P$_2$O$_5$.

2-Methylthioadenosine [66 mg, (200 μmol)] is added to 1 ml dry triethyl phosphate at 0° and 120 mg (800 μmol) freshly distilled phosphoryl chloride.[9] The mixture is stirred on ice in a tightly capped reaction vessel for 3.5 hr. Then the reaction mixture is poured into peroxide-free diethyl ether (100 ml) and the precipitate is collected by centrifugation and washed with diethyl ether. The precipitate is then dissolved in water and neutralized with 2 N NaOH. This solution is passed through a Dowex 50 (H$^+$) column (2.5-ml bed volume) followed by 50 ml H$_2$O. 2-Methylthioadenosine 5′-monophosphate (**V**) is then eluted from the column with 25 ml 45% acetic acid, which is evaporated with a rotary evaporator and coevaporated with H$_2$O to yield crystalline 2-methylthioadenosine 5′-monophosphate (**V**). Yield is 45 mg.

2-Methylthioadenosine 5′-monophosphate [1.2 mg (2 μmol)] and tributylamine [0.55 mg (2 μmol)] were dissolved in dry dimethylformamide and evaporated to dryness under reduced pressure in a vacuum desiccator. The residue is dissolved in 30 μl dry dimethylformamide, and N,N'-carbonyldiimidazole [2.4 mg (15 μmol)] is added, and the reaction mixture is allowed to stand with the exclusion of moisture to generate the phosphoramidate (**VI**).[10]

Separately, 10 mCi ortho[^{32}P]phosphate, carrier free, is added to 20 μl of 50 mM tributylammonium phosphate (1 μmol) and rendered anhydrous by coevaporation with dry dimethylformamide *in vacuo* (see Precautions, below).

The reaction mixture containing **VI** is treated with 1.2 μl methanol to

---

[9] M. Yoshikawa, T. Kato, and T. Takenishi, *Tetrahedron Lett.* **50,** 5065 (1967).
[10] D. E. Hoard and D. C. Ott, *J. Am. Chem. Soc.* **87,** 1785 (1965).

decompose the excess carbonyldiimidazole and diluted with 41 µl dry dimethylformamide. Then 8 µl of this reaction mixture (containing 0.33 µmol **VI**) is added to the radioactive tributylammonium phosphate (10,000 Ci/µmol) and allowed to react at room temperature for 20 hr with the exclusion of moisture. The reaction mixture can then either be applied to an anion-exchange high-performance liquid chromatography column (e.g., Partisil SAX; Whatman, Clifton, NJ) and eluted with a linear gradient of 10–700 mM $NH_4H_2PO_4$, or applied as a streak to a high-voltage paper electrophoresis apparatus (such as Shandon L24, using 50 mM sodium citrate, pH 4.05 as the electrolyte). In the latter case, the radioactive streak corresponding to the streak of UV-absorbing 2-methylthioadenosine 5'-diphosphate (**VII**) running slightly faster than ADP is cut out, and eluted by descending chromatography with 5% acetic acid (see Precautions, below).

The final product is assayed either by its UV absorption (λ_{max} 271, molar extinction 15,300) or by bioassay using platelet aggregation in comparison with authentic, nonradioactive 2-MeSADP.

Precautions

The above procedure requires handling potentially dangerous amounts of ^{32}P, especially as the reactions are carried out in small volumes. The reaction vessels should be covered with 1-cm thick clear plastic to absorb the high-energy β-particles, and steel clamps, etc., should not be used to avoid the generation of X-rays by Bremsstrahlung interaction. After each manipulation, a Geiger counter should be used to check for contamination of hands and instruments. Both phosphorylation steps require the strict exclusion of moisture, and it might be found convenient to perform them in a clear plastic glass box flushed with dry nitrogen. Several practice runs should be performed with a trace amount of ^{32}P before attempting a synthesis at high specific activity. Great care should be taken with handling the paper electrophoresis strips to ensure that the exposure of hands and eyes to the uncovered radioactivity is limited to a few seconds. Any stainless steel HPLC column that is used for purification of the final product will become heavily contaminated with ^{32}P. It should be used for no other purpose, and should be stored and used behind lead shields.

Binding Studies

The binding of [^{32}P]2-MeSADP to the exterior aspect of platelets can be observed using either platelet-rich plasma or washed platelet suspensions (see Fig. 1).[7] The majority of our work was done in platelet-rich

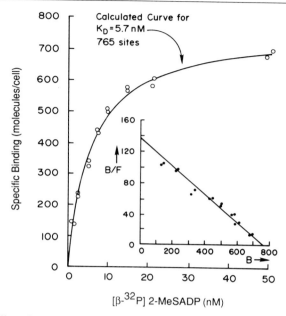

FIG. 1. Binding of 2-methylthioadenosine [β-^{32}P]diphosphate to intact washed platelets. The insert is a Scatchard plot of the data. From Ref. 6 with permission, copyright American Society of Clinical Investigation, Inc.

plasma, because the pharmacological responses of platelets (such as aggregation and the accumulation of cyclic AMP) are more reproducible when the platelets are suspended in their own plasma, and such platelets are less likely to have been stimulated to release their stored ADP than are washed platelets.

Platelets have 500–1200 binding sites for 2-MeSADP per platelet, all apparently presenting the same affinity for the ligand. We found no evidence for high- and low-affinity states. The rate of dissociation of the radioligand is rapid (about 1.4 min^{-1} at 23°), and thus some care is required when separating bound radioactivity from free radioactivity to prevent dissociation of specifically bound material.[7]

We found it convenient to use centrifugation through a mixture of silicone oils [Dow Corning (Midland, MI) #550 and #200, 1 cs] with density about 1.028. This results in a pellet of platelets carrying with them about 0.2 μl plasma/10^8 platelets. This volume is adequate to account for all of the observed "nonspecific" binding.[7]

Conclusion

[^{32}P]2-MeSADP is an excellent radioligand that identifies a single class chain of binding sites on the surface of the platelets. It is a powerful aggregating agent, and (like natural [^{32}P]ADP) it prevents the accumulation of cyclic AMP within platelets exposed to such agents as prostacyclin, presumably by activating a receptor that switches the adenylate cyclase off. The latter effect,[3] and the binding of 2-MeSADP,[7] is inhibited by incubating the platelets with *p*-chloromercuribenzene sulfonate, a sulfhydryl reagent. This reagent does not block the induction of the "shape change" by ADP, suggesting that this pharmacological effect of ADP is mediated by a receptor other than the one inhibiting the adenylate cyclase.

[^{32}P]2-MeSADP is unusual as a pharmacological radioligand in that it is a full agonist: it can be used with intact cells, it is highly polar, and its binding is very rapid. In the case of several receptors that activate the adenylate cyclase, studies with antagonist radioligands suggest that agonists initially bind to a high-affinity state of the receptor that is rapidly converted to a low-affinity form, perhaps as a prelude to receptor desensitization. This does not seem to occur with the ADP receptor on blood platelets, since the binding of the agonist [^{32}P]2-MeSADP yields straight Scatchard plots and the dissociation constant at equilibrium is not higher than the dissociation constant calculated from the initial rates of association and dissociation.[7] The absence of detectable high-affinity states of the receptor together with the finding of "receptor reserve" have been used to buttress the hypothesis that receptors in general act catalytically to switch the adenylate cyclase off or on.[11,12]

Acknowledgment

This work was supported by a Grant-in-Aid of the American Heart Association and grants from the Veterans Administration.

---

[11] D. E. Macfarlane, *Trends Pharmacol. Sci.* **5**, 11 (1984).
[12] D. E. Macfarlane, *Mol. Pharmacol.* **22**, 580 (1982).

[14] Interaction of Nucleotide Affinity Analog 5'-p-Fluorosulfonylbenzoyladenosine with Platelet ADP Receptor: Aggregin

By WILLIAM R. FIGURES, L. MARIE SCEARCE, ROBERTA F. COLMAN, and ROBERT W. COLMAN

Introduction

The initial response of human blood platelets to various agonists is a morphological change, in which the cells undergo a transformation from disks to spheres with the simultaneous appearance of cytoplasmic projections or pseudopodia. This transformation is visualized using scanning electron microscopy or detected optically as a decrease in the light transmission of stirred platelet suspensions. ADP, a potent inducer of platelet shape change as well as platelet aggregation, effects these responses presumably due to binding of the nucleotide to a cell surface receptor.

The identification of an ADP receptor responsible for the induction of ADP-mediated shape change and aggregation has been the subject of a number of investigations. Born and Feinberg[1] measured 88,000 "high-affinity sites" for [^{14}C]ADP on platelets in plasma. Because [^{14}C]ADP added to intact platelets can be converted to [^{14}C]ATP,[2] the results of these experiments are difficult to interpret as ATP is a known competitive antagonist of ADP. Binding of [^{14}C]ADP to isolated plasma membranes was measured by Nachman and Ferris.[3] A Scatchard plot of the binding data demonstrated 100,000 binding sites for ADP. Because at least two known nucleotide-binding proteins, actin and myosin, are located on the inner surface of the platelet membrane,[4] the actual binding to an exofacial ADP receptor would be difficult to extract from this data. Adler and Handin[5] have isolated an ADP-binding protein (M_r 60,000) from platelet membranes by freezing and thawing. No further studies have been reported relating this protein to ADP-induced shape change or aggregation

---

[1] G. V. R. Born and M. Feinberg, *J. Physiol. (London)* **251**, 803 (1975).
[2] J. F. Mustard, M. A. Packham, D .W. Perry, M. A. Guccione, and R. L. Kinlough-Rathbone, in "Biochemistry and Pharmacology of Platelets" (K. Elliot, ed.). p. 72. Elsevier, New York, 1975.
[3] R. L. Nachman and B. J. Ferris, *Biochemistry* **249**, 704 (1974).
[4] J. S. Bennett, G. Vilaire, R. F. Colman, and R. W. Colman, *J. Biol. Chem.* **256**, 1185 (1981).
[5] J. R. Adler and R. I. Handin, *J. Biol. Chem.* **254**, 3866 (1979).

or to an ADP receptor protein in intact platelets. Lipps et al.[6] demonstrated two classes of ADP-binding sites on the platelet. The binding constant of the high-affinity site was in the range of ADP concentration necessary to promote platelet shape change and aggregation. Most of these studies have rendered little information concerning the molecular nature of the receptor.

The nucleotide affinity analog, 5'-p-fluorosulfonylbenzoyladenosine (FSBA), has been demonstrated to modify covalently enzymatic and allosteric regulatory sites[7-17] requiring nucleotides as well as the ATP-binding sites of actin and myosin on the internal surface of the platelet membrane.[4] Bennett et al.[18] have presented evidence that FSBA can covalently incorporate into a single polypeptide (M_r 100,000) on the external surface of the platelet membrane. The labeling of this protein demonstrates considerable specificity as it is protected by ADP and its competitive inhibitor ATP, but not by adenosine, AMP, or epinephrine in experiments carried out with platelet membranes. In the same study, FSBA was shown to induce a rapid inhibition of ADP-induced shape change that was progressive with time of incubation with FSBA. Figures et al.[19] have demonstrated that incubation of intact washed platelets with FSBA resulted in the inhibition of ADP-mediated aggregation as well as exposure by ADP of latent fibrinogen-binding sites. A single site of covalent incorporation was observed concomitantly in the washed platelet preparations. The inhibition of ADP-induced platelet activation coupled with the modification of an externally oriented membrane protein makes this protein an attractive candidate for an adenine nucleotide receptor site that mediates shape change and aggregation as well as exposure of fibrinogen receptors in human platelets. The use of the technique outlined in this chapter should facilitate investigation of the interactions of this protein with other receptors. Furthermore,

---

[6] J. P. M. Lipps, J. J. Sixma, and M. E. Schiphorst, *Biochim. Biophys. Acta* **628**, 451 (1980).
[7] R. F. Colman, P. K. Pal, and J. Wyatt, this series, Vol. 42, p. 240.
[8] P. K. Pal, W. J. Wechter, and R. F. Colman, *J. Biol. Chem.* **250**, 8140 (1975).
[9] J. L. Wyatt and R. F. Colman, *Biochemistry* **16**, 1333 (1977).
[10] S. Roy and R. F. Colman, *Biochemistry* **18**, 4683 (1979).
[11] F. S. Esch and W. S. Allison, *J. Biol. Chem.* **253**, 6100 (1978).
[12] T. E. Mansour and R. F. Colman, *Biochem. Biophys. Res. Commun.* **41**, 1370 (1978).
[13] D. W. Pettigrew and C. Frieden, *J. Biol. Chem.* **253**, 3623 (1978).
[14] M. J. Zoller and S. S. Taylor, *Fed. Proc., Fed. Am. Soc. Exp. Biol.* **38**, 316 (1979).
[15] C. S. Hixon and E. G. Krebs, *J. Biol. Chem.* **254**, 7509 (1979).
[16] K. V. Saradambal, R. A. Bednar, and R. F. Colman, *J. Biol. Chem.* **256**, 11866 (1981).
[17] A. E. Annamalai and R. F. Colman, *J. Biol. Chem.* **256**, 10276 (1981).
[18] J. S. Bennett, R. F. Colman, and R. W. Colman, *J. Biol. Chem.* **253**, 7346 (1978).
[19] W. R. Figures, S. Niewiarowski, T. A. Morinelli, R. F. Colman, and R. W. Colman, *J. Biol. Chem.* **256**, 7789 (1981).

the enhanced ability to study this receptor protein-mediating platelet response to ADP will more than likely lead to an increased understanding of stimulus–response coupling at the molecular level.

Methods and Results

Isolation of Platelets from Blood

Platelet-rich plasma (PRP) is prepared from whole human blood anticoagulated with ACD [sodium citrate (0.085 M), citric acid (0.079 M), dextrose (0.180 M)] by differential centrifugation in 50-ml tubes (120 g, 15 min, 37°). The top layer containing platelet-rich plasma is separated from the red cell layer and used as a source of platelets in all studies.

Removal of Plasma from Platelets

Potato apyrase is prepared by extraction of potatoes using the method of Molnar and Lorand.[20] These preparations contain both ADPase and ATPase activities as well as a platelet-aggregating lectin. The lectin was separated from the apyrase using fetuin–agarose affinity chromatography.[21] Washed platelets are prepared from platelet-rich plasma by the method of Mustard et al.[22] Platelets are sedimented from platelet-rich plasma (1100 g, 20 min) and resuspended in Tyrode's–albumin buffer [NaH_2PO_4 (0.02 mM), NaCl (136 mM), KCl (2.68 mM), $NaHCO_3$ (11.9 mM), dextrose (5.4 mM), bovine serum albumin (Cohn fraction V; 0.35%), pH 7.35] containing heparin (25 U/ml), apyrase (50 U/ml), calcium chloride (2.0 mM), and magnesium chloride (1.0 mM). The use of apyrase prevents the action of ADP on platelets during preparation and the use of heparin minimizes exposure to thrombin. After a 20-min incubation at 37°, the cells are sedimented (1100 g, 20 min) and resuspended in the original buffer without the addition of heparin. After incubation (20 min, 37°) the cells are resedimented and suspended in the same buffer in the absence of apyrase and heparin.

Preparation of 5'-p-Fluorosulfonylbenzoyladenosine

5'-p-Fluorosulfonylbenzoyladenosine is prepared by the method of Wyatt and Colman.[9] The synthesis involves the condensation of adenosine with fluorosulfonylbenzoyl chloride. For the synthesis of radiolabeled

---

[20] J. Molnar and L. Lorand, *Arch. Biochem. Biophys.* **93**, 353 (1961).
[21] K. A. E. Whigham, A. H. Drummond, W. Edgard, and C. R. M. Prentice, *Thromb. Haemostasis* **36**, 652 (1976).
[22] J. F. Mustard, D. W. Perry, N. G. Ardlie, and M. A. Packham, *Br. J. Haematol.* **22**, 103 (1972).

FSBA, [2-³H]adenosine (New England Nuclear, Boston, MA) was utilized. Typical radiolabeled preparations have specific radioactivities of 20 Ci/mol. Once isolated, the compound is stored in solid form at $-70°$ in a desiccated environment. When needed, the stored sample is thawed and a small amount (~1 mg) dissolved in dimethylformamide (DMF). The concentration of FSBA is then determined by absorbance measurement of an aliquot of the DMF solution diluted into ethanol (1 : 3000). The absorbance (259 nm) is read versus a blank sample containing the same volume of pure DMF diluted into ethanol. The molar extinction coefficient for FSBA is $1.35 \times 10^4 \, M^{-1} \, cm^{-1}$ at 259 nm.

Labeling of Platelets with 5'-p-Fluorosulfonylbenzoyladenosine

A stock solution of FSBA dissolved in DMF is added to suspensions of washed platelets to give a desired final concentration of FSBA. All control samples are treated with DMF alone. The DMF concentration should never exceed 1.0% of the volume of the platelet suspension (5×10^8 cells/ml). To block ADP-induced activation of platelets completely, 40 μM FSBA is incubated with the platelets for 40 min at 37°. At the conclusion of the incubation period, the cells are sedimented (1200 g, 20 min) to separate them from the excess FSBA and then resuspended in fresh Tyrode's buffer. Because FSBA inhibits ADP aggregation of platelets, even gel-filtered platelets can be sedimented in this fashion without clumping of the cells caused by centrifugation-induced ADP leakage. All incubations are carried out in the presence of adenosine deaminase (2 U/ml) to prevent the effects of any contaminating adenosine in the preparation.[23]

Preparation of ¹²⁵I-Labeled Fibrinogen and Binding of ¹²⁵I-Labeled Fibrinogen to Platelets

Fibrinogen (Kabi, Stockholm, Sweden) is purified by ammonium sulfate precipitation as described by Niewiarowski et al.[24] Prior to the purification, solutions of fibrinogen (10 mg/ml) are treated with diisopropyl fluorophosphate (10 mM, 1 hr) to inhibit any protease activity associated with the fibrinogen. The clotability of fibrinogen prepared by this method is >93%. The purified fibrinogen is labeled with ¹²⁵I by the iodine monochloride technique.[25] Excess ¹²⁵I is removed by gel filtration of the sample

---

[23] D. C. B. Mills, W.R. Figures, L. M. Scearce, R. F. Colman, and R. W. Colman, *J. Biol. Chem.* **260**, 8078 (1985).

[24] S. Niewiarowski, A. Z. Budzynski, T. A. Morinelli, T. M. Budzynski, and G. J. Stewart, *J. Biol.Chem.* **256**, 917 (1980).

[25] A. S. McFarlane, *Biochem. J.* **62**, 135 (1956).

on BioGel P-2. The clotability of the labeled fibrinogen is usually >90% and the specific activity ranges from 3.1 to 50 mCi/nmol fibrinogen.

Effect of 5'-p-Fluorosulfonylbenzoyladenosine on ADP-Mediated Platelet Shape Change and Aggregation

For measurement of shape change platelet suspensions (0.5 ml, 2 × 10^8 cells/ml) are incubated at 37° in siliconized glass cuvettes in a dual-channel Chronolog Lumi-Aggregometer (Chronolog Corp., Havertown, PA). The reference channel contains platelets that have been incubated with 10 μM ADP. Platelet suspensions for shape change measurements are incubated with FSBA or carrier solvent (DMF) alone. After the addition of ADP (10 μM), the change in light transmittance is recorded on a chart recorder.

The greatest decrease in light transmittance is defined as 100% of maximum shape change. All other measurements on similar platelet suspensions are given in relation to this value. In addition to the extent, the rate of the decrease in light transmittance is measured with the slope of the tracings. The greatest rate of change is defined as 100% of maximum rate of shape change and all other rates for a specific compound and platelet suspension are expressed in terms of the maximum. Platelet aggregation is measured in the same instrument as an increase in light transmittance of the platelet suspension.

When platelets are prelabeled with FSBA as described above, shape change in response to ADP is inhibited. Figure 1 demonstrates the concentration dependence of inhibition by FSBA. Platelets that are preincubated with various concentrations of FSBA (20 min, 37°) are tested for their ability to undergo shape change in response to ADP. Control cells are incubated with DMF (carrier solvent) alone. As shown in Fig. 1, increasing concentrations of FSBA inhibit shape change progressively. Figure 2 shows electron micrographs of resting platelets, control cells (shape change induced by ADP), and cells that are inhibited by FSBA. These data confirm the observations of inhibition measured optically. Resting cells (Fig. 2A) appear mostly discoidal in appearance, as do the cells that are inhibited by FSBA (Fig. 2D). The control cells appear as spheroids with multiple pseudopodia (Fig. 2B), as do cells treated with the carrier solvent (DMF) alone (Fig. 2C). Aggregation of platelets is similarly inhibited by pretreatment with FSBA as shown in Fig. 3 (inset). The platelets are pretreated with FSBA (40 μM, 20 min, 37°). Control cells are treated with carrier solvent (DMF) only. Aggregation of the platelets is induced by the addition of fibrinogen (0.1%) and ADP (10 μM). Under these conditions, aggregation of the platelets is totally inhibited by FSBA. As a

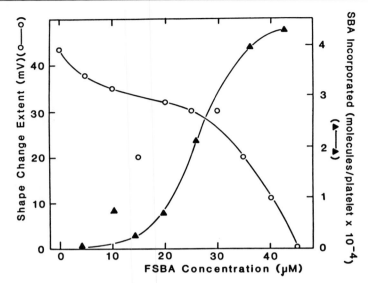

FIG. 1. Inhibition of shape change of washed platelets by FSBA and incorporation of labeled FSBA into platelet proteins. Samples of washed platelets were incubated for 1 hr at 37° with FSBA or with the solvent DMF at 0.5% (v/v) in the presence of adenosine deaminase (2 U/ml). Shape change (○) was induced by addition of ADP (2 μM) and the maximum extent of recorder deflection was taken as the extent of shape change. In a separate experiment, washed platelets were incubated with [^{3}H]FSBA under similar conditions and the amount of specific incorporation into nondialyzable material (▲) was determined.

test of specificity of the inhibition, the ability of FSBA-treated cells to respond to azo-PGH_2, a prostaglandin endoperoxide analog and a potent platelet agonist,[26] is determined. As demonstrated in Table I, the shape change mediated by azo-PGH_2 is not inhibited by pretreatment of the cells with FSBA. Table I compares the effects of FSBA with apyrase and creatine phosphate/creatine phosphokinase (CP/CPK). These two enzyme systems inhibit platelet activation by metabolizing ADP to AMP and ATP, respectively. Note that in every case tested, FSBA works identically to the ADP-depleting enzymes.

Measurement of Fibrinogen Binding to Platelet Receptors

The measurement of fibrinogen binding to platelet receptors is carried out using a silicone oil centrifugation technique in which the binding of ^{125}I-labeled fibrinogen to these sites is measured. Platelets (3 × 10^8

[26] T. A. Morinelli, S. Niewiarowski, E. Kornecki, W. R. Figures, Y. Wachtfogel, and R. W. Colman, *Blood* **61,** 41 (1983).

Fig. 2. Scanning electron micrographs of washed platelets demonstrating shape change. The platelets were fixed in glutaraldehyde (2.5%) and deposited with gentle suction onto Nucleopore membranes (1 μm). All samples contained adenosine deaminase (2 U/ml). (A) Control: Platelets were stirred in the aggregometer for 5 min and then fixed (96% disk shaped, optical change signal, 0 mV). (B) Control platelets fixed 30 sec after the addition of 2 μM ADP (4% disk shaped, optical shape change, 85 mV). (C) Platelets incubated for 45 min with 0.5% (v/v) DMF and then for 30 sec with ADP (2 μM) (5% disk shaped, optical change signal, 84 mV). (D) Platelets incubated for 45 min with FSBA (49 μM) and then with ADP (2 μM) (87% disk shaped, optical change signal, 10 mV).

cells/ml) are incubated with ADP (10 μM) and fibrinogen at concentrations ranging from 0 to 50 μg/ml. The cells are incubated for 20 min, at which time an aliquot of the reaction mixture is layered on top of silicone oil in an Eppendorf centrifuge tube with a constricted tip. The silicone oil consists of a mixture of DC550 (eight parts) and DC250 (two parts) (William

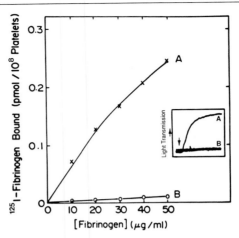

FIG. 3. Effect of FSBA on ADP-induced aggregation and fibrinogen binding of washed platelets. Washed platelets were pretreated with FSBA (100 μM) or carrier solvent (DMF) for 40 min. The ability of the cells to bind ^{125}I-labeled fibrinogen specifically in the presence of ADP (100 μM) over a range of ^{125}I-labeled fibrinogen concentrations is shown. (A) DMF-treated cells; (B) FSBA-treated cells. *Inset:* The cells were then tested for aggregation in response to ADP (100 μM) and fibrinogen (0.1%).

F. Nye, Inc., New Bedford, MA). The tubes are then centrifuged for 2 min in the Eppendorf centrifuge. The pellets are then removed by severing the constricted tip with dog-nail clippers. These pellets are then counted for ^{125}I in an LKB (Rockville, MD) γ counter. In order to determine specific binding of ^{125}I-labeled fibrinogen, identical samples containing an excess of unlabeled fibrinogen (1%) are evaluated in the same system. Specific binding is then calculated as the total binding minus nonspecific binding.

When the platelets are pretreated with FSBA, the fibrinogen binding mediated by ADP is severely inhibited (Fig. 3). We have also noted the ability of FSBA to reverse the fibrinogen binding mediated by ADP in a time-dependent manner similar to the effect of apyrase shown in Fig. 4. As previously observed with apyrase, the longer the time of incubation of platelets with the fibrinogen with ADP, the less fibrinogen is chased by addition of FSBA (Fig. 4).

Determination of Covalent Incorporation of [³H]SBA into Platelet Surface Protein

The incorporation of the affinity label into platelet membrane protein can be determined using two methods. The first method involves the labeling of the cells with FSBA as described above followed by the isola-

TABLE I
EFFECT OF VARIOUS INHIBITORS OF SHAPE CHANGE ON SHAPE
CHANGE MEDIATED BY ADP AND AZO-PROSTAGLANDIN H_2[a]

	Inhibitors			
Agonist	FSBA	Apyrase	CP/CPK	CTA_2
ADP	−	−	−	+
Azo-PGH_2	+	+	+	−

[a] Platelet shape change was induced by either ADP (10 μM) or azo-PGH_2 (10 nM) in the presence of various inhibitors. −, Inhibition observed; +, no inhibition. Note that the thromboxane inhibitor CTA_2 had no effect on the ability of ADP to mediate platelet shape change whereas it inhibited shape change by azo-PGH_2. The inhibition of ADP-mediated shape change by FSBA, CP/CPK, and apyrase was not observed when using azo-PGH_2 as an agonist.

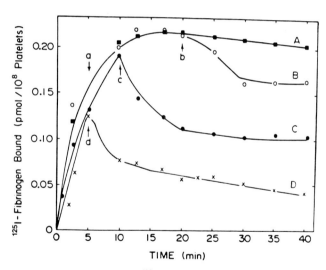

FIG. 4. Dissociation of platelet-bound ^{125}I-labeled fibrinogen by FSBA. Washed platelets were incubated with ^{125}I-labeled fibrinogen (40 μg/ml) and ADP (100 μM) and fibrinogen binding to the cells was measured over 40 min. An experiment representative of three similar time courses is presented. Maximal fibrinogen binding to the cells occurred after 10 min of incubation with ^{125}I-labeled fibrinogen and ADP (curve A). The effect of FSBA (100 μM) on ^{125}I-labeled fibrinogen binding was measured. Addition of FSBA at 5 min (point d) induced 66% of the bound fibrinogen to dissociate from the cells (curve D). Addition of FSBA at 10 min (point c) induced 48% of the bound fibrinogen to dissociate from the cells (curve C). Addition of FSBA at 20 min (point b) induced 13% of the bound fibrinogen to dissociate from the cells (curve B). Addition of carrier solvent DMF (0.25%) at any time during the time course was without effect (point a).

tion of platelet membranes, solubilization of the membrane proteins, and separation of the proteins by sodium dodecyl sulfate–polyacrylamide gel electrophoresis (SDS–PAGE).[18,19] The glycerol lysis method of Barber and Jamieson[27] is used for the lysis of the cells and the isolation of platelet membranes. The membrane proteins are then solubilized in SDS (10%), urea (8 M), dithiothreitol (DTT; 0.2 M), and diisopropylfluorophosphate (DFP; 10 mM). It is important to include DFP in the solubilization mixture as the labeled protein is extremely susceptible to hydrolytic breakdown by proteases. The solubilized platelet membrane proteins are then separated by SDS–PAGE using the method of Weber and Osborn.[28] The system is used because of the neutral pH of the running gel, thus stabilizing the ester bond at the 5′ position of the ribose ring in FSBA, which is quite susceptible to hydrolysis at pH levels <6.0 and >8.0. The gels are sliced; the slices are counted, extracted into a mixture of toluene–water–Protosol (New England Nuclear) (9 : 1 : 10), and then counted in a liquid scintillation cocktail (Econofluor; New England Nuclear) to localize the radiolabeled protein. The FSBA has been shown to label a single membrane polypeptide in platelet membranes with M_r 100,000 (Fig. 5). This method renders reproducible results; however, the time and amount of material required does not allow for the determination of more than four values for incorporation of [^{3}H]SBA per unit of platelets.

An alternate method is used to measure the effect of FSBA concentration on covalent incorporation of FSBA into the cell surface receptor. In these assays the cells are incubated at 37° for 40 min. To stop the reaction, cells are then added to DTT (200 mM final concentration). DTT causes rapid hydrolysis of the sulfonyl fluoride, thus rendering any residual FSBA incapable of further reaction with the platelets. To measure the extent of incorporation into the receptor, the cells are removed by centrifugation (8000 g, 20 min) in an Eppendorf microfuge and the supernatants carefully aspirated. The pellets are then dissolved in phosphate buffer (0.1 M, pH 7.2) containing SDS (10%), urea (8 M), and EDTA (10 mM). The dissolution of the cells is allowed to proceed overnight (18 hr, 23°). The solutions of solubilized platelet proteins are then dialyzed in a multiple-chamber continuous flow dialysis chamber (Bethesda Research Laboratories, Gaithersburg, MD, #1200) for 3 days. The dialysis buffer consists of phosphate (0.05 M, pH 7.2) containing SDS (0.1%) and EDTA (10 mM). The dialysis step is necessary to remove any radiolabeled material that has not covalently bound to membrane protein. To assure complete removal of unbound radiolabel, a number of samples at the highest concentration are

[27] A. J. Barber and G. A. Jamieson, *J. Biol. Chem.* **245**, 6357 (1970).
[28] K. Weber and M. Osborn, *J. Biol. Chem.* **244**, 4406 (1968).

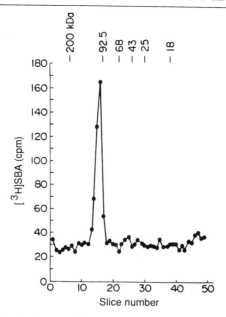

FIG. 5. Sodium dodecyl sulfate–polyacrylamide gel electrophoresis of platelet membrane polypeptide that incorporates [^{3}H]SBA. Washed platelets were treated with [^{3}H]FSBA (100 μM) for 20 min at 37°. Platelet membranes were prepared by the glycerol lysis method. The membrane vesicles were then solubilized, reduced, and subjected to electrophoresis in the presence of SDS. The gel was then sliced and the radioactivity in each slice was measured. A single membrane polypeptide on the platelet surface is covalently modified by FSBA.

included and tested on a daily basis. As mentioned above, it is important to maintain proteins labeled with SBA (SBA–protein) in a pH range between 7.0 and 8.0 because the ester bond at the 5' position is quite susceptible to hydrolysis at pH values outside these ranges. Dialysis is considered complete when the amount of radiolabel in a sample remains constant for 24 hr. Samples are then removed from the dialysis chamber and counted in a liquid scintillation cocktail (ACSII; Amersham, Arlington Heights, IL). The amount of radiolabel in the sample was evaluated and the specific activity calculated in terms of molecules incorporated per platelet. Nonspecific binding to nonnucleotide-binding sites is determined by use of an ADP (10 mM) chase. Specific binding can then be calculated by the difference between total binding and nonspecific binding. Using this method, a concentration dependence of the end point incorporation of FSBA into platelets may be determined (Fig. 1). The platelets are incubated with [^{3}H]FSBA for 20 min and the above method is utilized to determine the extent of the covalent incorporation ([^{3}H]SBA–protein). Note that the extent of covalent incorporation correlates very well with the observed inhibition of shape change.

Conclusions

The result of using FSBA to label nucleotide-binding sites on the external surface of the platelet membrane is consistent with the results expected if the reagent covalently incorporated into an ADP receptor mediating shape change and aggregation. Cells treated with FSBA for 20–40 min fail to respond to ADP such that both shape change and aggregation responses produced by this nucleotide in control cells is inhibited. This procedure therefore can be used to render platelets permanently incapable of inhibiting ADP-induced activation. In addition, the inhibition of ADP-mediated fibrinogen binding to fibrinogen receptors is also inhibited by pretreatment of the platelets with the nucleotide analog. The specificity of this inhibition is demonstrated by the lack of inhibition of shape change induced by the prostaglandin endoperoxide analog azo-PGH$_2$.[26] It is interesting to note that in all the systems tested the effect of FSBA on platelets is identical to that of the ADP-metabolizing enzymes apyrase and CP/CPK.

Further specificity was demonstrated by the finding that FSBA inhibited epinephrine-induced platelet aggregation and fibrinogen binding, indicating an ADP requirement without affecting binding to the α_2-adrenergic receptor.[29] In addition, we have demonstrated that the guanosine derivative 5'-p-fluorosulfonylbenzoylguanosine (FSBG) has no such effect,[23] and that FSBA acts over short periods of time as a partial agonist.[30] The use of this affinity analog has indicated the site of action to be a cell surface protein with M_r 100,000, distinct from glycoprotein IIIa,[31] which is designated *aggregin*. Aggregin is the platelet ADP receptor mediating shape change, aggregation, and fibrinogen binding.[32] A study by Mills *et al*.[23] has demonstrated that this site is distinct from the ADP-binding site on the platelet surface that controls the cyclic AMP (cAMP) levels in platelets. In this study, pretreatment of the platelets with FSBA under conditions known to inhibit ADP-mediated shape change and aggregation failed to

[29] W. R. Figures, L. M. Scearce, Y. Wachtfogel, J. Chen, R. F. Colman, and R. W. Colman, *J. Biol. Chem.* **261,** 5981 (1986).
[30] W. R. Figures, L. M. Scearce, P. DeFeo, G. Stewart, F. Zhou, J. Chen, J. Daniel, R. F. Colman, and R. W. Colman, *Blood* **70,** 796 (1987).
[31] R. W. Colman, W. R. Figures, Q. X. Wu, S. Y. Chung, T. A. Morinelli, G. P. Tuszynski, R. F. Colman, and R. W. Colman, *Arch. Biochem. Biophys.* **262,** 298 (1988).
[32] R. W. Colman, *FASEB J*. **4,** 1425 (1990).

impair the ability of ADP to lower cAMP levels. This observation strongly suggests that there are two types of binding sites on the cell surface for ADP: one site controlling aggregation and shape change, and the other controlling cAMP levels in the cell. It appears that FSBA interacts with the former site only. The ability to place a specific radiolabeled tag on this protein should facilitate its purification and allow investigators to define its modulation and control in physiologic and pathophysiologic states.

Acknowledgment

This work was supported by Program Project Grant HL 36579 from the National Heart, Lung and Blood Institute of the National Institutes of Health, an American Heart Association Grant-in-Aid #58424101 from the North Central Pennsylvania Chapter, an American Heart Association Special Investigatorship #58424301 from the South Eastern Pennsylvania Chapter, and NSF Grant DMB-9105116.

[15] Thrombin Receptors on Human Platelets

By ELIZABETH R. SIMONS, THERESA A. DAVIES, SHERYL M. GREENBERG, and NANCY E. LARSEN

The ability of thrombin to stimulate platelets has been known for many years and the subject has been thoroughly summarized and reviewed.[1-4] The stimulation is mediated through specific thrombin receptors on the

---

[1] G. C. White, II, E. F. Workman, Jr., and R. L. Lundblad, *in* "Chemistry and Biology of Thrombin" (R. L. Lundblad, J. W. Fenton, II, and K. G. Mann, eds.), p. 479. Ann Arbor Sci. Publ., Ann Arbor, Michigan, 1977.
[2] W. H. Seegers, *in* "Chemistry and Biology of Thrombin" (R. L. Lundblad, J. W. Fenton, II, and K. G. Mann, eds.), p. 1. Ann Arbor Sci. Publ., Ann Arbor, Michigan, 1977.
[3] T. C. Detwiler, *Ann. N.Y. Acad. Sci.* **370**, 67 (1981).
[4] M. B. Zucker and V. T. Nachmias, *Arteriosclerosis* **5**, 2 (1985).

surface of the platelet membrane.[2,5-22] This process requires active α-thrombin, and involves proteolysis of one of the receptors with release of a large glycoprotein fragment into the peripheral medium,[15,19,23-26] thus ruling out glycoprotein GPIb, which is not processed by thrombin, as one of the receptors.[17] Binding of thrombin to the platelet surface is very rapid, being maximal within 30 sec.[9,14,17,21-23,27-30]

Two classes of platelet thrombin receptors have been identified by equilibrium binding studies: a small number (450–750) of high-affinity (K_D of approximately 10^{-9}) and a larger number (50,000–70,000) of 100-fold lower affinity (K_D of approximately 10^{-7}) sites.[12,14,31] Even smaller numbers (150 high-affinity and 16,000 low-affinity sites) with comparable dissociation constants have been reported[20] (Table I). When the active site of thrombin has been blocked with tosyllysyl chloromethyl ketone (TLCK)[11] or diisopropyl fluorophosphate,[2,7,8] the modified thrombin binds equally

---

[5] T. C. Detwiler and R. D. Feinman, *Biochemistry* **12**, 282 (1973).
[6] T. C. Detwiler and R. D. Feinman, *Biochemistry* **12**, 2462 (1973).
[7] D. M. Tollefsen, J. R. Feagler, and P. W. Majerus, *J. Biol. Chem.* **249**, 2646 (1974).
[8] P. Ganguly, *Nature (London)* **247**, 306 (1974).
[9] B. M. Martin, R. D. Feinman, and T. C. Detwiler, *Biochemistry* **14**, 1308 (1975).
[10] B. M. Martin, W. W. Wasiewski, J. W. Fenton, II, and T. C. Detwiler, *Biochemistry* **15**, 4886 (1976).
[11] P. Ganguly and W. J. Sonnichsen, *Br. J. Haematol.* **34**, 291 (1976).
[12] D. M. Tollefson and P. W. Majerus, *Biochemistry* **15**, 2144 (1976).
[13] E. F. Workman, Jr., G. C. White, II, and R. L. Lundblad, *Thromb. Res.* **9**, 491 (1976).
[14] S. W. Tam and T. C. Detwiler, *Biochim. Biophys. Acta* **543**, 194 (1978).
[15] D. F. Mosher, A. Vaheri, J. J. Choate, and L. G. Gahmbery, *Blood* **53**, 437 (1979).
[16] W. F. Bennett and K. C. Glen, *Cell (Cambridge, Mass.)* **22**, 621 (1980).
[17] S. W. Tam, J. W. Fenton, and T. C. Detwiler, *J. Biol. Chem.* **225**, 6626 (1980).
[18] H. Holmsen, C. A. Dangelmaier, and H. Holmsen, *J. Biol. Chem.* **256**, 9393 (1981).
[19] N. E. Larsen and E. R. Simons, *Biochemistry* **20**, 4141 (1981).
[20] N. Tandon, J. T. Harmon, D. Rodbard, and G. A. Jamieson, *J. Biol. Chem.* **258**, 11840 (1983).
[21] R. J. Alexander, J. W. Fenton II, and T. C. Detwiler, *Arch. Biochem. Biophys.* **222**, 266 (1983).
[22] H. Holmsen, C. A. Dangelmaier, and S. Rongved, *Biochem. J.* **222**, 157 (1984).
[23] N. L. Baenziger, G. N. Broche, and P. W. Majerus, *Proc. Natl. Acad. Sci. U.S.A.* **68**, 240 (1971).
[24] D. R. Phillips and P. P. Agin, *Ser. Haematol.* **6**, 292 (1973).
[25] D. R. Phillips and P. P. Agin, *Biochem. Biophys. Res. Commun.* **352**, 218 (1974).
[26] D. R. Phillips and P. P. Agin, *Biochem. Biophys. Res. Commun.* **75**, 940 (1977).
[27] W. C. Horne and E. R. Simons, *Blood* **51**, 741 (1978).
[28] S. M. Greenberg-Sepersky and E. R. Simons, *J. Biol. Chem.* **259**, 1502 (1984).
[29] W. C. Horne, N. Norman, D. Swartz, and E. R. Simons, *Eur. J. Biochem.* **120**, 285 (1981).
[30] G. A. Jamieson, S. M. Jung, and A. Ordinas, *Ann. N.Y. Acad. Sci.* **370**, 96 (1981).
[31] E. F. Workman, Jr., G. C. White, II, and R. L. Lundblad, *Thromb. Res.* **9**, 491 (1976).

TABLE I
Number and Affinity of Platelet Thrombin Receptors

Type	Number per platelet	K_D (nM)	
		α-Thrombin	TLCK-thrombin
High affinity	450–750	0.2^a	1.2^a
		1.4^b	
	150	1.2^c	
Low affinity	50,000–70,000	30.0^a	600^b
	16,000	15.6^c	

[a] D. M. Tollefsen, J. R. Feagler, and P. W. Majerus, *J. Biol. Chem.* **249**, 2646 (1974).
[b] E. F. Workman, G. C. White II, and R. L. Lundblad, *J. Biol. Chem.* **252**, 7118 (1977).
[c] N. Tandon, J. T. Harmon, D. Rodbard, and G. A. Jamieson, *J. Biol. Chem.* **258**, 11840 (1983).

well to the same number of sites but does not activate the platelets or cleave the receptors.

There have also been some reports on thrombin blocked at the active site with D-phenylalanyl-L-prolyl-L-arginine chloromethyl ketone (PPACK-thrombin).[32–34] According to work from the same laboratory, PPACK-thrombin has a lower affinity for the platelet thrombin-binding sites (IC_{50} = 170 nM[35] in contrast to K_D = 1 nM for TLCK-thrombin[7,12,13]); after prolonged incubation with platelets (5 min) PPACK-thrombin inhibited the aggregation, secretion, cytoplasmic acidification, and Ca^{2+} rise normally initiated by 0.3–0.5 nM α-thrombin, but not the shape change. These blocked thrombins do, however, sensitize the platelets to subsequent stimulation by active thrombin[16,36–39] or by other stimuli,[40,41] appar-

[32] J. T. Harmon and G. A. Jamieson, *J. Biol. Chem.* **261**, 15298 (1986).
[33] J. T. Harmon and G. A. Jamieson, *Biochemistry* **27**, 2151 (1988).
[34] G. A. Jamieson, "Platelet Membrane Receptors: Molecular Biology, Immunology and Pathology," Alan R. Liss, New York, 1988.
[35] N. J. Greco, T. E. Tenner, Jr., N. N. Tandon, and G. A. Jamieson (personal communication).
[36] D. R. Phillips, *Thromb. Diath. Haemorrh.* **32**, 207 (1974).
[37] D. R. Phillips and P. P. Agin, *Biochem. Biophys. Res. Commun.* **352**, 218 (1974).
[38] T. A. Davies, S. M. Greenberg-Sepersky, and E. R. Simons, *J. Cell Biol.* **97**, 1522 (1983).
[39] T. A. Davies, S. M. Greenberg-Sepersky, and E. R. Simons, unpublished results.
[40] R. L. Kinlough-Rathbone, H. J. Reimers, and J. F. Mustard, *Science* **192**, 1011 (1976).
[41] H. J. Reimers, R. L. Kinlough-Rathbone, J. P. Cazenave, A. F. Senyi, J. Hirsch, P. Ganguly, and N. L. Gould, *Br. J. Haematol.* **42**, 137 (1979).

TABLE II
EFFECT OF PREEXPOSURE TO TOSYLLYSYL CHLOROMETHYL KETONE–THROMBIN

	TLCK-thrombin (U/ml)		Number of sites bound[a]		Percentage change attributable to preexposure to TLCK-thrombin			
						In rate of depolarization[c]		
							α-Thrombin doses	
Donor type	First exposure	Second addition	First exposure	Second addition	In number of sites[b]	0.005	0.01	0.5
Enhancer	0.0025	—	719	—	—	+20.7	−13.6	0
	0.05	—	6500	—	—	−37.9	−47.5	−9.5
	0.0025	0.05	798	8184	+25.9%	−86.6	−52.6	−31.1
Inhibitor	0.0025	—	100	—	—			
	0.05	—	6700	—	—			
	0.0025	0.05	100	5500	−17.9%			

[a] Calculated number of bound sites (each the average of two samples in a typical experiment) (one of four similar series of experiments with an enhancer platelet donor).
[b] Percentage of saturating control sites cross-linked when platelets are previously cross-linked with a subsaturating TLCK-thrombin dose. ^{14}C and/or ^{3}H derivatives were used.
[c] Percentage change in rate of depolarization attributable to preexposure and covalent binding of TLCK-thrombin on subsequent α-thrombin stimulation (control ≡ 0).

ently by increasing the number of receptors bound at a given low dose of active site-blocked thrombin[38,39] (Table II).

Original quantitation of the platelet thrombin receptors was accomplished by equilibrium binding of radioactively labeled active and/or active site-blocked thrombin to platelets[11,12,42]; these receptors were identified by the disappearance or decreased concentration of a Coomassie blue-stained membrane glycoprotein band on sodium dodecyl sulfate (SDS)-polyacrylamide gels[24–26] and by the appearance of a receptor fragment in the platelet supernatant after prolonged incubation with relatively high concentrations of thrombin.[15,17,23–26,43] These methods all involve noncovalent binding; although they have permitted identification of membrane proteins that have been proteolyzed by thrombin, they suffer from several limitations: (1) in the absence of covalent binding, dissolution of the membrane and its stimulus-bound receptor in detergents detaches the thrombin, making it necessary to detect the receptor by its disappearance as it is cleaved by the stimulus; (2) in determining the number of binding sites, nonspecific binding is difficult to avoid since hirudin (which has a high affinity for

[42] C. L. Knupp and G. C. White, II, *Blood* **65,** 578 (1985).
[43] J. W. Lawler, F. C. Chao, and P. H. Feng, *Thromb. Haemostasis* **37,** 355 (1977).

thrombin) cannot be used because it will remove any noncovalently bound thrombin; (3) there is evidence that the high-affinity receptor, being present in a small number of copies per cell, does not give rise to a detectable Coomassie blue band in platelet membrane gel electrophoresis[17,19]; and (4) the length of exposure of platelets to thrombin and/or the thrombin dose required is generally high, making it difficult to single out the intact initial high-affinity binding site.

It therefore seemed imperative to design a technique that would identify a specific binding site under nonequilibrium conditions and within seconds after exposure of the platelet to its stimulus. The application of photoreactive ligands was therefore inviting: they can be activated rapidly (within the millisecond-to-second time range), and form a stable bond that is impervious to detergent solubilization. Photoreactive groups can either be attached directly to a ligand without affecting its properties or can be prepared as nonspecific bifunctional cross-linking reagents[44-46]; if they contain a disulfide bond, these ligands also provide a convenient locus at which, by reduction, the ligand can be separated from its receptor once the complex has been identified. Therefore we devised, and reported in 1981,[19] the first determination of the platelet thrombin-binding sites via covalent binding of the thrombin to its receptors within seconds after exposure (i.e., under nonequilibrium conditions), and the identification of the thrombin–receptor complex by gel electrophoresis or chromatography of detergent-solubilized membranes. As summarized in Table III a single complex (200,000 Da), containing thrombin linked to a membrane glycoprotein (approximately 160,000 Da), believed to be the high-affinity binding site, was found when subsaturating doses of tosyllysyl chloromethyl thrombin (TLCK–thrombin) were covalently linked to washed human platelets. If the active α-thrombin derivative was substituted for the blocked enzyme, the size of the complex was reduced to approximately 120,000 Da, accounting thereby for the previously identified activation fragment.[15,43] In contrast the additional two platelet membrane glycoproteins found covalently linked when high concentrations of the thrombin derivatives were photoactivated, exhibiting molecular weights above 400,000 and approximately 46,000, respectively, were identical for the active and the blocked enzyme derivatives, indicating that the low-affinity receptors are not enzymatically processed by α-thrombin during the platelet activation process. We have reconfirmed these results in connection

---

[44] K. Peters and F. M. Richards, *Annu. Rev. Biochem.* **46**, 523 (1977).
[45] V. Chowdhry and F. H. Westheimer, *Annu. Rev. Biochem.* **48**, 293 (1979).
[46] A. E. Ruoho, A. Rashidbaigi, and P. E. Roeder, in "Membranes, Detergents and Receptor Solubilization," p. 119. Alan R. Liss, Inc., New York, 1984.

TABLE III
APPROXIMATE MOLECULAR SIZE OF PLATELET RECEPTOR–THROMBIN COMPLEXES AND THEIR REDUCED COMPONENTS

Receptor complexed with	Thrombin concentration[a] (U/ml)	Approximate size (Da) of receptor complexes							
		Unreduced[b,c]				Reduced[b,d]			
		Band 1	Band 2	Band 3	Band 4	Band 1	Band 2	Band 3	Band 4
DNCO-TLCK-thrombin	0.01		400,000	**200,000**					
DNCO-α-thrombin	0.01			**120,000**					
DNCO-TLCK-thrombin	0.10	600,000	400,000	**200,000**	46,000				
DNCO-α-thrombin	0.10		400,000	**120,000**	46,000				
ANPH-TLCK-thrombin	0.05	600,000	400,000	**200,000**					
DNCO-TLCK-thrombin	—			200,000			120,000	42,000	
DNCO-α-thrombin	—			120,000			60,000	42,000	
DNCO-TLCK-thrombin	0.10		400,000			210,000	160,000	145,000	56,000
DNCO-α-thrombin	0.01		400,000			210,000	160,000	145,000	56,000

[a] Saturation corresponds to 0.02 U/ml.
[b] Estimated by SDS–PAGE calibrated with platelets and with erythrocyte membrane proteins.
[c] Receptor coupled covalently to thrombin or TLCK-thrombin. Predominant band is indicated in **bold** numbers.
[d] Components of reduced and alkylated receptor complex, excluding the B-chain of thrombin, which is present in all cases.

with studies of the roles of these receptors in the platelet activation process, and expanded these studies by means of a different photoreactive coupling agent.[38,39,47] Yet a third photoreactive coupling agent has also yielded a 210,000-Da complex (i.e., a 173,000-Da platelet thrombin receptor plus a 37,500-Da thrombin).[48] The following sections will review the rationale of covalent ligand–receptor formation as well as its potential utility in delineating the role of these receptors in initiating the platelet response to thrombin.

In order to identify a membrane receptor by isolating the covalent complex formed between it and the specific ligand it recognizes, one may either attach a photoreactive cross-linking agent (e.g., a phenylazide or nitrophenylazide moiety) to the ligand, or one may allow the underivatized ligand to bind noncovalently to the cell surface and then cross-link it by a nonspecific bifunctional reagent. The former method has the advantage that it is specific, one can easily detect the ligand–receptor complex, one can determine exactly how many bonds will be formed between the ligand and its receptor, and one obtains little nonspecific binding. The disadvantage is that one must first be sure the photoreactive derivative has retained all of the biological functions and specificity of the parent compound. Bifunctional cross-linking agents, usually nitrophenylazo-derivatized imides (some also containing a reducible disulfide bond), have been used successfully for identification of specific platelet receptors when the ligand proved hard to derivatize, for example, in studies of the binding sites for collagen[49] and for fibrinogen,[50,51] but have not been applied to thrombin and will therefore not be discussed in detail here. Attachment of the photoreactive groups to thrombin without perturbing its ability to activate platelets or hydrolyze fibrinogen or the synthetic substrate tosylarginyl methyl ester, can be effected on the single carbohydrate chain that thrombin possesses,[19] or on one or more of the 22 lysyl residues of thrombin by reaction of an imide or an aldehyde with their ε-amine groups.[47,48] The two approaches have yielded the same results (Table I), a single high-affinity platelet membrane receptor (160,000–170,000 Da), and two to three low-affinity ones (46,000, >400,000, and >600,000),[19,47] implying that no other nearby membrane proteins are involved in any of these complexes since the covalent bonds have been made to different portions of the thrombin molecule.

[47] S. M. Greenberg-Sepersky and E. R. Simons, *Anal. Biochem.* **147**, 57 (1985).
[48] K. J. Danishefsky and T. C. Detwiler, *Biochim. Biophys. Acta* **801**, 48 (1984).
[49] J. Lahav, M. A. Schwartz, and R. O. Hynes, *Cell* **31**, 253 (1982).
[50] J. S. Bennett, G. Vilaire, and D. B. Cines, *J. Biol. Chem.* **257**, 8049 (1982).
[51] G. A. Marguerie, N. Thomas-Maison, M. H. Ginsberg, and E. F. Plow, *Eur. J. Biochem.* **139**, 5 (1984).

Experimental Procedures

A schematic diagram of the derivatization and cross-linking procedure, adapted from our original method,[19] is shown in Fig. 1 and is applicable to either active α-thrombin or tosyllysyl chloromethyl ketone-treated and hence enzymatically inactive TLCK-thrombin. Figure 2 shows a comparable scheme for preparation and utilization of the N-hydroxysuccinimidyl-6-(4'-azido-2'-nitrophenylamino)hexanoate (NHS-ANPH) derivative of these thrombins. The retention of platelet-activating and fibrinogen-hydrolyzing activity by the derivatized α-thrombin can and must be checked directly, and the retention of full platelet-binding capability by derivatized TLCK-thrombin must be verified via its inhibition of α-thrombin platelet stimulation.[19,38,39,47,48]

N,N'-Bis(2-nitro-4-azidophenyl)cystamine S,S-dioxide (DNCO) is not commercially available and was synthesized by us as described by its originators[52] from unlabeled precursors. The same compound was prepared by Hynes and co-workers[49] from [^{35}S]cystamine. It can be kept indefinitely in the dark or dissolved in pyridine, but it should be checked functionally before use for its ability to cross-link aldolase,[52] and spectroscopically for its ability to shift its absorbance maximum from 480 to 450 nm when photoirradiated (J. C. Whitin and E. R. Simons, unpublished). The other photocoupling reagent, NHS-ANPH, is commercially available (Pierce Co., Rockford, IL). The removal of excess free DNCO or NHS-ANPH after thrombin derivatization, which is best accomplished by Sephadex G-10 or G-25 chromatography or by dialysis, is essential as unbound reagent inhibits platelet activation and increases nonspecific covalent interactions.

For some applications, such as our study of positive cooperativity exhibited when platelets are pretreated with low concentrations of TLCK-thrombin before stimulation with active thrombin,[38,39] it has proven advantageous to use two different thrombin derivatives as ligands, one labeled with ^{3}H, the other with ^{14}C (Table II). Tritium labeling is shown in our original publication[19] as well as in Fig. 1, being achieved by reducing the Schiff base formed between periodate-treated thrombin and 2-mercaptoethylamine with sodium borotritide. ^{14}C labeling is accomplished by reductive methylation, an adaptation of the method of Tack et al.,[53] using [^{14}C]formaldehyde followed by sodium borohydride. If care is taken (short incubations and low concentrations of formaldehyde), the resultant thrombin remains fully active. An average extent of labeling, in

---

[52] C. K. Huang and F. M. Richards, *J. Biol. Chem.* **252**, 5514 (1977).
[53] B. F. Tack, J. Dean, D. Filat, P. E. Lorenz, and A. N. Schechter, *J. Biol. Chem.* **255**, 88421 (1980).

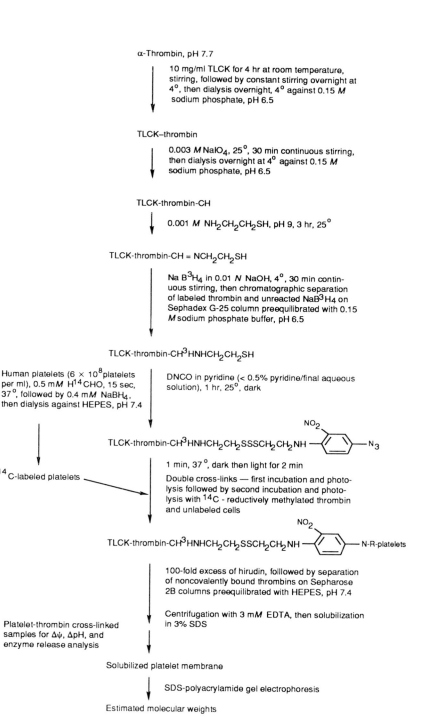

FIG. 1. Schematic diagram of typical thrombin derivatization and cross-link.

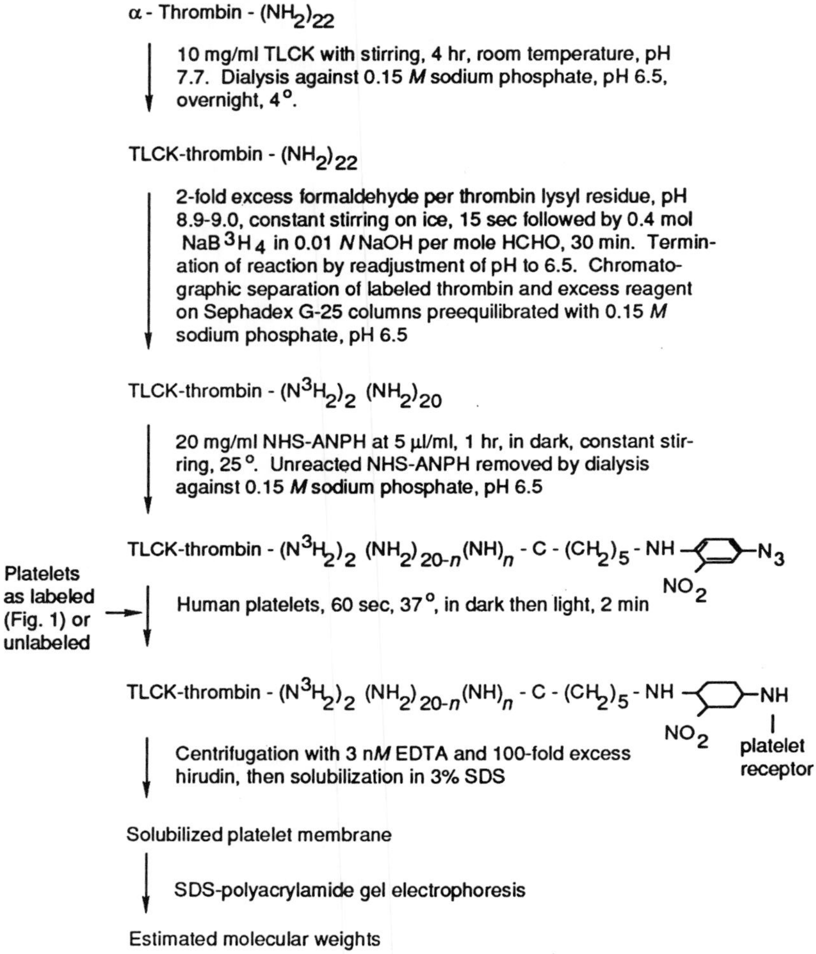

α - Thrombin - $(NH_2)_{22}$

  ↓ 10 mg/ml TLCK with stirring, 4 hr, room temperature, pH 7.7. Dialysis against 0.15 M sodium phosphate, pH 6.5, overnight, 4°.

TLCK-thrombin - $(NH_2)_{22}$

  ↓ 2-fold excess formaldehyde per thrombin lysyl residue, pH 8.9-9.0, constant stirring on ice, 15 sec followed by 0.4 mol NaB^{3}H$_4$ in 0.01 N NaOH per mole HCHO, 30 min. Termination of reaction by readjustment of pH to 6.5. Chromatographic separation of labeled thrombin and excess reagent on Sephadex G-25 columns preequilibrated with 0.15 M sodium phosphate, pH 6.5

TLCK-thrombin - $(N^3H_2)_2$ $(NH_2)_{20}$

  ↓ 20 mg/ml NHS-ANPH at 5 μl/ml, 1 hr, in dark, constant stirring, 25°. Unreacted NHS-ANPH removed by dialysis against 0.15 M sodium phosphate, pH 6.5

Platelets as labeled (Fig. 1) or unlabeled → TLCK-thrombin - $(N^3H_2)_2$ $(NH_2)_{20-n}(NH)_n$ - C - $(CH_2)_5$ - NH -⟨⟩- N$_3$ / NO$_2$

  ↓ Human platelets, 60 sec, 37°, in dark then light, 2 min

TLCK-thrombin - $(N^3H_2)_2$ $(NH_2)_{20-n}(NH)_n$ - C - $(CH_2)_5$ - NH -⟨⟩- NH / NO$_2$ | platelet receptor

  ↓ Centrifugation with 3 nM EDTA and 100-fold excess hirudin, then solubilization in 3% SDS

Solubilized platelet membrane

  ↓ SDS-polyacrylamide gel electrophoresis

Estimated molecular weights

FIG. 2. Typical NHS-ANPH thrombin derivatization.

our hands, approximates 10^8 cpm/mg for the ^{3}H-labeled and 10^6 cpm/mg for the ^{14}C-labeled thrombins. For other applications it has proven desirable to utilize platelets labeled on the plasma membrane proteins with one of these isotopes and DNCO-derivatized thrombins labeled with the other isotope so that the thrombin–receptor complex is the only (for the low doses identifying the high-affinity sites) or one of few (for the higher thrombin doses and hence low-affinity thrombin binding sites) proteins that are doubly labeled. The same reductive methylation technique is applicable to cell surface labeling,[54,55] and appears to be much less detrimental to platelet function than iodination or tritiation by reduction after enzymatic oxidation.[26,56]

Photocoupling of thrombin derivatives to platelets is carried out as depicted in Fig. 1, utilizing short thrombin–platelet incubation times (in the dark), followed by rapid photoactivation with a high-intensity lamp fitted with a filter to cut off wavelengths below 320 nm. Although bond formation is complete within seconds,[45] photoactivation can be carried on for 60 to 120 sec without causing any irradiation damage to the samples. The treatment of the ensuant DNCO-α-thrombin or DNCO-TLCK-thrombin cross-linked platelets, after photoactivation, with a large excess of hirudin, removes noncovalently bound thrombin. Failure to perform this step may lead to high nonspecific noncovalent binding. Requisite controls are photoactivation of platelets alone, and photoactivated DNCO- or ANPH-thrombin to which platelets were added only after the photoactivation. Hirudin can be added to an aliquot of the latter to verify that no labeled thrombin remains bound to the platelets when no covalent bonds between ligand and cells exist.

Identification of Covalent Thrombin–Receptor Complexes

When low (subsaturating, <2.5 nM) and high (saturating, >10 nM) concentrations of tritiated DNCO-TLCK-thrombin or DNCO-α-thrombin are exposed to platelets for 30 or 60 sec, photoactivated, and then processed as summarized in Fig. 1, the resultant platelet pellets can be solubilized in boiling sodium dodecyl sulfate (SDS) and applied to a polyacrylamide gel, or to a Sephacryl S-200 column, to obtain the complexes indicated in Table I.[19,38,39] (Since TLCK-thrombin binds more slowly than thrombin, albeit with the same K_D, 60 instead of 30 sec of incubation is used. All other conditions are the same for the active and active site-blocked derivatives.) The bands corresponding to the complex can be detected by their radioactivity. Utilizing this technique we previously identified a single radioactive complex, approximately 200,000 Da, as containing the high-affinity site and its covalently bound TLCK-thrombin, and two additional radioactive complexes, >400,000 and 46,000 Da, respectively, as the low-affinity sites coupled to saturating thrombin doses.[19] Newer electrophoretic techniques and better gel calibration have now allowed us to determine that an additional complex (possibly a dimer) at

[54] D. Mark and E. R. Simons, *Clin. Res.* **31**, 692a (1983).
[55] D. Mark, Ph.D. dissertation Boston University, Boston, MA (1984).
[56] L. G. Gahmbery, *J. Biol. Chem.* **251**, 510 (1976).

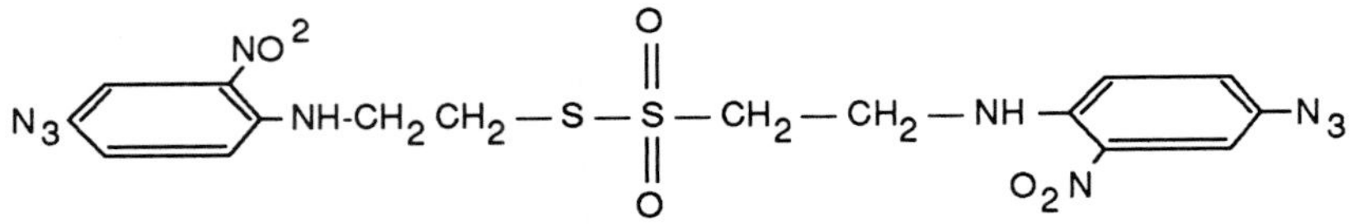

FIG. 3. Structure of DNCO.

400,000 Da is formed with low concentrations of TLCK thrombin while the >400,000-Da complex we reported earlier is, in reality, nearer 600,000.[39,47] If one uses active enzyme, i.e., DNCO-α-thrombin, the high-affinity receptor is proteolyzed, as noted in the introductory section of this chapter. While the native receptors all give positive PAS (periodic acid–Schiff) stains, i.e., are glycoproteins, the proteolyzed high-affinity receptor that is found bound to DNCO thrombin no longer does so; that is, the major portion of its carbohydrate moieties is attached to the released fragment described previously. One obtains very comparable results if the photoreactive agent is the succinimide NHS-ANPH and if covalent thrombin binding is thereby achieved at its lysyl instead of carbohydrate residues (Table I). This would not have been the anticipated result if the platelet thrombin receptor had been an assembly of noncovalently interacting polypeptide chains and led us to conclude[19,47] that each receptor is a single glycoprotein moiety.

Structure of Platelet Thrombin Receptors

The polypeptide chain composition of a receptor is initially studied by identifying the number and size of its component disulfide-linked polypeptide chains. One advantage of DNCO, shown in Fig. 3, is that it is cleavable by reduction whereas NHS-ANPH is not. It is therefore possible to isolate the DNCO-thrombin–receptor complex, either by electrophoresis or by gel chromatography,[19] and then to reduce the purified complex with dithiothreitol. If only the thrombin has been labeled, as occurs when the schemes in Figs. 1 or 2 are followed, reduced fragments of the receptor will not be labeled and must be detected by protein stain or assay; for this reason, it is preferable to label the cell surface proteins as well with a different isotope so that the source of the various reduced fragments is clear.[54,55] To date only studies on the ^{3}H-labeled DNCO-thrombin complexes have been performed (Table I).

As indicated in Table I (N. E. Larsen, unpublished), reduction of the high-affinity DNCO-TLCK-thrombin–receptor complex yielded three protein-positive bands at approximately 120,000, 42,000, and 28,000 Da and the complete disappearance of the original complex band at 200,000, indicating that a single receptor entity is involved. The 28,000-Da band was the only radioactive one and corresponded to the B chain of thrombin; the A chain is small (8,000 Da) and has never been detectable in our hands. When the complex had been formed with active DNCO-α-thrombin, the 120,000-Da band could no longer be detected but one at 60,000 Da appeared; the remaining two bands were unchanged (N. E. Larsen, unpublished). Thus the previously noted lytic fragment[15,43] has been released. The two disulfide-linked chains constituting the high-affinity thrombin receptor of the human platelet therefore apparently serve a dual role as recognition site[17] and as locus of actual thrombin stimulation since the latter cannot proceed unless the enzyme is active.

Reduction of the isolated large, low-affinity receptor complex (600,000 Da), which is identical for the complexes formed with TLCK and with α-thrombin, also leads to a disappearance of the original protein (Table III). Five bands appear, of which the smallest is again the B chain of thrombin. The others, at approximately 210,000, 160,000, 145,000, and 56,000 Da, have not yet been further identified. The other low-affinity receptor, also identical for both types of complexes, does not fully disappear on reduction; the B-thrombin band and another, at 42,000 Da, do appear, but no further characterization of this receptor has yet been performed.

Methods

Reagents

N-p-Tosyl-L-lysyl chloromethyl ketone (TLCK), N-tosyl-L-arginine methyl ester (TAME), sodium dodecyl sulfate (SDS), acrylamide, and N,N,N',N'-tetramethylethylenediamine (TEMED) can be purchased from Fisher (Pittsburgh, PA) or from Sigma Chemical Co. (St. Louis, MO); dithiothreitol (DTT) from Eastman (Rochester, NY); Sepharose 2B, SP-Sephadex C-50, and Sephacryl S-200 from Pharmacia (Piscataway, NJ); 4-fluoro-3-nitrophenylazide (F-NAP) from Pierce; sodium borotritide (50 Ci/mmol, >500 mCi/mg) from Amersham (Arlington Heights, IL); Protosol and Econofluor from New England Nuclear (Boston, MA); Ultrafluor and Soluscint-O liquid scintillation cocktail from National Diagnostics (Manville, NJ); fibrinogen from Kabi (Stockholm, Sweden); and dithiocarbodicyanine (diSC$_3$(5)) from Molecular Probes, Inc. (Eugene, OR). All other chemicals should be of reagent grade.

Thrombin Purification

Parke, Davis (Detroit, MI) topical thrombin must be purified[57] by ion-exchange chromatography on SP-Sephadex C-50 and affinity chromatography through a Sepharose 2B-lysine column to remove contaminating plasminogen. Fibrinogen clotting activity is determined by measuring the amount of time required for 0.1 ml of thrombin to clot 0.2 ml of fibrinogen (5 mg/ml of 0.15 M NaCl) at 37°, 1 U/ml being defined as the concentration (181 nM) that is capable of clotting 1 mg/ml fibrinogen in 0.25 min.[58] The ability to hydrolyze the small ester, TAME, is determined spectroscopically according to Hummel.[59] Protein concentrations can be determined by the method of Lowry *et al.*,[60] using bovine serum albumin for the standard curve.

Active-Site Blocking

For the preparation of active site-inhibited thrombin, a modification of the method of Workman *et al.*,[61] using TLCK, should be followed[19,27,62]: Purified α-thrombin is reacted with 10 mg/ml TLCK at pH 7.7 for 4 hr at room temperature, then overnight at 4° with constant stirring. TLCK-thrombin is then dialyzed against 4 liters of 0.15 M sodium phosphate buffer, pH 6.5, for 24 hr. The inactivated enzyme must be tested for esterase[59] and fibrinogen clotting activities,[58] which should be completely inhibited. The protein concentration can be measured as indicated above, and the relative purity verified via 10% (w/v) SDS–polyacrylamide tube gel electrophoresis. The modified thrombin is stable when stored at −20° in small aliquots.

DNCO Synthesis

The photolabel DNCO, N,N'-bis(2-nitro-4-azidophenyl)-cystamine S,S-dioxide, is synthesized according to Huang and Richards,[52] using 4-fluoro-3-nitrophenylazide (F-NAP) as the starting material. All handling of the azidophenyl compounds must be performed in the dark or under a red safe light. DNCO can be recrystallized from warm pyridine as red

---

[57] R. L. Lundblad, L. C. Uhteg, C. N. Vogel, H. S. Kindon, and K. G. Mann, *Biochem. Biophys. Res. Commun.* **66**, 482 (1975).
[58] R. L. Lundblad and J. H. Harrison, *Biochem. Biophys. Res. Commun.* **45**, 1344 (1977).
[59] B. C. Hummel, *Can. J. Biochem. Pharmacol.* **37**, 1393 (1959).
[60] O. H. Lowry, N. J. Rosebrough, A. L. Farr, and R. J. Randall, *J. Biol. Chem.* **193**, 265 (1951).
[61] E. F. Workman, Jr., G. C. White, II, and R. L. Lundblad, *J. Biol. Chem.* **252**, 7118 (1977).
[62] N. E. Larsen, W. C. Horne, and E. R. Simons, *Biochem. Biophys. Res. Commun.* **87**, 403 (1979).

needles (observed mp, 124–125°; reported mp, 127–128°; λ_{max} 459 nm. DNCO can be stored indefinitely in crystal form at 4° in the dark. For thrombin derivatization, a stock solution of 20 mg/ml DNCO in pyridine is prepared and addition to thrombin calculated so that the final pyridine concentration in the thrombin sample is less than 0.5% (v/v).

Platelet Preparation

Blood is drawn by antecubital venipuncture from normal volunteers and mixed with 3.8% (w/v) sodium citrate in plastic tubes for anticoagulation purposes, to produce a final concentration of 0.38% citrate. The blood is then centrifuged to produce a platelet-rich plasma and gel filtered on Sepharose 2B, as previously described,[27] in HEPES buffer (0.137 M NaCl, 0.0038 M HEPES, 0.0056 M D-glucose, 0.0038 M monobasic sodium phosphate, 0.0027 M KCl, 0.001 M $MgCl_2 \cdot 6H_2O$, at pH 7.35).[63,64] Apyrase is prepared from potatoes by the method of Molnar and Lorand,[65] or purchased from Sigma, and used at a final concentration of about 0.15 U/ml in the buffer. The concentration of gel-filtered platelets (GFP) is conveniently determined turbidimetrically on a Zeiss spectrophotometer at 436 nm, and evaluated by means of a standard calibration curve prepared from Coulter counter measurements.

Thrombin Derivatization by Reductive Amination

The thrombin analog is prepared by derivatizing α-thrombin via the carbohydrate residues on its B chain. As reported earlier,[66] these carbohydrate residues are not involved in any of the three above-mentioned activities of thrombin. For detection of the thrombin and, hence, the thrombin–membrane–protein complex, the photoreactive derivative must contain a tritium label of high specific activity. In order to couple DNCO to thrombin, a free SH group must be available. Both of these requirements can be met by incorporating 2-mercaptoethylamine in the derivative via Schiff base formation with oxidized carbohydrate residues followed by reduction with sodium borotritide (Fig. 1).

As Fig. 1 indicates, purified thrombin or TLCK-thrombin can be oxidized with 0.1 ml of 0.02 M $NaIO_4$ per milliliter with continuous stirring at room temperature for 30 min. Excess periodate is removed by overnight dialysis at 4° against 0.15 M sodium phosphate, pH 6.5. Full conver-

---

[63] S. Timmons and J. Hawiger, *Thromb. Res.* **12**, 298 (1978).
[64] S. Timmons and J. Hawiger, this series, Vol. 169, p. 11.
[65] J. Molnar and L. Lorand, *Arch. Biochem. Biophys.* **93**, 353 (1961).
[66] T. C. Hageman, G. F. Endres, and H. A. Scheraga, *Arch. Biochem. Biophys.* **171**, 327 (1975).

sion of carbohydrates to aldehydes is verified by means of the N-methylbenzothiazolone hydrazone (MBTH) assay.[67] Fibrinogen clotting, TAME hydrolysis, and platelet stimulation assays should be performed and should show that the oxidized thrombin retains its biological activity before the derivatization is continued.

The oxidized α-thrombin in sodium phosphate (pH 6.5) is then reacted with 10 μl of 0.1 M 2-mercaptoethylamine per ml of thrombin solution at pH 9.0 for 3 hr to form the Schiff base. A 10-fold molar excess of sodium borotritide (>50 Ci/mmol) in cold 0.01 M NaOH, is then added and the reduction allowed to proceed in an ice bath, with constant stirring, for 30 min. The radiolabeled thrombin can be separated chromatographically from the unreacted reagent and its hydrolysate on Sephadex G-25 columns preequilibrated with sodium phosphate buffer, pH 6.5 or extensively dialyzed at 4° against 0.15 M sodium phosphate, pH 6.5, until only low background counts remain. As above, the product, tritiated thrombin, must be tested for fibrinogen clotting and TAME esterase activity and counted to determine specific activity. It can be stored at $-20°$ without detectable loss of enzymatic activity.

Radiolabeling by Reductive Methylation

Some of the free amino groups of α- and TLCK-thrombin can be converted to the monomethyl derivatives, having almost the same pK, and without affecting the thrombin activity, by reaction with formaldehyde and the radiolabeled reducing agent, sodium borotritide, using a modification of the procedure of Tack et al.[53] α-Thrombin has 22 lysines per molecule, some of which are buried in the protein structure and are unavailable to the labeling procedure without denaturing the protein, while others are readily available to modification. The goal of this modification procedure is hence to derivatize a small portion of the exposed ε-amino groups so that the modification does not intererfere with the binding and hydrolytic functions of thrombin: an α- or TLCK-thrombin stock solution at approximately 0.5 mg/ml in phosphate buffer is adjusted to pH 8.9–9.0, and a two-fold excess of formaldehyde ($H^{14}CHO$) per thrombin lysyl residue is added with constant stirring. The thrombin solution is placed on ice, and all subsequent operations are carried out in a well-ventilated fume hood, while stirring. The sodium borohydride is dissolved in 0.01 N NaOH; 0.4 mol NaBH$_4$ of this stock solution per mole HCOH is then added to the reaction vials. The reaction is terminated after 30 min by readjustment of the pH to 6.5. The radiolabeled thrombin is then chromatographically

---

[67] M. S. Paz, O. O. Blumenfield, M. Rejkind, E. Henson, C. Furfine, and P. M. Gallop, *Arch. Biochem. Biophys.* **109,** 548 (1965).

separated from the unreacted reagent and its hydrolysate on Sephadex G-25 columns preequilibrated with 0.15 M sodium phosphate buffer, pH 6.5. The resultant radiolabeled thrombin is again assayed for protein concentration, and fibrinogen clotting, esterase, and biological activities (or competition with α-thrombin for these activities if the derivative is an active site-blocked TLCK-thrombin). Small aliquots of the thrombin are stored at $-20°$. Relative purity can be evaluated on SDS–polyacrylamide gels.

Thrombin–Platelet Coupling

Immediately before use, an aliquot of the tritiated or ^{14}C-labeled mercaptoethylamine modified α- or TLCK-thrombin is reacted with 5 μl/ml of DNCO (20 mg/ml of pyridine) or N-hydroxysuccinimidyl-6-(4'-azido-2'-nitrophenylamino)hexanoate (NHS-ANPH) for 1 hr at room temperature. The unreacted NHS-ANPH or DNCO is removed by dialysis against HEPES buffer, pH 7.4. The resulting DNCO-thrombin is tested as before for fibrinogen clotting, TAME hydrolysis, and platelet-stimulating activity. Since TLCK competes with thrombin for the platelet-binding sites, this competition is used to ensure identity of binding characteristics of DNCO-TLCK-thrombin to those of α-thrombin.

Fresh human platelets, prepared and washed on Sepharose 2B as previously described,[27,62] are concentrated to 6 $\times$ 10^8/ml of HEPES buffer by being placed in a sealed dialysis bag surrounded with dry polyethylene glycol (Carbowax, Fisher, MA). These platelets are then incubated with constant stirring with the desired quantity of radiolabeled DNCO-thrombin, ANPH-thrombin, DNCO-TLCK-thrombin, or ANPH-TLCK-thrombin (0.375 μg thrombin/ml = 0.05 U/ml) in the dark and incubated for 30 or 60 sec. The mixture is then poured into shallow petri dishes and photoactivated at 37° with constant stirring using a 200-W mercury lamp and a 320-nm cutoff filter. Photolysis is essentially immediate. A 100-fold molar excess of hirudin must then be added to remove noncovalently bound thrombin; the platelets are passed over a Sepharose 2B column preequilibrated with HEPES buffer, pH 7.4, to separate free hirudin-bound TLCK-thrombin (or any form of cross-linked thrombin) from the platelets. Aliquots of the thrombin-cross-linked platelets are saved for gel electrophoresis; for others, the platelet concentration is adjusted to 6 $\times$ 10^7 platelets/ml and the residual rate of membrane potential change in response to α-thrombin is measured. In a control experiment performed in the absence of platelets, DNCO-α-thrombin exhibited no loss in biological activity as a result of this coupling procedure.

The photoactivated platelet–thrombin mixture for gel electrophoresis is then centrifuged at 4000 g at 4°, and washed twice with 3 mM EDTA, pH 7.4. The pellet is solubilized in 3% SDS at 100° (boiling H_2O) for 10 min. We have compared this solubilized total platelet pellet with that obtained by solubilization of platelet membranes, isolated according to Barber and Jamieson[68] after photocoupling.[19] There is no difference in the thrombin–protein complex (as detected by column chromatography or by electrophoresis), but a smaller yield. In either case, the solubilized material should be electrophoresed immediately or stored at $-20°$ for no longer than 24 hr.

Sodium Dodecyl Sulfate-Polyacrylamide Gel Electrophoresis

Sodium dodecyl sulfate-polyacrylamide gel electrophoresis is performed by the method of Weber and Osborn.[69] Gels are scanned at 550 nm for protein and 560 nm for carbohydrate. For detection of radioactive bands, the gels are sliced into 1.0-mm sections and incubated overnight at 37° in 7 ml of 3% Protosol in Ultrafluor or Soluscint-O, and then counted for tritium on a Packard (Rockville, MD) liquid scintillation counter for 5 min/vial.

Determination of Number of Thrombin Molecules Covalently Bound to Receptors

The radioactivity in an aliquot of washed cross-linked platelets can be used to evaluate the incorporation of radiolabeled TLCK-thrombin, and the number of sites cross-linked per platelet calculated from the specific activity of the α- or TLCK-thrombin derivative. This method can be applied either to singly cross-linked platelets or to platelets that have been exposed sequentially to a subsaturating dose and to a saturating dose of, respectively, ^{3}H-labeled and ^{14}C-labeled TLCK-thrombin.

Gel Filtration and Isolation of Receptor Complex

Pellets from 2×10^{10} solubilized platelets can be counted for tritium, concentrated, mixed with blue dextran (1 mg/ml), and applied to a Sephacryl S-200 column (65 $\times$ 2.5 cm) (void volume 95 ml). The column should be preequilibrated with 0.04 M sodium phosphate, 0.1 M NaCl, 0.002 M disodium EDTA, 0.5% (w/v) SDS, 0.002% (w/v) sodium azide, pH 7.2, and eluted with the same buffer at room temperature at approximately 25 ml/hr and the fractions collected. Radioactivity can be detected by counting 50 μl of each fraction in 5 ml of Ultrafluor for 1 min/vial (usually

[68] A. L. Barber and G. A. Jamieson, *J. Biol. Chem.* **245,** 6357 (1970).
[69] K. Weber and M. Osborn, *J. Biol. Chem.* **244,** 4406 (1969).

80–95% recovery). The OD_{280} of the radioactive peak(s) is generally too low to be detectable until the corresponding eluate fractions are concentrated; it can then be estimated via their extinction coefficient, $\varepsilon = 10/$ mg/ml at 280 nm,[42] and by their radioactivity.

Reduction and Alkylation of Thrombin–Receptor Complex

By pooling samples from corresponding peaks of several isolations of the thrombin–receptor complex, as described above, one can obtain a sufficient amount (20 µg) for reduction with an equal volume of 1 M dithiothreitol in 1 M Tris-HCl, pH 7.5 at 37° for 3 hr. The pH is then adjusted to approximately 9.0 and 0.2 ml of a solution of 10 mg/ml iodoacetate in 2 N NaOH is added per milliliter of sample. The tube is then flushed with N_2 and kept in the dark for 30 min. The reaction is stopped by the addition of excess 2-mercaptoethanol (3%). An aliquot of this solution can be retained for SDS–PAGE electrophoresis. The remainder of the sample can, if desired, be applied to a small Sephacryl S-200 column (13 × 0.625 cm) previously equilibrated with the same buffer used for gel filtration but with the addition of 1% 2-mercaptoethanol. Eluted protein can be detected by absorbance at 280 nm and by radioactivity.

Fluorescence Measurements of Membrane Potential Change

Changes in the fluorescence of platelet suspensions (60 × 10^6 platelets/ml in HEPES buffer, 37°) are measured as described[27,62] with a Perkin-Elmer (Norwalk, CT) 650-10S spectrofluorimeter equipped with a thermostatted cuvette holder and a stirrer. Platelets, either fresh or after preincubation (and, in some cases, cross-linking) with TLCK-thrombin, are adjusted to the appropriate concentration (6 × 10^7/ml) and incubated with 2 µM $diSC_3(5)$ for 3 min in the spectrofluorimeter cuvettes. The stimulus is then added by remote injection through a microliter syringe equipped with polyethylene tubing. We have previously shown[27,62] that such treatment with this probe does not perturb the thrombin-induced platelet depolarization and subsequent aggregation. The initial response, depolarization of the membrane potential, is monitored continuously ($\lambda_{exc} = 620$ nm, $\lambda_{em} = 670$ nm). The rate of depolarization is determined by measuring the initial slope of the thrombin-induced fluorescence change (Δcm/min). We have shown this to be a more reproducible quantity than the relative maximal fluorescence change or the change after 30 sec, the parameters used in our previous studies.[28] The total $diSC_3(5)$ uptake by the platelet suspension can be determined from the maximal fluorescence of the suspension after addition of Triton X-100 to a final concentration of 0.1%.

Membrane potential changes, in platelets as well as in other cells, were one of the simplest and easiest parameters whose change paralleled cellular activation and therefore became standard measures of such activation. With the advent of new probes that permit continuous evaluation of platelet activation via changes in cytoplasmic Ca^{2+} or H^+ concentrations, which are probably directly involved in the signaling process,[29,70-78] we have found that the resting Ca^{2+} levels and the magnitude and time of appearance of a thrombin-induced Ca^{2+} transient are even better and more reproducible measures of platelet activation than the resting membrane potential and its stimulus-induced changes.[77,78]

Nature of High-Affinity Receptor for α-Thrombin: Current Status

The nature of the high-affinity thrombin receptor on human platelets is still controversial. Initially shown to be processed by thrombin[7,10-13,19,61] and therefore thought unlikely to be GPIb,[26] it was then thought to be a small number of copies of that plentiful platelet membrane glycoprotein.[33,35] There is no doubt that glycocalicin, a component of the GPIb α chain that is liberated from that protein by calpain (a calcium-dependent platelet protease), has a binding site for thrombin and that patients who lack GPIb have a thrombin-activation defect.[33,34,79-89]

Part of the controversy may rest on the individual definition of a receptor. That is, for thrombin, a platelet membrane binding site, no matter how high an affinity it exhibits, is not necessarily a functional receptor,

---

[70] T. J. Rink and R. J. Hallam, *Trends Biochem. Sci.* **376,** 215 (1984).
[71] T. J. Rink and S. O. Sage, *Physiol. Soc.* **369,** 115 (1985).
[72] T. J. Rink, S. W. Smith, and R. Y. Tsien, *FEBS Lett.* **148,** 21 (1982).
[73] T. J. Rink, R. Y. Tsien, and T. Pozzan, *J. Cell. Biol.* **95,** 189 (1982).
[74] S. O. Sage and T. J. Rink, *Biochem. Biophys. Res. Commun.* **136,** 1124 (1986).
[75] S. O. Sage and T. J. Rink, *J. Biol. Chem.* **262,** 16364 (1987).
[76] G. D. Jones and A. R. L. Gear, *Blood* **71,** 1539 (1988).
[77] T. A. Davies, D. Drotts, G. J. Weil, and E. R. Simons, *Cytometry* **9,** 138 (1988).
[78] T. A. Davies, D. Drotts, G. J. Weil, and E. R. Simons, *J. Biol. Chem.* **264,** 19600 (1989).
[79] S. M. Jung and M. Moroi, *Biochim. Biophys. Acta* **761,** 152 (1983).
[80] J. L. McGregor, J. Brochier, F. Wild, G. Follee, M-C. Trzeciak, E. James, M. Dechavanne, L. McGregor, and K. J. Clemetson, *Eur. J. Biochem.* **131,** 427 (1983).
[81] K. Yamamoto, H. Kitagawa, K. Tanoue, and H. Yamazaki, *Thromb. Res.* **39,** 751 (1985).
[82] M. C. Berndt, B. H. Chong, H. A. Bull, H. Zola, and P. A. Castaldi, *Blood* **66,** 1292 (1985).
[83] J. Takamatsu, M. K. Horne, and H. R. Gralnick, *J. Clin. Invest.* **77,** 362 (1986).
[84] D. Bienz, W. Schnippering, and K. J. Clemetson, *Blood* **68,** 720 (1986).
[85] B. Adelman, A. D. Michelson, R. I. Handin, and K. A. Ault, *Blood,* **66,** 423 (1985).
[86] A. N. Wicki and K. J. Clemetson, *Eur. J. Biochem.* **163,** 43 (1987).
[87] A. D. Michelson and M. R. Barnard, *Blood* **70,** 1673 (1987).
[88] G. E. Marti, L. Magruder, W. E. Schuette, and H. R. Gralnick, *Cytometry* **9,** 448 (1988).
[89] J. E. B. Fox and M. C. Berndt, *J. Biol. Chem.* **264,** 9520 (1989).

unless its ligand–receptor complex elicits a functional response. Since thrombin is one of the most potent signal-generating agonists for platelets, the elicited response includes generation of an activation signal. In this context, the proposed identification of GPIb as a functional high-affinity platelet thrombin receptor on platelet membrane remains controversial. Although a number of monoclonal (and one polyclonal) antibodies to GPIb exist, only their effect on von Willebrand factor binding has been well documented.[81–83,85,86,90,91] There are, at this writing, still contradictory statements in the literature with respect to the effect of these antibodies on α-thrombin binding, and on platelet activation, with conclusions ranging from partial inhibition to a total lack of effect.[81–83,85,86,90] The situation is complicated by the fact that partial effects can be due either to a partial inhibition of the response of each platelet or to a full inhibition of a portion of the platelets in the suspension being tested. Although this creates a fundamental difference in interpretation, the experimental differentiation between the two possibilities is achievable only by individual cell studies, i.e., by flow cytometry or enhanced microscopic imaging. Another difficulty in reaching a unanimity of opinion on the identity of the high-affinity α-thrombin receptor on human platelets may lie with the parameter whose inhibition is considered a measure of anti-receptor antibody efficacy. Such a parameter has usually been aggregation (but this is thought to be mediated by a number of pathways), with high or low doses of α-thrombin being used in different studies. The question awaits, for its resolution, the availability of methods that will distinguish between thrombin binding and thrombin-induced activation of platelets over short (<5 sec) time intervals. The now documented existence[77,78] of responding subpopulations when α-thrombin doses below 4.5 nM are used, even though thrombin has bound uniformly to all the platelets, indicates that the responding and nonresponding platelets are all capable of activation when higher doses are employed. Thus, the role of the high-affinity thrombin receptor on human platelets in stimulus–response coupling is more complex than previously thought.

Final Note

The cloning and expression of a functional thrombin receptor has been reported.[92] The same receptor, a protein of 425 amino acid residues (approximately 50,000 Da) and exhibiting seven transmembrane regions,

---

[90] B. S. Coller, E. I. Peerscke, L. E. Scudder, and C. A. Sullivan, *Blood* **61**, 99 (1983).
[91] K. Yamamoto, N. Yamamoto, H. Kitagawa, K. Tanoue, G. Kosaki, and H. Yamazaki, *Thromb. Haemostas.* **65**, 162 (1986).
[92] T. K. H. Vu, D. T. Hung, V. I. Wheaton, and S. R. Coughlin, *Cell* **62**, 1057 (1991).

appears to be present in platelets and in endothelial cells. Site-directed mutagenesis has shown this to be a functional receptor whose mechanism of action involves a thrombin-mediated cleavage that then exposes a cell-activating thrombin-binding site. Thus the properties of this newly isolated and characterized receptor fit the previously published facts. These include (1) the report by Tam and Detwiler of a thrombin-induced platelet activation site separate from the thrombin-binding site on the same receptor[14] and (2) reports from several investigators[13,15,19,23] that the receptor is proteolytically cleaved by thrombin as part of the activation process, and that active site-inhibited thrombin binds to the receptor but does not activate the platelet. Conversely, since there is ample evidence that GPIb is not cleaved on platelet activation, and because the sequence of the receptor does not match that of GPIb, this recent publication reinforces the view that GPIb is not the high-affinity functional platelet thrombin receptor, as the concentrations used were <10 nM. The size of the receptor, approximately 50,000 Da, is one-fourth that of the previously published values (approximately 200,000 Da); it remains to be seen whether this means that there is an assembly of these proteins in the membrane that is not dissociated by prolonged boiling in SDS (e.g., disulfide cross-linking).

[16] Platelet Membrane Glycoprotein V Purification

By DAVID R. PHILLIPS and MICHAEL C. BERNDT

Introduction

Glycoprotein (GP) V is a relatively minor glycoprotein of the platelet plasma membrane. Glycoprotein V (M_r 82,000) can be labeled on intact platelets by either the periodate- or the galactose oxidase-labeling procedures and has the distinguishing feature of being the only thrombin substrate yet detected on the platelet surface.[1,2] Thrombin hydrolysis of GPV yields a soluble fragment termed GPV$_{fl}$ (M_r 69,500). The thrombin susceptibility of GPV indicates that it interacts with thrombin during platelet activation.[3] GPV does not appear to be involved in thrombin-induced platelet activation, however, because the liberation of GPV$_{fl}$ does not

[1] D. R. Phillips and P. P. Agin, *Biochem. Biophys. Res. Commun.* **75**, 940 (1977).
[2] D. F. Mosher, A. Vaheri, J. J. Choate, and L. G. Gahmbery, *Blood* **53**, 437 (1979).
[3] M. C. Berndt and D. R. Phillips, *in* "Platelets in Biology and Pathology" (J. L. Gordon, ed.), Vol. 2, p. 43. Elsevier/North-Holland Biomedical Press, Amsterdam, 1981.

correlate with platelet stimulation[4] and GPV antibodies fail to block platelet stimulation.[5] GPV may be either physically or metabolically associated with GPIb–IX, as these glycoproteins are all deficient in Bernard–Soulier platelets.[6,7] Amino acid sequencing of purified GPV has shown that this glycoprotein contains at least 11 leucine-rich repeats, each composed of the consensus sequence PXXLLXXXXXLXXLXLSXNXLXXL, which is also observed in $GPIb_\alpha$, $GPIb_\beta$, and GPIX.[8] This chapter describes a procedure for isolation of homogeneous GPV from fresh platelets.[9] The purification procedure uses fresh washed platelets as the starting material because platelet membranes prepared by sonication or sucrose gradient centrifugation lack GPV, and because clinically expired platelets (>72 hr from venipuncture) contain little or no detectable GPV, as determined by periodate/sodium boro[^{3}H]hydride labeling.

Detection of Glycoprotein V

The electrophoretic mobilities of GPV and GPV_{fl} on sodium dodecyl sulfate (SDS)–polyacrylamide gels are used to detect GPV during its purification. Platelets labeled by the periodate/sodium boro[^{3}H]hydride procedure are used for these determinations. Glycoprotein V has an apparent molecular weight of 82,000 as determined by SDS–polyacrylamide gel electrophoresis of the labeled platelets and is identifiable by its disappearance when platelets are incubated with 2 units/ml of thrombin for 15 min at 37°. The hydrolytic fragment GPV_{fl} has an apparent molecular weight of 69,500 and is the new band that appears in solution after thrombin hydrolysis.

Glycoprotein V-containing fractions are identified during purification by analysis of fractions using SDS–polyacrylamide gel electrophoresis before and after thrombin hydrolysis. Fractions (50 μl) are incubated in the absence and presence of α-thrombin (2.5 units/ml, final concentration) for 90 min at 22°. The reaction is terminated by the addition of SDS [2% (w/v), final concentration], and the samples are subjected to SDS–polyacrylamide gel electrophoresis. Glycoprotein V is identified with thrombin treatment by the disappearance of one protein band and the appearance of another that corresponds in mobility to GPV and GPV_{fl}, respectively, from periodate/sodium boro[^{3}H]hydride-labeled platelets. Because GPV

[4] E. B. McGowan, A. Ding, and T. C. Detwiler, *J. Biol. Chem.* **258**, 11243 (1983).
[5] D. Bienz, W. Schnippering, and K. J. Clemetson, *Blood* **68**, 720 (1986).
[6] K. J. Clemetson, J. L. McGregor, E. James, *et al., J. Clin. Invest.* **70**, 304 (1982).
[7] A. T. Nurden, D. Dupuis, T. J. Kunicki, and J. P. Caen, *J. Clin. Invest.* **67**, 1431 (1981).
[8] T. Shimomura, K. Fujimura, S. Maehama, *et al., Blood* **75**, 2349 (1990).
[9] M. C. Berndt and D. R. Phillips, *J. Biol. Chem.* **256**, 59 (1981).

TABLE I
PURIFICATION OF HUMAN PLATELET GLYCOPROTEIN V

Step	Total protein[a] (A_{280})	A_{280}/A_{260}
Washed platelets	12,000	
Extract	960	1.46
Ammonium sulfate fraction (40 to 50%)	91.5	1.69
Sephacryl S-200	10.5	1.45
Hydroxylapatite	4.45	1.38
DEAE-cellulose	1.86	1.28
CM-cellulose	0.45–1.00[b]	1.29

[a] Values are per 100 units of platelet concentrate.
[b] Values correspond to 0.45 to 1.00 mg of purified GPV based on an $E_{1\,cm}^{1\%}$ value to 10.0 (range in yield for four preparations).

stains poorly for protein, GPV-positive fractions are conveniently identified in the early steps of purification (i.e., before the Sephacryl S-200 column step, described below) by staining the gels for carbohydrate using the sensitive dansylhydrazine procedure.[10]

Purification Procedure

Platelets from freshly drawn platelet concentrates (within 18 hr of venipuncture) are washed at ambient temperature as described,[11] except that the final washing solution contains 0.15 M sodium chloride, 1 mM EDTA, and 0.01 M HEPES, pH 7.6. Glycoprotein V is eluted from the platelet plasma membrane by equilibrating the platelets (4 × 10^9/ml) at 37° for 24 hr in 0.3 M NaCl, 1 mM EDTA, and 0.01 M HEPES, pH 7.6. All subsequent stages of the purification procedure are performed at 4°.

The supernatant obtained from centrifugation of the salt-extracted platelets at 800 g for 30 min is clarified further by centrifugation at 35,000 g for 60 min. The clear extract is dialyzed versus a buffer containing 1 mM EDTA and 0.05 M potassium phosphate, pH 6.8. The dialyzed extract is brought to 40% saturation with solid ammonium sulfate, stirred for 30 min, and centrifuged at 13,000 g for 60 min. The supernatant is brought to 50% saturation with ammonium sulfate and treated similarly. The pellet is taken up in a minimum of the 0.05 M potassium phosphate buffer and dialyzed versus the same buffer.

[10] A. E. Eckhardt, C. E. Hayes, and I. J. Goldstein, *Anal. Biochem.* **73**, 192 (1976).
[11] D. R. Phillips, L. A. Fitzgerald, L. V. Parise, and B. Steiner, this volume [22].

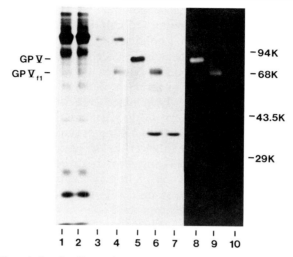

FIG. 1. Sodium dodecyl sulfate-polyacrylamide gel electrophoresis of purified GPV and the thrombin hydrolytic product GPV$_{f1}$. Lanes 1 to 4, periodate-labeled control and thrombin-treated platelet pellets and supernatants: lane 1, control platelet pellet; lane 2, platelet pellet after thrombin treatment; lane 3, control platelet supernatant; lane 4, platelet supernatant after thrombin treatment. Lanes 5 to 10, GPV in 0.1 M sodium phosphate, pH 7.4, treated with buffer or an equal concentration of α-thrombin for 1 min at 22° (GPV = 40 μg/ml; α-thrombin = 50 U/ml): lanes 5 to 7, protein stained with Coomassie Brilliant blue; lanes 8 to 10, carbohydrate stained with dansylhydrazine; lanes 5 and 8, 2 μg of GPV; lanes 6 and 9, thrombin digest of 2 μg GPV; lanes 7 and 10, α-thrombin. Protein standards in decreasing order of molecular weight: phosphorylase a, bovine serum albumin, ovalbumin, and carbonate dehydratase. (Reproduced with permission from Berndt and Phillips.[9])

The dialyzed sample is centrifuged at 35,000 g for 60 min to remove the flocculent precipitate that forms during dialysis. The clear supernatant is concentrated to about 10 ml by ultrafiltration using a Diaflo YM10 ultrafiltration membrane (Amicon, Danvers, MA). The concentrated sample is applied to a column (5 × 75 cm) of Sephacryl S-200 (Pharmacia, Piscataway, NJ), equilibrated with the 0.05 M potassium phosphate buffer, and eluted at a flow rate of 20 to 30 ml/hr. All GPV-positive fractions are pooled and dialyzed versus a buffer containing 0.2 mM EDTA and 5 mM potassium phosphate, pH 6.8, and then loaded onto a column (1 × 20 cm) of hydroxylapatite (Bio-Rad, Richmond, CA) equilibrated with the same buffer. The protein is eluted with a 300-ml linear gradient of 0.005 to 0.2 M potassium phosphate, pH 6.8 (0.1 mM in EDTA), at a flow rate of 12 ml/hr. Glycoprotein V-positive fractions are pooled, dialyzed versus a buffer containing 0.02 mM EDTA and 0.02 M sodium phosphate, pH 7.5, and loaded onto a column (0.7 × 14 cm) of DE-52 cellulose (Whatman,

Clifton, NJ) equilibrated with the same buffer. The flow-through is pooled, dialyzed versus a buffer containing 0.2 mM EDTA and 0.01 M sodium acetate, pH 4.5, and loaded onto a column (0.7 × 7 cm) of CM-52 cellulose (Whatman) equilibrated with the same buffer. The protein is eluted with a 100-ml linear gradient of 0 to 0.5 M NaCl at a flow rate of 12 ml/hr. The peak fractions containing GPV are pooled, dialyzed exhaustively against 0.05 M ammonium bicarbonate, lyophilized, and stored at $-20°$.

The purification procedure for GPV is summarized in Table I. This procedure reproducibly yields 0.45 to 1.0 mg of purified glycoprotein per 100 units of platelet concentrate (about 6×10^{12} platelets). Figure 1 shows purified GPV as detected by SDS–polyacrylamide gel electrophoresis and illustrates that the purified glycoprotein has electrophoretic properties identical to those of the glycoprotein in detergent-solubilized, periodate-labeled platelets and that the glycoprotein stains by Coomassie Brilliant blue and dansylhydrazine. Some physical properties of purified GPV have also been determined.[9] Glycoprotein V has an $E_{1cm}^{1\%}$ value of 10.0 at 280 nm. It contains ~48% carbohydrate by weight and is composed of neutral hexose, amino hexose, and sialic acid in a molar ratio of ~8 : 2 : 1. Electrophoresis of the purified glycoprotein according to O'Farrell[12] showed that it contains at least eight distinct isoelectric forms with isoelectric points (pI) ranging from 5.86 to 6.55; the four major forms have pI values of 6.28, 6.20, 6.12, and 6.04.

Acknowledgments

The authors thank James X. Warger and Norma Jean Gargasz for graphics assistance, Michele Prator and Kate Sholly for preparation of the manuscript, and Barbara Allen and Sally Gullatt Seehafer for editorial assistance.

[12] P. H. O'Farrell, *J. Biol. Chem.* **250**, 4007 (1975).

[17] Synthesis of a Yohimbine–Agarose Matrix Useful for Large-Scale and Micropurification of Multiple α_2-Receptor Subtypes

By STEVEN E. DOMINO, MARY G. REPASKE, CAROL ANN BONNER, MATTHEW E. KENNEDY, AMY L. WILSON, SUZANNE BRANDON, and LEE E. LIMBIRD

Introduction and Background

Epinephrine-provoked platelet aggregation and secretion of dense granules from human platelets are mediated by α_2-adrenergic receptors. This pharmacological characterization is based on the observation that platelet stimulation by epinephrine is blocked more potently by the α_2-selective antagonist, yohimbine, than by the α_1-selective antagonist, prazosin.[1] α-Adrenergic agonists are less useful for discriminating between subtypes. In fact, the prototypic α_2-selective agonist, clonidine, is actually a partial agonist (mixed agonist/antagonist) and its activity as an agonist in platelets can be demonstrated only by its ability to enhance synergistically ADP-induced platelet activation.[2]

α_2-Adrenergic receptor activation results in an inhibition of platelet cAMP accumulation due to inhibition of adenylate cyclase activity. In fact, the human platelet has been an important model system for elucidating the mechanism(s) of α_2-adrenergic receptor-mediated inhibition of cAMP synthesis.[3,4] Ironically, however, a number of lines of experimental evidence suggest that epinephrine-provoked secretion of dense granule contents occurs independent of decreases in intraplatelet cAMP levels. For example, Haslam *et al.* have shown that agonists linked to inhibition of cyclase can evoke platelet aggregation even in the presence of elevated levels of cyclic AMP.[5] Furthermore, we have shown that the removal of extraplatelet sodium blocks epinephrine-induced platelet secretion without affecting the ability of epinephrine to decrease cyclic AMP levels or to evoke primary aggregation.[6] These findings indicate that decreases

---
[1] J. A. Grant and M. C. Scrutton, *Br. J. Pharmacol.* **71**, 121 (1980).
[2] J. A. Grant and M. C. Scrutton, *Nature (London)* **227**, 659 (1979).
[3] T. Katada, J. K. Northup, G. M. Bokoch, M. Ui, and A. G. Gilman, *J. Biol. Chem.* **259**, 3578 (1984).
[4] A. G. Gilman, *Cell (Cambridge, Mass.)* **36**, 577 (1984).
[5] R. J. Haslam, M. M. L. Davidson, T. Davies, J. A. Lynnam, and M. D. McClenaghan, *Adv. Cyclic Nucleotide Res.* **9**, 533 (1978).
[6] T. M. Connolly and L. E. Limbird, *J. Biol. Chem.* **258**, 3907 (1983).

in cyclic AMP alone cannot account for epinephrine-provoked platelet secretion.

Because of the observation that receptor-mediated decreases in cyclic AMP production can be dissociated from receptor-induced platelet activation, we were interested in determining the role of other effector systems in α_2-adrenergic receptor-induced platelet secretion. Epinephrine activation of platelet secretion occurs subsequent to activation of the phospholipase C pathway in human platelets, as evidenced by agonist-induced increases in inositol phosphate and diacylglycerol production.[7] However, effects of epinephrine, as well as of ADP, on inositol phosphate production are indirect, as they are blocked by the cyclooxygenase inhibitor, indomethacin, and the endoperoxide/thromboxane A_2 antagonist, SQ 29,538. These data suggest that arachidonic acid must first be relased from another pool of lipids and, after conversion to cyclooxygenase products, elicit activation of the phospholipase C pathway. Release of this prior pool of arachidonic acid appears to be controlled by an Na^+/H^+ exchange mechanism, since manipulations that block Na^+/H^+ exchange also block the indomethacin-sensitive pathway of phospholipase C activation.[7,8] In contrast, thrombin (≥ 0.1 U/ml) and micromolar concentrations of the Ca^{2+} ionophores A23187 and ionomycin activate inositol phosphate production and platelet activation independent of the operation of an Na^+/H^+ exchange mechanism.[8] We demonstrated that a phosphatidylinositol-hydrolyzing phospholipase A_2 (PLA_2) activity is stimulated by epinephrine in human platelets. Furthermore, epinephrine or ADP stimulation of this PLA_2 activity is blocked by perturbants that block Na^+/H^+ exchange, suggesting that this enzyme might be responsible for mobilizing the pool of arachidonic acid that activates the phospholipase C pathway and resultant platelet secretion in response to epinephrine.[7] These findings suggest that the human platelet α_2-adrenergic receptor may be linked to multiple effector systems: (1) inhibition of adenylate cyclase activity and, either directly or indirectly, (2) stimulation of phospholipase A_2 activity.

The functional consequences of α_2-adrenergic receptor occupancy described above are not the focus of the present chapter, but serve as background for exploiting the various methodologies described. The present chapter will summarize methods for identification of α_2-adrenergic receptors on the human platelet and provide detailed protocols for solubilization of human platelet α_2 receptors and isolation of these receptors using

---

[7] J. D. Sweatt, I. A. Blair, E. J. Cragoe, Jr., and L. E. Limbird, *J. Biol. Chem.* **261**, 8660 (1986).

[8] J.D. Sweatt, S. L. Johnson, E. J. Cragoe, Jr., and L. E. Limbird, *J. Biol. Chem.* **260**, 12910 (1985).

yohimbine-agarose chromatography. Once obtained, the purified receptor provides the appropriate starting material for reconstitution experiments aimed at clarifying the multiple possible effector systems with which the α_2 receptor is capable of interacting. Thus, this resin will also be useful in determining the molecular events involved in receptor processing and turnover. In addition, we observed that this yohimbine-agarose resin not only permits purification of the α_{2A} subtype of α_2 receptors, which is the α_2 receptor subtype expressed in the human platelet, but also of other α_2 receptor subtypes (see below). Furthermore, this resin is applicable not only to large-scale purification of the receptor but also to micropurification of the α_2 receptor from metabolically labeled cells.

Identification of Human Platelet α_2-Adrenergic Receptors

Table I summarizes the commercially available radioligands that have been used for identifying human platelet α_2-adrenergic receptors and some of the advantages and disadvantages in using each of the ligands.[9-28] It is useful to keep in mind whether or not the radioligand is an antagonist,

[9] K. D. Newman, L. T. Williams, N. M. Bishopric, and R. J. Lefkowitz, *J. Clin. Invest.* **61,** 395 (1978).
[10] H. J. Motulsky and P. A. Insel, *Biochem. Pharmacol.* **31,** 2591 (1982).
[11] B.-S. Tsai and R. J. Lefkowitz, *Mol. Pharmacol.* **16,** 61 (1979).
[12] R. W. Alexander, B. Cooper, and R. I. Handin, *J. Clin. Invest.* **61,** 1136 (1977).
[13] T. W. Burns, P. E. Langley, B. E. Terry, D. B. Bylund, B. B. Hoffman, M. D. Tharp, R. J. Lefkowitz, A. Garcia-Sainz, and J. N. Fain, *J. Clin. Invest.* **67,** 467 (1981).
[14] K.-H. Jakobs and R. Raushek, *Klin. Wochenschr.* **56,** 139 (1978).
[15] M. L. Steer, J. Khoranagand, and B. Galgoci, *Mol. Pharmacol.* **16,** 719 (1979).
[16] C. J. Lynch and M. L. Steer, *J. Biol. Chem.* **256,** 3298 (1981).
[17] H. Motulsky, S. J. Shattil, and P. A. Insel, *Biochem. Biophys. Res. Commun.* **97,** 1561 (1981).
[18] D. E. MacFarlane, B. L. Wright, and D. C. Stump, *Thromb. Res.* **24,** 31 (1981).
[19] J. A. Garcia-Sevilla, P. J. Hollingsworth, and C. B. Smith, *Eur. J. Pharmacol.* **74,** 329 (1981).
[20] M. Daijiyi, H. Y. Meltze, and D. C. U'Prichard, *Life Sci.* **28,** 2705 (1981).
[21] S. K. Smith and L. E. Limbird, *Proc. Natl. Acad. Sci. U.S.A.* **78,** 4026 (1981).
[22] B. B. Hoffman, T. Michel, D. Mullikin-Kilpatrick, R. J. Lefkowitz, M. E. M. Tolbert, H. Gilman, and J. N. Fain, *Proc. Natl. Acad. Sci. U.S.A.* **77,** 4569 (1980).
[23] S. J. Shattil, M. McDonough, J. Turnbull, and P. A. Insel, *Mol. Pharmacol.* **19,** 179 (1981).
[24] J. J. Mooney, W. C. Horne, R. I. Handin, J. J. Schildkraut, and R. W. Alexander, *Mol. Pharmacol.* **21,** 600 (1982).
[25] R. R. Neubig, R. D. Gantzos, and R. S. Brasier, *Mol. Pharmacol.* (1985).
[26] J. T. Turner, C. Ray-Prenger, and D. B. Bylund, *Mol. Pharmacol.* (1985).
[27] J. W. Regan, R. M. DeMarinis, M. G. Caron, and R. J. Lefkowitz, *J. Biol. Chem.* **259,** 7864 (1984).
[28] S. M. Shreeve, C. M. Fraser, and J. C. Venter, *Proc. Natl. Acad. Sci. U.S.A.* **82,** 4842 (1985).

TABLE I
COMMERCIALLY AVAILABLE RADIOLIGANDS USED FOR IDENTIFICATION OF
α_2-ADRENERGIC RECEPTORS IN HUMAN PLATELET PREPARATIONS

Radioligands	Advantages or disadvantages	References for Intact platelets	References for Membrane preparations
Antagonists			
[^{3}H]Dihydroeryocryptine	High specific radioactivity; relatively high nonspecific binding	9, 10	11–13
[^{3}H]Dihydroergonine	High specific radioactivity; relatively high nonspecific binding	14	14
[^{3}H]Phentolamine	Relatively high nonspecific binding		15, 16
[^{3}H]Yohimbine	High affinity, high specific radioactivity (>80 Ci/mmol), low nonspecific binding	17, 18	16, 19–22
[^{3}H]Rauwolscine	Stereoisomer of α-yohimbine, but higher nonspecific binding		
Partial agonists			
[^{3}H]Clonidine	Probably labels solely "high-affinity state" for receptor–agonist interactions, therefore not useful for determination of total receptor density		19, 23
p-Amino[^{3}H]clonidine	Probably labels solely "high-affinity state" for receptor–agonist interactions, therefore not useful for determination of total receptor density		24
Agonists			
[^{3}H]Epinephrine	High affinity and specific radioactivity (60–80 Ci/mmol), but relatively high nonspecific binding that increases on storage; needs frequent repurification; identifies primarily "high-affinity state" for receptor–agonist interactions		21, 22

partial agonist, or full agonist in terms of inhibiting adenylate cyclase activity when selecting a radioligand to accomplish certain experimental goals. This is because receptor–agonist and receptor–partial agonist interactions are modulated by a number of regulatory agents that, when present, influence the ability to detect receptors in various preparations of the human platelet (cf. Table II). Thus, studies aimed at quantitating the

TABLE I (continued)

Radioligands	Advantages or disadvantages	References for Intact platelets	References for Membrane preparations
[³H]Norepinephrine	As above for [³H]epinephrine; additional limitation is receptor affinity 10-fold less for [³H]norepinephrine than [³H]epinephrine		16
[³H]UK14,304	High affinity, high specific radioactivity (>80 Ci/mmol); more stable to long-term storage; under appropriate conditions may identify multiple "affinity states" for receptor–agonist interactions		25, 26
Irreversible ligands[a]			
[³H]Phenoxybenzamine	30–60 Ci/mmol; commercial availability variable; extraordinarily high nonspecific binding; to date, only useful for identifying highly purified receptor preparations (30–60 Ci/mmol)		27
p-Azido[³H]clonidine	Specific activity 30–40 Ci/mmol; low yield of incorporation and relatively high nonspecific binding		28

[a] Useful for isolated and partially purified receptors only.

density of the receptors on the human platelet are best accomplished using antagonist radioligands. [³H]Yohimbine and [³H]rauwolscine are clearly the radiolabeled antagonist ligands of choice for identifying α_2 receptors on either intact platelets or in particulate preparations obtained from platelet lysates, because these ligands are available at high specific radioactivity and demonstrate very low nonspecific binding. The antagonist [³H]phenoxybenzamine (PBZ) is not useful for the routine detection of α_2 receptors because this ligand binds to a number of other receptors and nonreceptor proteins. However, this ligand is an alkylating agent and has been useful in covalently labeling highly purified preparations of the α_2 receptor for detection following sodium dodecyl sulfate (SDS)–polyacrylamide gel electrophoresis.[27]

As indicated in Table I, radiolabeled agonist and partial agonist agents have also been used to identify platelet α_2 receptors. However, as summa-

TABLE II
EFFECTORS THAT MODULATE α_2-ADRENERGIC RECEPTOR AFFINITY FOR AGONIST AND ANTAGONIST AGENTS IN HUMAN PLATELET PARTICULATE PREPARATIONS

Effector	Affinity of α_2 receptor for			
	Antagonists	Ref.	Agonists	Ref.
Guanine nucleotides	↑ (2-fold)	34	↓ (10- to 20-fold)	36
Na$^+$ (EC$_{50}$ 5–15 mM)	↑ (2-fold)	33	↓ (10- to 20-fold)	32, 33, 37
H$^+$ (pH 7.35 → pH 6.8)	↑ (2-fold)	35	↓ (10-fold)	35
Mg^{2+}	↓	34	↑ [a]	36

[a] The presence of Mg^{2+} may be essential for detecting the high-affinity state for α_2 receptor interactions, presumed to reflect a ternary complex of agonist · α_2R · G$_i$ (see text).

rized in Table II, a number of effectors influence α_2 receptor–agonist interactions. The interpretation of radiolabeled agonist binding to human platelet particulate preparations must be considered in light of the understanding that α_2 receptor–agonist interactions manifest multiple affinity states. The "high-affinity state" of the receptor (R) for agonists (Ag) has an equilibrium dissociation constant (K_D value) in the nanomolar range, whereas the "low-affinity state" possesses a K_D value for agonists in the micromolar range.[29] By analogy with receptors linked to stimulation of cyclase,[30] the high-affinity state of the α_2 receptor has been interpreted to represent a ternary complex of agonist–receptor–G$_i$, where G$_i$ is the GTP-binding protein linking receptor occupancy to inhibition of cyclase. GTP or hydrolysis-resistant guanine nucleotides appear to dissociate this high-affinity ternary complex (Ag · R · G$_i$) and promote the formation of the low-affinity state (Ag · R). The rapid dissociation of ligand from this lower affinity state means that radiolabeled agonist binding to this state is not "trapped" using the vacuum filtration methods routinely employed to terminate radioligand-binding assays. Consequently, one probably identifies exclusively the high-affinity state for agonists when [^{3}H]epinephrine or [^{3}H]norepinephrine are used as radioligands. This may or may not be true using [^{3}H]UK14,304, which possesses a significantly higher affinity than epinephrine at α_2 receptors.[25,26] Inherent in the above discussion is that the binding of radiolabeled agonists and partial agonists can only be studied using well-washed particulate preparations where endogenous GTP and Na$^+$ can be removed and where the concentrations of the GTP, Na$^+$, and Mg^{2+} effectors can be intentionally manipulated.

[29] B. B. Hoffman, T. Michel, T. Brenneman, and R. J. Lefkowitz, *Endocrinology (Baltimore)* **110**, 1926 (1982).
[30] A. DeLean, J. M. Stadel, and R. J. Lefkowitz, *J. Biol. Chem.* **255**, 7108 (1980).

As shown in Table II, Na^+ shares with GTP the ability to decrease receptor affinity for agonists[31-33] (also see Refs. 34–37). The slight increase in receptor affinity for antagonists caused by Na^+ is observed even in the presence of Mg^{2+}, whereas the effects of GTP on receptor–antagonist interactions appear to be masked by increasing concentrations of Mg^{2+}. Although the phenomenological effects of Na^+ and GTP on α_2 receptor–ligand interactions are similar, they are synergistic (or at least additive), suggesting distinct sites of action. Furthermore, the observation that alkylation of intact platelets with N-ethylmaleimide, exposure of platelet membranes to elevated temperatures (45° for 15 min or 60° for 5 min), or solubilization of unoccupied human platelet α_2 receptors with digitonin eliminates the regulatory effects of guanine nucleotides, but not those of Na^+, also suggests that the effects of Na^+ are not mediated by the GTP-binding α subunit of G_i.[34] In fact, it has been established that the effects of Na^+ on α_2 receptor–ligand interactions are due to an allosteric site for monovalent cations on the receptor itself.[38-40] The functional relevance of this site on α_2 receptors, or on other G protein-coupled receptors,[40] remains to be established.

Isolation of Human Platelets and Preparation of Washed Lysates (Particulate Preparations)

Although methods are available for preparing highly enriched surface membrane preparations from human platelets,[25,41,42] the yield of membrane protein is low using these procedures. Furthermore, highly purified membrane fractions do not appear to be necessary to detect [^{3}H]yohimbine binding or α_2 receptor-mediated inhibition of adenylate cyclase activity, although detection of radiolabeled agonist binding may be improved in

[31] B.-S. Tsai and R. J. Lefkowitz, *Mol. Pharmacol.* **14**, 540 (1978).
[32] T. Michel, B. B. Hoffman, and R. J. Lefkowitz, *Nature (London)* **288**, 709 (1980).
[33] L. E. Limbird, J. L. Speck, and S. K. Smith, *Mol. Pharmacol.* **21**, 609 (1982).
[34] Y.-D. Cheung, D. B. Barnett, and S. R. Nahorski, *Eur. J. Pharmacol.* **84**, 79 (1982).
[35] H. J. Motulsky and P. A. Insel, *J. Biol. Chem.* **258**, 3913 (1983).
[36] B.-S. Tsai and R. J. Lefkowitz, *Mol. Pharmacol.* **16**, 61 (1979).
[37] L. E. Limbird and J. L. Speck, *J. Cyclic Nucleotide Protein Phosphorylation Res.* **9**, 191 (1983).
[38] M. G. Repaske, J. M. Nunnari, and L. E. Limbird, *J. Biol. Chem.* **262**, 12381 (1987).
[39] J. N. Nunnari, M. G. Repaske, S. Brandon, and L. E. Limbird, *J. Biol. Chem.* **262**, 12387 (1987).
[40] D. A. Horstman, S. Brandon, A. L. Wilson, C. A. Guyer, and L. E. Limbird, *J. Biol. Chem.* **265**, 17307 (1990).
[41] A. J. Barber and G. A. Jamieson, *J. Biol. Chem.* **245**, 6357 (1970).
[42] D. M. F. Cooper and M. Rodbell, *Nature (London)* **282**, 517 (1979).

these preparations.[25] The procedure outlined below for preparing particulate preparations from freshly obtained whole blood or from outdated platelet packs has been useful for analysis of radioligand binding, α_2-adrenergic receptor-mediated inhibition of adenylate cyclase, or as starting material for detergent solubilization of α_2-adrenergic receptors.[21,36]

1. Human blood is obtained by venipuncture into 21-gauge butterfly needles and drawn into 60-ml syringes containing 6 ml of acid–citrate–dextrose (ACD: combine 12.5 g dextrose, 11.0 g of sodium citrate, 4.0 g citric acid, bring pH to 5.0 with citric acid, and bring to 500-ml final volume) as the anticoagulant.

2. Platelet-rich plasma (PRP) is obtained by centrifuging whole blood (30 ml/50-ml conical centrifuge tube) for 15 min at 1200 rpm in a Sorvall (Du Pont/Sorvall, Newtown, CT) GLC table-top centrifuge maintained at room temperature. The platelet-rich plasma is removed by aspiration using a *siliconized* Pasteur pipette, taking care not to remove the buffy coat.

3. The pH of the PRP is reduced to 6.5 by adding one-tenth volume of ACD. This manipulation prevents aggregation of the platelets during the subsequent centrifugation step. The PRP is centrifuged in 50-ml conical tubes (25 ml PRP/tube) for 20 min at room temperature at 2400 rpm in a Sorvall GLC table-top centrifuge.

3a. When platelets are obtained as outdated platelet packs from the Red Cross, the isolation procedure begins at step 3, except that the pH of the platelet pack does not have to be adjusted with ACD, as it has already been reduced prior to platelet storage.

4. The supernatant from the centrifugation in step 3, above (platelet-poor plasma) is removed by aspiration. The platelet pellet is washed once in 150 mM NaCl, 50 mM Tris-HCl, 20 mM EDTA, pH 7.5 by gentle resuspension in a siliconized Pasteur pipette. If contaminating red cells are present, care is taken to avoid resuspending these, and the milky white platelet suspension is transferred to a clean, siliconized 15-ml conical centrifuge tube. This washed platelet pellet is collected by centrifugation at room temperature for 20 min at 2400 rpm in a Sorvall GLC table-top centrifuge.

5. The platelets in the pellet are lysed by resuspending 1 ml (or less) of packed platelets into a 4-ml final volume of ice-cold lysing buffer (5 mM Tris-HCl, 5 mM EDTA, pH 7.5 with NaOH or Tris-HCl) to which is immediately added phenylmethylsulfonyl fluoride (PMSF) at a final concentration of 10 μM. This mixture is frozen in an ethanol/dry ice bath for 5 min. If platelet preparations are to be stored, this can be done at this point by freezing the platelets in liquid N_2 storage vials. Outdated platelet

preparations have been stored up to 2 years in liquid N_2 without apparent change in receptor–antagonist interactions.

6. The frozen platelets are thawed by standing at room temperature. The platelets are further lysed by transfer to a 40-ml Sorvall centrifuge tube and, after bringing the platelet volume to 10 ml with lysing buffer, by exposure to a Polytron for two 5-sec bursts interrupted by 30 sec on ice. (The tube is maintained in an ice water bath even during the 5-sec Polytron step.) Care is taken to wash the Polytron probe well to obtain all of the platelet preparation.

7. The platelet particulate preparation is isolated by centrifugation in a 12-ml Sorvall centrifuge tube for 10 min at 18,000 rpm (39,000 g) in a Sorvall 2B refrigerated centrifuge maintained at 4°. The pellet is washed two or four times by resuspension in lysing buffer followed by centrifugation at 39,000 g for 10 min. The number of washes depends on whether it is essential to remove endogenous guanine nucleotides (four washes) for the anticipated experiments.

8. The final wash of the platelet lysate is performed in the buffer into which the particulate preparation will ultimately be suspended. Recommendations for buffer conditions are given below:

   a. For radioligand-binding studies in which radiolabeled antagonist binding is to be optimized (K_D for [^{3}H]yohimbine is 0.6 nM under these conditions), the buffer is 25 mM glycylglycine, 2 mM EGTA, 120 mM NaCl, pH adjusted to 7.65 with NaOH.

   b. When agonist competition for radiolabeled antagonist binding is to be assessed, or direct radiolabeled agonist binding is to be performed, the resuspension buffer contains 25 mM glycylglycine, 4 mM MgCl$_2$, 2 mM EGTA, pH adjusted to 7.65 with N-methyl-D-glucamine. This buffer is prepared in the absence of sodium ion, e.g., the EGTA acid is employed rather than the sodium salt.

Solubilization of Particulate Preparation

The human platelet α_2 receptor can be solubilized into the mild detergent, digitonin, in such a way that receptor recognition of α-adrenergic agents retains the specificity characteristic of intact membrane preparations. The procedures outlined below describe (1) the final washes of the particulate preparations prior to exposure to digitonin-containing buffers, (2) procedures for solubilization of the α_2 receptor into digitonin-containing solutions, and (3) methods for identification of α_2 receptors in digitonin-containing preparations. A discussion of the characteristics of detergent-solubilized α_2 receptors follows the description of the methodologies for solubilizing and assaying these preparations.

1. Prior to detergent extraction, the particulate preparation can be washed by resuspension using a siliconized Pasteur pipette into 0.5 M KCl, 5 mM Tris-HCl, 5 mM EDTA, pH 7.5. This procedure is intended to remove peripheral membrane proteins and does not decrease α_2 receptor-binding activity. If a KCl extraction is employed, the particulate preparation is subsequently washed in 25 mM glycylglycine, 120 mM NaCl, and 2 mM EGTA, pH 7.65 by resuspension using a Pasteur pipette and centrifugation at 18,000 rpm for 10 min at 4° in a Sorvall 2B refrigerated centrifuge.

2. The pellet from the above centrifugation is resuspended in 10–20 up-and-down strokes in a Teflon–glass homogenizer into a buffer containing 0.08% digitonin, 25 mM glycylglycine, 120 mM NaCl, 2 mM EGTA, pH 7.65. Approximately 1.5–2 mg of membrane protein is suspended into 1 ml of this digitonin-containing buffer, but the homogenization is carried out such that the membrane pellet is homogenized into one-third of the final volume, and the rest of the volume is used to wash the pestle and homogenizer. The 0.08% digitonin-containing solution is stirred on ice for 15 min prior to centrifugation at 18,000 rpm (39,000 g) for 30 min at 4° in a Sorvall 2B centrifuge. This step does *not* solubilize α_2 receptors from the platelet membrane, but removes some protein and improves the subsequent solubilization of α_2 receptors on exposure to higher concentrations of digitonin.

3. The pellet from the above centrifugation is homogenized by Teflon–glass homogenizer into 0.5% digitonin, 25 mM glycylglycine, 120 mM NaCl, 2 mM EGTA, pH 7.65. Significantly better yields of receptor solubilization are obtained if the platelet pellet is homogenized into one-third of the final volume for the digitonin solubilization, as described in step 2 above. For example, approximately 50 mg of particulate preparation is solubilized into a final volume of 25 ml, but the pellet is transferred to the Teflon/glass homogenizer and homogenized with 10 up-and-down strokes in approximately 8 ml of 0.5% digitonin-containing buffer and the remaining 17 ml of detergent solution is used to wash extensively the homogenizer and pestle. The 0.5% digitonin-containing incubation is stirred on ice for 30 min prior to centrifugation at 105,000 g for 60 min at 4°. The supernatant from this centrifugation is defined as the "solubilized preparation," and has been demonstrated to pass through 0.22-μm filters, be included in Sepharose 2B columns, and be devoid of particulate material under electron microscopy,[21] thus meeting the operational criteria for a solubilized receptor.

It is important to note that digitonin is not readily soluble in aqueous solutions. Concentrated digitonin stock solutions are prepared by adding digitonin powder (Gerhard–Schlessinger, New York, NY) to just-boiled water. Once this digitonin solution has clarified, the solution is filtered through a Millipore (Bedford, MA) 0.45-μm HAWP filter. Digitonin stock solutions (up to 20%, w/v) prepared in this manner are then stored at 4°. Digitonin falls out of solution after long standing at 4°, particularly if Mg^{2+} is present. By storing as a concentrated solution in water, the digitonin may be redissolved at a later date, if necessary, by briefly reheating the solution to clarity. All digitonin-containing buffers for solubilization are made fresh the day of use. It is important to prepare buffers for column chromatography or sucrose gradient centrifugation well in advance of use, so that insoluble contaminants in the commercially extracted digitonin preparation can be removed by multiple filtrations through Millipore 0.45-μm HAWP filters, if necessary, until sedimentation of digitonin no longer occurs. Digitonin obtained from Gerhard–Schlessinger does not typically require multiple filtrations; material from other manufacturers does. The yield of receptor solubilization with digitonin definitely varies from lot to lot of digitonin but is routinely 35–40%.

Identification of α_2-Adrenergic Receptors in
Digitonin-Solubilized Preparations

Exchange of Solubilized Preparations into 0.1% Digitonin

Binding to digitonin-solubilized receptors can be detected in the 0.5% digitonin-containing extract, but is improved if the preparation is exchanged into 0.1% digitonin-containing buffers by Sephadex G-50 chromatography. Two-milliliter aliquots of solubilized preparation are applied to 1 × 20–22 cm Sephadex G-50 columns preequilibrated with 0.1% digitonin-containing buffers. The void volume, defined as the peak elution position of blue dextran 2000 (2.4–2.7 ml), is collected and used for subsequent receptor binding. Usually 85% or more of the solubilized receptors binding activity is obtained in the void volume of these columns. Although the Sephadex G-50 column is a "desalting" column, the height-to-volume ratio is critical for the purposes of exchange into lower digitonin concentrations, because the detergent micelles elute just after the void volume, not in the salt volume of the column.

Binding to unoccupied α_2 receptors in digitonin-solubilized receptors is best accomplished using [^{3}H]yohimbine; [^{3}H]rauwolscine demonstrates higher nonspecific binding to detergent-solubilized preparations. [^{3}H]Dihydroeryocryptine is absolutely unsuitable as a radioligand under these circumstances because of its binding to and uptake into digitonin-containing micelles.

[³H]Yohimbine-Binding Assays to Solubilized Preparations

[³H]Yohimbine binding to digitonin is assayed for 90 min at 15° in the presence of 7.5 nM [³H]yohimbine (unless saturation analysis as a function of increasing radioligand concentrations is being performed) in the absence or presence of competing agents in a final volume of 0.5 ml. In unpurified preparations, the solubilized receptor preparation is typically added as 0.4 ml of the 0.5-ml incubation. Separation of receptor-bound ligand from free ligand is accomplished at 4–10° by chromatography of the 0.5-ml incubation on 0.6 × 14 cm Sephadex G-50 columns equilibrated and eluted with 0.025% digitonin-containing buffers. The 0.9- to 1.1-ml volume corresponding to the elution volume for blue dextran 2000 (defined as the void volume and representing receptor-bound ligand) is collected into scintillation vials containing 10 ml of NEN-963 scintillation fluor (New England Nuclear, Boston, MA). Specific binding, defined as that binding competed for by 10 μM phentolamine, is ≥75% of total binding.

Although recovery of receptor-bound ligand from the 0.6 × 14 cm Sephadex G-50 columns described above is not quantitative, we have observed that the sensitivity and repeatability of detection of radioligand binding to detergent-solubilized α_2 receptors from human platelets is nonetheless significantly better using Sephadex G-50 chromatography to separate bound and free ligand at the end of the incubation than using alternative procedures, such as filtration through polyethyleneimine-coated glass fiber filters, precipitation with polyethylene glycol, or adsorption of radioligand onto dextran-coated charcoal (M. Repaske, 1985, unpublished findings).

Characteristics of Digitonin-Solubilized Human Platelet α_2-Adrenergic Receptor

Exposure of human platelet particulate preparations to digitonin results in the solubilization of an α_2 receptor ligand binding site, which retains its recognition properties for α_2-adrenergic agents.[21] The solubilized receptor possesses an affinity for [³H]yohimbine indistinguishable from that observed for binding to platelet membranes under comparable incubation conditions. Thus, the K_D for [³H]yohimbine is 6.0 nM in the presence of 50 mM Tris-HCl, 5 mM MgCl$_2$, 2 mM EDTA, pH 7.65 and 0.6 nM in the presence of 25 mM glycylglycine, 120 nM NaCl, 2 mM EGTA, pH 7.65. (Tris-HCl actually suppresses detectable [³H]yohimbine binding.) Furthermore, a number of antagonist competitors demonstrate potency (K_D value) in digitonin-solubilized preparations identical to that in native membranes.[21] Agonists also compete for [^{3}H]yohimbine binding in digitonin-solubilized preparations with the same order of potency as that observed in native membranes. However, the EC_{50} values for agonists are considerably (10- to 20-fold) increased after solubilization, indicating a selective loss in receptor affinity for agonists on membrane disruption.[21]

Altered Properties of α_2-Adrenergic Receptor Agonist Interactions in Detergent-Solubilized Preparations

A number of lines of experimental evidence suggest that the selective decrease in α_2 receptor affinity for agonist agents on solubilization occurs because digitonin disrupts α_2 receptor interaction with the GTP-binding protein, which, in the native membrane, is responsible for conferring the high-affinity state for receptor–agonist interactions. First, guanine nucleotides do not modulate receptor affinity for agonists in solubilized preparations in which the receptors are unoccupied at the time of solubilization. In fact, the EC_{50} for epinephrine for competing with [^{3}H]yohimbine for binding to solubilized receptors (EC_{50}, 6 μM) resembles that for epinephrine competing for [^{3}H]yohimbine binding to native membrane preparations in the presence of guanine nucleotides (EC_{50}, 3 μM), whereas the EC_{50} for epinephrine in competing with [^{3}H]yohimbine binding to membrane preparations is 0.2 μM in the absence of added guanine nucleotides. Second, α_2 receptor occupancy by the radiolabeled agonist, [^{3}H]epinephrine, prior to digitonin solubilization stabilizes receptor interactions with the effector component(s) conferring sensitivity to guanine nucleotides, as evidenced by the ability of Gpp(NH)p to facilitate [^{3}H]epinephrine dissociation from prelabeled, solubilized [^{3}H]epinephrine–receptor complexes, in a manner analogous to guanine nucleotide-facilitated [^{3}H]epinephrine dissociation from the receptors in intact membrane preparations.[21] In addition, [^{3}H]epinephrine occupancy of platelet α_2 receptors prior to digitonin solubilization induces or stabilizes a molecular complex of larger apparent size that sediments more rapidly in sucrose gradients than α_2 receptors, which are unoccupied[21] or antagonist occupied[21,43] at the time of solubilization. Finally, the presence of guanine nucleotides with [^{3}H]epinephrine during the membrane-prelabeling incubation prevents or reverses the agonist-stabilized increase in receptor size.[21] Taken together, the above findings suggest that digitonin solubilization of human platelet membranes extracts the α_2-adrenergic receptor without perturbing the

[43] T. Michel, B. B. Hoffman, R. J. Lefkowitz, and M. G. Caron, *Biochem. Biophys. Res. Commun.* **100**, 1131 (1981).

recognition properties of the receptor. Solubilization, however, does disrupt putative α_2 receptor–GTP-binding protein interactions unless the receptor is occupied by an agonist at the time of solubilization. The practical consequence of this mechanistic understanding is that identification of α_2 receptors in digitonin-solubilized preparations requires that the antagonist [^{3}H]yohimbine be used as the radioligand, as the binding of [^{3}H]epinephrine to the detergent-solubilized low-affinity state of the α_2 receptor would not be detected.

Synthesis of Yohimbine–Agarose Matrix

The present protocol for synthesis of the yohimbine-agarose matrix for purification of the α_2 receptor has been streamlined since our initial report[38] to increase the ease of synthesis, decrease the quantity of reagents used and hence reduce synthetic costs, and improve the overall coupling efficiency. The procedure is straightforward and has been replicated many times and by many laboratory colleagues with comparable yield.

Routinely, 150 ml of yohimbine-agarose is prepared. The resin is stable (in terms of capacity, degree of purification, and ultimate yield of purified α_2-adrenergic receptor) for at least 9 months. The synthesis is performed using 15-ml aliquots of the diaminodipropylamine agarose backbone.

The diaminodipropylamine agarose (15 ml packed resin volume; Pierce Biochemicals, Rockford, IL) is aliquoted in a 1 : 1 aqueous slurry into 30-ml glass syringes. The resin is retained in the column using a single piece of nylon mesh and an O-ring to secure the mesh at the base of the flat-bottomed syringe.

Because yohimbinic acid is not readily soluble in aqueous buffers at the concentrations used in this protocol, we chose to perform the synthesis in dimethyl sulfoxide. A contributing consideration to this choice was our knowledge that the dicyclohexylcarbodiimide reagent used for coupling and, more importantly, the dicyclohexylurea by-product of the reaction both are soluble in dimethyl sulfoxide (DMSO).

Prior to initiating the synthetic steps, the resin is saturated with DMSO by six washes. For each wash, the syringe is filled with DMSO, the resin resuspended by end-over-end mixing on a rotator, and the DMSO eluted from the syringe dropwise through the syringe outflow. For end-over-end rotation the syringe is stoppered with a rubber stopper fitted with a 21-guage needle and a three-way stopcock, which allows degassing of the resin mixture prior to removing the stopper. When the resin has become saturated with DMSO, it will have shrunk to two-thirds its original volume and is translucent yellow in appearance.

Yohimbinic acid monohydrate (Aldrich Chemicals, Milwaukee, WI), dissolved in DMSO (1.2 g/15 ml resin), is added to each aliquot of resin and the resin suspended in a 20-ml final volume with DMSO, again by end-over-end rotation. The dicyclohexylcarbodiimide (DCCD; Aldrich Chemicals; 1.25 g/15 ml resin) is dissolved in 2–3 ml of DMSO by heating at 60° and injected into the yohimbinic acid–resin slurry using a heated glass pipette. The syringe is quickly rotated to allow rapid mixing of the reagents. The total volume of the reaction is brought to 30 ml before end-over-end rotation of the syringe. The final concentration of reactants added is 100 mM yohimbinic acid and 200 mM DCCD, representing a 10- and 20-fold molar excess, respectively, of these reactants over the amino group capacity of the diaminopropylamino-agarose backbone, as quoted by the manufacturer.

Each syringe is covered with foil to protect yohimbinic acid, which reportedly is light sensitive, from inactivation. After 4 hr of end-over-end rotation, triethylamine (TEA) is added to the mixture to minimize the possibility that the amino groups on the diaminodipropylamine agarose matrix become protonated, which would render these groups unavailable for reaction in the DCCD-catalyzed coupling (cf. Fig. 1). Triethylamine addition is as follows: 8.4 μl TEA after 45 min of rotation, 12.6 μl TEA after 90 min, 12.6 μl after 135 min, and 21 μl after 180 min. After the final TEA addition, the resin is rotated overnight.

The next day (day 2), the reactants are removed by filtration through the resin and the resin is washed six times with DMSO as before. The DCCD-catalyzed yohimbinic acid coupling and the TEA additions are repeated as before. In contrast to our initial methodology,[38] we have found it unnecessary to repeat the coupling a third or fourth time.

On day 3, the resins from each individual column are pooled in an acid-washed scintered glass funnel. It is not uncommon for each of the resins to have a unique color ranging from pale yellow to dark brown. The origin of this color difference is unknown, although we suspect it is due, at least in part, to reactants derived from the rubber O-ring that fastens the nylon mesh to the bottom of each column. Nonetheless, differences in color at this stage do *not* correlate with differences in yohimbinic acid coupling or, ultimately, differences in the binding or elution of α_2-adrenergic receptors. Hence, the differences can be ignored. The pooled resin is washed with DMSO until the absorbance of the DMSO eluate at 300 nm is less than 0.05, with zero defined as the absorbance of DMSO at 300 nm. This low absorbance assures that yohimbinic acid, dicyclohexylurea, and other side products have been effectively removed from the resin. It is important to transfer the DMSO washes to the quartz cuvette with glass pipettes and to avoid Tygon and plastic tubing, since passing DMSO through these materials profoundly affects the UV absorbance of the DMSO. The DMSO

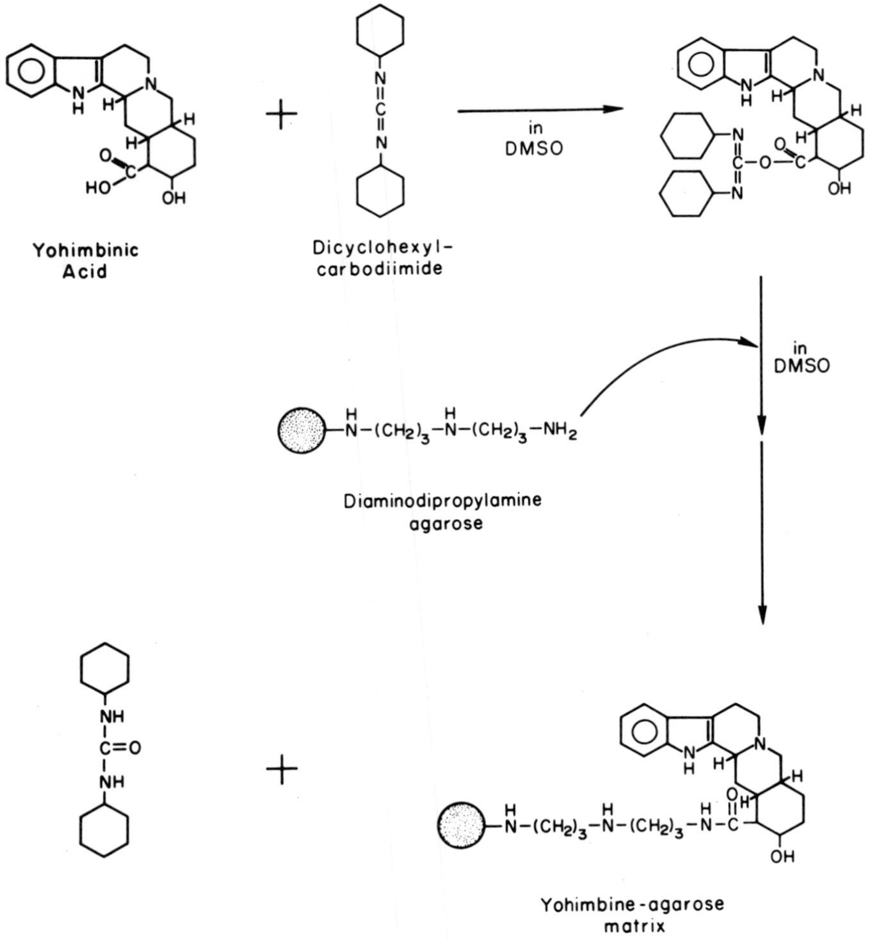

FIG. 1. Schematic overview of the synthetic route for preparing the yohimbine-agarose affinity resin. After two dicyclohexylcarbodiimide-catalyzed reactions to couple yohimbinic acid to diaminodipropylamine (DAPA)-agarose, the unconjugated amino groups are blocked by coupling acetate to the DAPA-agarose using a water-soluble carbodiimide (see text).

washing of the resin takes a minimum of 4 liters of DMSO. For economy, later washes are saved and used for the early washes of the yohimbinic acid-agarose affinity matrix during a subsequent resin preparation.

The yohimbine-agarose resin is then reswelled into water and aliquoted at 15 ml/column into freshly methanol-washed glass syringes secured with new nylon mesh and O-rings. The amino groups that were not coupled with yohimbinic acid during the DCCD-catalyzed reaction in DMSO are now blocked by acetylation using the water-soluble carbodiimide, 1-ethyl-3-(3-dimethylaminopropyl)carbodiimide (EDAC). In our experience, this step is critical for allowing the yohimbine-agarose matrix to behave as an affinity-purification tool step rather than an ion-exchange column. The resin is equilibrated with 100 mM sodium acetate, pH 4.5, and EDAC (1.15 g/15 ml resin) is added in 1–2 ml of H_2O to the resin slurry. The resin is then mixed by rapid rotation and the pH of the reaction is maintained between pH 4.7 and 5.0 by dropwise addition of 1.0 N acetic acid over the first 15 min of the reaction (this usually requires 10–12 drops over the first 2–3 min and a total of 15–18 drops over the first 15 min). After the pH stabilizes (usually at 15 min), the mixture is brought to 30 ml with 100 mM sodium acetate, pH 4.5, and rotated overnight at room temperature. This reaction is performed twice, although the second reaction may be carried out for 4 hr rather than overnight. After extensive washing with deionized water, the resin is stored at 6–10° in 20 mM EDTA in foil-wrapped glass syringes. Prior to later use for affinity chromatography, the aliquot of resin needed was again subjected to acetylation as described above.

The amount of yohimbinic acid coupled to the agarose resin can be estimated spectrophotometrically by suspending 0.125 ml resin 1 : 1 (w/v) with water and adding 1.75 ml glycerol. Using a similar suspension of diaminodipropylamine agarose as a blank, a UV spectrum is obtained. The spectrum obtained more closely resembles that of yohimbine than yohimbinic acid. Assuming that the extinction coefficient of yohimbinic acid coupled to the amino group of the spacer arm is similar to that for yohimbine, the above coupling protocol results in the conjugation of 3–5 μmol yohimbinic acid per 1 ml packed bed volume, compared with a primary amino group capacity of 16–20 μmol/ml, as quoted by the manufacturer for diaminodipropylamine agarose. However, it is important to note that varying capacity (i.e., 3–5 μmol yohimbinic acid/ml packed resin volume) does *not* result in a detectably different capacity for the binding or elution of α_2-adrenergic receptors.

Yohimbine-Agarose Affinity Chromatography of Human Platelet or Porcine Brain α_2-Adrenergic Receptors: Large-Scale Purification

Two side-by-side 15-ml yohimbine-agarose columns are adsorbed with approximately 750 ml each of digitonin-solubilized preparation in the following manner. One day prior to chromatography, two 15-ml resin aliquots are treated with EDAC and sodium acetate as described above to modify any unreacted amino groups with the acetate moiety prior to affinity chromatography, thus minimizing the ion-exchange properties of the resin.

The resin is then prewashed with digitonin solubilization buffer (above) in an effort to coat and equilibrate the resin with this 1% digitonin-containing solution. The resin is packed at a flow rate of 25 ml/min. Typically, the first 50–100 ml of digitonin-solubilized preparation applied to the resin represents the last column effluents from previous affinity chromatography protocols.

The NaCl concentration of the digitonin-solubilized preparation is increased to 500 mM prior to yohimbine-agarose chromatography in an effort to reduce nonspecific adsorption to the matrix due to electrostatic interactions with the matrix. One volume of diluent buffer is added to each 3 vol of digitonin-solubilized preparation to bring the NaCl concentration to 500 mM, reduce the digitonin concentration to 0.75%, and retain the HEPES buffer concentration at 50 mM. α_2-Adrenergic receptor-containing preparations are adsorbed to the resin until the capacity of adsorption is reduced from the 80–90% characteristic of the outset of chromatography to less than 70% of applied receptor retained by the resin. Typically, a 15-ml resin will accommodate 750–1000 ml of diluted digitonin-solubilized preparation. Adsorption typically takes 3.5 to 4 days. Solubilized preparations are thawed as needed every 24 hr. After the adsorption phase is terminated, the resin is washed for half the duration of the adsorption phase (approximately 2 days) with a wash buffer containing 0.2% digitonin, 50 mM HEPES, 300 mM NaCl, 2 mM EGTA, pH 7.6. Column washing is performed at a flow rate of 10–12 ml/hr.

Prior to elution, the yohimbine-agarose affinity resin is equilibrated with 0.1% digitonin, 40 mM HEPES, 0 mM NaCl, 2 mM EGTA, pH 7.6. The resin is then eluted in the presence of 10^{-4} or 10^{-5} M phentolamine in the presence of a gradient of 0–600 mM NaCl or, when concentrated receptor is desired, in the presence of 600 mM NaCl. When NaCl gradient elution is performed in the presence of phentolamine, α_2-adrenergic receptor-binding activity elutes at a salt concentration of approximately 150–220 mM NaCl.

The elution position of α_2-adrenergic receptor-binding activity is determined by assay of [^{3}H]yohimbine binding to fractions exchanged over Sephadex G-50 to remove phentolamine. Thus, phentolamine removal is accomplished by dilution of 25 μl of column eluate to a final volume of 150 μl with a buffer containing 0.2% digitonin, 0.1% bovine serum albumin, 40 mM HEPES, 25 mM glycylglycine, 100 mM NaCl, 5 mM EGTA, pH 7.6. This 150-μl sample is exchanged over a 0.75-ml Sephadex G-50 column prepared in a cut-off 1.0-ml polypropylene disposable pipette as previously described.[38] The 450-μl eluate containing the void volume (substantially depleted of phentolamine) is incubated with 75 nM [^{3}H]yohimbine and receptor-binding activity is assayed as described earlier in the text.

Applicability of Yohimbine-Agarose Chromatography to Multiple α_2-Adrenergic Receptor Subtypes

The existence of subtypes of α_2-adrenergic receptors has been implicated in a variety of physiological and pharmacological studies over the last decade and, more recently, has been confirmed by molecular cloning strategies. Bylund[44] has suggested the following nomenclature for these α_2 receptor subtypes. α_2-Adrenergic receptors of the α_{2A} subtype, like those expressed in human platelet or the porcine brain preparations described above, are characterized by antagonist selectivity of yohimbine (nM) $\ggg$ prazosin (10 μM) and an agonist selectivity of oxymetazoline > UK14304 (a clonidine analog) > epinephrine. In contrast, α_2 receptors of the α_{2B} subtype interact with prazosin with a relatively higher affinity such that yohimbine (nM) > prazosin (10–100 nM). α_{2B} Receptors possess an agonist selectivity of UK14304 > epinephrine > oxymetazoline. The α_2-adrenergic receptor identified in opossum kidney (OK) cells has an intermediate specificity between the more widely distributed α_{2A} and α_{2B} subtypes, and has been referred to by Bylund as the α_{2C} subtype. Continually accruing data suggest that other subtypes will be identified. In all cases, however, the α_2 receptor subtypes possess a relatively high affinity (K_D in the nanomolar range) for yohimbine, indicating the likely utility of the yohimbine-agarose matrix for purification of multiple α_2 receptor subtypes. As an example, the yohimbine-agarose matrix has been used to isolate the α_{2B} receptor subtype expressed in NG108-15 neuroblastoma × glioma cultured cells.[45]

Effectiveness of Yohimbine-Agarose Affinity Matrix for Microscale Isolation of α_2 Receptors from Metabolically Labeled Cells

Another situation where the yohimbine-agarose matrix has proven useful is in the microisolation of α_2-adrenergic receptors from metabolically labeled cells. Typically, identification of receptor maturation in [^{35}S]methionine-labeled cells or the characterization of posttranslational modification in appropriately labeled cells is monitored following immunoprecipitation of the receptor of interest. However, available antibodies against the α_2-adrenergic receptor do not quantitatively immunoprecipitate the receptor protein. Consequently, we evaluated whether micropurification of the α_2 receptor using the yohimbine-agarose matrix might suffi-

---

[44] D. Bylund, *Trends Pharmacol. Sci.* **9**, 356 (1990).
[45] A. L. Wilson, K. Seibert, S. Brandon, E. J. Cragoe, Jr., and L. E. Limbird, *Mol. Pharmacol.* **39**, 481 (1991).

ciently resolve the α_2-adrenergic receptor from the other radiolabeled proteins to detect this entity in metabolically labeled cells.

The α_{2A}-adrenergic receptor possesses a cysteine residue near its C terminus that is predicted to be a candidate for acylation by palmitic acid. Labeling of two independent permanent transformant clones of porcine kidney LLC-Pk1 cells with [^{3}H]palmitic acid results in the covalent incorporation of this fatty acid into a variety of digitonin-extractable proteins. However, batchwise adsorption of this preparation with the yohimbine-agarose matrix results in the affinity elution of a single radiolabeled species, as identified on SDS-PAGE, which migrates at the molecular weight characteristic of the α_2-adrenergic receptor expressed within these LLC-Pk1 cells as identified by photoaffinity labeling with [^{125}I]iodoazidorauwolscine carboxamide.[46] These findings suggest that the properties of functional α_2-adrenergic receptors, as revealed by pulse–chase experiments in metabolically labeled cells, can be studied even in the absence of immunoprecipitating antibodies simply by microaffinity adsorption/elution using the yohimbine-agarose matrix.

Summary

We have provided a detailed protocol for the synthesis of a yohimbine-agarose matrix that has been shown to be effective for isolation of the α_{2A}-adrenergic receptor from human platelet and purification of the α_{2A}-adrenergic receptor to apparent homogeneity from porcine brain cortex using chromatography on only two sequential yohimbine-agarose columns.[38] In addition, this affinity matrix also interacts with α_2 receptors of the α_{2B} subtype extracted from cultured NG108-15 cells.[45] Finally, this affinity matrix has proven useful for monitoring posttranslational modifications of the receptor in digitonin extracts of metabolically labeled cells.[46] Thus, this affinity matrix can be exploited for the purification of multiple α_2-adrenergic receptor subtypes on both a macro- and microscale and should be of value to any laboratory exploring the molecular basis for α_2-adrenergic functions.

---

[46] M. E. Kennedy, A. L. Wilson, and L. E. Limbird, *Fed. Proc., Fed. Am. Soc. Exp. Biol.* A1584 (1991).

[18] 5-[^{3}H]Hydroxytryptamine and [^{3}H]Lysergic Acid Diethylamide Binding to Human Platelets

By JOHN R. PETERS, DAVID P. GEANEY, and D. G. GRAHAME-SMITH

The interactions of 5-hydroxytryptamine (5-HT) with human platelets *in vitro* are complex. First, the amine induces the related processes of shape change and aggregation[1] and second, it is concentrated within the cell by an energy-requiring uptake process.[2] The original observation that signals for each of these cellular events were mediated via separate receptors was made indirectly some years ago on the basis of structure–activity relationships.[3] However, the direct demonstration of high-affinity binding associated with the receptor for shape change and aggregation has required the application of radioligand-binding techniques for tritiated 5-hydroxytryptamine ([^{3}H]5-HT) using intact platelets,[4] and for tritiated lysergic acid diethylamide ([^{3}H]LSD) using human platelet membranes.[5]

The two methods as described involve different approaches to the definition of specific high-affinity binding. Not only is the agonist ligand [^{3}H]5-HT bound to both receptor sites for aggregation and uptake, but the active uptake process contributes to its apparent nonspecific binding. High-affinity (aggregation) receptor-associated [^{3}H]5-HT is therefore distinguished from lower affinity (uptake) receptor-associated [^{3}H]5-HT by the use of different concentrations of unlabeled 5-HT in the displacement of radioligand from each site. On the other hand, definition of binding of [^{3}H]LSD to the shape change/aggregation site relies on the pharmacological specificity of the ligand for this receptor, which has a very low affinity for the 5-HT uptake recognition site.[6]

5-[^{3}H]Hydroxytryptamine Binding to Intact Human Platelets

The intact cell method has the advantages of dealing with receptor populations as close to the physiological state as possible, of using the natural agonist as ligand, and the potential for simultaneous estimation of

---

[1] J. R. A. Mitchell and A. A. Sharp, *Br. J. Haematol.* **10**, 78 (1964).
[2] R. S. Stacy, *Br. J. Pharmacol.* **16**, 284 (1961).
[3] G. V.R. Born, K. Juengjaroen, and F. Michal, *Br. J. Pharmacol.* **44**, 117 (1972).
[4] J. R. Peters and D. G. Grahame-Smith, *Eur. J. Pharmacol.* **68**, 243 (1980).
[5] D. P. Geaney, M. Schächter, J. M. Elliott, and D. G. Grahame-Smith, *Eur. J. Pharmacol.* **97**, 87 (1984).
[6] A. Laubscher and A. Pletscher, *Life Sci.* **24**, 1833 (1979).

active uptake, as well as for use in a radioreceptor assay for circulating drugs (see below). The disadvantages include the relatively low affinity and low specific activity of the ligand, the complexities of calculation of high-affinity binding, and the errors introduced by possible leakage of endogenous 5-HT from the cells.

Preparation of Platelets

Whole blood is taken from an antecubital vein through a large (at least 19-gauge) needle. To minimize local trauma it should ideally be allowed to flow without applied negative pressure into 1% (w/v) EDTA in 0.7% NaCl (w/v) (9:1) as anticoagulant. After centrifugation at 170 g for 12 min at 37°, the resulting supernatant, platelet-rich plasma (PRP), is removed with a Pasteur pipette, the hematocrit having been recorded. Platelet-rich plasma is stored on ice, and prior to assay is diluted by 50% with 0.1% EDTA in 0.9% NaCl (pH 7.4) to give a final platelet density of between 1 and 3×10^8 platelets per milliliter as counted by a Coulter counter (industrial model D; Coulter, Hialeah, FL).

5-[^{3}H]Hydroxytryptamine-Binding Assay

Receptor binding of [^{3}H]5-HT is measured by a centrifugation assay at 2°. Diluted PRP as above is added in aliquots of 450 μl to 1.5-ml conical polypropylene tubes (Eppendorf GmbH, Hamburg, Germany) containing 50 μl [^{3}H]5-HT to produce final concentrations in the range 0.5–10 nmol [^{3}H]5-HT. Incubation is for 2–2.5 min at 2° to reach equilibrium binding, at which point 50 μl of unlabeled 5-HT or buffer is added to parallel samples at each radioactive concentration, resulting in free 5-HT concentrations of 10^{-4} M, 10^{-8} M, or zero. All tubes are vortexed briefly and incubated for a further 2 min prior to centrifugation. Alternatively, if only high-affinity binding is to be quantitated, incubations are begun by adding diluted PRP to tubes containing both [^{3}H]5-HT and a specific antagonist, such as pizotifen or ketanserin, at a final concentration of 10^{-7} M. Both types of incubation are terminated at 4 min by rapid (90 sec) centrifugation at 2700 g. The supernatant is then decanted and the surface of the platelet pellet and the walls of the tube washed times with 800 μl ice-cold normal saline. Platelet pellets are then fragmented using a soniprobe converter and suspended in distilled water. Aliquots of this suspension are counted in scintillation cocktail using standard B counting with quench correction.

Relevant practical points are that (1) care not to traumatize the platelets either at venipuncture or during preparation of PRP, will, in the presence of Ca^{2+}-free buffer, minimize unlabeled 5-HT release and the consequent artifactual apparent reduction in total binding, (2) the procedures are all

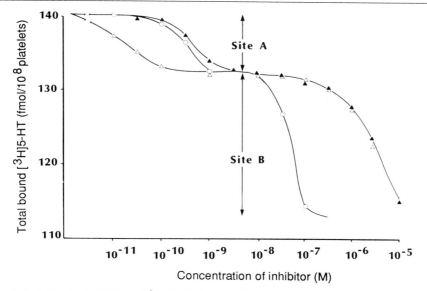

FIG. 1. *In vitro* inhibition of [^{3}H]5-HT binding to intact human platelets. Site A is stereospecifically inhibited by the α ($\triangle$) and β ($\blacktriangle$) forms of flupenthixol. ($\bigcirc$), 5-HT. The isomers are equally ineffective against site B binding. For K_i values see Table I.

performed on ice to decrease active uptake of [^{3}H]5-HT and so minimize the apparent nonspecific binding, and (3) timing of assays is consistent, to decrease the intersubject variation in nonspecific binding.

Calculation of Components of 5-[^{3}H]Hydroxytryptamine Binding

Total receptor-bound [^{3}H]5-HT is displaced by increasing concentrations on unlabeled 5-HT, resulting in a bisigmoidal curve with a plateau between 10^{-9} and 10^{-8} M that indicates a two-site system. The equilibrium dissociation constants for the two sites of this experimentally derived data, calculated by log probit analysis and the Cheng–Prusoff[7] equation, are 0.5 and 35 nM. Thus measurement of bound [^{3}H]5-HT in the absence and in the presence of 10^{-8} or 10^{-4} M unlabeled 5-HT will yield data for calculating by subtraction the absolute values of specific bound [^{3}H]5-HT for both the high-affinity, low-capacity site (A) and the lower affinity, higher capacity site (B) at that free [^{3}H]5-HT concentration. Similarly, the incubation with specific antagonists, for example, α-flupenthixol at 10^{-7} M, will result in inhibition of binding to site A only compared with total [^{3}H]5-HT bound at that concentration (Fig. 1).

[7] Y. Cheng and W. H. Prusoff, *Biochem. Pharmacol.* **22**, 3099 (1973).

The adsorption isotherm of specific [^{3}H]5-HT binding displaced by 10^{-4} M unlabeled 5-HT is biphasic, in the range 0.5–10 nM free [^{3}H]5-HT, and transformation into the Scatchard analysis reveals a curvilinear plot. By using parallel incubations, displacing with 10^{-8} M cold 5-HT at each free concentration of [^{3}H]5-HT, the high-affinity component of binding may be calculated. Treating each site separately in this manner produces two linear plots with values for site A of K_d = 0.7 nM, B_{max} = 7.9 fmol/10^8 platelets and for site B K_d = 24 nM, B_{max} = 180 fmol/10^8 platelets.

Kinetics of 5-[^{3}H]Hydroxytryptamine Binding

Total specific binding of [^{3}H]5-HT to intact human platelets is rapid, association half-time being approximately 70 sec, with equilibrium by 2 min. Displacement of specific [^{3}H]5-HT bound by unlabeled 5-HT shows a half-time of 45 sec. The experimentally determined association and dissociation rate constants from pseudo-first-order plots[8] of specific binding are k_{ob} = 1.77 min^{-1} and k_2 = 0.55 min^{-1}, giving a derived equilibrium dissociation constant of 4.4 nM. This is, as expected, intermediate between the K_d values of sites A and B, as determined independently.

Specific [^{3}H]5-HT binding is directly proportional to the platelet count over the working range and represents 25–40% of the total bound [^{3}H]5-HT across the free concentration range 0.5–10 nM. Total bound [^{3}H]5-HT is less than 4% of the free [^{3}H]5-HT at all incubation concentrations.

Uptake of 5-[^{3}H]Hydroxytryptamine and Nonspecific Binding

The nonspecific or nondisplaceable component of binding is conventionally defined in radioligand-binding assays as linear with time and concentration of ligand. With regard to [^{3}H]5-HT and platelets, this nonspecific binding is nonlinear at low free ligand concentrations (below 10 nM), and a saturable component with a K_d of 15 nM may be inhibited by preincubation of platelets at 37° with 10^{-3} M N-ethylmaleimide, sodium cyanide, and ouabain for 30 min prior to cooling for the receptor assay. Because the receptor density is unaffected and the residual nonspecific binding becomes linear after such treatment, the proportion sensitive to metabolic inhibition is likely to represent the active transport of [^{3}H]5-HT into the platelets still active at 2°, albeit with a reduced B_{max}. Thus, the rate of 5-HT uptake into platelets may be calculated in the same experiments by correcting total nonspecific binding, at the same free ligand concentration,

---

[8] L. T. Williams and R. J. Lefkowitz, "Receptor Binding Studies in Adrenergic Pharmacology." Raven Press, New York, 1978.

TABLE I
INHIBITORY CONSTANTS FOR 5-[^{3}H]HYDROXYTRYPTAMINE
RECEPTOR BINDING AND ACTIVE UPTAKE OF
5-[^{3}H]HYDROXYTRYPTAMINE[a]

Inhibitor	K_i site B [^{3}H]5-HT binding (μM)	K_i uptake [^{3}H]5-HT (μM)
Fluoxetine	0.004	0.003
5-HT	0.015	0.015
Chlorimipramine	0.13	0.08
Nortriptyline	0.50	0.25
Chlorpromazine	1.25	1.2
Mianserin	30.00	17.00
Pizotifen	85.00	70.00
Methysergide	250.00	75.00

[a] Under the same conditions.

for the nonmetabolically inhibitable values derived from parallel incubations.

Specific Inhibition of 5-[^{3}H]Hydroxytryptamine-Binding Sites and Function

The above relationship between lower affinity site B [^{3}H]5-HT binding and the nonsaturable (uptake) component of nonspecific binding, as suggested by their similar K_d values, may be confirmed using specific inhibitors of the uptake process. Classical uptake inhibitors chlorpromazine, chlorimipramine, and nortriptyline show K_i values for the two processes that are identical, and that agree closely with published data (Table I). In addition compounds known to be specific inhibitors of 5-HT-induced aggregation, for example, methysergide, pizotifen, and the active isomers of flupenthixol, are equally ineffective against site B binding and uptake (Table I, Fig. 1).

Use of inhibitors to displace [^{3}H]5-HT binding from the high-affinity site A shows that stereospecific effects are seen with the α and β isomers of the neuroleptic flupenthixol (Fig. 1). The specific inhibitors of aggregation, methysergide and pizotifen, are also competitive inhibitors at this site. The rank order of inhibitor potency (IC$_{50}$) of these agents, in terms of 5-HT-induced platelet aggregation, is given in Table II. The kinetics of the inhibiton may be examined using single-dose inhibitor studies with variable-dose free [^{3}H]5-HT both *in vitro* or *in vivo* (Fig. 2), where in-

TABLE II
INHIBITORY CONSTANTS K_i FOR 5-[^{3}H]HYDROXYTRYPTAMINE
RECEPTOR BINDING AND IC_{50} VALUES FOR INHIBITION OF
5-HYDROXYTRYPTAMINE INDUCED PLATELET AGGREGATION[a]

Inhibitor	K_i [^{3}H]5-HT site A binding (nM)	IC_{50} 5-HT aggregation (μM)
α-Flupenthixol	0.02	0.05
Methysergide	0.2	0.07
Mianserin	0.8	1.2
Pizotifen	1.1	2.3
β-Flupenthixol	2.0	11.0
5-HT	0.5	0.8

[a] See Ref. 4.

creased equilibrium dissociation constant and unchanged B_{max} confirm the inhibition as competitive in type.

Application of 5-[^{3}H]Hydroxytryptamine Binding to Human Platelets

The methodology described may be used to quantitate directly the two receptor sites for [^{3}H]5-HT with or without the parameters for active uptake of the amine under the same conditions. The characteristics of the two receptor sites have been shown to alter under the influence of administered[9] or endogenous[10] steroid hormones. Alternatively, the assay may be used appropriately to assess the potency of pharmacologic agents known to affect either platelet 5-HT-induced shape change/aggregation or 5-HT uptake. This may be undertaken either *in vitro* by competition binding or by estimation of decreased binding of [^{3}H]5-HT to platelets from patients with circulating levels of such agents (Fig. 2). Because the assay is performed in plasma, comparison of [^{3}H]5-HT binding at a single free ligand concentration using PRP from a patient taking a given drug, with values obtained from an *in vitro* dose inhibition curve for the same agent, will indicate the effective circulating plasma level of that agent—a true radioreceptor assay. The example in Fig. 2 illustrates a migraine prophylactic agent affecting site A binding. Equally, the effect of circulating concentrations of, for example, tricyclic antidepressant agents could be assayed using competition for site B [^{3}H]5-HT binding as the indicator.

[9] J. M. Elliott, J. R. Peters, and D. G. Grahame-Smith, *Eur. J. Pharmacol.* **66**, 21 (1980).
[10] J. R. Peters, J. M. Elliott, and D. G. Grahame-Smith, *Lancet* **2**, 933 (1979).

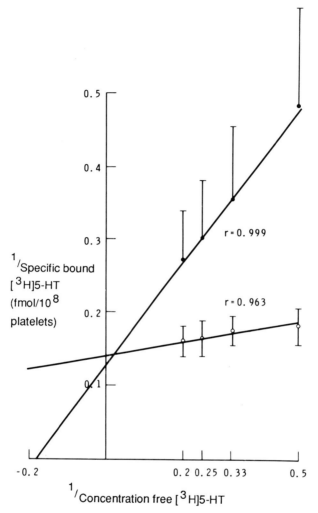

FIG. 2. High-affinity [^{3}H]5-HT binding to human platelets. Double-reciprocal analysis of means of site A [^{3}H]5-HT binding in controls ($n = 16$, $K_d = 4.7 \pm 0.1 \times 10^{-10}$ M, $B_{max} = 6.97 \pm 0.7$ fmol/10^8 platelets) and pizotifen-treated patients ($n = 13$, $K_d = 6.84 \pm 0.2 \times 10^{-9}$ M, $B_{max} = 7.83 \pm 0.9$ fmol/10^8 platelets). $p < 0.001$ for K_d, not significant (NS) for B_{max}. K_d and B_{max} for site B binding were identical for the two groups.

[³H]Lysergic Acid Diethylamide Binding to Human Platelet Membranes

D-Lysergic acid diethylamide (LSD) is a potent inhibitor of 5-HT-induced shape change and aggregation but is virtually inactive against 5-HT uptake in resuspended human platelets.[6] [³H]Lysergic acid diethylamide has been used to label 5-HT receptors in rat brain,[11,12] human brain,[13] and rabbit platelets.[14] In these systems, classical inhibitors of 5-HT uptake are weak. The binding of [³H]LSD to human platelet membranes described here characterizes the specific receptor responsible for 5-HT-induced shape change and aggregation.

Membrane Preparation

The sequence of preparation of human platelet membranes is as follows.

1. Blood is drawn by venipuncture from the antecubital vein using a 19-gauge needle. Anticoagulation is ensured by adding 1% EDTA in normal saline (9:1).
2. The sample is centrifuged at 180 g for 15 min at 20°, and the platelet-rich plasma is aspirated.
3. Platelet-rich plasma is centrifuged at 1200 g for 7.5 min at 10° and the pellet is resuspended in hypotonic Tris–EDTA [5 mM Tris, 0.1% (v/v) EDTA, pH 7.5].
4. Suspension is homogenized with 20 strokes in motor-driven Teflon/glass homogenizer and centrifuged at 30,000 g for 15 min at 4°.
5. The membrane pellet is resuspended and homogenized a second time in hypotonic buffer and, following a second centrifugation, in incubation buffer [50 mM Tris, 120 mM sodium chloride, 5 mM potassium chloride, 2 mM magnesium chloride, and 0.05% (w/v) ascorbic acid, pH 7.3].
6. Final centrifugation (30,000 g for 15 min at 4°) and resuspension in incubation buffer is at a protein content of 0.15–0.35 mg/ml.

[³H]Lysergic Acid Diethylamide-Binding Assay

Receptor binding of [³H]LSD to human platelet membranes is measured by filtration separation of bound and free ligand at equilibrium. The incubation tubes are inoculated with 50 μl [³H]LSD (final concentration 0.25–2.5 nM) and either 50 μl spiperone (final concentration 300 nM) in

[11] J. P. Bennett and S. H. Snyder, *Brain Res.* **94**, 523 (1975).
[12] S. J. Peroutka and S. H. Snyder, *Mol. Pharmacol.* **16**, 687 (1979).
[13] A. J. Cross, *Eur. J. Pharmacol.* **82**, 77 (1982).
[14] A. Laubscher, A. Pletscher, and H. J. Noll, *Pharmacol. Exp. Ther.* **216**, 385 (1981).

incubation buffer, or buffer alone. Spiperone is used as the displacing agent because, like LSD, it is a potent inhibitor of 5-HT-induced shape change and aggregation and is a weak inhibitor of 5-HT uptake,[6] but is structurally dissimilar to LSD. The incubation is begun by addition of 400 μl of platelet membrane suspension in buffer, continues at 37° to equilibrium at 4 hr, and is stopped by the addition of 5 ml ice-cold buffer [50 mM Tris, 0.01% (v/v) bovine serum albumin, pH 7.7]. Separation is by filtration under reduced pressure through Whatman (Clifton, NJ) GF/F filters. These are washed twice more with 5 ml of the same buffer, dried in air at 25°, and counted in 10 ml Instagel (Packard, Rockville, MD) scintillation fluid at 40% efficiency. Aliquots of membrane suspension are stored at $-20°$ for later assay of protein by the method of Lowry.[15]

For inhibition studies, the competing ligand is added to incubates without spiperone to determine the inhibition of total binding, and also to incubates with 300 nM spiperone to determine any inhibition of nonspecific binding that may occur. By subtracting any inhibition of nonspecific binding from the inhibition of total binding the degree of inhibition of specific binding can be accurately determined for each inhibitor concentration. Between 4 and 12 concentrations of each ligand are studied to determine the IC_{50} value.

Characteristics of [³H]Lysergic Acid Diethylamide Binding to Human Platelet Membranes

The total binding of [³H]LSD is inhibited by spiperone over approximately two orders of magnitude to reach a plateau at 300 nM spiperone. We therefore use this concentration of spiperone to define the specific binding of [³H]LSD. A plateau occurs at the same level for inhibition of binding by D-LSD between 10^{-8} and 10^{-7} M.

Specific binding of [³H]LSD reaches equilibrium by 4 hr and remains constant for 10 hr at 37°. Thus, 4 hr is used as the routine incubation time. Dissociation of the specific binding occurs on addition of 300 nM spiperone. At a free concentration of 0.5 nM [³H]LSD, the half-time for association is 56 min with an association rate constant of 0.017 ± 0.004 nM^{-1} min^{-1} (mean $k_1 \pm$ SEM, $n = 7$). The half-time for dissociation is 173 min with a dissociation rate constant of 0.0040 ± 0.0003 min^{-1} (mean $k_2 \pm$ SEM, $n = 4$). The kinetically derived value for the equilibrium dissociation constant is therefore $K_d = k_2/k_1 = 0.0040/0.017 = 0.24$ nM.

Total bound [³H]LSD does not exceed 1% of total radioactivity at any free concentration. The specific binding of [³H]LSD to human platelet

---

[15] O. H. Lowry, N. J. Rosebrough, A. L. Farr, and R. J. Randall, *J. Biol. Chem.* **193**, 265 (1951).

membranes is saturable in the range 0.25–2.5 nM. No further specific binding sites are observed at concentrations up to 15 nM. Scatchard analysis of the binding curve reveals a single site. Specific binding represents 30–60% of the total radioactivity bound to the platelet membranes. In 19 control subjects (10 males, 9 females, aged 17–60 years) the binding affinity (K_d) was 0.53 ± 0.02 nM (mean ± SEM) and the capacity (B_{max}) was 57.1 ± 5.6 fmol/mg protein. Binding studies in a single male subject repeated on four separate occasions during a 10-day period showed a coefficient of variation of 16% for the K_d and 11% for the B_{max}.

Specific binding of [^{3}H]LSD was related linearly to protein concentration in the range 0.05–0.50 mg/ml.

5-Hydroxytryptamine antagonists are all potent inhibitors of [^{3}H]LSD binding to human platelet membranes (Table III). Like spiperone, haloperidol is a butyrophenone, but with predominantly dopamine antagonist properties. It is a weaker inhibitor of binding than spiperone by approximately two orders of magnitude. The 5-HT uptake inhibitors are also weak binding inhibitors, as are histamine and α- and β-adrenergic antagonists. The isomers of butaclamol display a marked stereospecificity, with the (+)-isomer, the active form, being over two orders of magnitude more potent than the (−)-isomer. Similar stereospecificity is displayed by the α- and β-isomers of flupenthixol. Of the biogenic amines tested, 5-HT was by far the most potent inhibitor, although 5-HT is much less potent than its antagonists. 5-Hydroxytryptamine agonists quipazine and 5-methoxydimethyltryptamine (5-MDMT) have a similar affinity for 5-HT.

Hill plots of the inhibition of specific [^{3}H]LSD binding reveal slopes close to unity for LSD, spiperone, and nearly all the other 5-HT antagonists tested, but 5-HT and the other 5-HT agonists are all considerably lower. Inhibition by 5-HT does not appear to be altered by the presence of 10 μM guanosine triphosphate (GTP), or 100 μM 5'-guanylylimidodiphosphate [Gpp(NH)p].

There is no correlation between the inhibition of [^{3}H]LSD binding to human platelet membranes and the inhibition of the active uptake of 5-HT by human platelets. However, there is a highly significant correlation between the inhibition of [^{3}H]LSD binding and the inhibition of 5-HT-induced shape change in resuspended human platelets (Fig. 3).

Application of [^{3}H]Lysergic Acid Diethylamide Binding to Human Platelet Membranes

The specific binding of [^{3}H]LSD fulfills the criteria for binding to a receptor in that it is saturable and of high affinity, kinetically follows the law of mass action for a bimolecular interaction, and demonstrates the

TABLE III
INHIBITION OF SPECIFIC BINDING OF [^{3}H]LYSERGIC ACID DIETHYLAMIDE TO HUMAN PLATELET MEMBRANES[a]

	Concentration (nM)	n_H		Concentration (nM)	n_H
IC$_{50}$			IC$_{50}$		
5-HT antagonists			Endogenous amines		
D-LSD	1.5	0.99	5-HT	185	0.60
Methiothepin	2.1	0.77	Dopamine	>100,000	
Bromo-LSD	2.7	0.99	Histamine	>100,000	
Metergoline	2.8	0.85	Isoprenaline	>100,000	
Pirenperone	7.3	0.95	Noradrenaline	>100,000	
Spiperone	13.2	0.98			
Ketanserin	13.5	0.97			
Pizotifen	15.3	0.76	5-HT agonists		
Ergotamine	18.3	1.08	Quipazine	80.1	0.60
Cyproheptadine	23.7	0.72	5-MDMT	116	0.48
(+)-Butaclamol	26.3	1.11	Mescaline	5,929	0.54
Methysergide	30.7	0.93			
Mianserin	38.4	0.90			
α-Flupenthixol	38.8	0.81	5-HT uptake blockers		
β-Flupenthixol	4,748	1.28			
(−)-Butaclamol	>10,000		Amitriptyline	321	
			Fluoxetine	1,256	
			Imipramine	1,324	

[a] Human platelet membranes were incubated with [^{3}H]LSD (0.5 nM) ± spiperone (300 nM) + inhibitor at 4–12 concentrations per assay. The inhibition of specific binding was determined by subtracting any inhibition of nonspecific binding (incubates containing 300 nM spiperone) from the inhibition of total binding (incubates without spiperone) at each inhibitor concentration. The IC$_{50}$ was estimated from log-probit analysis of the inhibition curves. The Hill coefficient (n_H) is given for 5-HT antagonists and agonists.

specificity and stereospecificity of a 5-HT receptor. Functionally, this binding site correlates with the 5-HT receptor responsible for shape change and aggregation, but not that associated with the active uptake of 5-HT.

The inhibitory profile of the platelet 5-HT receptor for shape change and aggregation as described here is intermediate between that of the proposed[12] central 5-HT$_1$ and 5-HT$_2$ sites. It more closely resembles the 5-HT$_2$ site in that 5-HT antagonists are much more potent inhibitors of platelet [^{3}H]LSD binding than agonists. However, the 5-HT agonists are relatively more potent inhibitors of binding to platelets than of binding to central 5-HT$_2$ sites. Accordingly, there is a close correlation between [^{3}H]LSD binding to human platelet membranes and [^{3}H]LSD binding to human frontal cortex defined by 1 μM LSD (Fig. 4). The platelet has

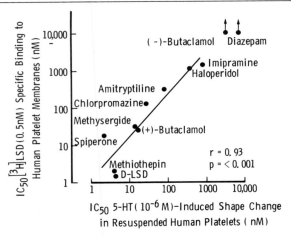

FIG. 3. Correlation of affinity of inhibitors of [^{3}H]LSD binding to human platelet membranes and of 5-HT-induced shape change in resuspended human platelets.[6] Linear regression analysis by the method of least squares.

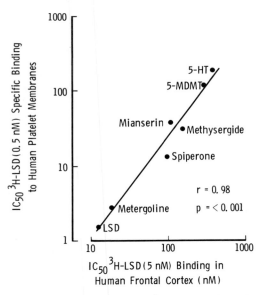

FIG. 4. Correlation of affinity of inhibitors of [^{3}H]LSD binding to human platelet membranes and [^{3}H]LSD binding in human frontal cortex.[13] Linear regression analysis by the method of least squares.

previously been proposed as a model for the central 5-HT neuron[16,17] and the correlation between [³H]LSD binding to human platelet membranes and binding to human frontal cortex supports this analogy.

Thus [³H]LSD binding to human platelet membranes may prove useful in studying pathological and drug-induced changes in 5-HT receptors in humans, both as a direct assessment of the platelet 5-HT receptor for shape change and aggregation and as a correlate for central 5-HT receptors. For example, [³H]LSD-binding capacity is not increased in patients with classical migraine, suggesting that the increased platelet aggregation to 5-HT reported in migraineurs is not due to a primary platelet receptor change.[18] Conversely, [³H]LSD-binding capacity is increased in patients on chronic neuroleptic treatment. This may explain the enhancement of aggregation to 5-HT observed in these patients and raises the possibility that similar changes may occur in the brain.[19]

[16] J. M. Sneddon, *Prog. Neurobiol.* **1**, Part 2, 151 (1973).
[17] M. Graf and A. Pletscher, *Br. J. Pharmacol.* **65**, 601 (1979).
[18] D. P. Geaney, M. G. Rutterford, J. M. Elliott, M. Schächter, K. M. S. Peet, and D. G. Grahame-Smith, *J. Neurol., Neurosurg. Psychiatry* **47**, 720 (1984).
[19] M. Schächter, D. P. Geaney, D. G. Grahame-Smith, P. J. Cowen, and J. M. Elliot, *Br. J. Pharmacol.* **19**, 453 (1985).

[19] Platelet Serotonin Transporter

By GARY RUDNICK and CYNTHIA J. HUMPHREYS

Introduction

Blood platelets contain two distinct transport systems for serotonin.[1] The first of these systems moves serotonin from the plasma into the cytoplasm and the second transports cytoplasmic serotonin into the storage organelle, or dense granule. Both systems move serotonin uphill, against a concentration gradient, and are therefore coupled to the input of metabolic energy. In each case this energy comes from hydrolysis of cytoplasmic ATP, which is used to generate transmembrane ion gradients.

At the plasma membrane the Na^+,K^+-ATPase directly creates Na^+ (out > in) and K^+ (in > out) gradients and, indirectly, through generation of a transmembrane electrical potential ($\Delta\psi$, inside negative), creates a Cl^- gradient (out > in). The transmembrane gradients of Na^+, K^+, and

[1] G. Rudnick, H. Fishkes, P. J. Nelson, and S. Schuldiner, *J. Biol. Chem.* **255**, 3638 (1980).

Cl⁻ serve, in turn, as driving forces for serotonin transport. A second component of the plasma membrane, the serotonin transporter, couples the inward flux of Na⁺ and Cl⁻, and the outward flux of K⁺, to the entry of serotonin.[2]

A similar situation is found in the dense granule membrane, where two membrane components are responsible for serotonin accumulation. The first component is an H⁺-pumping ATPase that appears to be related to similar enzymes in other secretory granules, endosomes, lysosomes, and coated vesicles.[3] This ATPase generates a transmembrane pH difference (ΔpH, acid inside) and $\Delta\Psi$ (positive inside). The ΔpH and $\Delta\Psi$ are then utilized by a second component, the amine transporter, which exchanges cytoplasmic serotonin for one or more intragranular hydrogen ions.[4,5]

Assay of Serotonin Transport into Intact Platelets

Principle

In intact platelets, both systems work in series to concentrate serotonin in the dense granule. Analysis of the individual transport systems is difficult because the ultimate platelet capacity for accumulated serotonin is primarily a function of storage in dense granules, whereas the rate of accumulation is determined by the plasma membrane system. The alkaloid reserpine competitively inhibits serotonin transport into dense granules and has relatively little effect at the plasma membrane.[1] Consequently, reserpine-treated platelets have been used to measure the activity of the plasma membrane transporter in intact cells. At high serotonin concentrations and long incubation times, there is a marked difference between normal and reserpinized platelets.[6]

The differential sensitivity of plasma and granule membranes to externally added gramicidin provides another way to distinguish between the two systems in intact platelets. Gramicidin is an ionophore that renders the membrane freely permeable to Na⁺, K⁺, H⁺, and other monovalent cations. Platelets that have accumulated serotonin to a steady state level lose relatively little serotonin when gramicidin is added, but reserpinized platelets lose almost all of their accumulated serotonin.[6] This effect of gramicidin is due to its ability to dissipate Na⁺ and K⁺ gradients across

[2] G. Rudnick, *J. Biol. Chem.* **252**, 2170 (1977).
[3] G. Rudnick, in "Physiology of Membrane Disorders" (T. E. Andreoli, D. D. Fanestil, J. F. Hoffman, and S. G. Schultz, eds.), 2nd ed., p. 409. Plenum, New York, 1986.
[4] H. Fishkes and G. Rudnick, *J. Biol. Chem.* **257**, 5671 (1982).
[5] S. E. Carty, R. G. Johnson, and A. Scarpa, *J. Biol. Chem.* **256**, 11244 (1981).
[6] G. Rudnick, R. Bencuya, P. J. Nelson, and R. A. Zito, Jr., *Mol. Pharmacol.* **20**, 118 (1981).

the plasma membrane, which removes the driving force for serotonin transport. In reserpinized platelets all of the serotonin is cytoplasmic, and exits when the driving forces are removed. In normal platelets most of the serotonin is granular, and is retained because the granule membrane is less affected by gramicidin. The two systems can also be separated by measuring how much serotonin is released by platelet aggregating agents that cause secretion of dense granule contents.[7]

To examine transport of serotonin into intact platelets, freshly isolated platelets are suspended in a buffer that provides the ions Na^+, K^+, and Cl^-, whose fluxes are coupled to serotonin transport. This buffer also provides glucose, which platelets use as an energy source to generate transmembrane ion gradients. Transport is initiated when [^{3}H]serotonin is added to the platelet suspension and is allowed to continue at 37° for varying periods of time. When this assay is used to study initial rates of serotonin transport, it is necessary, initially, to carry out a time course to ensure that the time point chosen for the determination of initial rates is appropriate. Transport is terminated by diluting the suspension with ice-cold medium. External serotonin is removed by filtering the suspension under vacuum and rinsing the filters. To avoid loss of [^{3}H]serotonin from the platelets by diffusion during this process, the filters are immediately removed from the vacuum filtration apparatus and dried. The process of dilution, filtration, and washing typically takes less than 15 sec.

Reagents

> Platelets: Prepare by differential centrifugation from freshly collected blood (human or porcine) anticoagulated with acid–citrate–dextrose[8]
> Platelet buffer: 66 mM NaCl, 25 mM Na_2HPO_4, 6 mM KCl, 6 mM D-glucose, and 1 mM $MgSO_4$, adjusted to pH 6.7 with H_3PO_4
> [1,2-^{3}H]Serotonin: 14,000 to 17,000 counts per minute (cpm)/pmol
> NaCl: 0.1 M
> Nitrocellulose filters, 0.45-gmm pore size (GN-6; Gelman Sciences, Ann Arbor, MI): Either 25-mm circles or 25-mm squares cut from larger sheets

Procedure

Transport is initiated by the addition of [^{3}H]serotonin to 0.2 ml of a suspension of platelets in platelet buffer at 37°, bringing platelet protein and [^{3}H]serotonin concentrations to approximately 1 mg/ml and 0.1 μM,

---
[7] H. Affolter and A. Pletscher, *Mol. Pharmacol.* **22**, 94 (1982).
[8] R. H. Aster and J. H. Jandl, *J. Clin. Invest.* **43**, 843 (1964).

respectively. The suspensions are incubated at 37° for varying periods of time and then transport is terminated. Two milliliters of ice-cold NaCl is rapidly added to the transport mixture, which is then filtered under vacuum through a nitrocellullose filter. The reaction tube and filter are immediately rinsed with an additional 2 ml of cold NaCl. The filters are immediately removed from the vacuum, dried, and counted in scintillation vials using scintillation fluid suitable for nonaqueous samples.

Plasma Membrane Vesicles

Although transport studies in intact platelets can yield valuable information, their usefulness is limited by our inability to control the cytoplasmic and intragranular ion composition and the possibility of metabolism by monoamine oxidase in platelet mitochondria. To circumvent these difficulties, we have used preparations of vesicles derived from plasma membrane and dense granule membrane.[2,4]

Serotonin Transport Assay

Principle

To demonstrate serotonin transport in plasma membrane vesicles, ion gradients must be imposed across the vesicle membrane. This is accomplished by first incubating a dilute vesicle suspension in potassium phosphate buffer, collecting the vesicles by centrifugation, and resuspending them in the same buffer. This process equilibrates the vesicle interior with K^+. The equilibrated vesicle suspension is diluted into isosmotic NaCl containing [^{3}H]serotonin, and incubated at 25–37°. After a given time, part or all of the reaction mixture is diluted with ice-cold buffer, filtered on a nitrocellulose filter, and the filter is washed and counted for vesicular serotonin.

Ionic Requirements. External Na^+ and Cl^- are both required for serotonin transport into the vesicles.[2] The transporter catalyzes cotransport of one Na^+ and one Cl^- with each serotonin molecule[9,10] and although Cl^- is not strictly required for binding,[9,11] both ions are necessary for the translocation of bound substrate. Following dissociation of internal serotonin, the transporter must return to a form that binds external serotonin. This return step requires translocation of internal K^+ under physiological conditions.[12] If internal K^+ is absent, internal H^+ replaces K^+ in

[9] J. Talvenheimo, H. Fishkes, P. J. Nelson, and G. Rudnick, *J. Biol. Chem.* **258**, 6115 (1983).
[10] P. J. Nelson and G. Rudnick, *J. Biol. Chem.* **257**, 6151 (1982).
[11] J. Talvenheimo, P. J. Nelson, and G. Rudnick, *J. Biol. Chem.* **254**, 4631 (1979).
[12] P. J. Nelson and G. Rudnick, *J. Biol. Chem.* **254**, 10084 (1979).

the return step.[13] Thus, in membrane vesicles devoid of K^+, serotonin transport is strongly dependent on internal pH (being maximal at low pH values), but the internal pH of K^+-containing vesicles has little effect on transport. External H^+ inhibits (as does external K^+) and the transport rate decreases below external pH 6.5.[13]

Electrogenicity and Stoichiometry. The ionic gradients imposed across the vesicle membrane in the normal transport assay do not generate a significant transmembrane electrical potential ($\Delta\Psi$).[12] In the presence of valinomycin, a K^+-specific ionophore, the K^+ gradient (in > out) generates a diffusion potential of 40–50 mV (inside negative). This potential has negligible effects on serotonin transport, suggesting that the transport process is electroneutral. Thus, the number of positively and negatively charged ions moving across the membrane with each turnover should add up to zero. Since one Na^+ is cotransported with serotonin[9] and one K^+ is countertransported,[2] the charge movements accompanying serotonin and Cl^- transport should cancel out. Serotonin is transported as a cation under normal circumstances,[13] requiring cotransport of one Cl^- to render the entire process electroneutral.

Reversibility of Transport System. The serotonin transporter is quite capable of catalyzing efflux as well as influx. Efflux is stimulated by internal Na^+ and Cl^- and by external K^+,[10,12] and is inhibited by the tricyclic antidepressant imipramine,[10] which also inhibits serotonin influx into intact platelets,[14] membrane vesicles,[2] and synaptosomes.[15] This property renders ambiguous the question of whether the transport assay measures transport in the physiological direction. From the experimental design of the serotonin transport assay, the directionality of serotonin transport is determined by artificially imposed ion gradients, and not by the orientation of the vesicles. The preparation of plasma membrane vesicles is vesicular, as evidenced by its ability to retain serotonin, and at least 70% right side out from studies of glycoprotein accessibility to hydrolytic enzymes.[16] Nevertheless, it is still difficult to state with certainty that serotonin is transported only into right-side-out vesicles.

Reagents

Platelet plasma membrane vesicles in 0.25 M sucrose, 10 mM Tris-HCl, pH 7.5, are prepared by the method of Jamieson and co-workers, as described in [4] in this volume, with the following exception. The lysis medium in which the glycerol-loaded platelet pellet is resuspended

---

[13] S. Keyes and G. Rudnick, *J. Biol. Chem.* **257**, 1172 (1982).
[14] A. Todrick and A. T. Tait, *J. Pharm. Pharmacol.* **21**, 751 (1969).
[15] M. J. Kuhar, R. H. Roth, and G. K. Aghajanian, *J. Pharmacol. Exp. Ther.* **181**, 36 (1972).
[16] A. J. Barber and G. A. Jamieson, *J. Biol. Chem.* **245**, 6357 (1970).

additionally contains 5 mM MgSO$_4$ and 33 µg/ml deoxyribonuclease I. Lysis is performed at 37° and the lysis mixture is incubated for 10 min at this temperature, with occasional mixing, before layering on sucrose gradients.

>Potassium phosphate: 0.1 M, pH 6.7, containing 1 mM MgSO$_4$
>[1,2-^{3}H]Serotonin: 8000–12,000 cpm/pmol
>NaCl (0.1 M), 1 mM MgSO$_4$
>Nitrocellulose filters: 0.45-µm pore size (GN-6; Gelman)

Procedure

Platelet plasma membrane vesicles are diluted 10-fold or more into potassium phosphate buffer and incubated for 15 min at 37°. The membranes are then collected by centrifugation at 48,000 g for 20 min at 4° and resuspended to a final concentration of 2 to 5 mg membrane protein per milliliter. Serotonin transport is initiated by diluting 5 to 10 µl of the vesicle suspension (20 to 30 µg protein) into 0.2 ml of 0.1 M NaCl, 1 mM MgSO$_4$ containing 0.1 µM [^{3}H]serotonin, and the samples are incubated for varying lengths of time at 37°. Transport is terminated by dilution with cold NaCl and subsequent filtration as described above in the assay for transport into intact platelets.

Imipramine Binding

Imipramine markedly inhibits serotonin transport at nanomolar concentrations.[14] This avidity allows imipramine binding to be used as a measure of the serotonin transporter. Although the serotonin transport system may be the site of action for tricyclic antidepressants, the imipramine-binding site has not, as yet, been demonstrated to reside on the serotonin transporter. Nevertheless, much of the experimental evidence on imipramine binding and inhibition of serotonin transport is consistent with transport and binding being properties of the same macromolecule.[17] With this assumption, our laboratory has used imipramine to measure the serotonin transporter. Imipramine is not a substrate for the transport system, and is not accumulated within the vesicles under the conditions of the serotonin transport assay.[11] Maximal imipramine binding requires the presence of external Na$^+$ and Cl$^-$. However, ion gradients have no detectable effect on vesicle-associated imipramine because imipramine is not transported. Just as imipramine is a competitive inhibitor of serotonin transport, serotonin competitively inhibits imipramine binding. This inter-

---

[17] G. Rudnick, J. Talvenheimo, H. Fishkes, and P. J. Nelson, *Psychopharmacol. Bull.* **19**, 545 (1982).

action between serotonin and imipramine sites persists even after solubilization of the plasma membrane in detergent.[18] Although solubilized transporters have no transport function, they retain, for binding, the same substrate and ionic specificities as in the native membranes.

Imipramine-Binding Assays

Principle

The simplest imipramine-binding assay is a filter assay, in which free imipramine is removed by filtering a vesicle suspension. Unfortunately, some imipramine always binds to the filter, and raises the blank values. To reduce this complication, we use glass fiber filters pretreated with polyethyleneimine, which bind much less imipramine. A second method is to separate free imipramine from vesicles by centrifugation through small Sephadex columns. This technique, although capable of removing free imipramine from detergent-solubilized transporter and giving very low blanks, requires more time than the filter assay. To maintain the same free concentration in assays of native and solubilized vesicles, more imipramine is added in assays that contain detergent. This compensates for imipramine partitioning into detergent micelles that otherwise significantly decreases the free imipramine concentration. The solubilized transporter can also be measured in the filtration assay by first adsorbing it to calcium phosphate gel prior to filtration. The bound transporter retains similar binding characteristics to the native and solubilized forms and its association with calcium phosphate allows rapid removal of unbound imipramine.

Equipment

Clinical centrifuge (International Equipment Company, Needham Heights, MA)

Disposable 1-ml syringes fitted at the bottom with 1.6-mm-thick porous polyethylene disks (Bel-Art Products, Pequannock, NJ)

Reagents

Platelet plasma membrane vesicles (5 to 10 mg membrane protein per milliliter) in 0.25 M sucrose, 10 mM Tris-HCl, pH 7.5 containing 1 mM MgSO$_4$

Digitonin-Solubilized Platelet Membranes. Membrane vesicles prepared as described above are diluted 20-fold with 0.2 M sodium phosphate, pH 6.7, containing 1 mM MgSO$_4$ collected by centrifugation at 48,000 g for 20 min at 4° and resuspended to a volume of 1.2 ml in the same buffer.

---

[18] J. Talvenheimo and G. Rudnick, *J. Biol. Chem.* **255**, 8606 (1980).

Digitonin (1.43%; 2.8 ml) in 0.29 M NaCl, 1 mM MgSO$_4$ is added to the membrane suspension at 4°, bringing the final volume to 4.0 ml and the final concentrations to 0.06 M sodium phosphate, pH 6.7, 1 mM MgSO$_4$, 0.2 M NaCl, and 1% (w/v) digitonin. A ratio of 5 to 8 mg digitonin per milligram of membrane protein is always maintained. After mixing the suspension thoroughly, the insoluble material is collected by centrifugation at 226,000 g at 4°. The clear, colorless supernatant fraction, which contains about 0.9% digitonin and 1 mg of protein per milliliter, is carefully decanted, frozen in liquid nitrogen, and stored at −70°. Specific imipramine-binding activity remains stable for at least 2 months when stored at this temperature.

[^{3}H]Imipramine: 6000–80,000 cpm/pmol
Glass fiber filters (GF/B; Whatman, Clifton, NJ): Presoak in 0.3% polyethyleneimine
CaPO$_4$ Gel: Prepare by the method of Keilin and Hartree[19]
NaCl: 0.2 M, containing 1 mM MgSO$_4$
LiCl: 0.2 M, containing 1 mM MgSO$_4$
Sephadex G-50 (fine) preswollen in 0.2 M NaCl or LiCl

Procedures

Filtration Assay. Membrane vesicles are washed and resuspended in potassium phosphate buffer as described for the serotonin transport assay. To initiate binding, 5- to 10-μl samples of the vesicle suspension (20 to 30 μg) are diluted into 0.2 ml of 30 nM [^{3}H]imipramine in either 0.2 M NaCl or 0.2 M LiCl, containing 1 mM MgSO$_4$, 4°. After the samples are incubated for 30 to 60 min at 4°, binding is terminated. The reaction mixture is quickly diluted with 4 ml ice-cold NaCl and filtered under low vacuum through a glass fiber filter. The vacuum should be just high enough to pull the diluted reaction mixture through the filter in 2 to 5 sec. The assay tube and filter are rinsed three times with 4 ml cold NaCl and then the filter is counted in Optifluor (Packard, Meriden, CT). Specific imipramine binding is determined by subtracting binding values measured in the absence of Na$^+$ (Li$^+$ substituted for Na$^+$) or in the presence of excess serotonin (100 μM).

Calcium Phosphate Assay. In the assay modified for use with detergent-solubilized membrane, samples containing 0.05 to 0.10 mg of solubilized protein in a volume of 50 to 200 μl are mixed thoroughly with 20 μl of calcium phosphate gel and incubated for 30 min on ice. Following centrifugation at 500 g for 10 min to sediment the gel and adsorbed protein,

---

[19] D. Keilin and E. F. Hartree, *Proc. R. Soc. London, Ser. B.* **124**, 397 (1938).

the supernatant fractions are discarded. The gel and adsorbed protein are gently resuspended in 0.2 ml of 0.2 M NaCl or 0.2 M LiCl containing 100 nM [^{3}H]imipramine and 1 mM MgSO$_4$ and incubated at 4° for 30 min. Imipramine binding to the adsorbed protein is assayed as described above for intact vesicles. Each sample is diluted rapidly with 4 ml of ice-cold 0.2 M NaCl, filtered through a cellulose triacetate filter, and rinsed three times with 4 ml of cold NaCl. Once the filters are counted, specific imipramine-binding activity is calculated, assuming quantitative binding of transporter to the gel.[20]

Gel-Filtration Assay. Sephadex centrifuge columns are prepared by adding to the 1-ml mark swollen Sephadex G-50 preequilibrated in NaCl or LiCl. The columns are placed in small test tubes on ice for 2 hr to allow drainage of excess fluid. Samples for the binding assay are prepared by diluting membrane vesicles (0.04 to 0.10 mg protein) into 100 μl of ice-cold 0.2 M NaCl or 0.2 M LiCl containing [^{3}H]imipramine at a final concentration of 30 nM. The samples are placed on ice and incubated for 30 min. Each sample is assayed by applying 60 μl to the top of an ice-cold column, and then centrifuging the sample immediately for 1 min at half-maximal speed in the clinical centrifuge (International Equipment Company). The delay between sample application and the start of centrifugation is kept to less than 15 sec. The column effluents are collected directly into 1.5-ml conical plastic tubes during the centrifugation. An aliquot from each sample is counted in scintillation fluid suitable for aqueous samples to measure the concentration of bound [^{3}H]imipramine; a second aliquot is used to determine the protein concentration. Again, specific imipramine binding is determined by subtracting binding values measured in the absence of Na$^+$ or in the presence of excess serotonin (100 μM).

This gel-filtration technique is also used to measure specific imipramine binding to digitonin-solubilized membranes. Samples containing up to 1% (w/v) detergent in 0.2 ml NaCl are incubated with a final concentration of 100 nM [^{3}H]imipramine for 15 min at 4°. At the end of the incubation period, 60-μl aliquots of the sample are centrifuged through Sephadex G-50 columns as described above.

Dense Granule Membrane Vesicles

The platelet lysate from which plasma membrane vesicles are isolated also contains vesicles derived from dense granules. The two independent serotonin transport systems are both active in the crude lysate, but by choosing the appropriate conditions, it is possible to measure either plasma

[20] J. Talvenheimo, Ph.D. thesis, Yale University, New Haven, Connecticut (1981).

membrane or granule membrane transport without interference from the other system.[1] In low Na^+ medium the plasma membrane serotonin transporter is inactive, while transport into granule membrane vesicles is driven by external ATP. Thus, the plasma membrane transporter is measured by imposing an Na^+ gradient as described above in the absence of ATP. The granule amine transporter is driven by ATP-dependent H^+ pumping and does not require Na^+. Thus, transport into granule-derived membrane vesicles is assayed in low-Na^+ medium containing ATP. The specificity of the approach is demonstrated by the fact that ATP-dependent serotonin transport in low Na^+ is inhibited by reserpine and not by imipramine, while imipramine but not reserpine blocks Na^+ gradient-driven serotonin accumulation in the absence of ATP.[1]

Assay for Serotonin Transport into Dense Granule Membrane Vesicles

Principle

The platelet dense granule amine transporter couples serotonin transport to the transmembrane ΔpH (acid inside) and $\Delta \Psi$ (positive inside) generated by the ATP-driven H^+ pump; one molecule of cytoplasmic serotonin is exchanged for one or more intragranular protons.[4,5] Transport of serotonin into dense granule membrane vesicles, then, can be driven either by supplying ATP to the H^+-pumping ATPase or by artificially imposing an electrochemical H^+ potential ($\Delta \bar{\mu}_{H^+}$).[21] In the absence of ATP, a $\Delta \bar{\mu}_{H^+}$ (inside acid and positive) is imposed by first preloading vesicles with K^+-containing, low pH medium and then transferring them to K^+-free, higher pH medium containing nigericin, which exchanges H^+ and K^+. This manipulation generates a ΔpH (interior acid) across the vesicle membrane that is maintained by nigericin-catalyzed H^+ influx driven by the K^+ gradient. Efflux of the permeant anions Cl^- and SCN^- generates a $\Delta \Psi$ (interior positive) that also drives transport.[4]

Reagents

Platelet lysate (5 to 10 mg lysate protein per milliliter) in 0.25 M sucrose containing 10 mM Tris-HCl, pH 7.5

Purified Dense Granule Membrane Vesicles. Five milliliters of the platelet lysate is layered onto the tops of each of six ice-cold linear gradients prepared from 17 ml 0.3 M sucrose containing 10 mM K^+-HEPES, pH 7.4, and 17 ml 40% (w/v) meglumine diatrizoate (Hypaque meglumine, 60%; Winthrop Laboratories, Div. Sterling Drug, New York, NY) containing 10 mM K^+-HEPES, pH 7.4. The gradients are then centrifuged at

---

[21] G. E. Dean, H. Fishkes, P. J. Nelson, and G. Rudnick, *J. Biol. Chem.* **259**, 9569 (1984).

83,000 g for 30 min at 4° in a Beckman (Palo Alto, CA) SW 28 swinging bucket rotor. The upper band in the gradients contains membrane vesicles derived from dense granules and is aspirated with a Pasteur pipette. The upper bands from all six gradients are combined and washed once by diluting them 20-fold into 0.3 M sucrose containing 10 mM K$^+$-HEPES, pH 7.4 and sedimenting at 48,000 g for 20 min at 4°. The sedimented membranes are resuspended with 0.3 M sucrose containing 10 mM K$^+$-HEPES, pH 7.4, to about 7 mg of membrane protein per milliliter and then frozen in liquid nitrogen. Transport activity is stable in these granule membrane vesicles for at least a year if they are stored at $-70°$.

[1,2-^{3}H]Serotonin
K$^+$-HEPES: 10 mM, pH 7.4
K$^+$-HEPES: 10 mM, pH 7.4, containing 0.3 M sucrose
RM-A buffer: 10 mM K$^+$-HEPES, pH 8.5, containing 0.3 M sucrose, 2.5 mM MgSO$_4$, 5 mM disodium ATP adjusted to pH 8.5 with KOH, and 5 mM KCl
HEPES stop solution: 10 mM K$^+$-HEPES, pH 8.5, containing 0.3 M sucrose
Nitrocellulose filters: 0.45-μm pore size (GN-6; Gelman)
KN buffer: 0.125 M KCl containing 25 mM KSCN and 10 mM K$^+$-HEPES, pH 5.6
RM-B buffer: 10 mM Tris–acetate, pH 8.5, containing 0.3 M sucrose, 2.5 mM MgSO$_4$, and 1 μM nigericin
Tris stop solution: 10 mM Tris–HEPPS, pH 8.5, containing 0.3 M sucrose

Procedure

To measure ATP-driven serotonin accumulation by dense granule membrane vesicles, a suspension of platelet lysate or purified dense granule membrane vesicles is diluted 80-fold into K$^+$-HEPES buffer, pH 7.4. Following a 5-min incubation at 37° the vesicles are collected by centrifugation at 48,000 g for 30 min at 4°, and resuspended to a concentration of about 20 mg of protein per milliliter in K$^+$-HEPES buffer, pH 7.4, containing sucrose. Transport is initiated by adding 0.05 to 0.10 mg washed lysate to 0.2 ml of RM-A buffer containing 0.18 μM [^{3}H]serotonin, 37°. After incubation for varying periods of time at 37°, transport is terminated by rapidly adding 2 ml of ice-cold HEPES stop solution and filtering through a nitrocellulose filter. After the reaction tube and filter are washed with an additional 2 ml of the same buffer, the filter is dried and counted as above.

To measure serotonin transport driven by $\Delta\bar{\mu}_{H^+}$, a suspension of platelet lysate or membrane vesicles is diluted 35-fold into KN buffer and incubated for 5 min at 37°. Following centrifugation at 48,000 g for 20 min at 4°, the sedimented membranes are resuspended in the same buffer to a concentration of about 15 mg of membrane protein per milliliter. Transport is initiated when a sample of this suspension (20 to 50 μg) is diluted into 200 μl of RM-B buffer containing 0.18 μM [^{3}H]serotonin. At the end of the incubation period, transport is terminated by the method described above for ATP-driven serotonin uptake except that an ice-cold Tris stop solution is used.

[20] Binding of Platelet-Activating Factor 1-O-Alkyl-2-acetyl-sn-glycero-3-phosphorylcholine to Intact Platelets and Platelet Membranes

By FRANK H. VALONE

Platelet-activating factor (PAF) binding is quantitated by incubating [^{3}H]PAF with washed platelets for 30 min at room temperature or with platelet membranes for 120 min on ice. Platelet-bound PAF is separated from unbound PAF by centrifugation, whereas membrane-bound PAF is separated by filtration on glass fiber filters.

Assay Method

Reagents

[^{3}H]PAF (alkyl-2-acetyl-*sn*-glyceryl-3-phosphorylcholine, 1-O-[alkyl-1′,2′-^{3}H]: 150 Ci/mmol (Du Pont/New England Nuclear, Boston, MA)
PAF: L-α-lecithin, β-acetyl-, γ-O-alkyl- (Bachem, Torrance, CA)
PAF analogs and antagonists
Citrate anticoagulant: 0.076 M citric acid, 0.15 M sodium citrate, pH 5.2
Phosphate-buffered saline (PBS): 0.076 M sodium/potassium phosphate, pH 7.2, 0.1 M NaCl
Tris–EDTA–Mg^{2+} (TEM): 10 mM Tris, pH 7.5, 1 mM EDTA, 5 mM MgCl$_2$
Tris–Mg^{2+}–K$^+$–BSA (TMKB): 10 mM Tris, pH 7.5, 5 mM MgCl$_2$, 10 mM KCl, 0.25 delipidated bovine serum albumin (BSA)
Glass fiber filters (GL/C; Whatman, Clifton, NJ)

Procedures

Preparation of Platelets. Analysis of PAF binding requires fresh platelets that are washed to remove plasma acylhydrolases that metabolize PAF. The platelets may be prepared by any method that yields platelets that are highly sensitive to PAF (half-maximal platelet aggregation at 5–20 nM PAF for human platelets, and 5–20 pM PAF for rabbit platelets). Human platelet-rich plasma is prepared from fresh blood that was collected into sodium citrate anticoagulant (20% final volume). The platelets are sedimented by centrifugation at 1500 g for 15 min at 20° onto cushions of autologous erythrocytes (2 ml packed erythrocytes/50 ml plasma).[1,2] The platelet/erythrocyte pellet is resuspended in 50 ml PBS, containing 20% (v/v) citrate anticoagulant, final pH 6.8, and then washed three times by repeat centrifugation. After the third wash, the platelet/erythrocyte pellet is resuspended in PBS containing 1.8 mM CaCl$_2$ and 0.1% (w/v) human serum albumin, and the erythrocytes are removed by centrifugation at 150 g for 10 min at 20°. The platelet-rich supernatant is then removed and the concentration of platelets is adjusted to 2×10^8/ml. Rabbit platelets may be prepared in a similar fashion for PAF-binding studies except that the platelets are washed without using erythrocyte cushions.

Platelet-Activating Factor Binding to Intact Platelets. Platelet-activating factor binding assays are performed routinely at room temperature or at 4° because higher temperatures increase nonspecific binding and PAF metabolism, whereas lower temperatures diminish the kinetics of binding to an impractical rate.[1] ^{3}H-Labeled platelet-activating factor (4000 cpm = 0.05 pmol) in 2 μl of PBS containing 1.8 mM CaCl$_2$ and 0.1% (w/v) human serum albumin is added to 1.5-ml microfuge tubes. Aliquots of competitive binding inhibitors, such as unlabeled PAF and PAF analogs, or buffer are then added to the tubes. Buffer (0.25 ml) is then added to suspend the tritiated PAF and the competitive inhibitors, after which 0.25 ml of platelet suspension (1×10^8 platelets) in the same buffer is added to the tubes. The suspension is mixed gently and then incubated at room temperature without further mixing. After 30 min, the platelets are sedimented by centrifugation for 1 min in a microfuge. The supernatants are carefully aspirated and transferred to scintillation vials for quantitation of unbound [^{3}H]PAF. The platelet pellets are resuspended in 0.5 ml of distilled water by repeated aspiration with a pipette and transferred to scintillation vials for quantitation of bound [^{3}H]PAF.

To quantitate nonspecific PAF binding to the plasticware, control tubes in each experiment should contain all reagents except platelets and should

[1] F. H. Valone, E. Coles, V. R. Reinhold, and E. J. Goetzl, *J. Immunol.* **129**, 1637 (1982).
[2] F. H. Valone, *Thromb. Res.* **50**, 103 (1988).

be processed as described above. Because nonspecific PAF binding to platelets may constitute up to 50% of the total PAF binding, specific receptor-mediated binding should be determined in each experiment. A full binding competition curve between radiolabeled and unlabeled PAF is required to be certain that the receptor is saturated, thereby allowing determination of the percentage of nonspecific binding (see below). In routine practice, however, a 2000-fold molar excess of unlabeled PAF (100 pmol) fully saturates the PAF receptor and can be used to estimate specific binding with the formula: % specific binding = % [^{3}H]PAF binding in the absence of unlabeled PAF minus % [^{3}H]PAF binding in the presence of a 2000-fold molar excess of unlabeled PAF.

Platelets and other cells may metabolize PAF extensively, thereby increasing apparent nonspecific binding.[2] To quantify PAF metabolism, platelet-bound PAF should be extracted by a modified Bligh–Dyer method using chloroform : methanol : water (1.1 : 1.1 : 1, v/v/v).[3] The extracts are dried under a stream of nitrogen, resuspended in 0.5 ml methanol, and spotted on silica gel thin-layer plates that are developed with chloroform : methanol : acetic acid : water (50 : 25 : 8 : 4, v/v/v/v). Standards on each plate include [^{3}H]PAF, [^{3}H]lyso-PAF (1-*O*-alkyl-*sn*-glycero-3-phosphorylcholine), and [^{3}H]phosphatidylcholine. Platelet-activating factor is metabolized mainly to 1-*O*-alkyl-2-acyl-*sn*-glycero-3-phosphocholine, which migrates with the [^{3}H]phosphatidylcholine standard.[2] An alternative method to avoid PAF metabolism is to quantify binding of a nonmetabolizable PAF analog such as [^{3}H]WEB 2086 (Amersham, Arlington Heights, IL) using methods described for PAF binding. These analogs bind effectively to PAF receptors on platelets but may not bind to all classes of PAF receptors on other cells.

Platelet-Activating Factor Binding to Platelet Membranes. Platelet-activating factor binding to platelet membranes is quantitated by the method of Hwang *et al.*[4] This method yields membranes from a variety of cells that reliably bind PAF.[4,5] Washed platelets are resuspended at a concentration of 2×10^9/ml in 10 m*M* Tris, pH 7.5, 1 m*M* EDTA, 5 m*M* MgCl$_2$ (TEM) and are then disrupted by three cycles of freezing in liquid nitrogen followed by slow thawing at room temperature. Alternatively, platelets may be disrupted in a nitrogen cavitation bomb.[5] Plasma membranes are isolated by centrifugation of the lysed platelet suspension at 63,500 *g* for 5 hr on a discontinuous sucrose density gradient of 12% (w/v) and 27% (w/v) sucrose in TEM buffer. The membrane fraction that

---

[3] F. H. Valone and L. B. Epstein, *J. Immunol.* **141**, 3945 (1988).
[4] S.-B. Hwang, M.-H. Lam, and S.-S. Pong, *J. Biol. Chem.* **261**, 532 (1986).
[5] F. H. Valone and N. M. Ruis, *Biotechnol. Appl. Biochem.* **8**, 465 (1986).

contains the bulk of the PAF-binding activity forms a band between the 12 and 27% sucrose layers. Unbroken platelets, granules, and other debris are sedimented to the bottom of the tube.

The membrane fraction is sedimented by centrifugation at 100,000 g for 60 min and resuspended in 10 mM Tris, pH 7.5, 5 mM MgCl$_2$, 10 mM KCl. The isolated membrane fraction retains binding activity for at least 2 weeks when stored at $-70°$. Platelet-activating factor binding to platelet membranes is assessed by a filtration technique using a vacuum filtration manifold. Between 50 and 200 μg of membrane protein in 1 ml TEM, 10 mM KCl, and 0.25% (w/v) bovine serum albumin (TMKB) is added to tubes containing 0.5 pmol (50,000 cpm) of [^{3}H]PAF without or with unlabeled PAF or PAF antagonists. Aliquots of each sample (50 μl) are transferred to scintillation vials for determination of the total radioactivity in each tube. The tubes are incubated for 120 min on ice, after which membrane-bound PAF is separated from unbound PAF by filtration on Whatman GL/C glass fiber filters that were presoaked for 1 hr in TMKB. The filters are then dried and placed in scintillation vials for quantitation of bound [^{3}H]PAF. Platelet-activating factor metabolism is not a major problem during PAF binding to membranes. Metabolism can be quantified by thin-layer chromatography as described above.

Quantification and Characterization of Platelet-Activating Factor Binding. Platelet-activating factor binding is characterized in terms of the competition for binding between [^{3}H]PAF and either unlabeled PAF, PAF antagonists, or PAF analogs.[6] The percentage binding of [^{3}H]PAF in the absence or presence of a range of concentrations of inhibitors is determined and used to draw binding competition curves or saturation isotherms. These curves provide estimates of the binding affinity of PAF analogs and antagonists relative to that of PAF itself. The plateau representing the minimal percentage of tritiated PAF binding induced by a 100- to 1000-fold molar excess of unlabeled PAF defines the nonsaturable, nonspecific component of PAF binding. Platelet-activating factor binding must reach equilibrium in order to draw saturation isotherms. To assure that equilibrium has been reached, the time course of binding of the lowest ligand concentration used in each isotherm should be followed until the percentage specifically bound is stable. The number of binding sites and their binding affinity may be estimated graphically by the method of Scatchard,[7] in which the quantity of bound PAF divided by the quantity of unbound PAF is plotted on the ordinate, and the quantity of bound PAF is plotted on the abscissa. The dissociation constant is determined from

---

[6] G. A. Weiland and P. B. Molinoff, *Life Sci.* **29**, 313 (1981).
[7] G. Scatchard, *Ann. N.Y. Acad. Sci.* **51**, 660 (1949).

the slope of the curve and the maximal binding capacity (number of binding sites) from the x intercept. Curve fitting and calculation of the binding constants are improved substantially by using a computer program based on the least squares method.[8]

Comments

Platelet-activating factor is a phospholipid and therefore requires special handling. Because PAF is insoluble in aqueous solutions, buffers must contain 0.1% (w/v) human serum albumin or 0.25% BSA as carrier proteins. Thus, the binding data obtained in these studies represent an equilibrium between receptor bound- and albumin-bound PAF. Platelet-activating factor is stable for at least 6 months when stored at $-20°$ in methanol or 2.5% BSA in PBS under an atmosphere of air. Stock solutions of tritiated PAF and unlabeled PAF may be used repeatedly in binding assays.

[8] P. J. Munson and D. Rodbard, *Anal Biochem.* **107,** 220 (1980).

[21] Binding of Fibrinogen and von Willebrand Factor to Platelet Glycoprotein IIb–IIIa Complex

By JACEK HAWIGER and SHEILA TIMMONS

The mechanism through which a platelet hemostatic plug is formed involves at least two adhesive macromolecules: von Willebrand factor (vWF) and fibrinogen.[1] von Willebrand factor is thought to provide a molecular anchor between the subendothelium and platelets, because in von Willebrand disease formation of a platelet hemostatic plug is impaired. On the other hand, fibrinogen is considered to provide interplatelet linkages after activation of platelets with ADP and exposure of the glycoprotein IIb–IIIa (GPIIb–IIIa) receptor for fibrinogen.[2]

The involvement of fibrinogen in platelet aggregation remained unexplained until it was demonstrated that aggregation of platelets induced by ADP is accompanied by their binding of fibrinogen. Subsequent experi-

[1] J. Hawiger, *in* "Hemostasis and Thrombosis: Basic Principles and Clinical Practice" (R. W. Colman, J. Hirsh, V. J. Marder, and E. W. Salzman, eds.), Chapter 12, p. 182. Lippincott, Philadelphia, 1987.
[2] J. Hawiger, *Ann. N.Y. Acad. Sci.* **614,** 270 (1991).

ments indicated that the main role of platelet agonists such as ADP, epinephrine, or thrombin is to induce exposure of binding sites for fibrinogen on the platelet membrane. After stimulation with thrombin and inactivation of free thrombin with hirudin, we observed specific binding of ^{125}I-labeled fibrinogen with remarkably similar characteristics to those after stimulation with ADP.[3] The interaction of fibrinogen with human platelets stimulated with thrombin (0.05 U/ml) fulfills the criteria for specific, saturable, and reversible binding. It is mediated by receptors that bind approximately 44,000 molecules of fibrinogen per platelet with an apparent dissociation constant (K_D) of 1.8×10^{-7} M. Because the plasma concentration of fibrinogen is approximately 50 times higher, it is apparent that even when its plasma level decreases to 5% of the normal value the fibrinogen level is still sufficient to saturate the available binding sites. The binding of fibrinogen to platelets requires calcium, because it is prevented or reversed by ethylenediaminetetraacetic acid (EDTA). The platelet membrane glycoprotein IIb–IIIa complex that is involved in the binding of fibrinogen belongs to the β_3 family of the integrin receptor gene superfamily.[4] About 40,000 molecules of the glycoprotein IIb–IIIa complex (integrin $\alpha_{IIb}\beta_3$) are available on one platelet.[5] This suggests a 1:1 stoichiometry between fibrinogen and the glycoprotein IIb–IIIa complex on the membrane of activated platelets. We have shown that this process is either prevented or reversed by cyclic AMP-mediated reactions, inasmuch as an increase in cAMP levels induced by prostacyclin (PGI$_2$) or a nonprostanoid inhibitor, forskolin, correlates with inhibition of binding of fibrinogen to activated platelets, which is paralleled by inhibition of platelet aggregation.[3,6] Isolated glycoproteins IIb–IIIa in the form of a calcium-held heterodimer bind fibrinogen attached to solid support.[7] Isolated GPIIb–IIIa complexes can be incorporated into phospholipid vesicles (liposomes) and used for binding studies. Their binding capacity, however, is much lower.[8] The primary structure of GPIIb and GPIIIa was elucidated by cDNA cloning and sequencing.[9,10]

[3] J. Hawiger, S. Parkinson, and S. Timmons, *Nature (London)* **283**, 195 (1980).
[4] E. Ruoslahti and M. D. Pierschbacher, *Science* **238**, 491 (1987).
[5] B. S. Coller, E. I. Peerschke, L. E. Scuder, and C. A. Sullivan, *J. Clin. Invest.* **72**, 325 (1983).
[6] S. Graber and J. Hawiger, *J. Biol. Chem.* **257**, 14606 (1982).
[7] R. L. Nachman, L. L. K. Leung, M. Kloczewiak, and J. Hawiger, *J. Biol. Chem.* **259**, 8584 (1984).
[8] L. Parise and D. R. Phillips, *J. Biol. Chem.* **261**, 14011 (1986).
[9] M. Poncz, R. Eisman, R. Heidenreich, S. M. Silver, G. Vilaire, S. Surrey, E. Schwartz, and J. S. Bennett, *J. Biol. Chem.* **262**, 8467 (1987).
[10] L. A. Fitzgerald, B. Steiner, S. C. Rall, Jr., S. S. Lo, and D. R. Phillips, *J. Biol. Chem.* **262**, 3936 (1987).

Human fibrinogen interacts with binding sites exposed on GPIIb–IIIa of stimulated platelets through the tentacles present on γ and α chains.[11] The 12-residue carboxyl-terminal segment of the γ chain, encompassing residues 400–411, was pinpointed by us as the platelet receptor recognition domain.[12–14] Following our findings that isolated α chains from human fibrinogen were reactive with ADP-activated platelets,[11] we showed that the sequences RGDF (α95–98) and RGDS (α572–575) are involved in the interaction of human fibrinogen α chain with receptors on activated platelets.[15] Both domains contain the sequence RGD, identified previously as "the cell recognition site" of fibronectin.[16]

It is apparent then that one molecule of fibrinogen bridging two platelets bears six binding domains, resulting in a higher affinity due to the higher than one valency level.[1] Thus, the presence of three domains on each half of the fibrinogen molecule, potentially able to interact with platelet receptors, provides optimal conditions for tighter binding of fibrinogen to platelets and for their subsequent aggregation (Fig. 1). In solid-phase binding of fibrinogen to human platelets *in vitro,* the α chain domain, encompassing the sequence RGDF (α95–98), did not seem to be involved.[17] All three domains are prone to enzymatic attack by plasmin, which can cleave the bond Lys^{406}–Gln^{407} in the γ chain and clip off sequences containing RGD in the α chain.[18]

In addition to fibrinogen, vWF is another adhesive protein in plasma that can bind to human platelets and mediate their aggregation or agglutination. Two distinct mechanisms of vWF binding to platelets are known. The first involves the nonintegrin receptor, GPIb–IX complex, which binds human vWF in the presence of the antibiotic ristocetin. A deficiency of the platelet membrane GPIb–IX complex in giant platelet syndrome (Bernard–Soulier syndrome) results in an inability of platelets to interact

---

[11] J. Hawiger, S. Timmons, M. Kloczewiak, D. D. Strong, and R. F. Doolittle, *Proc. Natl. Acad. Sci. U.S.A.* **79**, 2068 (1982).
[12] M. Kloczewiak, S. Timmons, and J. Hawiger, *Biochem. Biophys. Res. Commun.* **107**, 181 (1982).
[13] M. Kloczewiak, S. Timmons, T. Lukas, and J. Hawiger, *Biochemistry* **23**, 1767 (1984).
[14] M. Kloczewiak, S. Timmons, M. A. Bednarek, M. Sakon, and J. Hawiger, *Biochemistry* **28**, 2915 (1989).
[15] J. Hawiger, M. Kloczewiak, M. A. Bednarek, and S. Timmons, *Biochemistry* **28**, 2909 (1989).
[16] M. D. Pierschbacher and E. Ruoslahti, *Nature (London)* **309**, 30 (1984).
[17] D. A. Cheresh, S. A. Berliner, V. Vincente, and Z. M. Ruggeri, *Cell (Cambridge, Mass.)* **58**, 945 (1989).
[18] R. F. Doolittle, *in* "Hemostasis and Thrombosis" (A. L. Bloom and D. P. Thomas, eds.), Chapter 11, p. 163. Churchill-Livingstone, Edinburgh, 1981.

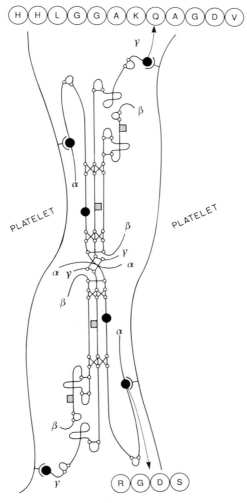

FIG. 1. Proposed model of interaction of human fibrinogen with human platelets. The fibrinogen molecule, composed of three pairs of nonidentical chains (α, β, and γ), is arranged in an antiparallel configuration. The shaded boxes represent carbohydrates on the β and α chains. The platelet receptor recognition domains marked as black dots encompass sequences 95–98 and 572–575 on the α chain and sequence 400–411 on the γ chain. One molecule of fibrinogen can be engaged in trans and cis interaction with platelet receptors made of GPIIb–IIIa.

with vWF in the subendothelial matrix[19] and to bind vWF in the presence of ristocetin.[20] Such binding results in aggregation of intact platelets or agglutination of formalin-fixed, metabolically inert platelets.[21] The binding domain of the GPIb–IX complex, which interacts with vWF, encompasses the amino-terminal segment of glycocalicin, a heavily glycosylated, extraplatelet region of $GPIb_\alpha$.[22–24] The complementary binding domain on human vWF encompasses the amino-terminal segment between residues Val^{449}–Lys^{728} containing two receptor recognition domains for the platelet nonintegrin receptor, glycoprotein Ib (Cys^{474}–Pro^{488} and Leu^{694}–Pro^{708}), which are separated by a 205-residue segment.[25] Another domain, containing the RGD motif (residues 1744–1746) recognized by integrins, is localized in the carboxyl-terminal "module" of the vWF subunit.

The second mechanism of binding of vWF to human platelets operates when thrombin or ADP is used to stimulate platelets.[26,27] Thrombin-induced binding of vWF is not observed in platelets from patients with Glanzmann's thrombasthenia, known to be deficient in GPIIb–IIIa.[28] Such binding is blocked by monoclonal antibodies against GPIIb–IIIa.[29] This binding of vWF, induced by thrombin and ADP but not ristocetin, is blocked by the synthetic peptide RGDS, which is an analog of the sequence 1744–1747 present in vWF.[30,31] Surprisingly, the synthetic dodecapeptide corresponding to the carboxyl-terminal segment of human fibrinogen γ chain (γ400–411), and not present in vWF, also blocked binding of vWF to ADP- and thrombin-stimulated platelets.[32] Thus, vWF can bind to platelets via GPIb–IX complex (ristocetin-dependent mechanism) and via the GPIIb–IIIa complex (thrombin-, ADP-dependent mechanism).

Binding of fibrinogen and vWF to platelet receptors is either prevented

---

[19] H. J. Weiss, V. T. Turitto, and H. R. Baumgartner, *J. Lab. Clin. Med.* **72**, 750 (1978).
[20] J. L. Moake and J. D. Olson, *Thromb. Res.* **19**, 21 (1980).
[21] K. M. Brinkhous and M. S. Read, this series, Vol. 169, p. 149.
[22] T. Okamura, C. Lombart, and G. A. Jamieson, *J. Biol. Chem.* **251**, 5950 (1976).
[23] V. Vincente, P. J. Kostel, and Z. M. Ruggeri, *J. Biol. Chem.* **263**, 18473 (1988).
[24] V. Vincente, R. A. Houghten, and Z. M. Ruggeri, *J. Biol. Chem.* **265**, 274 (1990).
[25] H. Mohri, Y. Fujimura, M. Shima, A. Yoshioka, R. A. Houghten, Z. M. Ruggeri, and T. S. Zimmerman, *J. Biol. Chem.* **263**, 17901 (1988).
[26] T. Fujimoto, S. Ohara, and J. Hawiger, *J. Clin. Invest.* **69**, 1212 (1982).
[27] T. Fujimoto and J. Hawiger, *Nature (London)* **297**, 154 (1982).
[28] Z. M. Ruggeri, R. Bader, and L. DeMarco, *Proc. Natl. Acad. Sci. U.S.A.* **79**, 6038 (1983).
[29] Z. M. Ruggeri, L. DeMarco, L. Gatti, R. Bader, and R. R. Montgomery, *J. Clin. Invest.* **72**, 1 (1983).
[30] E. F. Plow, M. D. Pierschbacher, E. Ruoslahti, G.A. Marguerie, and M. H. Ginsberg, *Proc. Natl. Acad. Sci. U.S.A.* **82**, 8057 (1985).
[31] S. Timmons and J. Hawiger, *Trans. Assoc. Am. Physicians* **99**, 226 (1986).
[32] S. Timmons, M. Kloczewiak, and J. Hawiger, *Proc. Natl. Acad. Sci. U.S.A.* **81**, 4935 (1984).

or reversed by cyclic AMP-mediated reactions. An increase in cAMP levels induced by prostacyclin (PGI_2) or a nonprostanoid inhibitor, forskolin, correlates with inhibition of binding of fibrinogen and vWF to activated platelets, and is paralleled by inhibition of platelet aggregation.[3,6,26] Clearly, tight binding of adhesive proteins to the glycoprotein IIb–IIIa complex cannot take place when platelet cyclic AMP rises. In addition to a cyclic AMP-dependent kinase phosphorylating a number of proteins (see above), platelets have a cyclic GMP-dependent kinase that also exerts a negative effect on the receptor function of the GPIIb–IIIa complex.[33]

Measuring the binding of purified fibrinogen or vWF to human platelets requires that the platelets be isolated from other plasma cells and proteins. The method for separation of human platelets from plasma proteins using albumin gradient centrifugation and Sepharose 2B gel filtration (outlined below) has been described in Volume 169 in this series.[34] The ligands used in binding studies, fibrinogen and vWF, must be purified in such a way that they are freed of each other. The purification procedure described below fulfills this objective. Finally, what follows is the experimental protocol for binding of fibrinogen and vWF to their receptors exposed on the GPIIb–IIIa complex during the activation of platelets with thrombin and ADP.

Materials and Methods

A large-scale method of isolating human fibrinogen from plasma by cold ethanol precipitation has been described by Doolittle and colleagues.[35] In our studies on the binding of fibrinogen to human platelets we used both cold ethanol-precipitated fibrinogen and commercially available Kabi fibrinogen (Kabi, Stockholm, Sweden) that was further purified by ammonium sulfate precipitation to remove highly soluble and partially degraded fibrinogen.[32,36] In both cases the fibrinogen was separated from vWF and other high molecular weight proteins by column chromatography. One should avoid methods employing ε-aminocaproic acid to inhibit plasmin because this inhibitor interferes with the binding of fibrinogen.

Human vWF was isolated from fibrinogen and other plasma proteins by our modification of the method by Martin *et al.*[31,37] This method utilizes

---

[33] M. E. Mendelsohn, S. O'Neill, D. George, and J. Loscalzo, *J. Biol. Chem.* **265**, 19028 (1990).
[34] S. Timmons and J. Hawiger, this series, Vol. 169, p. 11.
[35] R. F. Doolittle, D. Schubert, and S. A. Schwartz, *Arch. Biochem. Biophys.* **118**, 456 (1976).
[36] I. B. Lipinska, B. Lipinski, and V. Gurewich, *J. Lab. Clin. Med.* **84**, 509 (1974).
[37] S. E. Martin, V. J. Marder, L. S. Loftus, and G. H. Barlow, *Blood* **55**, 848 (1980).

cold ethanol precipitation, differential fractionation with polyethylene glycol (PEG 3500), and column chromatography.

Caution: Because human blood products are used, precautions should be undertaken, e.g., double protective gloves should be worn during handling of blood products that are biohazardous, irrespective of routine testing of such products for human immunodeficiency virus (HIV) and hepatitis viruses.

Purification of Fibrinogen. Kabi fibrinogen is supplied as a lyophilized powder containing protein (1 g) with 1 g of sodium citrate and 0.4 g sodium chloride. One gram of this material is dissolved by layering it onto 10 ml of deionized H_2O containing 10^{-5} M PPACK (D-phenylalanyl-L-propyl-L-arginine chloromethyl ketone; Calbiochem, La Jolla, CA). The fibrinogen should be allowed to slowly absorb; it usually takes about 2 hr. Since this is a concentrated solution it should not be mixed manually or on a Vortex until dissolved. The sample is then applied to a Sepharose 4B-CL column (2.5 × 90 cm) that has been equilibrated with the following buffer: 0.13 M sodium citrate, 0.0125 M sodium phosphate, 0.15 M NaCl, 1 mM EDTA, 10^{-5} M PPACK, pH 7.6. The material is eluted with the same buffer and 4-ml fractions are collected into polypropylene tubes. The fractions rich in fibrinogen in the second peak are pooled (about 120 ml) and dialyzed against 1 liter of 0.05 M NaCl for 2 hr with two changes of the dialysate. It should not be dialyzed longer at this step because the fibrinogen will precipitate, or polymerize, and cannot be redissolved. After dialysis, the fibrinogen solution is collected into a polypropylene centrifuge bottle (200 ml) and saturated ammonium sulfate to a final concentration of 16% (v/v) is added dropwise with stirring.[32] When all of the ammonium sulfate solution has been added, the mixing is stopped and the mixture is allowed to sit at 4° for 30 min. The mixture is then centrifuged in the same bottle at 1200 g at 4° for 10 min. The supernatant is decanted and discarded, and the pellet is dissolved in citrate/phosphate/NaCl buffer, pH 7.6, by gently stirring with a Teflon rod. To remove residual ammonium sulfate, the dissolved fibrinogen is then dialyzed against citrate/phosphate/NaCl buffer, pH 7.6, for 2 hr with three changes of buffer. The protein concentration is determined by measuring the absorbance at 280 nm and using the extinction coefficient 1.5 to calculate the concentration, which is usually 10–15 mg/ml. This fibrinogen preparation is stored at −40° in 1-ml aliquots. It should not be frozen and thawed more than one time.

Purification of von Willebrand Factor. Cryoprecipitated human antihemophilic factor from the American Red Cross is thawed at 37° for 5 min. A batch of 10 bags of cryoprecipitate yields about 10 mg of purified vWF. The contents of 10 bags are pooled into a siliconized glass beaker and chilled to −5°. While the solution is stirred, 95% ethanol at −20° is added dropwise to a final ethanol concentration of 4% (v/v). The mixture is centrifuged at 5000 g for 5 min in plastic bottles. The supernatant is discarded and the surface of the pellet is washed with 8% ethanol at 0°. After transferring the tube to room temperature, the pellet is dissolved in 50 ml of Tris-HCl buffer, pH 7.0. Sodium citrate, 0.5 M, is added to give a final concentration of 0.02 M and then the pH is adjusted to 6.3 by gradual addition of 0.02 M citric acid. Contaminating plasma proteins are precipitated at room temperature by slowly adding 40% (w/v) polyethylene glycol (PEG 3500) with continuous stirring to a final concentration of 4% (w/v). The mixture is allowed to sit without stirring at room temperature for 15 min and then is centrifuged at 5000 g at room temperature for 15 min. The pellet is discarded and 40% (v/v) PEG 3500 is slowly added to the supernatant at room temperature while stirring to have a final concentration of 10% for precipitation of the vWF-rich fraction. After centrifugation, the supernatant is discarded and the pellet chilled to 0° on ice and the surface washed once with cold 8% ethanol in 0.02 M Tris-HCl, pH 7.0, to remove PEG and once with cold 0.02 M Tris-HCl, pH 7.0 to remove ethanol. The pellet is dissolved in 10 ml of 0.02 M Tris-HCl, 0.02 M sodium citrate, 0.1 M sodium chloride buffer, pH 7.3, containing 10^{-5} M PPACK. This material is applied to a Sepharose 4B-CL column (2.5 × 90 cm) that had been equilibrated with the same buffer without PPACK and 5-ml fractions are collected. The first peak eluted in the void volume (V_0) contains only vWF. It is pooled and concentrated by dialysis against 40% (w/v) Ficoll in 0.01 M sodium phosphate, 0.15 M sodium chloride, pH 7.4, for 16–20 hr at 4°. The concentrated vWF is then dialyzed against 0.01 M sodium phosphate, 0.15 M sodium chloride buffer, pH 7.4 for 3 hr at 4° with two changes of buffer. Purified vWF migrates as a single band with an apparent M_r 240,000 on sodium dodecyl sulfate-polyacrylamide gel electrophoresis (SDS–PAGE) under reducing conditions and shows a normal pattern of multimers on agarose electrophoresis. One milligram of purified vWF usually contains less than 55 ng of fibrinogen as determined by radioimmunoassay and it has 86 arbitrary units of ristocetin cofactor activity (1 unit equals that measured in 1 ml of normal pooled plasma). The protein concentration is determined by the method of Bradford.[38] Before concentration the purified vWF may be stored at −40° and thawed without loss or degradation. After concentration, it should not be frozen because there will be a considerable loss due to precipitation when it is thawed. Concentrated vWF is stable at 4° for about 2–3 weeks.

Iodination of Purified Fibrinogen and von Willebrand Factor. The iodine monochloride method was adapted because it is mild enough to

---

[38] M. M. Bradford, *Anal. Biochem.* **72**, 248 (1975).

retain the biologic properties of fibrinogen and vWF.[39] Incorporation of 0.5 atom of ^{125}I per molecule according to the following procedure results in a preparation that has not been denatured and is sufficiently radiolabeled to study its binding to platelets. A stock solution of iodine monochloride is prepared by dissolving 108 mg of sodium iodate in 2 ml H_2O and 150 mg of sodium iodide in 8 ml of 6 N HCl. The iodate solution should be forcibly injected (using a glass 5-ml pipette and rubber pipette filler) into the iodide–hydrochloric acid solution and quickly mixed with the pipette to avoid precipitation of the iodine. If no yellow color is detectable, the volume is adjusted with H_2O to reach 45 ml. This solution is stable for about 6 months if stored at 4° in a brown bottle. At the time of the iodination procedure the stock iodine monochloride is diluted 1 : 10 (v/v) in 2 M NaCl. Glycine (0.8 M) in 0.2 M NaCl adjusted to pH 8.5 must be prepared fresh on the day of use. A charcoal-filtered iodination hood is recommended for safety in handling the free ^{125}I. An example of the system for labeling of fibrinogen is as follows: 2 ml of purified fibrinogen in solution (15 mg/ml) and 0.2 ml of glycine buffer, pH 8.5, are mixed in a 15 × 100 mm polypropylene tube. Into another tube containing 0.015 ml of the iodine monochloride diluted in NaCl, the following solutions are added: 0.02 ml of ^{125}I (2 mCi), and 0.2 ml glycine buffer, pH 8.5. After mixing, the iodine monochloride mixture is forcibly injected into the fibrinogen solution using a 9-in glass Pasteur pipette and a pipette filler. To measure total radioactivity, 0.005 ml of the mixture is counted in a γ counter. The radiolabeled fibrinogen is then transferred into a dialysis membrane tubing and dialyzed against 250 ml of the citrate/phosphate/saline buffer, described above, for 4 hr with four changes of buffer. After dialysis and microcentrifugation, protein is measured at absorbance 280 nm. The proportion of bound versus free radioactivity is determined in a 0.005-ml sample of the ^{125}I-labeled fibrinogen mixed with 0.05 ml of 20% (w/v) albumin, and precipitated with 1 ml of 20% (w/v) trichloroacetic acid. When using the iodine monochloride method, 50% of the radioactivity is incorporated into the protein and 99.5% of this radioactivity is precipitated with trichloroacetic acid. The specific activity is about 4 × 10^7 cpm/mg of ^{125}I-labeled fibrinogen and 5 × 10^8 cpm/mg for ^{125}I-labeled vWF. The concentration of ^{125}I-labeled fibrinogen is adjusted to 10 mg/ml, divided into 0.1-ml aliquots, and stored at −40° for use up to 60 days. At the time of the binding assay, an aliquot is thawed and diluted to have 20 times the desired final concentration (i.e., 3.4 μM diluted 0.025 ml/0.5 ml in the

---

[39] A. S. McFarlane, *J. Clin. Invest.* **42**, 346 (1963).

binding assay will give a final concentration of 0.17 μM). The ^{125}I-labeled vWF is stored undiluted at 4° for 2 weeks.

Platelet Agonists. Platelet agonists such as ADP, thrombin, and epinephrine are dissolved in phosphate-buffered saline (PBS) (0.02 M sodium phosphate in 0.15 M NaCl, pH 7.35, is used). If an agonist, such as phorbol myristate acetate (PMA) and ionophore A23187, requires an organic solvent [dimethyl sulfoxide (DMSO) or ethanol, respectively] the concentration should be as low as possible, less than 0.1% (v/v), and appropriate controls of the diluent should be included in the protocol.

Separation of Platelets from Plasma Proteins. Blood is collected from healthy volunteers who have abstained from all medications for the prior 10 days and from caffeine-containing beverages for 10 hr. Human platelets are separated from plasma proteins by applying platelet-rich plasma to a stepwise albumin gradient and centrifugation at 180 g for 15 min at room temperature.[34] The platelet layers are collected and applied to a Sepharose 2B gel filtration column and the platelet-rich fraction eluted as described in detail in Volume 169 of this series.[34]

Binding Assay. The binding assay is performed with platelets suspended in 5 mM HEPES [4-(2-hydroxyethyl)-1-piperazineethanesulfonic acid] balanced salt buffer, pH 7.35, containing 0.1% (w/v) dextrose and 0.35% (w/v) albumin. The platelet suspension for use in the assay of ^{125}I-labeled fibrinogen or ^{125}I-labeled vWF binding should be separated from plasma proteins and used within 2 hr of blood drawing. Using 12 × 75 mm polypropylene tubes, 0.45 ml of the platelet suspension containing 1 × 10^8 platelets is added, followed by agonist in 0.025 ml and ^{125}I-labeled fibrinogen in 0.025 ml. If inhibitors of the binding of ^{125}I-labeled fibrinogen to platelets are tested, 0.025–0.05 ml inhibitor (e.g., peptide) is added for 5 min, or longer, according to the specific test system, to the platelet suspension before the agonist and radiolabeled fibrinogen.[15] The mixture is incubated at room temperature without stirring for a time that is determined by the requirements of the test system, usually 15 min. After incubation, the mixture is layered over 0.5 ml of 25% (w/v) sucrose in a microcentrifuge tube and centrifuged at 10,000 g for 5 min. Alternatively, the bound radioactivity can be separated from the free by layering the incubation mixture over a cushion of dibutyl phthalate/Apiezon oil C (9 : 1, v/v).

Samples of the supernatant (0.05 ml) are collected for measurement of unbound radiolabeled ligand. The remaining content of the tube is frozen in a dry ice–acetone mixture, and the tips of the tubes containing the immobilized pellets are severed using the hot blade of a utility knife. The pellet is transferred to a counting vial, and the radioactivity bound to the platelets is counted in a γ counter.

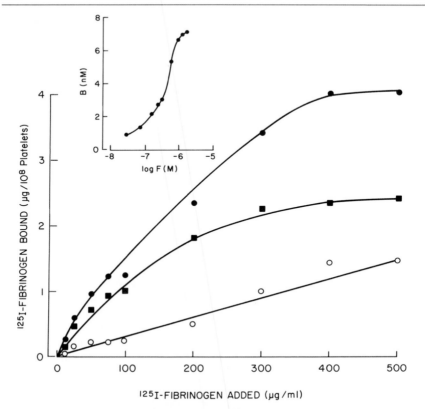

FIG. 2. Concentration-dependent binding of ^{125}I-labeled fibrinogen to human platelets (1×10^8) separated from plasma proteins and suspended in a final volume of 0.55 ml. Incubation was at room temperature for 20 min without stirring. Total observed binding (●) in the presence of 10 μM ADP, specific binding (■) calculated by subtracting observed nonspecific binding from total observed binding, and observed nonspecific (○) binding in the presence of ADP after addition of a 50-fold excess of unlabeled fibrinogen. *Inset:* Klotz plot of the binding data.

Comments

The method presented above allows measurement of binding of ^{125}I-labeled fibrinogen or ^{125}I-labeled vWF to human platelets separated from plasma products. The binding requires stimulation of platelets with agonists such as thrombin, ADP, epinephrine, or phorbol myristate acetate (PMA). The precise mechanism through which agonists change receptors for fibrinogen and vWF on GPIIb–IIIa from nonbinding to binding mode remains an enigma.[1] Nevertheless, the binding of ^{125}I-labeled fibrinogen

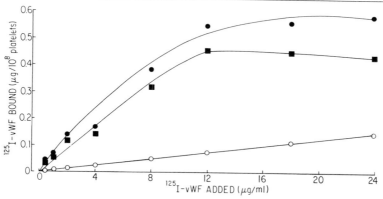

FIG. 3. Concentration-dependent binding of ^{125}I-labeled vWF to human platelets (1×10^8) separated from plasma proteins and suspended in a final volume of 0.55 ml. Incubation was at room temperature for 20 min without stirring. Total observed binding (●) in the presence of 10 μM ADP, specific binding (■) calculated by subtracting observed nonspecific binding from total observed binding, and observed nonspecific (○) binding in the presence of ADP after addition of a 50-fold excess of unlabeled vWF.

and of ^{125}I-labeled vWF to receptors on activated platelets fulfills the criteria for ligand–receptor interaction: it is concentration dependent, saturable, and specific (Figs. 2 and 3, respectively). To verify the saturability of binding, the binding isotherm that represents specific binding can be transformed into a Klotz plot,[40] as illustrated in Fig. 2 (inset). The inflection in the Klotz plot represents half-saturable binding and lies within the range of the dissociation constant (K_D). The latter provides a measure of the affinity of binding. The capacity of binding indicates the number (n) of molecules of ligand bound to one platelet. Both the affinity and the capacity of binding have been determined in several laboratories in regard to the fibrinogen–platelet interaction using the Scatchard plot or double-reciprocal plots.[41] The reported number of molecules of ^{125}I-labeled fibrinogen bound to one platelet ranges from 44,000 to 50,000 and the K_D is within the remarkably narrow range of 1.2 to 1.8×10^{-7} M. There are, however, two caveats posed by these data. First, when platelets are activated an additional pool of GPIIb–IIIa complexes, present in the α granules, is exposed as a result of fusion of the platelet membrane with the α granule membrane.[42] The α granule membrane becomes a part of the platelet membrane and an estimated increase in platelet surface area following

[40] I. M. Klotz, *Science* **217**, 1247 (1982).
[41] G. Scatchard, *Ann. N.Y. Acad. Sci.* **51**, 660 (1949).
[42] K. Niija, E. Hodson, R. Bader, V. Byers-Ward, J. A. Koziol, E. F. Plow, and Z. M. Ruggeri, *Blood* **70**, 475 (1987).

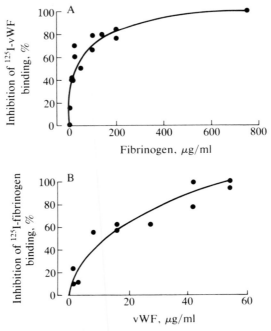

FIG. 4. Inhibition of binding of ^{125}I-labeled vWF by fibrinogen and binding of ^{125}I-labeled fibrinogen by vWF. (A) Binding of ^{125}I-labeled vWF (9 μg/ml to ADP-treated human platelets (1 × 10^8/0.55 ml) in the presence of various concentrations of human fibrinogen. Labeled and unlabeled ligands were added simultaneously to human platelets 5 min before addition of ADP (10 μM). (B) Binding of ^{125}I-labeled fibrinogen (60 μg/ml) to ADP-treated platelets in the presence of various concentrations of vWF. Other conditions for binding were as in (A). Each data set represents one of three experiments carried out with platelets from different donors. [S. Timmons and J. Hawiger, *Proc. Natl. Acad. Sci. U.S.A.* **81,** 4935–4939 (1984).]

activation can reach 60%.[43] Depending on the conditions employed for preparation of platelets and their activation, it is likely that the number of GPIIb–IIIa complexes that are able to bind fibrinogen or anti-GPIIb–IIIa monoclonal antibody Fab fragments will vary.[42] Predictably, receptors on the GPIIb–IIIa complexes can be occupied by unlabeled fibrinogen and/or vWF secreted from the α granules. Second, due to the dimeric structure of fibrinogen and the geometry of its interaction with platelet receptors (parallel rather than perpendicular) in reference to the plane of the platelet

[43] G. V. R. Born, *J. Physiol. (London)* **209,** 487 (1970).

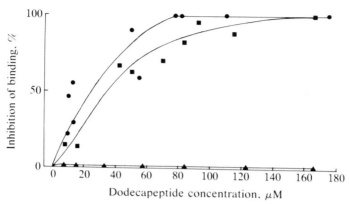

FIG. 5. Concentration-dependent inhibitory effect of the dodecapeptide (γ400–411) on ^{125}I-labeled vWF binding (■) as compared with ^{125}I-labeled fibrinogen binding (●) to human platelets (1 × 10^8/0.55 ml) 5 min prior to addition of ^{125}I-labeled vWF or ^{125}I-labeled fibrinogen (60 μg/ml) followed by ADP. The effect of dodecapeptide on ristocetin (0.6 mg/ml)-induced binding (▲) was done in parallel. Data represents one of three experiments carried out with platelets from different donors. [S. Timmons and J. Hawiger, *Proc. Natl. Acad. Sci. U.S.A.* **81**, 4935–4939 (1984).]

membrane, one molecule of fibrinogen can occupy two receptors on the same platelet (see Fig. 1).

The analysis of binding of vWF to thrombin- or ADP-stimulated platelets is hampered by its multimeric structure. Simply, vWF is a multivalent ligand akin to polymeric collagen or fibrin. In such a situation, it is inappropriate to apply methods of binding analysis based on the assumption that the binding of a monovalent ligand is being measured.

Fibrinogen and vWF compete for the same receptor on thrombin- or ADP-stimulated platelets. When ^{125}I-labeled vWF at 9 μg/ml was added to ADP-treated platelets in the presence of various concentrations of fibrinogen, inhibition of binding was observed (Fig. 4A). Fibrinogen at concentrations of 18 to 675 μg/ml inhibited 40–100% of ^{125}I-labeled vWF binding, respectively. The fibrinogen IC$_{50}$ value for ^{125}I-labeled vWF binding was 74 nM.

When the system was reversed and binding of ^{125}I-labeled fibrinogen (60 μg/ml) was measured in the presence of nonlabeled vWF, inhibition of binding of ^{125}I-labeled vWF was observed (Fig. 4B). The vWF IC$_{50}$ value for ^{125}I-labeled fibrinogen binding was 70 nM (calculated for vWF subunit of M_r 240,000). Thus, the two adhesive glycoproteins, vWF and fibrinogen, inhibited each other at the given concentrations in respect to their binding to the site(s) exposed on ADP-treated platelets. However, when the con-

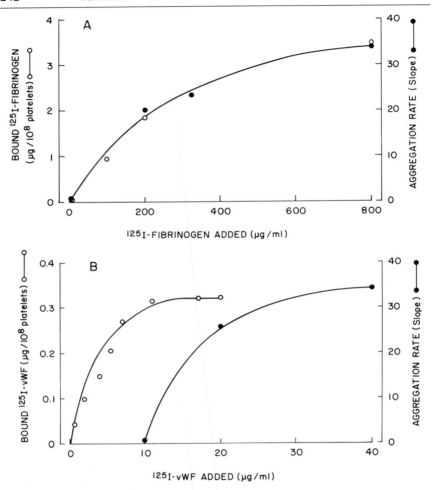

FIG. 6. Correlation of binding of ^{125}I-labeled fibrinogen to human platelets with their aggregation and the lack thereof in the case of vWF. (A) Binding of ^{125}I-labeled fibrinogen (○) and its mediation of aggregation (●) of ADP-treated human platelets. (B) Binding of ^{125}I-labeled vWF (○) and its mediation of aggregation (●) of ADP-treated platelets.

centration of ^{125}I-labeled fibrinogen was increased to 2 mg/ml, corresponding to its plasma level, inhibition by vWF (60 μg/ml) was not observed. Because of its limited solubility, higher concentrations of vWF cannot be used.

The synthetic dodecapeptide (M_r 1188), encompassing the sequence HHLGGAKQAGDV, corresponds to the carboxyl-terminal segment of the human fibrinogen γ chain (γ400–411). The synthetic dodecapeptide inhibited binding of ^{125}I-labeled vWF to ADP-treated platelets in a concentration-dependent manner. Inhibition of 50% of the binding of ^{125}I-labeled vWF (9 mg/ml) was observed at 35 μM dodecapeptide (Fig. 5). In contrast, the dodecapeptide did not inhibit ristocetin-induced binding of ^{125}I-labeled vWF. Thus, the dodecapeptide inhibited binding of ^{125}I-labeled vWF, as well as ^{125}I-labeled fibrinogen, to human platelets only when the activation-dependent GPIIb–IIIa receptor mechanism was involved.

Correlation of binding of ^{125}I-labeled fibrinogen or ^{125}I-labeled vWF to human platelets with their aggregation showed the following differences. The binding was done at room temperature without stirring to dampen subsequent aggregate formation, which will prevent meaningful measurement of bound ligand. The aggregation was monitored at 37° with constant stirring (1100 rpm) in a Payton aggregometer (Payton Associates, Buffalo, NY). Human platelets, separated from plasma proteins, and ^{125}I-labeled fibrinogen or ^{125}I-labeled vWF were used for binding and aggregation measurement following addition of ADP. Binding of ^{125}I-labeled fibrinogen correlated with aggregation rate (slope value) over the range of ligand concentrations used ($r = 0.955$). Both binding and aggregation reached maximum at 300–500 μg/ml (Fig. 6A). In contrast, the binding of ^{125}I-labeled vWF reached saturation at 10 μg/ml, whereas aggregation required at least 50 μg/ml to reach maximum (Fig. 6B). This pattern indicates that polymerization of vWF multimers bound to platelet GPIIb–IIIa is needed for formation of platelet aggregates. Polymerization of ^{125}I-labeled vWF bound to subendothelial matrix of human renal artery was noted by Sakariassen et al.[44] Because platelet aggregation was mediated by either ligand, the binding data support the notion that fibrinogen and/or vWF provide molecular bridges required for platelet aggregate formation. However, aggregation is a multistep process. It begins with stimulatory, agonist-induced signal transduction, cytoskeletal rearrangement, switching the receptors on the GPIIb–IIIa complex to the binding mode, binding of fibrinogen and vWF, formation of small aggregates (<10 platelets) not detectable by conventional aggregometer, secretory response, and formation of large aggregates (>10 platelets) detectable in an aggregometer. Therefore, measurement of binding of ^{125}I-labeled fibrinogen constitutes a useful and very sensitive assay for the receptor function of the GPIIb–IIIa complex and its blockers.

Acknowledgments

This work was supported by NIH Grants HL30647, HL30648, and HL45994.

[44] K. S. Sakariassen, P. A. Bolhuis, and J. J. Sixma, *Nature (London)* **279**, 636 (1979).

[22] Platelet Membrane Glycoprotein IIb–IIIa Complex: Purification, Characterization, and Reconstitution into Phospholipid Vesicles

By David R. Phillips, Laurence Fitzgerald, Leslie Parise, and Beat Steiner

Introduction

Glycoprotein (GP) IIb and GPIIIa are the major cell surface glycoproteins in the platelet plasma membrane. These glycoproteins were first identified by carbohydrate staining of platelet proteins separated on sodium dodecyl sulfate (SDS)–polyacrylamide gels and by radiolabeling of surface proteins of intact platelets.[1,2] The two glycoproteins were subsequently found to form a Ca^{2+}-dependent complex in nonionic detergent solution[3] with a molecular weight of 265,000 calculated from hydrodynamic parameters.[4] The mole ratio of the two glycoproteins is 1 : 1,[5] indicating that the complex (termed the GPIIb–IIIa complex) is a heterodimer. Glycoprotein IIb–IIIa has been purified, and much is known about its structure. Glycoprotein IIb consists of two disulfide-linked subunits, $GPIIb_\alpha$ (M_r 125,000) and $GPIIb_\beta$ (M_r 23,000), while GPIIIa has only one polypeptide chain (M_r 108,000, reduced).

The primary sequences of both GPIIIa and GPIIb have been determined by molecular cloning[6,7] and were used to establish the integrin superfamily of receptors that mediate a large number of cellular interactions.[8] Integrins are composed of α and β subunits that form a noncovalent heterodimer complex. Glycoprotein IIb is the α subunit of GPIIb–IIIa (also termed α_{IIb}), and GPIIIa is the β subunit (also termed β_3). As of this writing, seven distinct but homologous β subunits have been identified, termed β_1, β_2, β_3, β_4, β_5, β_6, and β_7. Each of these β subunits can associate with one of the 10 different α subunits now identified.[9] Although

---

[1] D. R. Phillips, *Biochemistry* **11**, 4582 (1972).
[2] R. L. Nachman and B. Ferris, *J. Biol. Chem.* **247**, 4468 (1972).
[3] T. J. Kunicki, D. Pidard, J. P. Rosa, and A. T. Nurden, *Blood* **58**, 268 (1981).
[4] L. K. Jennings and D. R. Phillips, *J. Biol. Chem.* **257**, 10458 (1982).
[5] N. A. Carrell, L. A. Fitzgerald, B. Steiner, H. P. Erickson, and D. R. Phillips, *J. Biol. Chem.* **260**, 1743 (1985).
[6] L. A. Fitzgerald, B. Steiner, S. C. Rall, *et al.*, *J. Biol. Chem.* **262**, 3936 (1987).
[7] M. Poncz, R. Eisman, R. Heidenreich, *et al.*, *J. Biol. Chem.* **262**, 8476 (1987).
[8] R. O. Hynes, *Cell (Cambridge, Mass.)* **48**, 549 (1987).
[9] E. Ruoslahti and F. G. Giancotti, *Cancer Cells* **1**, 119 (1989).

platelets contain several integrins, including $\alpha_2\beta_1$ (a collagen receptor), $\alpha_5\beta_1$ (a fibronectin receptor), $\alpha_v\beta_3$ (a vitronectin receptor), in addition to GPIIa–IIIa, GPIIb–IIIa appears to be unique in that it is the only integrin that is restricted to platelets and cells of megakaryoblastic potential. Glycoprotein IIIa is a single polypeptide of 762 amino acids with a large (692 amino acids) extracellular domain, a single transmembrane domain, and a short cytoplasmic tail.[5] There are 56 cysteines in the extracellular domain. Glycoprotein IIIa is approximately 40% homologous to β_1 and β_2. Glycoprotein IIb is composed of a light and a heavy chain, which are derived from a single mRNA, but are posttranslationally cleaved.[10] The heavy chain of GPIIb is entirely extracellular and is 871 amino acids in length. It contains four 65-amino acid repeating segments, each of which contains a 12-amino acid sequence characteristic of the Ca^{2+}-binding sequence of calmodulin.[7] The single transmembrane domain of GPIIb is found in the light chain. All the disulfides of GPIIb are disulfide linked and the linking pattern of GPIIb has been reported.[11]

The role of GPIIb–IIIa in platelet function was indicated by their absence in the platelets of patients with the inherited bleeding disorder, Glanzmann's thrombasthenia.[12,13] Platelets from these patients are defective in that they do not aggregate after exposure to the normal platelet agonists. The key function of GPIIb–IIIa in aggregation is that this glycoprotein complex serves as the receptor for adhesive proteins (fibrinogen, fibronectin, von Willebrand factor, and vitronectin) on activated platelets.[14] The binding of fibrinogen and von Willebrand factor are essential for platelet aggregation. Glycoprotein IIb–IIIa also appears to be involved in Ca^{2+} transport, and may be a carrier of several platelet-specific antigens.[15] Glycoprotein IIb–IIIa complexes in reconstituted phospholipid vesicles have fibrinogen-binding properties similar to the fibrinogen-binding properties of intact platelets,[16] which demonstrates that the GPIIb–IIIa complex is the fibrinogen receptor on the surface of activated platelets.

The aim of this chapter is to present the procedures currently in use in this laboratory for isolating the GPIIb–IIIa complex from outdated platelet concentrates and for characterizing the properties of the complex. The

[10] P. F. Bray, J. P. Rosa, V. R. Lingappa et al., Proc. Natl. Acad. Sci. U.S.A. **83**, 1480 (1986).
[11] J. J. Calvete, A. Henschen, and J. Gonzalez-Rodriguez, Biochem. J. **261**, 561 (1989).
[12] D. R. Phillips and P. P. Agin, J. Clin. Invest. **60**, 535 (1977).
[13] A. T. Nurden and J. P. Caen, Br. J. Haematol. **28**, 253 (1974).
[14] E. F. Plow and M.H. Ginsberg, Prog. Hemostasis Thromb. **9**, 117 (1988).
[15] D. R. Phillips, I. F. Charo, L. V. Parise, and L. A. Fitzgerald, Blood **71**, 831 (1988).
[16] L. V. Parise and D. R. Phillips, J. Biol. Chem. **260**, 10698 (1985).

purification procedure yields milligram quantities of the GPIIb–IIIa complex directly from lysates prepared from clinically outdated platelet concentrates. Methods will be described for determining whether isolated glycoproteins and glycoproteins in platelet membranes are complexed or dissociated. Conditions for reconstituting the GPIIb–IIIa complex into phospholipid vesicles and for examining the ligand-binding properties of the purified GPIIb–IIIa complex in a membranous system will be described. Finally, a method will be outlined for the isolation of functionally active subunits of the GPIIb–IIIa complex.

Isolation of Glycoprotein IIb–IIIa Complex

The reported methods for purification of GPIIb and GPIIIa include lectin affinity chromatography,[17] monoclonal antibody affinity chromatography,[18] selective extraction of isolated membranes followed by gel filtration,[4,19] and high-performance liquid chromatography of Triton X-114-extracted glycoproteins.[20] Although each of these methods has advantages for specific applications, we have found that the method described below gives a high yield of functionally active glycoprotein. This method has several advantages. First, because the method avoids extreme pH conditions, the isolated glycoproteins exist in their native, heterodimer form. Second, detergent extracts of whole platelets are used as the starting material, which eliminates the loss of glycoprotein during membrane isolation. Third, the method is simple and does not require the prior generation of monoclonal antibodies. The method described below is routinely used for the rapid isolation of the GPIIb–IIIa complex from Triton X-100 lysates of whole platelets with a 40% yield.

Preparation of Platelet Lysates

Outdated platelet concentrates, stored at 4° from the time of outdating, are obtained from blood banks and are used within 14 days of venipuncture. Platelet concentrates are pooled into 1-liter centrifuge bottles and are centrifuged at 300 g for 5 min at 4° to remove contaminating blood cells. An additional low-speed spin is sometimes used when erythrocytes are still evident in suspension. The supernatant containing the platelets is removed and centrifuged at 1800 g for 15 min at 4° to sediment the platelets. The platelets are dislodged and resuspended in a buffer containing 150

[17] L. L. K. Leung, T. Kinoshita, and R. L. Nachman, *J. Biol. Chem.* **256**, 1994 (1981).
[18] R. P. McEver, J. U. Baenziger, and P. W. Majerus, *Blood* **59**, 80 (1982).
[19] L. A. Fitzgerald, B. Leung, and D. R. Phillips, *Anal. Biochem.* **151**, 169 (1985).
[20] P. J. Newman and R. A. Kahn, *Anal. Biochem.* **132**, 215 (1983).

mM NaCl, 20 mM Tris-HCl, 1 mM EDTA, pH 7.4, using a plastic pipette attached to a rubber bulb, with care being taken not to disturb the erythrocytes, which should remain at the bottom of the tube. The platelets are washed three times at 4° using this procedure. The final platelet pellet is lysed by resuspending the cells in 5 to 10 vol of 1% (w/v) Triton X-100, 10 mM Tris-HCl, 150 mM NaCl, 1 mM $CaCl_2$, 10^{-5} M leupeptin (Vega Biotechnologies, Tucson, AZ), pH 7.2 at 4°. The lysates are centrifuged at 30,000 g for 15 min at 4° to remove cytoskeletal elements, rapidly frozen in dry ice–methanol, and stored at $-80°$ until further processing is required.

Chromatography on Concanavalin A-Sepharose

Triton X-100 lysates (containing 2 to 3 g protein) from 100 units of platelet concentrates are rapidly thawed by immersion of the tube into a 37° bath and centrifuged at 30,000 g at 4° to remove insoluble material. The solution is then applied at ambient temperature at a flow rate of 1 to 2 ml/min to a 1.5 × 10-cm column of concanavalin A (binding capacity of more than 20 mg glycoprotein/ml gel; Sigma Chemical Co., St. Louis, MO) equilibrated with buffer A [0.1% (w/v) Triton X-100, 10 mM Tris-HCl, 150 mM NaCl, 1 mM $CaCl_2$, and 0.05% (w/v) NaN_3, pH 7.4]. The bound glycoproteins are subsequently washed with several bed volumes of buffer A and eluted with buffer A containing 100 mM α-methyl-D-mannoside at a flow rate of 0.5 ml/min and collected into 2- to 2.5-ml fractions. Figure 1 presents a typical chromatogram and shows that measurements of absorbance at 280 nm of individual fractions is a simple and useful measurement of the relative protein concentration in the Triton X-100-containing solution used for this chromatogram.

Figure 2 shows the distribution of proteins that are eluted from the concanavalin A affinity column. The original Triton X-100 lysate (Fig. 2A, lane 1) contains many proteins in addition to faint bands corresponding to $GPIIb_\alpha$ and GPIIIa. These lysates from outdated concentrates have reduced amounts of actin-binding protein (M_r 250,000) and P235 (M_r 235,000) compared to platelets of freshly drawn blood. This is because these proteins are substrates for the endogenous Ca^{2+}-dependent protease and are hydrolyzed during platelet storage. However, storage of outdated platelets according to the conditions described above does not appear to affect GPIIb and GPIIIa qualitatively or quantitatively. Figure 2A (lane 2) shows the proteins that do not bind to the concanavalin A column while lane 3 shows the proteins that bind and are eluted. The primary glycoproteins present in the bound and eluted fractions are thrombospondin, GPIIb, GPIIIa, and fibrinogen. The thrombospondin in these fractions is removed

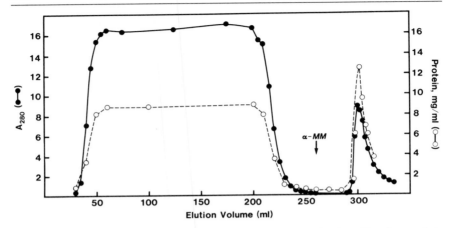

FIG. 1. Concanavalin A affinity chromatography of lysed platelets. The first peak is the concanavalin A flow-through; the second peak is the glycoproteins retained by the concanavalin A and eluted with 100 mM α-methyl-D-mannoside (α-MM; added at the arrow). Shown are both the absorbance at 280 nm (●) and the protein concentration (○). (Reproduced with permission from Fitzgerald et al.[19])

by heparin-Sepharose chromatography as described below. Most detectable fibrinogen is removed by subsequent chromatography on Sephacryl S-300, but it may also be useful to employ the wheat germ agglutinin affinity column procedure also described below to eliminate fibrinogen contamination.

Chromatography on Heparin-Sepharose

The thrombospondin that elutes with the GPIIb–IIIa complex from the concanavalin A column is removed by heparin-Sepharose affinity chromatography. The GPIIb–IIIa-containing fractions from the concanavalin A column are pooled and added at a flow rate of 1 to 2 ml/min to a column (1.5 × 10 cm) of heparin-Sepharose (prepared as described,[21] coupling 1 mg heparin/ml resin) preequilibrated with buffer A. The flow-through fractions of this column contain GPIIb, GPIIIa, and fibrinogen (Fig. 2A, lane 4). The addition of 0.5 M NaCl to the buffer elutes proteins of M_r 180,000 and 160,000 (Fig. 2A, lane 5), which are thrombospondin and a degradation product of this protein, respectively.[22]

[21] K. H. Weisgraber and R. W. Mahley, *J. Lipid Res.* **21**, 316 (1980).
[22] J. Lawler, F. C. Chao, and C. M. Cohen, *J. Biol. Chem.* **257**, 12257 (1982).

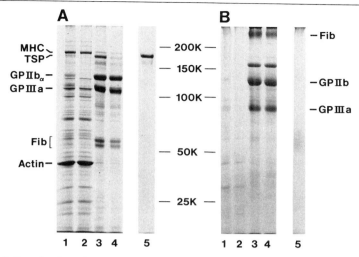

FIG. 2. Proteins from the concanavalin A affinity column in Fig. 1 and the heparin affinity column that were separated by SDS–polyacrylamide gel electrophoresis under reducing (A) and nonreducing (B) conditions. Lane 1 shows the total Triton X-100 extract. Lane 2 is the concanavalin A flow-through fraction. Lane 3 is the concanavalin A-retained fraction. Lane 4 is the heparin flow-through fraction. Lane 5 is the heparin-retained fraction. Fibrinogen (Fib) and thrombospondin (TSP) are also indicated. Approximately 20 mg of protein was electrophoresed in lanes 1 and 2, 5 to 10 μg in lanes 3–5. MHC, Myosin heavy chain. (Reproduced with permission from Fitzgerald et al.[19])

Chromatography of Sephacryl S-300

The proteins that do not bind to heparin are concentrated three- to fivefold by ultrafiltration (PM30 membrane; Amicon Corp., Danvers, MA) to a volume of 8 to 10 ml and chromatographed at ambient temperature on a column of Sephacryl S-300 (2.5 × 110 cm; Pharmacia, Piscataway, NJ) equilibrated with buffer A at a flow rate of 50 to 80 ml/hr; fractions of 5 to 5.5 ml are collected. Figure 3 shows an SDS–polyacrylamide gel of fractions obtained from such a column. Glycoproteins IIb and IIIa coelute at a relative elution volume (V_e/V_t) of ~0.4 to 0.45, mostly separated from fibrinogen and lower molecular weight contaminants. The GPIIb- and GPIIIa-containing fractions from such a column are >98% pure as shown by densitometry of SDS–polyacrylamide gels; a minor contaminant is detectable at M_r 180,000 to 200,000 (Fig. 3B). When fractions containing GPIIb and GPIIIa are pooled, concentrated, and again chromatographed on Sephacryl S-300, no contaminating bands are apparent on SDS–polyacrylamide gels (data not shown). The purified glycoprotein is rapidly frozen by immersion in a dry ice–methanol bath and stored at −80°.

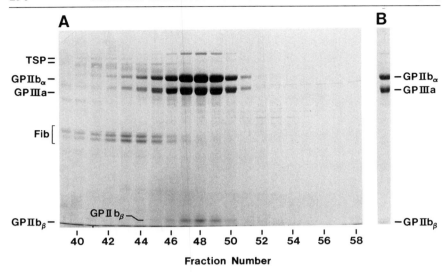

FIG. 3. Sephacryl S-300 gel filtration of the heparin flow-through fraction. (A) SDS–polyacrylamide gel electrophoresis was performed on equal aliquots of individual fractions from the S-300 column. Thrombospondin (TSP), GPIIb$_\alpha$, GPIIb$_\beta$, GPIIIa, and fibrinogen (Fib) bands are indicated. (B) Fractions 46–50 are pooled, and 5 μg was electrophoresed. Only a minor contaminant at M_r 180,000–200,000 is evident. (Reproduced with permission from Fitzgerald et al.[19])

Storage in a frostfree freezer should be avoided. The glycoprotein solution should be rapidly thawed at 37° to avoid dissociation of the GPIIb–IIIa complex. Table I shows the results for a typical isolation of GPIIb and GPIIIa.

Chromatography on Wheat Germ Agglutinin-Sepharose

For experiments requiring the complete removal of fibrinogen, the isolated glycoproteins are adsorbed and eluted from a wheat germ agglutinin affinity column. Because fibrinogen does not contain wheat germ agglutinin receptors, efficient separation of fibrinogen from the GPIIb–IIIa complex is achieved. The GPIIb–IIIa-containing fractions from the first Sephacryl S-300 column are pooled, applied at a flow rate of 0.2 ml/min to a 1 × 5-cm column of wheat germ agglutinin-Sepharose (binding capacity of 3–5 mg glycoprotein/ml gel; Sigma Chemical Co.) equilibrated with buffer A. The column is washed with 5 to 10 bed volumes of buffer A, and the bound GPIIb and GPIIIa are eluted with buffer A containing 100 mM N-acetyl-D-glucosamine. The wheat germ agglutinin affinity step is not routinely employed in GPIIb and GPIIIa preparations because most of the

TABLE I
GLYCOPROTEIN IIb–IIIa PURIFICATION STEPS[a]

Purification step	Total protein (mg)	Amount of GPIIb–IIIa[b] complex		
		Amount (mg)	Total protein[b] (%)	Yield[c] (%)
Triton X-100 lysate	2680	—	—	—
Concanavalin A retained	178	91	51	(100)
Heparin flow-through	112	61	56	68
Sephacryl S-300 pooled fractions	39	37	93	41

[a] Reproduced with permission from Fitzgerald et al.[19]
[b] Calculated from densitometric scans of SDS–polyacrylamide gels.
[c] The percentage yield was based on the content of GPIIb and GPIIIa in the concanavalin A-retained material (see text).

detectable fibrinogen (by SDS–polyacrylamide gels) is eliminated by the Sephacryl S-300 chromatography and because the wheat germ agglutinin has poor recovery (25 to 50%) of GPIIb and GPIIIa.

Radiolabeling of Purified Glycoprotein IIb–IIIa Complex

Iodination of the purified GPIIb–IIIa complex [by lactoperoxidase or chloramine-T (chloramine-T from Sigma)] proved inefficient when the protein was dissolved in Triton X-100 due to high labeling of Triton X-100. To circumvent this problem, the purified GPIIb–IIIa complex is adsorbed onto DEAE-Sephacel, washed to remove Triton X-100, and labeled by the chloramine-T method.[23] DEAE-Sephacel (~100 μl of packed resin) is first washed five times with a buffer containing 0.1% (w/v) Triton X-100, 1 mM CaCl$_2$, 20 mM Tris-HCl, pH 7.4 (buffer B). This is easily performed by pelleting the resin using 10-sec spins in a fixed-angle microfuge (Eppendorf, Hamburg, Germany) and resuspending the resin using a pipettor. The washed resin is then mixed with the purified GPIIb–IIIa complex (50 to 100 μg in 1 to 15 ml of buffer B) for 30 to 60 min at 4°. The cold temperature is required to minimize dissociation of the GPIIb–IIIa complex, which can occur under the low salt concentrations required for adsorption. The GPIIb–IIIa adsorption to DEAE-Sephacel requires an ionic strength below 75 mM NaCl (~5 to 7 mΩ). The resin, with bound glycoprotein, is washed five times with buffer B without Triton X-100 to remove the Triton X-100 from the resin. The resin is resuspended to 1 ml

[23] F. C. Greenwood, W. M. Hunter, and J. S. Glover, *Biochem. J.* **89**, 114 (1963).

TABLE II
PROPERTIES OF GLYCOPROTEINS IIb AND IIIa

Property	GPIIb	GPIIIa	GPIIb–IIIa complex
Sedimentation coefficient ($s_{20,w}$)	4.7s	3.2s	8.6s
Stokes radius (Å)	61	67	71
Frictional coefficient (f/f_0)	1.7	2.1	1.5
Molecular weight (M_r)			
Hydrodynamic parameters	125,000	93,000	265,000
SDS–polyacrylamide gels	136,000	95,000	
Amino-terminal sequence (GPIIb$_\alpha$ and GPIIIa)	(Leu-Asn-Leu-Asp)	(Gly-Pro-Asn-Ile)	
Percentage carbohydrate	14	16	

of this buffer and the glycoprotein is iodinated by adding 1 mCi of carrier-free Na^{125}I and 10 µl of 0.66 mM chloramine-T. The suspension is incubated for 1 to 2 min at room temperature, and iodination is quenched by adding 10 µl of 0.8 mM sodium metabisulfite. The chloramine-T and sodium metabisulfite are dissolved in distilled water immediately before use. The ^{125}I-labeled GPIIb–IIIa-containing resin is then washed five times with buffer B, and the labeled glycoprotein is released from the resin by incubating 15 min with buffer B containing 0.5 M NaCl. The ^{125}I-labeled GPIIb–IIIa complex can be eluted over a PD-10 column (Sigma) to further remove unbound ^{125}I. For this removal, a PD-10 column is washed according to the manufacturer's directions with 5 ml of buffer B containing 1 mg/ml bovine serum albumin and is reequilibrated with several bed volumes of buffer B. The efficiency of this iodination procedure is low, and ~1 to 2 × 10^6 cpm can be expected to be incorporated into the GPIIb–IIIa complex, which is about 0.1% incorporation with a specific activity of 10,000 to 20,000 cpm/µg.

Characterization of Glycoprotein IIb–IIIa Complex

When GPIIb and GPIIIa are isolated in the presence of Ca^{2+}, the two glycoproteins exist as the Ca^{2+}-dependent heterodimeric complex GPIIb–IIIa. Treatment of the isolated GPIIb–IIIa complex with divalent cation chelators will cause the two glycoproteins to dissociate. Other conditions that promote dissociation of the isolated GPIIb–IIIa complex include high pH, low ionic strength buffers, and repeated freezing and thawing. The dissociated glycoproteins will either remain monomeric (at 4°) or will polymerize into multimers (at 37°). Table II summarizes the properties of the GPIIb–IIIa complex and the dissociated glycoproteins. The hydrodynamic properties of glycoproteins in lysates of fresh platelets and those isolated from either fresh platelets or outdated platelet concentrates are indistinguishable, indicating that the glycoproteins are not hydrolyzed or grossly altered in structure during purification or platelet storage at 4°.

Because the fibrinogen receptor function of GPIIb and GPIIIa most likely requires that the two glycoproteins exist as a heterodimeric complex, it is of interest to determine the association state of the isolated glycoproteins. Three methods have been used to characterize the association state of isolated GPIIb and GPIIIa. One method relies on the sedimentation of isolated glycoproteins:[24] the GPIIb–IIIa complex sediments at 8.6S, which is clearly different from the sedimentation coefficients of dissociated glycoproteins (GPIIb at 4.7S and GPIIIa at 3.2S) or polymerized glycoproteins (sedimentation coefficients >8.6S). Another method is thrombin hydrolysis: when GPIIb is complexed with GPIIIa, GPIIb$_\alpha$ is not hydrolyzed by thrombin; GPIIb$_\alpha$ of monomeric GPIIb is a thrombin substrate.[25] The third method determines the binding activity of isolated glycoproteins with "complex-specific" monoclonal antibodies;[26–28] these antibodies bind to the GPIIb–IIIa complex but not to glycoproteins dissociated by chelation of divalent cations. The sedimentation method is routinely used in this laboratory and will be discussed below. A modification of this method has been used to determine the association state of these glycoproteins in intact platelets and will also be discussed.

The thrombin-induced hydrolysis and monoclonal antibody-binding methods both have limitations in determining whether GPIIb and GPIIIa are complexed or dissociated. Glycoprotein IIb and/or GPIIIa apparently undergo a change in structure when dissociated.[24] Accordingly, it is not known whether increased thrombin hydrolysis of GPIIb$_\alpha$ or loss of antibody binding occurs because of dissociation of the GPIIb–IIIa complex or because of a subsequent change in the structure of the dissociated glycoproteins. Thus, it is conceivable that if GPIIb and GPIIIa could exist as dissociable monomers prior to a change in structure, the monomers would still bind to "complex-specific" monoclonal antibodies and would be resistant to thrombin hydrolysis. Because of this uncertainty, we con-

[24] L. A. Fitzgerald and D. R. Phillips, *J. Biol. Chem.* **260**, 11366 (1985).
[25] K. Fujimura and D. R. Phillips, *J. Biol. Chem.* **258**, 10247 (1983).
[26] D. Pidard, R. R. Montgomery, J. S. Bennett, and T. J. Kunicki, *J. Biol. Chem.* **258**, 12582 (1983).
[27] R. P. McEver, E. M. Bennett, and M. N. Martin, *J. Biol. Chem.* **258**, 5269 (1983).
[28] J. S. Bennett, J. A. Hoxie, S. F. Leitman, G. Vilaire, and D. B. Cines, *Proc. Natl. Acad. Sci. U.S.A.* **80**, 2417 (1983).

clude that these two methods may not give a reliable indication of the association state of the glycoproteins and that sedimentation is the preferred method. Sedimentation has the added advantage of being able to detect polymerized forms of the glycoproteins.

Sucrose Density Centrifugation of the Isolated Glycoprotein IIb–IIIa Complex

A simple procedure for determining the sedimentation coefficient(s) of isolated glycoproteins is to sediment unlabeled GPIIb and GPIIIa with radiolabeled calibration proteins through a 5 to 25% sucrose gradient.[24] The sucrose gradients are drained into individual fractions and the sedimentation position(s) of GPIIb and GPIIIa are determined from SDS gels; the positions of the calibration proteins are determined by counting the radioactivity in each fraction.

Useful calibration proteins are bovine serum albumin $s_{20,w} = 4.6S$ (Sigma Chemical Co.) and bovine liver catalase, $s_{20,w} = 11.3S$ (Boehringer GmbH, Mannheim, Germany), which are iodinated by the chloramine-T method.[23] Approximately 20 nCi (containing <1 μg protein) of each ^{125}I-labeled protein is mixed with ~50 μg of GPIIb–IIIa complex in a total volume of 0.2 ml. This sample is then gently layered onto a 5 to 25% linear sucrose gradient in a buffer containing 1 mM EDTA, 1% Triton X-100, 150 mM NaCl, 25 mM Tris-HCl, pH 7.2 (5 ml, precooled to 4°), which is centrifuged for 7.5 hr at 55,000 rpm at 4°. This volume and centrifugation time are convenient for the SW 55 rotor (Beckman Instruments, Inc., Spinco Division); for conversion to other rotors, $\omega^2 t = 9 \times 10^{11}$. To drain the gradient, 15-drop fractions are collected by puncturing the tube with a gradient drainer (~20 fractions per gradient). The γ emissions from each fraction are measured to determine the sedimentation positions of the calibration proteins. Sodium dodecyl sulfate–polyacrylamide gel electrophoresis is used to determine the positions of GPIIb and GPIIIa. Samples are prepared for electrophoresis by precipitation of proteins with trichloroacetic acid, which concentrates dilute samples so that minor proteins can be detected by Coomassie Brilliant blue staining or autoradiography. This is accomplished by treating a 100- to 200-μl aliquot of each sucrose gradient fraction with 1.0 ml of 12% trichloroacetic acid at 4°. A precipitate immediately forms and is pelleted by centrifugation at 2000 g for 30 min at 0°. The trichloroacetic acid pellet is directly solubilized with 100 μl of a SDS-containing sample buffer,[29] to which NaOH is added to neutralize the residual trichloroacetic acid. The amounts of $GPIIb_\alpha$ and GPIIIa in each

---

[29] U. K. Laemmli, *Nature (London)* **227**, 680 (1970).

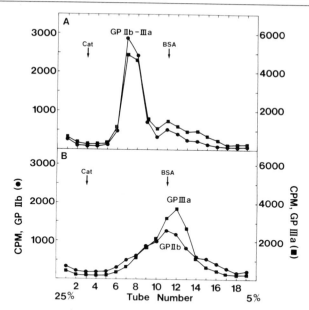

FIG. 4. Sedimentation of the GPIIb–IIIa complex and GPIIb and GPIIIa through 5 to 15% sucrose gradients containing 1% Triton X-100. (A) Sedimentation of the GPIIb–IIIa complex in 2 mM CaCl$_2$. (B) Sedimentation of dissociated GPIIb and GPIIIa in 5 mM EDTA. The amounts of GPIIb (●) and GPIIIa (■) were determined by counting the radioactivities in the GPIIb and GPIIIa bands after SDS–polyacrylamide gel electrophoresis using radiolabeled proteins. Similar data are obtained from densitometric scans of the stained gel (14). The sedimentation positions of bovine serum albumin (BSA) and catalase (Cat) are indicated.

fraction are determined from densitometric scans of the Coomassie Brilliant blue-stained gel. Figure 4 shows a typical sedimentation of the GPIIb–IIIa complex, GPIIb, and GPIIIa compared to the two reference proteins. Note that the 20 fractions collected are sufficient to discriminate between complexed and dissociated glycoproteins. For more precise evaluation, more fractions should be collected (~40) and more calibration proteins included (a total of five).

Quantitation of Glycoprotein IIb–IIIa Complex in Intact Platelets

The amount of GPIIb–IIIa complex in intact platelets can be determined either by physical studies, i.e., sucrose density gradient centrifugation of lysates of ^{125}I-labeled cells,[24] or by quantitating the binding of monoclonal antibodies that bind to a complex-dependent epitope on GPIIb–IIIa (see Ref. 30).

For physical studies, platelets are isolated by centrifugation, ^{125}I labeled, and washed as described.[31] The ^{125}I-labeled platelets are resuspended in a small volume at a concentration of 2 to 5 × 10^8 cells/ml and maintained at 37° for up to 1 hr before lysis or other treatment. The platelets must be maintained in Ca^{2+}-containing buffers (0.25 to 1 mM Ca^{2+}) to prevent dissociation of the GPIIb–IIIa complex. If EDTA buffers are used at 20–25° for the initial removal of plasma proteins, ~30% of the GPIIb–IIIa complex is dissociated at the time of lysis. Platelets can be incubated in EDTA-containing buffers at 4 to 25° for short periods (up to 15 min) without dissociation of the GPIIb–IIIa complex. High pH conditions and low salt concentrations should also be avoided as these conditions also favor dissociation.

^{125}I-Labeled platelets are directly lysed in a buffer that stabilizes the association state (i.e., the amounts of dissociated and complexed forms) of GPIIb and GPIIIa. Aliquots of platelets are pipetted into 4 vol of the Triton X-100 lysis buffer [1 mM EDTA, 1% (w/v) Triton X-100, 150 mM NaCl, 10 μM leupeptin, 1 mM phenylmethylsulfonyl fluoride, 25 mM Tris-HCl, pH 7.2] at 4°. The temperature of the Triton X-100 lysis buffer is equilibrated to 4° in polystyrene tubes immersed in melting ice before the addition of platelets. The EDTA in the Triton X-100 lysis buffer prevents any dissociated GPIIb and GPIIIa from forming complexes, while the cold temperature prevents any dissociation of GPIIb–IIIa complexes by the EDTA. Thus, no change in the amounts of complexed or dissociated glycoproteins will occur, preserving the association state of the glycoproteins in the intact cell at the time of lysis. The EDTA–Triton X-100 lysates can be maintained at 4° for up to 24 hr with no change in association state, but sucrose gradients are usually begun within 1 hr. Sucrose gradients are performed at 4° as described above. The proteins of individual gradient fractions are precipitated with trichloroacetic acid and analyzed by SDS–polyacrylamide gel electrophoresis. An aliquot of unlabeled, purified GPIIb–IIIa complex is added to each fraction before precipitation and electrophoresis so that GPIIb and GPIIIa can be identified on the dried gels and the ^{125}I-labeled glycoproteins can be cut out to quantitate their distribution in the gradient. The sedimentation position of GPIIIa is more useful than GPIIb$_\alpha$ because the sedimentation of the dissociated GPIIIa ($s_{20,w}$ = 3.2S) is more distinct from the GPIIb–IIIa complex ($s_{20,w}$ =

---

[30] A. T. Nurden, J. N. George, and D. R. Phillips, in "Biochemistry of Platelets" (D. R. Phillips and M. A. Shuman, eds.), p. 159. 1986.

[31] D. R. Phillips, this volume [35].

8.6S) and because other proteins coelectrophorese with GPIIb. Thus, the percentage of glycoprotein as the GPIIb–IIIa complex is calculated according to Eq. (1), in which the total radioactivity in the region of the gradient containing the complex (usually fractions 10–13, if 22–23 fractions are collected using the sedimentation conditions described earlier) is divided by the sum of the radioactivity in the regions containing complexed and dissociated glycoproteins (usually fractions 15–18).

$$\text{GPIIb–IIIa complex (\%)} = \frac{\text{total [}^{125}\text{I]GPIIIa (cpm) in GPIIb–IIIa complex}}{\text{[}^{125}\text{I]GPIIIa (cpm) in GPIIb–IIIa complex} + \text{[}^{125}\text{I]GPIIIa (cpm) in GPIIIa}} \quad (1)$$

There are usually one or two fractions that cannot be clearly described as having either complexed or dissociated glycoproteins, but the total number of counts in these fractions is only 5 to 10% of the total.

Glycoprotein IIb–IIIa complexes can also be quantitated by measuring the binding of a complex-dependent antibody. Several have been described, including 7E3, 10E5, A_2A_9, AP2, and PAC-1 (see Ref. 30). When GPIIb–IIIa complexes are dissociated, for example by calcium chelation, the epitope for these antibodies is lost. In contrast, many antibodies have been described that bind to epitopes on individual glycoproteins, for example, AP3 to GPIIIa, and TAB to GPIIb. These antibodies usually have the characteristic of binding to both GPIIb–IIIa complexes and to the dissociated glycoproteins. Differences in epitope expression can therefore be equated to the amount of GPIIb–IIIa complex. Methods for the detection of monoclonal antibody binding to platelets have been described (see Ref. 30), but are beyond the scope of this chapter.

Reconstitution of Glycoprotein IIb–IIIa Complex into Phospholipid Vesicles

Two methods have been used in this laboratory for reconstituting the GPIIb–IIIa complex into phospholipid vesicles. One method[32] produces small (40 to 50 nm), sided vesicles with the GPIIb–IIIa complex oriented exclusively to the outer surface of the vesicles. These vesicles have proven useful for studying glycoprotein morphology in membranes and phospholipid requirements for glycoprotein reconstitution. Another method[16] produces large or fused vesicles in which only 55% of the GPIIb–IIIa complex is exposed to the outside of the vesicle. These vesicles are readily sedi-

---

[32] L. V. Parise and D. R. Phillips, *J. Biol. Chem.* **260**, 1750 (1985).

mented at microfuge g forces (12,800 g) and are retained on filters (0.2-μm pore size), making them useful for measuring the amounts of radiolabeled ligands bound to the GPIIb–IIIa complex.

Incorporation of Glycoprotein IIb–IIIa Complex into Small, Sided Vesicles

To incorporate the GPIIb–IIIa complex (isolated in Triton X-100 as described above) into phospholipid vesicles, it is first necessary to replace the Triton X-100 with the dialyzable detergent octylglucoside. Triton X-100 is removed by adsorbing the purified GPIIb–IIIa complex onto a 5-ml concanavalin A affinity column at 4°.[32] The column is washed with 50 ml of a buffer containing 50 mM Tris-HCl, 0.5 mM CaCl$_2$, 0.1 M NaCl, 60 mM octylglucoside (Calbiochem, La Jolla, CA), and 0.02% (w/v) NaN$_3$, pH 7.3. The GPIIb–IIIa complex is eluted from the column with 50 ml of the same buffer containing 100 mM α-methylmannoside. The GPIIb–IIIa complex is concentrated by ultrafiltration (YM30 membrane; Amicon, Lexington, MA) to 3 to 6 ml and dialyzed at 4° in three changes of a 10-fold excess volume of the Tris-HCl buffer with octylglucoside, a step that removes the α-methylmannoside. The GPIIb–IIIa complex is then filtered through a 0.9 × 2.5-cm column of Sephadex G-75 to remove any potentially contaminating concanavalin A and is stored at $-70°$ in glass tubes until use.

The GPIIb–IIIa complex is incorporated into phospholipid vesicles by a modification of the procedure of Helenius *et al.*[33] Phosphatidylserine (bovine brain; Avanti Polar Lipids, Birmingham, AL) and phosphatidylcholine (egg yolk; Avanti) (70% phosphatidylserine, 30% phosphatidylcholine) are added to the bottom of a glass test tube. The organic solvents are removed by drying the phospholipids under a stream of filtered nitrogen and then twice redissolving them in ~0.2 ml of diethyl ether and redrying. The GPIIb–IIIa complex (100 μg GPIIb–IIIa/ml) in 60 mM octylglucoside is added to the dried phospholipids (glycoprotein : phospholipid, 1 : 2.9 to 1 : 3.6, w/w) at ambient temperature. The glycoprotein and phospholipids are mixed by drawing the solution 20 times through a fine glass Pasteur pipette and then by sonicating the solution for 5 min in a bath-type sonicator (Laboratory Supplies Co., Inc.). To form the vesicles, octylglucoside is removed from the mixture by dialysis for 17 to 20 hr against 2 changes of a 1000-fold excess of buffer without octylglucoside [50 mM Tris-HCl, 0.5 mM CaCl$_2$, 0.1 M NaCl complex, and 0.02% (w/v) NaN$_3$, pH 7.3]. If vesicles are initially formed with [^{3}H]GPIIb–IIIa complex and [^{14}C]phos-

[33] A. Helenius, E. Fries, and J. Kartenbeck, *J. Cell Biol.* **75**, 866 (1977).

phatidylcholine, >90% of the two labels float through a 10 to 40% sucrose gradient. The morphology of the GPIIb–IIIa complex in these vesicles has been observed by negative staining electron microscopy.[32]

Incorporation of Glycoprotein IIb–IIIa Complex into Large Vesicles

Many of the ligands that bind to the GPIIb–IIIa complex (e.g., fibrinogen, von Willebrand factor, and fibronectin) are comparable in size to the small vesicles (diameter 40 nm) described in the preceding section. With such vesicles, ligand-binding measurements are difficult because the ligand and the small vesicles have similar filtration and sedimentation properties. Because it is useful to measure the ligand-binding properties of membrane glycoproteins when they are in membranes, large vesicles containing the GPIIb–IIIa complex have been constructed and have proven useful for ligand-binding analysis.[16] To incorporate the GPIIb–IIIa complex into large vesicles suitable for ligand-binding studies, vesicles are formed as above with the following modifications: (1) the GPIIb–IIIa complex at a concentration of 400 µg/ml is added to the dried phospholipids with a glycoprotein : phospholipid weight ratio of 1 : 2.9; (2) a trace amount of ^{131}I-labeled GPIIb–IIIa complex is added to the mixture of phospholipids and unlabeled GPIIb–IIIa complex; and (3) octylglucoside is removed by dialysis into a buffer containing 3 mM CaCl$_2$. The larger phospholipid vesicles are isolated by centrifugation at 12,800 g for 15 min in an Eppendorf centrifuge (model 5414). The vesicles form a visible white pellet. The supernatant is gently removed with a 1-cm^3 tuberculin syringe. Between 50 and 60% of the GPIIb–IIIa complex is routinely recovered in the phospholipid pellet as determined by a trace amount of ^{131}I-labeled GPIIb–IIIa complex included in the vesicle preparation. The pellet is resuspended in the desired buffer and kept on ice until use.

Ligand Binding to Glycoprotein IIb–IIIa Containing Vesicles

In a typical binding experiment, phospholipid vesicles containing unlabeled GPIIb–IIIa complex and a trace amount of ^{131}I-labeled GPIIb–IIIa complex are incubated for 1 hr in a 23° water bath with ^{125}I-labeled fibrinogen in a buffer containing 50 mM Tris-HCl, 0.1 M NaCl, 0.5% (v/v) bovine serum albumin (fraction V; Calbiochem), and 0.02% (w/v) NaN$_3$ (pH 7.3) with or without excess unlabeled fibrinogen, at a final volume of 200 µl.[16] Incubations take place in 1.5-ml polypropylene Eppendorf test tubes that are precoated with a 0.5% bovine serum albumin solution. The ^{125}I-labeled fibrinogen bound to the phospholipid vesicles is separated from the unbound ^{125}I-labeled fibrinogen by filtration (using a filtration manifold from Hoefer) of 50-µl aliquots through 0.2-µm pore size, 25-mm diameter polycarbonate filters (Nuclepore Corp., Pleasanton, CA) that are presoaked overnight at 4° in a solution of 0.5% bovine serum albumin. Filters pretreated in this manner demonstrate minimal nonspecific ^{125}I-labeled fibrinogen binding. The filters are washed twice with 1.3 ml of ice-cold buffer containing 50 mM Tris-HCl, 0.1 M NaCl, 0.02% (w/v) NaN$_3$, and a CaCl$_2$ concentration equivalent to that in the incubation mixture. Filters are counted in a γ counter with two windows set to count simultaneously ^{125}I and ^{131}I. ^{125}I and ^{131}I spillovers are determined from samples containing the individual isotopes, and spillover corrections are made for each double-labeled sample. The specific activity of ^{131}I-labeled GPIIb–IIIa complex in each resuspended GPIIb–IIIa-containing vesicle preparation is determined by the protein microassay of Peterson,[34] using bovine serum albumin as the standard, and by ^{131}I radioactivity measurements. ^{125}I-Labeled GPIIb–IIIa complex is included in the phospholipid vesicles to quantitate the amount of protein retained on each filter. The variation in ^{131}I-labeled glycoprotein retention is usually insignificant within replicates of a given sample. However, it should be noted that the amount of ^{131}I-labeled GPIIb–IIIa complex retained per filter decreases slowly but significantly with time following resuspension of the vesicle pellet (from ~100 to 70% retention after several hours). Also, resuspension of vesicles in EDTA causes ~30% less ^{131}I-labeled GPIIb–IIIa complex to be retained per filter. Thus, by including the ^{131}I tracer in the phospholipid vesicles, it is possible to determine precise values for the moles of ^{125}I-labeled fibrinogen bound per milligram of GPIIb–IIIa complex. The fibrinogen binding to these GPIIb–IIIa-containing vesicles has many of the characteristics of ^{125}I-labeled fibrinogen binding to whole platelets or isolated platelet membranes.[16]

The fibrinogen-binding properties of the GPIIb–IIIa complex incorporated into phospholipid vesicles by this procedure are similar to the fibrinogen-binding properties of the fibrinogen receptor on platelets. Fibrinogen-binding activity is specific, saturable, reversible, time dependent, and Ca^{2+} dependent. The apparent dissociation constant for ^{125}I-labeled fibrinogen binding to GPIIb–IIIa-containing vesicles is 15 nM, and the maximal binding capacity is 0.1 mol of ^{125}I-labeled fibrinogen/mol of GPIIb–IIIa. This is in contrast to activated platelets, which bind up to 1 mol of fibrinogen per mole of GPIIb–IIIa.[35] This finding suggests that purified GPIIb–IIIa lacks a cofactor necessary for its optimal activity. ^{125}I-Labeled fibrinogen binding is inhibited by amino sugars, the anti-GPIIb–IIIa monoclonal antibody 10E5, and the decapeptide from the carboxyl terminus of

[34] G. L. Peterson, *Anal. Biochem.* **83**, 346 (1977).
[35] E. F. Plow and G. A. Marguerie, *J. Biol. Chem.* **255**, 10971 (1980).

the fibrinogen γ chain, and by peptides containing the RGD sequence. Furthermore, little or no ^{125}I-labeled fibrinogen binds to phospholipid vesicles lacking protein or containing proteins other than the GPIIb–IIIa complex (i.e., bacteriorhodopsin, apolipoprotein A-I, or glycophorin). Also, other ^{125}I-labeled plasma proteins (transferrin, orosomucoid) do not bind to the GPIIb–IIIa vesicles. This appears, therefore, to be an example of a reconstituted receptor system.

Isolation of Functionally Active Glycoproteins IIb and IIIa

Isolation of GPIIb and GPIIIa has been achieved either by column chromatography of the EDTA-treated GPIIb–IIIa complex[4] or by preparative polyacrylamide gel electrophoresis of the SDS-treated GPIIb–IIIa complex.[17,18] In both instances, however, the isolated glycoproteins are functionally inactive in that they fail to reform heterodimeric complexes when incubated with excess Ca^{2+}. This suggests that the dissociated glycoproteins have undergone a change in structure, perhaps because of the denaturing conditions used for glycoprotein purification. In this section, a method is described for the rapid separation of GPIIb from GPIIIa, under nondenaturing conditions, that produces functionally active glycoproteins in that they are capable of reforming GPIIb–IIIa complexes when mixed in the presence of Ca^{2+}.

Dissociation of Glycoprotein IIb–IIIa Complex

The purified GPIIb–IIIa complex, obtained from Sephacryl S-300 column chromatography, is adjusted to 2.8 mM EDTA and pH 8.9 at 4°. This solution is then incubated for 5 min at 37° (the pH changes to 8.5 under these conditions), immersed in an ice bath, and adjusted to pH 7.5. This procedure results in dissociation of the GPIIb–IIIa complex and in the formation of monomeric subunits as analyzed by sucrose gradient sedimentation.

High-Performance Liquid Chromatographic Separation of Glycoprotein IIb from Glycoprotein IIIa

Dissociated glycoproteins are separated at ambient temperature on a TSK-4000 gel-filtration column (300 × 7.5 mm). The column must first be preconditioned to separate GPIIb from GPIIIa effectively, which can be achieved by eluting it with a buffer containing 0.1% (w/v) Triton X-100, 0.1 M NaCl, 50 mM Tris-HCl, pH 7.0 for ~42 days at a flow rate of 0.1 ml/min. The column is then equilibrated for 3 hr with the same buffer but containing 0.05% (w/v) Triton X-100. An aliquot (250 μl) of the dissociated

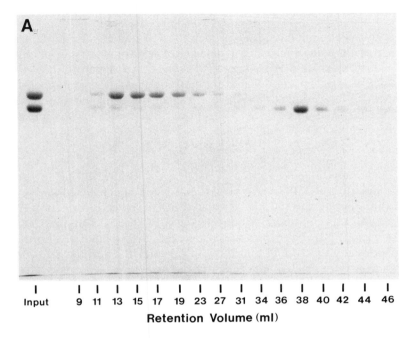
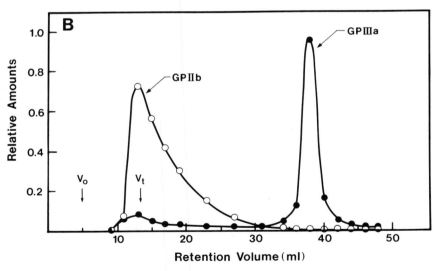

FIG. 5. Separation of GPIIb from GPIIIa by high-performance liquid chromatography in Triton X-100. (A) SDS–polyacrylamide gels of proteins eluted from a TSK-4000 column stained with Coomassie Brilliant blue. (B) Quantitative distributions of GPIIb (○) and GPIIIa (●) in the eluted fractions as determined from densitometric scans of the SDS gels. V_0, initial volume; V_t, total volume.

subunits is centrifuged for 10 min at 4° in an Eppendorf microcentrifuge, loaded onto the column, and eluted with the 0.05% Triton X-100 buffer. Figure 5 shows a typical separation of GPIIb from GPIIIa. Ten percent of the GPIIb peak is GPIIIa; 5% of the GPIIIa peak is GPIIb.

Reassociation of the isolated glycoproteins requires immediate readdition of Ca^{2+} (6 mM) to the fractions obtained from the high-performance liquid chromatography column. When isolated GPIIb is mixed with isolated GPIIIa, 10 to 15% of the dissociated glycoproteins reform the heterodimeric complex.

Acknowledgments

This work was supported in part by Grants HL 28947 and HL 32254 form the National Institutes of Health.

The authors thank Barbara Allen and Sally Gullatt Seehafer for editorial assistance, James X. Warger and Norma Jean Gargasz for graphics, and Kate Sholly, Michele Prator, Linda Harris Odumade, and Linda Parker for manuscript preparation.

[23] von Willebrand Factor Binding to Platelet Glycoprotein Ib Complex

By ZAVERIO M. RUGGERI, THEODORE S. ZIMMERMAN,[1] SUSAN RUSSELL, ROSSELLA BADER, and LUIGI DE MARCO

von Willebrand factor (vWF) is a complex multimeric glycoprotein that plays an essential role in platelet function.[1a] It is required for normal platelet adhesion to exposed subendothelium and for normal platelet plug formation at sites of vascular injury. The function of vWF is particularly important in vessels of small caliber, where conditions of high wall shear rate prevail. It is now recognized that the mechanisms underlying vWF function comprise interaction with components of the subendothelium as well as with specific receptors on the platelet membrane. Two distinct vWF-binding sites have been recognized so far on platelets, one related to the membrane glycoprotein (GP) Ib and the other to the heterodimeric GPIIb–IIIa complex.[2]

[1] Deceased.
[1a] Z. M. Ruggeri and T. S. Zimmerman, *Blood* **70**, 895 (1970).
[2] Z. M. Ruggeri, L. De Marco, L. Gatti, R. Bader, and R. R. Montgomery, *J. Clin. Invest.* **72**, 1 (1983).

This chapter deals with reviewing the methodology involved in studying the vWF interaction with GPIb. von Willebrand factor binds to GPIb in the presence of the antibiotic ristocetin. Such binding correlates with the occurrence of platelet aggregation induced by ristocetin in the presence of vWF.[3] The latter function of vWF is often referred to as the ristocetin cofactor activity. If platelets are metabolically inactive, as after fixation with formalin, they will nonetheless still respond to ristocetin in the presence of vWF with what is more properly called agglutination. Therefore, the ristocetin-dependent interaction of vWF with GPIb can be analyzed using either fresh, metabolically active platelets or fixed, metabolically inactive platelets.

Ristocetin, however, is not the only tool with which to study vWF binding to GPIb under experimental conditions. Following removal of sialic acid residues from the vWF glycoprotein, the resulting asialo-vWF exhibits the ability to interact directly with GPIb.[4] It is, therefore, apparent that GPIb-related binding sites for vWF are exposed on the membrane of unstimulated platelets and can interact with a modified form of the molecule. The concept that a modification of the vWF molecule can be the initial event promoting its interaction with platelets is a stimulating alternative, or perhaps addition, to the concept that a modification of the platelet surface or microenvironment, as possibly brought about by ristocetin,[5,6] is necessary to promote vWF binding to GPIb. Indeed, it has also been demonstrated that another substance with the property of inducing vWF binding to GPIb, the snake protein botrocetin,[7] exerts its function by forming a bimolecular complex with vWF that in turn binds to platelets[7,8]; botrocetin itself does not interact with the platelet surface.[8] Using asialo-vWF, it is also possible to demonstrate that receptors on GPIIb–IIIa are exposed following vWF binding to GPIb.[4] This constitutes a potentially important link between two major binding sites for adhesive glycoproteins on the platelet membrane.

The procedures described in this chapter relate to the general methodology for studying receptor–ligand interaction. The basic constituents of the experimental system are, in this case, highly purified vWF and platelets washed free, as much as possible, of other blood components. The vWF to be used as a tracer is radiolabeled with ^{125}I. As mentioned above,

---

[3] K.-J. Kao, S. V. Pizzo, and P. A. McKee, *Proc. Natl. Acad. Sci. U.S.A.* **76,** 5317 (1979).

[4] L. De Marco, A. Girolami, S. Russell, and Z. M. Ruggeri, *J. Clin. Invest.* **75,** 1198 (1985).

[5] S. E. Senogles and G. L. Nelsestuen, *J. Biol. Chem.* **258,** 12327 (1983).

[6] B. S. Coller, *J. Clin. Invest.* **61,** 1168 (1978).

[7] M. S. Read, S. V. Smith, M. A. Lamb, and K. M. Brinkhous, *Blood* **74,** 1031 (1989).

[8] M. Sugimoto, G. Ricca, M. E. Hrinda, A. B. Schreiber, G. H. Searfoss, E. Bottini, and Z. M. Ruggeri, *Biochemistry* **30,** 5202 (1991).

washed platelets can be either fresh or fixed. A detailed description of the methods used to prepare these reagents and to study ristocetin-induced binding of vWF to platelets is presented. The procedure used to prepare asialo-vWF and to study its interaction with platelets, as well as that to measure botrocetin-mediated vWF binding to platelets, will not be described here but have been the subjects of previous reports.[4,8,9] Botrocetin-induced platelet agglutination/aggregation is also described in [12] in Volume 169 of this series.

Purification of von Willebrand Factor

Reagents

Human plasma cryoprecipitate (the source of vWF): It can be obtained as such or, if blood bank facilities are available, it can be prepared from fresh plasma using the procedure of Pool and Shanon.[10]

Aluminum hydroxide (Rehsorptar from Armour Pharmaceutical Co., Kankakee, IL)

Bentonite (B-3378 from Sigma Chemical Co., St. Louis, MO)

Polyethylene glycol (P-2139, average M_r 8000; Sigma): A 40% (w/v) solution is prepared in imidazole buffer (see below)

Sepharose CL-4B (Pharmacia Fine Chemicals, Piscataway, NJ)

Citrate buffer: 0.02 M Tris, 0.02 M trisodium citrate, 0.02 M ε-aminocaproic acid, pH 7.0

Imidazole buffer: 0.02 M imidazole, 0.01 M trisodium citrate, 0.02 M ε-aminocaproic acid, 0.15 M NaCl, pH 6.5; sodium azide at a concentration of 0.02% (w/v) is added as a bacteriostatic agent

Tris buffer: 0.02 M Tris, 0.15 M NaCl, pH 7.35

Procedure

The method is a modification of two published procedures,[11,12] and begins with 1000 ml of cryoprecipitate, from which a final yield of 15 to 20 mg of purified vWF is expected. The cryoprecipitate is thawed at 0–4° (approximately 18 hr) and then centrifuged at 9800 g (r_{max}) for 45 min at 4°. The supernatant is discarded and the precipitate is dissolved with several additions of citrate buffer, up to a final volume of 300 ml. The

---

[9] L. De Marco and S. S. Shapiro, *J. Clin. Invest.* **168**, 321 (1981).
[10] J. G. Pool and A. E. Shanon, *N. Engl. J. Med.* **273**, 1443 (1965).
[11] J. Newman, A. J. Johnson, M. H. Karpatkin, and S. Puszkin, *Br. J. Haematol.* **21**, 1 (1971).
[12] M. E. Switzer and P. A. McKee, *J. Clin. Invest.* **57**, 925 (1976).

solution is transferred to a plastic beaker and Rehsorptar is added dropwise (5 ml/100 ml of redissolved precipitate) while stirring. The mixture is further stirred for 10 min at room temperature (22–25°), then transferred into plastic bottles and centrifuged at 7500 g (r_{max}) for 30 min at 20°.

In the meantime, bentonite (2.5 mg/ml of supernatant) is placed in clean plastic centrifuge bottles and the powder is mixed with a small amount of citrate buffer to obtain a slurry. At the end of the centrifugation, the supernatant is poured directly into the bottles containing the bentonite, quickly mixed, and the mixtrue rapidly centrifuged at 3800 g (r_{max}) for 5 min at 20°.

The supernatant from the last centrifugation is transferred to a plastic beaker and the pH is adjusted to 6.5 by slow addition of 0.02 M citric acid. A volume of the 40% polyethylene glycol solution is added dropwise, under continuous stirring, to give a final concentration of 3% (w/v) polyethylene glycol. The mixture is stirred for an additional 10 min at room temperature and then centrifuged at 3800 g (r_{max}) for 15 min at 20°. The supernatant is recovered and transferred to a plastic beaker and additional polyethylene glycol is added following the same procedure to give a final concentration of 15%. After stirring for an additional 20 min at room temperature, the mixture is centrifuged at 9800 g (r_{max}) for 30 min at room temperature.

At the end of centrifugation, the supernatant is discarded and the precipitate is redissolved in the smallest possible volume of imidazole buffer. The precipitate obtained from 1 liter of cryoprecipitate can usually be redissolved in a final volume of approximately 30–35 ml. Complete dissolution usually takes 90–180 min with continuous, but gentle, agitation on a shaker.

The dissolved precipitate is ultracentrifuged at 113,000 g (r_{max}) for 45 min at 20°, after which the lipids floating on top are discarded and the clear solution is separated from the insoluble material at the bottom of the tube. The clear solution is then applied onto a Sepharose CL-4B column, equilibrated with imidazole buffer. A 100-cm-long siliconized glass column, 5 cm in diameter and with a bed volume of approximately 1700 ml, is used (the actual agarose column is approximately 80 cm long). The column is run at a flow rate of 60 ml/hr at room temperature, usually overnight for convenience. The optical density of the effluent is continuously monitored at 280 nm and 10-ml fractions are collected. The asymmetric peak appearing at the void volume contains the purified vWF. The column is developed until all proteins are eluted. It is then washed with at least three bed volumes of buffer and finally stored with buffer containing 0.1% (w/v) sodium azide.

von Willebrand factor is a multimeric protein and consists of molecular forms of varying molecular weight. Figure 1 shows the distribution of

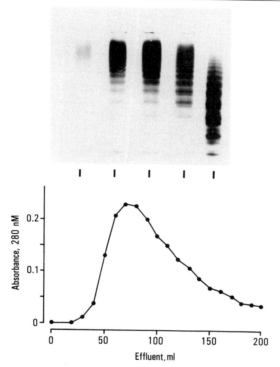

FIG. 1. Multimeric structure of vWF eluting from the gel-filtration column. *Lower:* Elution profile of the protein peak appearing at the void volume, detected by light absorbance at 280 nm. *Upper:* SDS-agarose gel electrophoresis of samples corresponding to different positions of the elution profile, as indicated by black bars. Note that the largest multimers appear first. Cathode at the top. Electrophoresis was performed as described in [21] in Volume 169 of this series.

multimers of different size across the vWF peak eluting from the Sepharose column. Usually, all the fractions corresponding to the ascending part and the first half of the descending part of the peak are pooled. Although the vWF peak is well separated from the subsequent one, it is preferable to discard the later eluting fractions to decrease the likelihood of contaminating proteins in the vWF preparation.

The protein concentration in the vWF pool obtained from the gel-filtration column is usually between 0.1 and 0.2 mg/ml. Because a higher concentration usually is needed, particularly for labeling the protein, the vWF must be concentrated. The most effective method is to transfer the vWF solution into regular dialysis tubing and then surround it with the hygroscopic agent Aquacide II, obtained from Calbiochem-Behring Corporation (La Jolla, CA). When sufficient water is removed to achieve the desired concentration, the dialysis tubing is rinsed with distilled water and the vWF is dialyzed against Tris buffer. The vWF preparation is then aliquoted and stored at $-70°$.

Comments

According to individual needs, smaller amounts of starting cryoprecipitate or plasma can be used. In that case, a gel-filtration column of smaller diameter can be employed, but its length should not be decreased.

If the protein concentration in the vWF preparation is derived from the optical density (OD), the following equation, which takes into account a correction for light scattering, should be used:

$$OD_{280} - (1.7 \times OD_{320})/0.7 = \text{vWF concentration (mg/ml)} \qquad (1)$$

The method used to analyze the multimeric structure of vWF is described in detail in [21] in Volume 169 of this series. In view of the heterogeneous nature of vWF and the fact that manipulation of the molecule can lead to loss of the largest multimer, it is advisable to analyze the multimeric composition of each purified vWF preparation before use. In fact, loss of the largest multimers affects the functional properties of vWF. If multimeric analysis cannot be performed, one alternative is to measure the ristocetin cofactor activity of the purified preparation. If no contaminants are present and the preparation has a normal multimeric distribution, the specific activity should be greater than 100 units of ristocetin cofactor per milligram of vWF.[4]

It is not clear whether the use of protease inhibitors during the purification procedure is of any help in achieving the goal of obtaining vWF with the full complement of multimers, including the largest. Nevertheless, protease inhibitors now are routinely added to the cryoprecipitate when it is first thawed out. The following inhibitors are used, all from Calbiochem-Behring, and all at a final concentration of 10 μM in the cryoprecipitate: D-phenylalanyl-L-prolyl-L-arginine chloromethyl ketone, dansyl-L-glutamyl-L-glycyl-L-arginine chloromethyl ketone, and D-phenylalanyl-L-phenylalanyl-L-arginine chloromethyl ketone, as well as leupeptin (Chemicon, Los Angeles, CA), at a final concentration of 15 μg/ml. The same inhibitors, but at one-tenth the final concentration, may also be added to the imidazole buffer used to resuspend the final precipitate and to equilibrate and run the gel-filtration column.

Labeling of von Willebrand Factor

Reagents

1,3,4,6-Tetrachloro-3α,6α-diphenylglycouril (Iodogen; Pierce Chemical Co., Rockford, IL): A 1-mg/ml solution is prepared in dichloromethane. It can be stored sealed and protected from light for several weeks at $-20°$

Sephadex G-25 Medium (Pharmacia)

Tris $(0.02\ M)$–$0.15\ M$ NaCl buffer, pH 7.35

KI: 10 mg/ml in distilled water

$Na^{125}I$, carrier free (Amersham, Arlington Heights, IL)

Procedure

The method described here is slightly modified from that previously published by Fraker and Speck.[13] The whole procedure is performed under a fume hood in a laboratory designated for protein iodination.

Between 100 and 400 μl of Iodogen solution is pipetted into a glass tube or vial, and the dichloromethane is evaporated under a stream of nitrogen. The tube is sealed, protected from light, and kept on ice until used (within a few hours).

A small column (for example, 0.9-cm diameter by 15-cm length) is packed with Sephadex G-25 medium and equilibrated with Tris buffer. It is possible to use prepacked disposable columns (PD-10, Pharmacia); in any case, the column should be discarded after each use. In order to increase the recovery of radiolabeled vWF, a 1-ml volume of normal plasma is filtered through the column before its use, followed by four to five bed volumes of equilibration buffer. The flow rate should be approximately 40 ml/hr. Immediately before use, the tube coated with Iodogen is repeatedly rinsed with Tris buffer to remove excess reagent. The purified vWF solution is then pipetted into the tube (not more than 500 μl), followed by the appropriate amount of $Na^{125}I$. We routinely use 0.5 to 1 mCi/mg of vWF. The mixture is kept on ice for 10–12 min. At the end of the incubation time, 500 μl of KI is applied onto the Sephadex column and, as soon as this solution has entered the gel, the iodination mixture is also applied. Fractions of 0.5 ml are collected and the radioactivity in a 5-μl sample from each fraction is counted. Radioactive fractions eluting in a peak at the void volume are pooled, the radioactivity of the pool is measured accurately, and the protein concentration is measured to calculate the

---

[13] D. J. Fraker and J. C. Speck, *Biochem. Biophys. Res. Commun.* **80**, 849 (1978).

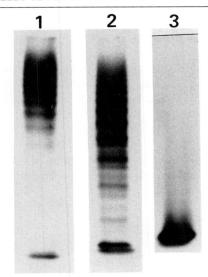

FIG. 2. Sodium dodecyl sulfate-agarose gel electrophoresis of radiolabeled vWF. Three samples labeled separately with ^{125}I are shown. Lane 1 shows a sample with intact maintained multimeric structure. Lane 2 shows a sample with only moderate reduction of the largest multimers. Lane 3 shows a sample with total loss of vWF multimers. Electrophoresis was performed as described above (see Fig. 1), with the exception that the dry gel was exposed to X-ray film directly, without incubating with antibodies specific for vWF.

specific activity, which is usually between 0.2 and 0.4 mCi/mg. ^{125}I-Labeled vWF is stored in aliquots frozen at $-70°$ until used.

Comments

von Willebrand factor is susceptible to degradation during and after the iodination procedure. As an example, the multimeric structure of three different preparations of ^{125}I-labeled vWF is shown in Fig. 2. To obtain meaningful results, only preparations containing all multimeric forms, including the largest, should be used. Therefore, it is advisable to check multimeric structure and/or ristocetin cofactor activity of ^{125}I-labeled vWF immediately after labeling and also at regular intervals if it is stored for prolonged periods of time.

Preparation of Washed Platelets

Reagents

Normal human blood: Obtain by clean venipuncture through 19-gauge needles, collect into polypropylene syringes, and transfer immediately into polypropylene tubes, mixing 5 vol into 1 vol of acid/citrate/dextrose anticoagulant, which is prepared as 85 mM trisodium citrate, 65 mM citric acid, 111 mM glucose

Bovine serum albumin solution (fraction V, Cat. No. 12659; Calbiochem-Behring Corp.): Aproximately 60% in distilled water, pH 6.5, osmolarity between 290 and 310 mOsm

Calcium-free Tyrode's buffer: Two different buffers are used. One is composed of 137 mM NaCl, 2 mM MgCl$_2$, 0.42 mM NaH$_2$PO$_4$, 11.9 mM NaHCO$_3$, 2.9 mM KCl, 5.5 mM glucose, 10 mM HEPES, pH 6.5. The other has the same composition, but the pH is adjusted to 7.35. Bovine serum albumin, 20 mg/ml, is added to part of the latter buffer

Apyrase (grade III; Sigma Chemical Co.)

Procedure

This method is a modification of that originally described by Walsh *et al.*[14] Platelet-rich plasma is prepared from the blood as soon as possible after collection by three successive centrifugation steps at 1200 g (r_{max}) for 60 sec at 22–25°. Each time the platelet-rich plasma is removed and the blood is recentrifuged without mixing. Apyrase is added to the platelet-rich plasma at a final concentration of 5 ATPase units/ml.

Albumin density gradients are prepared in flat-bottom polypropylene tubes (10-ml capacity). The albumin solution is used both undiluted and diluted 1:2 and 1:3 in Tyrode's buffer, pH 6.5. Aliquots of each concentration (300 μl) are pipetted into each tube and layered one over the other, starting with the most concentrated. The gradient is rendered less discontinuous by gently mixing at the interface between different layers. Platelet-rich plasma is then layered over the albumin gradient (not more than 6 ml in each tube). Platelets are then sedimented at 1200 g (r_{max}) for 15–20 min. At the end of the centrifugation the platelets should be collected in a narrow band that partly enters the albumin gradient and is separated from the supernatant plasma by a layer of albumin. The supernatant is discarded and most of the albumin underneath the platelets is gently removed by suction with a siliconized glass pipette. Platelets are gently resuspended in approximately half a volume of Tyrode's buffer, pH 6.5, and apyrase is added at 2 ATPase units/ml. The platelet suspension is again layered onto albumin gradients and the centrifugation repeated. The supernatant is then discarded, the albumin underneath the platelets is removed, the platelets are resuspended in

---

[14] P. N. Walsh, D. C. B. Mills, and J. G. White, *Br. J. Haematol.* **36**, 281 (1977).

the same volume of Tyrode's buffer, pH 6.5, and apyrase is added at 0.2 ATPase units/ml. Again the platelet suspension is layered onto albumin gradients and the centrifugation repeated for the last time. The supernatant is discarded, albumin underneath the platelets is removed by aspiration, and the platelets are resuspended in the desired volume of Tyrode's buffer, pH 7.35. Starting from 40 to 60 ml of blood, one can expect to have 1-2 ml of washed platelet suspension with a count of $1-2 \times 10^9$/ml, depending on the initial count in the platelet-rich plasma. The albumin concentration in the final suspension varies, depending on how carefully the excess has been removed during the washing procedure, but it should be between 3 and 5 mg/ml.

Comments

Platelets washed with this procedure routinely give irreversible aggregation when stimulated with ADP (4–10 μM) or epinephrine (10 μM) in the presence of fibrinogen and $CaCl_2$. They retain the ability to respond for about 2 hr or more.

Immediately after preparation, these platelets do not aggregate on addition of ristocetin, unless vWF is added to the mixture. Subsequently, some response to ristocetin may become evident even without addition of vWF, due to leakage of endogenous platelet vWF.

Fixation of Washed Platelets

Reagents

Tris (0.02 M), 0.15 M NaCl buffer, pH 7.35

Formaldehyde solution, 2% in Tris buffer, prepared from commercially available formaldehyde (which is usually a 37% solution): Dilute 2 ml of formaldehyde into 98 ml of Tris buffer

Procedure

Washed platelets are mixed 1 : 1 (v/v) with formaldehyde and incubated at 37° for 1 hr. The mixture is then kept at 4° for 12–18 hr. The platelets are then sedimented by centrifugation at 2000 g (r_{max}) for 15 min at room temperature, and washed twice in Tris buffer. They are finally resuspended in Tris buffer containing 0.02% sodium azide and adjusted to a count of $4-6 \times 10^8$/ml. They are stored at 4°.

Comments

Platelets fixed with this technique will agglutinate in response to ristocetin in the presence of vWF for at least 6–8 weeks. It is advisable to wash them once every week during storage, to change the buffer and remove possible products of bacterial contamination (proteases).

Ristocetin-Induced Binding of von Willebrand Factor to Platelets

Reagents

Washed platelets, fresh or fixed
^{125}I-Labeled vWF and unlabeled purified vWF
Ristocetin (Sigma Chemical Co.; >90% ristocetin A): Dissolve at a concentration of 15 mg/ml in 0.15 M NaCl
Tyrode's buffer, pH 7.35 (as described above), containing 20 mg/ml bovine serum albumin
Tris buffer, consisting of 0.02 M Tris, 0.15 M NaCl, pH 7.35
Sucrose: 20% solution in Tyrode's buffer containing albumin

Procedure

Platelets are used at a final count of 1×10^8/ml. A suitable volume is pipetted from the suspension of washed platelets into 1-ml plastic tubes, to give the desired final number after taking into account the addition of all other reagents. If the starting suspension must be diluted, this is done with Tyrode's buffer containing albumin. ^{125}I-Labeled vWF is then added. In a typical experiment, six concentrations are tested, with doubling amounts of ^{125}I-labeled vWF from 1 to 32 μg/ml (final concentration). Stock solutions are prepared in Tris buffer at 10 times the final concentration, and one-tenth the final volume is then added. Two sets of tubes are prepared. To a series of six tubes, unlabeled vWF is added at a 50-fold excess over the concentration of the ^{125}I-labeled vWF. An equivalent volume of Tris buffer is added to the other series of six tubes.

The same volume (one-tenth of final) of ristocetin solution is then pipetted into each tube, to give a final concentration of 1.5 mg/ml. The mixture is gently mixed and then left at room temperature for 30 min, without agitation.

In the meantime, sucrose solution (300 μl) is pipetted into 450-μl microcentrifuge tubes with capillary tips (No. 72.702, 47 × 7 mm; Sarstedt, Hayward, CA). Care must be taken that the capillary tips are filled with sucrose solution before applying the radioactive mixture; this can be

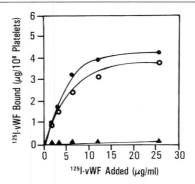

Fig. 3. Ristocetin-induced binding of ^{125}I-labeled vWF to platelets in the presence of monoclonal antibodies. Note that the binding of vWF (at the various concentrations indicated) is blocked by an anti-GPIb monoclonal antibody (▲), but not by an antibody against GPIIb–IIIa (●). Control (○).

achieved by centrifuging the tubes. At the end of the incubation, a 50-μl aliquot of each mixture is pipetted and carefully layered on top of the sucrose solution, in duplicate. The tubes are then centrifuged at 12,000 g (r_{max}) for 4 min. Platelets sediment through the sucrose solution and pellet at the bottom of the tube. Soluble components of the mixture remain on top of the sucrose solution well separated from the platelets. The portion of the tube tips containing the platelet pellet is cut off with a scalpel or cutter (because the capillary tips have small diameter, there is no spilling of solution after the cut) and the radioactivity associated with the platelet pellet is counted in a γ scintillation spectrometer (2 min should be sufficient). The remainder of the tube is also counted to have an objective measurement of the concentration of free vWF. Two 5-μl aliquots of each mixture are also counted *in toto* to determine the total radioactivity present.

Comments

The starting solutions of ^{125}I-labeled and unlabeled vWF should be as concentrated as possible, so that relatively small volumes can be used to achieve the final desired concentrations. A practical example is the preparation of a platelet suspension with 5×10^8 platelets/ml (25 μl is used), ^{125}I-labeled vWF in solutions from 10 to 320 μg/ml (12.5 μl is used), and unlabeled vWF in solutions from 0.125 to 4 mg/ml (50 μl is used, to give a 50-fold excess over ^{125}I-labeled vWF). Ristocetin solution (12.5 μl) and Tyrode's buffer with albumin (25 μl) complete the experimental mixture.

The calculation of the amount of ^{125}I-labeled vWF bound to platelets is done by subtracting the counts bound in the presence of excess unlabeled vWF (representing low-affinity, presumably nonspecific binding) from those bound in the absence of unlabeled vWF (total binding). Theoretically, a greater excess of unlabeled vWF (300-fold) should be used for the estimation of nonspecific binding; this is practically difficult because it requires large volumes of concentrated vWF solution, but can be done for the mixtures containing lower concentrations of radiolabeled vWF. From the known number of platelets in the pellet (5×10^6 if this procedure is followed) and the specific activity of ^{125}I-labeled vWF, one can calculate the amount of vWF bound (specific binding). This result is usually expressed as micrograms per 10^8 platelets. The amount of free ^{125}I-labeled vWF remaining in each mixture is calculated by determining the difference between the amount added and the amount bound to the platelets, and is also measured objectively as described above. The procedure described here can be modified to include other reagents, such as monoclonal antibodies that allow definition of the receptor specificity for ristocetin-induced binding of vWF to platelets. In Fig. 3, binding of ^{125}I-labeled vWF to platelets has been blocked by anti-GPIb monoclonal antibody, but not by antibody to GPIIb–IIIa. This is evidence for the existence of more than one binding site for vWF on platelets, a conclusion also supported by studies with platelets deficient in GPIb or GPIIb–IIIa.[2,15]

Because vWF is a multimeric glycoprotein and, therefore, a potential multivalent ligand, the interpretation of the results of binding studies using mathematical models described for monovalent ligands (Scatchard-type analysis) is potentially misleading. Nevertheless, if this type of analysis is applied to vWF-binding isotherms, the results demonstrate the existence of a single class of noninteracting binding sites with excellent linear fit.[16,17] This suggests that the binding of vWF to GPIb occurs through independent, even though potentially multiple, monovalent interactions. Scatchard-type analysis of binding isotherms can be performed with a computer-assisted program.[18,19] In this case, the nonsaturable (nonspecific) component of binding can be calculated from the total binding isotherm, avoiding the use of unlabeled vWF.

---

[15] Z. M. Ruggeri, R. Bader, and L. De Marco, *Proc. Natl. Acad. Sci. U.S.A.* **79,** 6038 (1982).
[16] A. B. Federici, R. Bader, S. Pagani, M. L. Colibretti, L. De Marco, and P. M. Mannucci, *Br. J. Haematol.* **73,** 93 (1989).
[17] L. De Marco, M. Mazzucato, D. De Roia, A. Casonato, A. B. Federici, A. Girolami, and Z. M. Ruggeri, *J. Clin. Invest.* **86,** 785 (1990).
[18] P. J. Munson and D. Rodbard, *Anal. Biochem.* **107,** 220 (1980).
[19] P. J. Munson, this series, Vol. 92, p. 542.

[24] Isolation and Characterization of Glycoprotein Ib

By ANDREAS N. WICKI, JEANNINE M. CLEMETSON, BEAT STEINER, WOLFGANG SCHNIPPERING, and KENNETH J. CLEMETSON

Glycoprotein Ib is one of the major components of the outer surface of the platelet plasma membrane and contains most of the sialic acid of the platelet membrane contributing to the surface charge.[1] Because of its absence in platelets in the inherited bleeding disorder Bernard–Soulier syndrome, it was one of the first platelet glycoproteins to which a functional role was ascribed.[2,3] It is thus clear that the GPIb–IX complex plays an essential role in normal platelet adhesion and activation, particularly at high shear rates. The first structural information came with the isolation and partial characterization of the proteolytic degradation products "macroglycopeptide"[4] and glycocalicin,[5] which were later shown to be derived from GPIb. Glycocalicin is described in [25] in this volume. Methods have been established for the isolation of amounts of intact, purified GPIb–IX adequate for biochemical characterization. All three of the constituent chains of the GPIb–IX complex, $GPIb_\alpha$, $GPIb_\beta$, and GPIX have now been cloned and sequenced[6-9] so that a great deal of structural information is now available.

Glycoprotein Ib–IX, in common with most other membrane glycoproteins, requires the presence of a detergent for its solubilization from the membrane. Several different detergents have been used.[10] When nondenaturing detergents are employed to maintain the molecule in a biologically

---

[1] N. O. Solum, I. Hagen, C. Filion-Myklebust, and T. Stabaek, *Biochim. Biophys. Acta* **597,** 235 (1980).
[2] C. S. P. Jenkins, D. R. Phillips, K. J. Clemetson, D. Meyer, M.-J. Larrieu, and E. F. Lüscher, *J. Clin. Invest.* **57,** 112 (1976).
[3] A. T. Nurden and J. P. Caen, *Nature (London)* **255,** 720 (1975).
[4] D. S. Pepper and G. A. Jamieson, *Biochemistry* **9,** 3706 (1970).
[5] T. Okamura, C. Lombart, and G. A. Jamieson, *J. Biol. Chem.* **251,** 5950 (1976).
[6] J. A. Lopez, D. W. Chung, K. Fujikawa, F. S. Hagen, T. Papayannopoulou, and G. J. Roth, *Proc. Natl. Acad. Sci. U.S.A.* **84,** 5615 (1987).
[7] K. Titani, K. Takio, M. Handa, and Z. M. Ruggeri, *Proc. Natl. Acad. Sci. U.S.A.* **84,** 5610 (1987).
[8] J. A. Lopez, D. W. Chung, K. Fujikawa, F. S. Hagen, E. W. Davie, and G. J. Roth, *Proc. Natl. Acad. Sci. U.S.A.* **85,** 2135 (1988).
[9] M. J. Hickey, S. A. Williams, and G. J. Roth, *Proc. Natl. Acad. Sci. U.S.A.* **86,** 6773 (1989).
[10] H. A. Cooper, K. J. Clemetson, and E. F. Lüscher, *Proc. Natl. Acad. Sci. U.S.A.* **76,** 1069 (1979).

active state, a second problem is encountered, because GPIb–IX remains specifically associated with a group of cytoskeletal components. In contrast to GPIIb–IIIa, which is not generally linked to the cytoskeleton in resting platelets,[11] GPIb–IX seems to be largely associated with the cytoskeleton, which influences the amount of GPIb–IX that can be solubilized. This linkage to actin-binding protein can be disrupted by treating the platelets with N-ethylmaleimide[12] before preparation of membranes or Triton X-114 phase separation. Although N-ethylmaleimide is clearly a cysteine-blocking reagent, the mechanism of action in this case is not yet understood. The linkage of GPIb to the cytoskeleton is via actin-binding protein,[13] which is rapidly degraded by calpain. It is retained in the presence of the inhibitors of this enzyme.[14] The optimal conditions for this first solubilization stage to avoid loss of GPIb–IX in the cytoskeleton fraction are still poorly defined but affect the yield of GPIb–IX. Although it is always possible to have platelets in a defined state by working with fresh blood from individual donors, the large amount of blood bank platelets required for preparative work often means that storage and handling conditions cannot be defined before isolation starts. Use of N-ethylmaleimide to uncouple GPIb–IX from the cytoskeleton plus efficient calpain inhibitors to prevent degradation to glycocalicin may solve these problems.

Although platelets contain several proteases, the principal activity is due to calcium-activated neutral thiol protease (calpain).[15] When released by lysed platelets calpain cleaves GPIb, forming glycocalicin. Proteases from other cells may also be a problem. In preparative-scale isolations it is difficult to remove all contaminating leukocytes that are rich in enzymes that degrade platelet glycoproteins.[16] Thus, it is essential to add a cocktail of protease inhibitors to the platelets before solubilization to avoid the degradation of GPIb and the production of cleavage products that might interfere at later purification stages.

Lectin affinity chromatography on wheat germ agglutinin was an early method that was shown to be of value in purifying GPIb–IX.[17] Wheat germ agglutinin interacts predominantly with glycoproteins rich in sialic acid

---

[11] D. R. Phillips, L. K. Jennings, and H. H. Edwards, *J. Cell Biol.* **86**, 77 (1980).
[12] J. E. B. Fox, L. P. Aggerbeck, and M. C. Berndt, *J. Biol. Chem.* **263**, 4882 (1988).
[13] J. R. Okita, D. Pidard, P. J. Newman, R. R. Montgomery, and T. J. Kunicki, *J. Cell Biol.* **100**, 317 (1985).
[14] N. O. Solum, T. M. Olsen, G. O. Gogstad, I. Hagen, and F. Brosstad, *Biochim. Biophys. Acta* **729**, 53 (1983).
[15] D. R. Phillips and M. Jakábová, *J. Biol. Chem.* **252**, 5602 (1977).
[16] K. Bykowska, J. Kaczanowska, M. Karpowicz, J. Strachurska, and M. Kopec, *Thromb. Haemostasis* **50**, 768 (1983).
[17] K. J. Clemetson, S. L. Pfueller, E. F. Lüscher, and C. S. P. Jenkins, *Biochim. Biophys. Acta* **464**, 493 (1977).

present on O-linked oligosaccharides. On platelets these are relatively few and include GPIb, GPIIIb (GPIV), and GPV.[17,18] A more selective lectin is peanut agglutinin, which binds fairly specifically to GPIb, although only after sialic acid residues are removed with neuraminidase.[19] The problems of specificity are overcome by using the Triton X-114 phase separation method. Glycoprotein Ib displays exotic behavior by partitioning into the aqueous phase, rather than in the Triton phase, as would be expected for an integral hydrophobic glycoprotein.[6] Most of the other glycoproteins, including GPIIIb and GPV, partition into the detergent phase. Thus, together with wheat germ agglutinin affinity chromatography, the Triton X-114 phase separation method provides the basis for an isolation procedure for GPIb–IX that is relatively simple and effective. Still further purification can be effected by ion-exchange chromatography on Q-Sepharose or by taking advantage of the fact that GPIb contains a thrombin-binding site by affinity chromatography on thrombin-Sepharose.

Monoclonal antibodies to GPIb–IX also provide a method for isolating the GPIb–IX complex.[20,21] Questions of availability and expense, however, necessarily restrict the generalization of this approach.

Experimental Procedures

Isolation and Washing of Platelets

Platelets are isolated from citrate-treated blood within 20 hr of collection. The buffy coats from 100 units are transferred into one-quarter of their volume of 9.6 mM glucose, 3 mM KCl, 100 mM NaCl, 10 mM EDTA, 30 mM sodium citrate, pH 6.5, to a final concentration of about 3×10^9 platelets/ml. The platelets are collected by centrifugation at 1500 g for 10 min and washed twice in 30 mM glucose, 120 mM NaCl, 10 mM EDTA, 30 mM sodium citrate, pH 6.5, and once in 134 mM NaCl, 10 mM EDTA, 10 mM Tris-HCl buffer, pH 7.4, giving about 75 ml of platelet pellet.

Solubilization and Triton X-114 Separation

The pellet is suspended in 100 ml of 124 mM NaCl, 20 mM EDTA, 2 mM N-ethylmaleimide (omit this if GPIb–IX linked to actin-binding protein is required), 10 mM Tris-HCl buffer, pH 7.4. The mixture is cooled to 4° and

---

[18] M. Moroi and S. M. Jung, *Biochim. Biophys. Acta* **798**, 295 (1984).
[19] P. S. Judland and D. J. Anstee, *Protides Biol. Fluids* **27**, 871 (1979).
[20] M. C. Berndt, C. Gregory, A. Kabral, H. Zola, D. Fournier, and P. A. Castaldi, *Eur. J. Biochem.* **151**, 637 (1985).
[21] V. Vicente, R. A. Houghten, and Z. M. Ruggeri, *J. Biol. Chem.* **265**, 274 (1990).

mixed with 100 ml of 2% (w/v) Triton X-114 at 4°. Phenylmethylsulfonyl fluoride (PMSF) in methanol is added to give a final concentration of 2 mM, and the mixture is stirred for 15 min and centrifuged at 10,000 g for 30 min at 4°. The supernatant is further centrifuged at 100,000 g for 1 hr at 4°; the supernatant from this second centrifugation (220–230 ml) should be clear. The clear supernatant (20–25 ml) is carefully overlayered on 20-ml cushions of 6% (w/v) sucrose, 154 mM NaCl, 1 mM EDTA, 0.06% (w/v) Triton X-114, 10 mM Tris-HCl, pH 7.4, in 50-ml centrifuge tubes. The tubes are heated to 37° for 5 min in a water bath. The upper phase becomes opaque as clouding of the Triton X-114 occurs. The tubes are then centrifuged at 1000 g for 10 min in an uncooled centrifuge. To avoid mixing the phases, it is important that the centrifuge is well balanced and that no braking is used. The upper phase is carefully removed, and Triton X-114 is added to give a final concentration of 1% (w/v). The mixture is stirred at 4° until a clear solution is obtained, again layered above the sucrose solution in the centrifuge tubes, heated to 37° for 5 min, and again centrifuged at 1000 g for 10 min. The upper phases are pooled and Brij 99 (Sigma, St. Louis, MO) is added to a final concentration of 0.5% (w/v). The pH is checked, adjusted (if necessary) to 7.4 with 1 N HCl, and the solution is left standing at 4° for 2–3 hr. Generally a precipitate forms at this stage, which is removed by centrifuging at 10,000 g for 10 min. If the preparation must be interrupted and the material stored frozen, then the best point is just before this centrifugation, which is then carried out on thawing.

Wheat Germ Agglutinin-Sepharose 4B Affinity Chromatography

The clear supernatant is applied to a column of wheat germ agglutinin-Sepharose 4B [15 × 3 cm; 1 mg wheat germ agglutinin (WGA)/ml Sepharose], which is equilibrated in 0.5% (w/v) Brij 99, 154 mM NaCl, 20 mM Tris-HCl, pH 7.4, and washed through with the same buffer until the optical density (OD) of the eluate returns to baseline. The bound material is then eluted using 0.5% Brij 99, 30 mM NaCl, 20 mM Tris-HCl, pH 7.4, plus 2.5% (w/v) N-acetylglucosamine. A silver-stained, two-dimensional polyacrylamide gel electrophoresis separation of the bound material is shown in Fig. 1.

Q-Sepharose Ion-Exchange Chromatography

The fractions containing material that bound to the wheat germ agglutinin-Sepharose 4B are pooled, adjusted to pH 7, and applied to a 10 × 2.5-cm Q-Sepharose (Pharmacia, Piscataway, NJ) column equilibrated with 0.5% Brij 99, 30 mM NaCl, 20 mM Tris-HCl, pH 7.0, and washed through with the same buffer until the flow-through has eluted. A 200-ml gradient

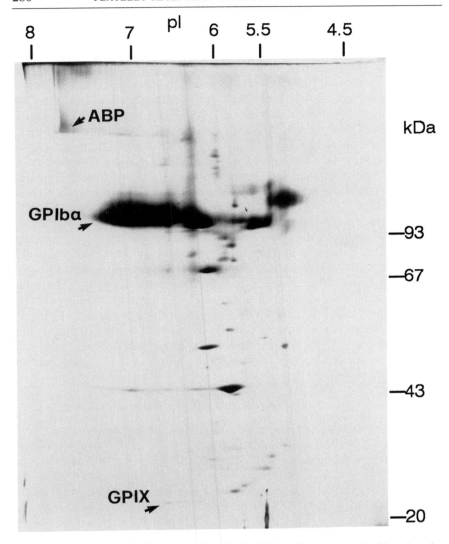

FIG. 1. Two-dimensional isoelectric focusing/5–17% gradient polyacrylamide gel, under reducing conditions, of the platelet components present in the aqueous phase of a Triton X-114 phase separation that bind to wheat germ agglutinin-Sepharose. Glycoprotein Ib$_\beta$ and GPIX stain relatively weakly with the silver method. ABP, Actin-binding protein.

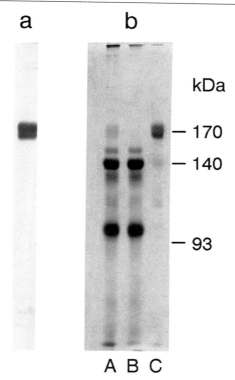

FIG. 2. Sodium dodecyl sulfate-polyacrylamide gel electrophoresis (6% gel), under nonreducing conditions, of GPIb purifications. (a) Glycoprotein Ib purified by Triton X-114 phase separation, wheat germ agglutinin chromatography, and Q-Sepharose ion-exchange chromatography. Note the presence of two, barely separated bands due to size polymorphism. (b) Lane A, glycoproteins from Triton X-100-solubilized platelets that bind to wheat germ agglutinin; lane B, flow-through on thrombin-Sepharose chromatography of the material shown in lane A; lane C, thrombin-Sepharose-bound material from lane A eluted with 0.5 M NaCl.

from 0.25 to 0.5 M NaCl in 0.5% Brij 99, 20 mM Tris-HCl, pH 7.0, is then applied to the column to elute the bound material. The fractions containing protein (OD$_{280}$) are examined by SDS–polyacrylamide gel electrophoresis[22] with silver staining[23] (Fig. 2a) and those containing GPIb complex (a low, broad peak eluting between 0.38 and 0.48 M NaCl) are pooled, dialyzed against water, and lyophilized.

[22] U. K. Laemmli, *Nature (London)* **227**, 680 (1970).
[23] J. H. Morrissey, *Anal. Biochem.* **117**, 307 (1981).

Preparation of Thrombin-Sepharose 4B

Bovine thrombin is purified by the method of Fenton et al.[24] and stored at $-80°$ in phosphate-buffered saline containing 1 M NaCl. The affinity matrix is prepared by reacting 45 mg thrombin with 10 ml CNBr-activated Sepharose 4B (Pharmacia) according to standard procedures. To block the enzymatic activity of thrombin, the thrombin-Sepharose is then incubated with 0.32 mM D-phenylalanine-L-prolyl-L-arginine chloromethyl ketone (PPACK; Calbiochem, La Jolla, CA) in 150 mM NaCl, 20 mM Tris-HCl buffer, pH 7.0, for 1 hr at room temperature. The column is then washed and stored at 4° in the same buffer containing 0.05% (w/v) NaN_3.

Thrombin-Sepharose Affinity Chromatography

Washed, outdated platelets are lysed at 4° for 15 hr in a buffer containing 1% (w/v) Triton X-100, 150 mM NaCl, 1 mM $CaCl_2$, 1 mM $MgCl_2$, 0.02% (w/v) NaN_3, 10 μM leupeptin, 0.5 mM PMSF, 1 mM N-ethylmaleimide, 20 mM Tris-HCl, pH 7.3. The lysate is centrifuged for 30 min at 30,000 g to remove insoluble material. The supernatant is stored at $-80°$ until use or applied directly at room temperature to a wheat germ agglutinin-Sepharose column preincubated in 0.1% (w/v) Triton X-100, 150 mM NaCl, 1 mM $CaCl_2$, 1 mM $MgCl_2$, 0.02% (w/v) NaN_3, 20 mM Tris-HCl, pH 7.0. The column is washed with the same buffer until the optical density of the flow-through is <0.05. The bound glycoproteins are then eluted with the same buffer containing 25 mg/ml N-acetylglucosamine, 10 μM leupeptin, 0.5 mM PMSF, 1 mM N-ethylmaleimide and directly applied to the thrombin-Sepharose, which was already equilibrated in the same buffer at room temperature. The column is washed until the optical density of the flow-through is below 0.02. The bound proteins are eluted with the same buffer containing 0.5 M NaCl. The purity of the GPIb–IX complex obtained and the contents of the flow-through were checked by sodium dodecyl sulfate (SDS)–polyacrylamide gel electrophoresis (Fig. 2b).

Gel Filtration in Sodium Dodecyl Sulfate

The lyophilized pool is dissolved in 3–4 ml of 0.1% (w/v) SDS, 100 mM NH_4HCO_3 and applied to a 90 × 2.5-cm AcA 34 (IBF, Villeneuve la Garenne, France) column equilibrated in the same buffer. One main peak and two minor peaks are obtained, the larger containing GPIb and other high molecular weight proteins, and the minor peaks containing GPIX and

---

[24] J. W. Fenton, II, M. J. Fasco, and A. B. Stackrow, *J. Biol. Chem.* **252**, 3587 (1977).

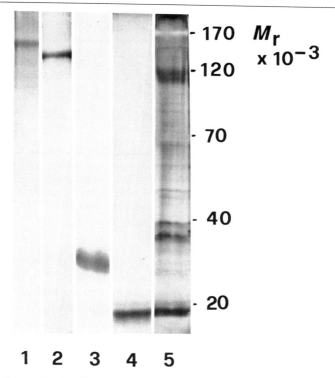

Fig. 3. Sodium dodecyl sulfate-polyacrylamide gel electrophoresis (7–17% gradient) of GPIb–IX complex components. Lane 1, GPIb, nonreduced; lane 2, GPIb$_\alpha$ chain; lane 3, GPIb$_\beta$ chain; lane 4, GPIX; lane 5, GPIb complex in Q-Sepharose eluate.

proteins in the 30K–40K region. To obtain pure GPIX, the GPIX peaks are pooled, lyophilized, and rechromatographed on a 90 × 2.5-cm AcA 44 column in the same buffer system (Fig. 3, lane 4). The fractions containing GPIb are pooled, lyophilized, and rechromatographed on a 90 × 2.5-cm AcA 34 column in 0.1% (w/v) SDS, 100 mM NH$_4$HCO$_3$ to separate intact GPIb (Fig. 3, lane 1). Fractions are examined by SDS–polyacrylamide gel electrophoresis with silver staining and pooled accordingly.

Isolation of α and β Subunits of Glycoprotein Ib

Purified GPIb from which GPIX has been removed is dissolved in 3–4 ml of 0.1% (w/v) SDS, 100 mM NH$_4$HCO$_3$, and dithiothreitol (DTT) is added to an end concentration of 1% (w/v). The α and β chains are separated by gel filtration on a 90 × 2.5-cm AcA 44 column. The fractions

TABLE I
PURIFICATION OF GPIb–IX

Purification step	Volume (ml)	Total protein[a] (mg)	GPIb[b] (mg)
Triton X-114 solubilization	230	8000	80
WGA loading	180	4100	54
WGA eluate	125	100	25
Q-Sepharose eluate	150	60	18
GPIb fractions, AcA 34	6	5	5

[a] Total protein was determined with the BCA protein assay (Pierce Chem. Corp., Rockford, IL) with BSA as standard.
[b] Glycoprotein Ib was determined using Laurell rocket electrophoresis against rabbit anti-glycocalicin antibodies with glycocalicin as standard.

containing the β subunit are pooled, lyophilized, redissolved, and again passed through the AcA 44 column to obtain pure $GPIb_\beta$ (Fig. 3, lane 3). The fractions containing the α subunit are pooled, lyophilized, redissolved, and separated on a 90 × 2.5-cm AcA 34 column. Fractions containing pure $GPIb_\alpha$ are pooled (Fig. 3, lane 2).

Determination of Yield

The yield at various purification stages can be determined by using specific antibodies. It is relatively easy to isolate glycocalicin (see [25] in this volume) and to prepare polyclonal, monospecific antibodies to this fragment of GPIb that can be used in Laurell-type assays. Alternatively, monoclonal antibodies to GPIb are available and radioimmunoassays using them have been described.[25] The yields obtained by the Laurell technique from a typical preparation using rabbit anti-glycocalicin antibodies are shown in Table I.

Characterization and Properties of Glycoprotein Ib–IX Complex Molecules

The GPIb–IX complex as defined here consists of those molecular species that are reproducibly found associated with GPIb when the latter is isolated in the presence of nondenaturing detergents. These include GPIX, actin-binding protein, and actin. The primary structure of all these components is known,[6–9] as well as a good deal about their functional

[25] B. S. Coller, E. Kalomiris, M. Steinberg, and L. E. Scudder, *J. Clin. Invest.* **73**, 794 (1984).

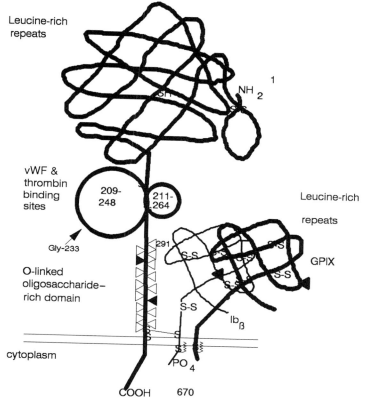

FIG. 4. Schematic drawing of the GPIb–IX complex, showing the various domains and the relation between the components. The numbers alongside the GPIb$_\alpha$ chain indicate the amino acid positions. The arrow indicating Gly^{233} is the position of the mutation in one family of "platelet-type" von Willebrand's disease.

domains. A schematic drawing of the GPIb–IX complex is shown in Fig. 4, with the main features indicated. Glycoprotein Ib consists of two chains, α (140 kDa) and β (27 kDa), linked by a disulfide bond just above the membrane surface. Even after reduction of the disulfide bond the subunits can be dissociated only under denaturing conditions, suggesting a strong noncovalent binding. Glycoprotein IX is noncovalently tightly associated with GPIb in a 1:1 ratio.[26] All three chains cross the plasma membrane and contain transmembrane and cytoplasmic sequences.[6–8] Glycocalicin

[26] X. Du, L. Beutler, C. Ruan, P. A. Castaldi, and M. C. Berndt, *Blood* **69**, 1524 (1987).

consists of the extracellular part of the α chain of GPIb and is easily removed from the platelet surface by treatment with various proteases, in particular endogenous platelet calpain. Glycocalicin contains the von Willebrand factor- and thrombin-binding sites.[5] All these subunits contain leucine-rich domains, motifs originally found in leucine-rich α_2-glycoprotein, which are now known in a wide range of proteins from different species. The function of this motif remains obscure but it seems to be associated with adhesive and protein–protein interactions. The α chain of GPIb contains several distinct domains; these are illustrated in Fig. 4: an N-terminal region containing a disulfide loop; a domain with six or seven leucine-rich repeats; a region with two interlinked disulfide-loops, probably critical for receptor function; a region containing threonine-rich repeats, highly O-glycosylated (40 oligosaccharides at a density of 1 per 3–4 amino acids), which produce a semistiff, rodlike structure, probably holding the receptor domains out from the platelet surface; and finally, transmembrane and cytoplasmic domains, the latter probably involved in linkage to actin-binding protein and hence to the submembranous cytoskeleton. The outer domain also contains two N-linked oligosaccharide chains. The structure of the N- and O-linked oligosaccharide chains has been determined.[27,28]

Molecular variants of GPIbα have been reported. Four different polymorphic alleles termed A, B, C, and D were first reported from Japan,[29] leading to different molecular mass species (Fig. 2a). The domain containing the polymorphism was localized to the O-glycosylated region[30] and has been reported to be caused by doubling and tripling of a 13-amino acid sequence (Ser^{399}–Thr^{411})[31] containing five putative O-glycosylation sites. In addition, a single amino acid polymorphism, Thr^{145} (89%)–Met^{145} (11%) (ACG–ATG), has been reported[31] that may be responsible for the Sib^a alloantigen.

The bleeding disorder "platelet-type" von Willebrand's disease is caused by a defect in GPIb leading to spontaneous binding of von Willebrand factor. The name comes from the observation that one of the symptoms is a deficiency of high molecular weight multimers of von Willebrand factor in plasma. In one such case a point mutation was found involving a change at residue 233 of the GPIb$_\alpha$ chain from Gly to Val.[32] This lies in

---

[27] T. Tsuji and T. Osawa, *J. Biochem.* (*Tokyo*) **101**, 241 (1987).
[28] S. A. M. Korrel, K. J. Clemetson, H. van Halbeek, J. P. Kamerling, and J. J. Sixma, *Eur. J. Biochem.* **140**, 571 (1984).
[29] M. Moroi, S. M. Jung, and N. Yoshida, *Blood* **64**, 622 (1984).
[30] M. Meyer and I. Schellenberg, *Thromb. Res.* **58**, 233 (1990).
[31] J. Ware, S. Russell, and Z. M. Ruggeri, *Thromb. Haemostasis* **65**, 770 (1991).
[32] J. L. Miller, D. Cunningham, V. A. Lyle, and C. N. Finch, *Proc. Natl. Acad. Sci. U.S.A.* **88**, 4761 (1991).

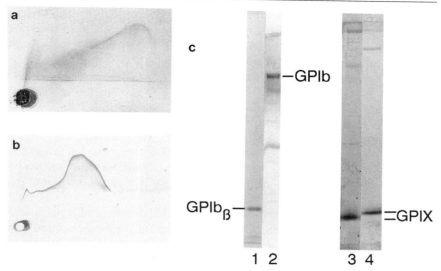

FIG. 5. (a) Crossed immunoelectrophoresis of whole platelets, solubilized in Triton X-100 (100 μg protein) against rabbit anti-GPIb$_\beta$ antiserum. (b) Crossed immunoelectrophoresis of whole platelets, solubilized in Triton X-100 (100 μg protein) against rabbit anti-GPIX antiserum. (c) Western blots on Immobilon of gel electrophoresis of whole platelets, using rabbit anti-GPIb$_\beta$ antiserum (lanes 1 and 2) or rabbit anti-GPIX antiserum (lanes 3 and 4) as primary antibody and goat anti-rabbit antibodies coupled to alkaline phosphatase (Bio-Rad, Richmond, CA) as second antibody. Bands were visualized using nitro blue tetrazolium and 5-bromo-4-chloro-3-indolyl phosphate (Sigma, St. Louis, MO) as substrate. Lanes 1 and 3: platelets solubilized under reducing conditions; lanes 2 and 4: platelets solubilized under nonreducing conditions.

the disulfide-linked double-loop structure[33] that is a region of GPIb$_\alpha$ already strongly implicated as involved in the von Willebrand factor-binding site.

The GPIb$_\beta$ chain consists of an N-terminal region, a leucine-rich repeat, and transmembrane and cytoplasmic domains. The cytoplasmic domain is phosphorylated[34] on Ser^{166} by cAMP-dependent kinase[35] and has a cysteine residue that is palmitoylated.[36] Phosphorylation of GPIb$_\beta$ via cAMP-dependent protein kinase is thought to control actin polymerization in response to platelet stimulation by collagen.[37] Glycoprotein IX has a very

[33] D. Hess, J. Schaller, E. E. Rickli, and K. J. Clemetson, *Eur. J. Biochem.* **199**, 389 (1991).
[34] B. Wyler, D. Bienz, K. J. Clemetson, and E. F. Lüscher, *Biochem. J.* **234**, 373 (1986).
[35] M. R. Wardell, C. C. Reynolds, M. C. Berndt, W. Wallace, and J. E. B. Fox, *J. Biol. Chem.* **264**, 15656 (1989).
[36] L. Muszbek and M. Laposata, *J. Biol. Chem.* **264**, 9716 (1989).
[37] J. E. B. Fox and M. C. Berndt, *J. Biol. Chem.* **264**, 9520 (1989).

similar overall structure to $GPIb_\beta$ but the cytoplasmic domain, although also containing a cysteine residue that is palmitoylated,[36] lacks the phosphorylation site. Both $GPIb_\beta$ and GPIX have single N-linked oligosaccharides of the biantennary lactosamine type.[38]

Considerable circumstantial evidence points to a relationship between GPV and the GPIb complex. Together with GPIb–IX, it is also absent or affected in the inherited bleeding disorder Bernard–Soulier syndrome,[39] although there is no direct evidence so far for a physical association between it and GPIb–IX. On the other hand, GPV also belongs to the leucine-rich repeat family of proteins and has a structure (as far as it is known) with considerable similarities to $GPIb_\alpha$.[40] Glycoprotein V is readily cleaved from the platelet surface by thrombin but is not involved as a thrombin receptor in platelet activation.[41]

Actin-binding protein is a relatively well-characterized cytoskeletal component that has been shown by several groups to be associated with GPIb.[13,42] It can be dissociated from GPIb in intact platelets by treatment with N-ethylmaleimide or, as with other components, dissociated from the complex in the presence of SDS.

Preparation and Characterization of Antibodies to Glycoproteins Ib_α, Ib_β, and IX

Rabbits are immunized at 2-week intervals for 8–10 weeks with $GPIb_\alpha$, $GPIb_\beta$, or GPIX (150–300 µg/ml) in physiological saline containing 0.1% Brij 99, mixed 1 : 1 with Freund's complete (for the first immunization) or incomplete (for successive immunizations) adjuvant. After 10–12 weeks, booster immunizations are given. The antiserum titer is checked and high-titer bleedings are pooled. Figures 5a and 5b show crossed immunoelectrophoresis results obtained with antibodies to $GPIb_\beta$ and GPIX. As shown in Fig. 5c, the antibodies reacted specifically with their antigen.

Acknowledgment

This work was supported by the Swiss National Science Foundation, Grants #3.302.082, 3.232.085, and 31-25633.88.

[38] A. N. Wicki and K. J. Clemetson, *Eur. J. Biochem.* **163,** 43 (1987).
[39] K. J. Clemetson, J. L. McGregor, E. James, M. Dechavanne, and E. F. Lüscher, *J. Clin. Invest.* **70,** 304 (1982).
[40] T. Shimomura, K. Fujimura, S. Maehama, M. Takemoto, K. Oda, T. Fujimoto, R. Oyama, M. Suzuki, K. Ichihara-Tanaka, K. Titani, and A. Kuramoto, *Blood* **75,** 2349 (1990).
[41] D. Bienz, W. Schnippering, and K. J. Clemetson, *Blood* **68,** 720 (1986).
[42] J. E. B. Fox, *J. Clin. Invest.* **76,** 1673 (1985).

[25] Platelet Glycocalicin

By JOSEPH LOSCALZO and ROBERT I. HANDIN

Glycocalicin is an M_r 110,000, hydrophilic, proteolytic fragment of the leucine-rich two-chain integral membrane protein of the platelet glycoprotein Ib. This heavily glycosylated surface protein serves as a receptor for ristocetin-dependent von Willebrand factor binding,[1] has been implicated as a thrombin receptor[2] and as a quinidine-dependent antibody receptor,[3] and comprises the IgG Fc receptor[4] on the platelet, although more recent data call into question the physiologic importance of these last three functions. These binding functions are believed to be localized to the glycocalicin domain. Platelets from individuals with Bernard–Soulier syndrome are deficient in glycoprotein Ib[5,6] and show defects in these binding functions.

The topological relationship between glycocalicin and glycoprotein Ib is indicated in Fig. 1. Glycoprotein Ib consists of two glycoproteins, the α and β subunits, linked by one or more disulfide bonds.[7] In the platelet membrane, glycoprotein Ib forms a noncovalent complex with a second smaller glycoprotein, glycoprotein IX. Glycocalicin is derived from the larger α chain and itself consists of several domains as defined by proteolytic cleavage with trypsin.[8] A large, trypsin-resistant domain with an apparent molecular weight of 65,000 contains most of the glycosylation sites (the so-called macroglycopeptide), while an M_r 45,000 trypsin-sensitive domain containing two internal disulfide bonds (Cys^{209}–Cys^{248} and Cys^{211}–Cys^{264},[9] comprises the thrombin-binding site and the von Willebrand factor-binding region.

Glycocalicin is 56.5% (w/w) carbohydrate, and these sugar residues are confined largely to the trypsin-resistant macroglycopeptide domain in which the major oligosaccharide is a hexasaccharide O-linked to one in

---

[1] T. Okumura and G. A. Jamieson, *Thromb. Res.* **8**, 701 (1976).
[2] P. Ganguly and N. L. Gould, *Br. J. Haematol.* **42**, 137 (1979).
[3] T. J. Kunicki, M. M. Johnson, and R. H. Aster, *J. Clin. Invest.* **62**, 716 (1978).
[4] A. Moore, G. D. Ross, and R. L. Nachman, *J. Clin. Invest.* **62**, 1053 (1978).
[5] A. T. Nurden and J. P. Caen, *Nature (London)* **255**, 720 (1975).
[6] C. S. P. Jenkins, D. R. Phillips, K. J. Clemetson, D. Meyer, M.-J. Larrieu, and E. F. Luscher, *J. Clin. Invest.* **57**, 112 (1976).
[7] D. R. Phillips and P. Poh Agin, *J. Biol. Chem.* **252**, 2121 (1977).
[8] T. Okumura, M. Hasitz, and G. A. Jamieson, *J. Biol. Chem.* **253**, 3435 (1978).
[9] D. Hess, J. Schaller, E. E. Rickli, and K. J. Clemetson, *Eur. J. Biochem.* **199**, 389 (1991).

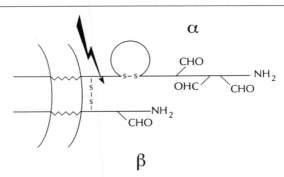

FIG. 1. Structure of platelet glycoprotein Ib and derived glycocalicin structure. Arrow indicates the site of calpain cleavage on the α subunit of glycoprotein Ib with the resulting release of glycocalicin from the membrane.

four amino acids.[10] The hexasaccharide is composed of two sialic acid, two galactose, one N-acetylglucosamine, and one N-acetylgalactosaminitol residues organized in the sequence shown in Fig. 2.[11] Capping of these side chains with sialic acid is responsible for much of the negative charge on the platelet surface. In addition to these O-linked oligosaccharides, four N-linked glycosylation sites have been identified.[12]

Glycocalicin was first purified by Lombart and colleagues[13,14] and subsequently by Solum and co-workers,[15] exploiting the intrinsic calcium-dependent protease (calpain) of platelets[16] to which glycoprotein Ib is exquisitely sensitive.[17,18] The former group of investigators used sonication, while the latter group used both freeze-thawing and high ionic strength to release stored platelet calcium and thereby activate the protease. Glycocalicin may also be hydrolyzed from glycoprotein Ib by extracel-

[10] P. A. Judson, D. J. Anstee, and J. R. Clamp, *Biochem. J.* **205**, 81 (1982).
[11] T. Tsuji, S. Tsunehisa, Y. Watanabe, K. Yamamoto, and T. Osawa, *J. Biol. Chem.* **258**, 6335 (1983).
[12] J. A. Lopez, D. W. Chung, K. Fujikawa, F. S. Hagen, T. Papayannopoulou, and G. J. Roth, *Proc. Natl. Acad. Sci. U.S.A.* **84**, 5616 (1987).
[13] C. Lombart, T. Okumura, and G. A. Jamieson, *FEBS Lett.* **41**, 30 (1974).
[14] T. Okumura, C. Lombart, and G. A. Jamieson, *J. Biol. Chem.* **251**, 5950 (1976).
[15] N. O. Solum, I. Hagen, C. Filion-Myklebust, and T. Stabaek, *Biochim. Biophys. Acta* **597**, 235 (1980).
[16] D. R. Phillips and M. Jakábová, *J. Biol. Chem.* **252**, 5602 (1977).
[17] K. J. Clemetson, *Blood Cells* **9**, 319 (1983).
[18] E. B. McGowan, K.-T. Yeo, and T. C. Detwiler, *Arch. Biochem. Biophys.* **227**, 287 (1983).

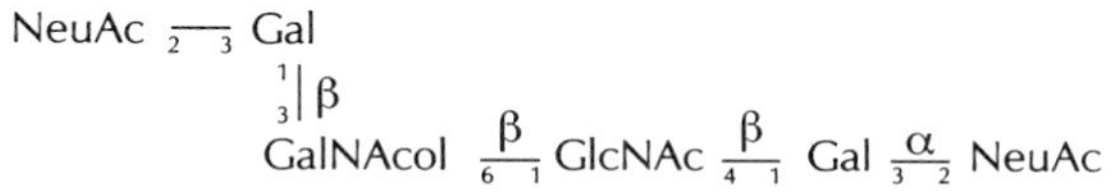

$$\text{NeuAc} \xrightarrow[2\longrightarrow 3]{} \text{Gal}$$
$$\,_3^1|\beta$$
$$\text{GalNAcol} \xrightarrow[6\longrightarrow 1]{\beta} \text{GlcNAc} \xrightarrow[4\longrightarrow 1]{\beta} \text{Gal} \xrightarrow[3\longrightarrow 2]{\alpha} \text{NeuAc}$$

FIG. 2. Structure of the principal oligosaccharide of glycocalicin. NeuAc, sialic acid; Gal, galactose; GalNAcol, N-acetylgalactosaminitol; and GlcNAc, N-acetylglucosamine.

lular metalloproteases produced by gram-negative bacteria[19] (e.g., *Serratia marcescens* metalloprotease) and by plasmin.[20] Subsequent purification steps have employed chromatography on insolubilized lectins such as wheat germ lectin-Sepharose 6-MB,[14] thrombin-Sepharose 4B,[21] conventional gel-filtration columns, hydroxyapatite, or phenyl boronate followed by ion-exchange high-performance liquid chromatography (HPLC).[22]

We have found that the following relatively simple procedure consistently yields sufficient quantities of purified glycocalicin for use in biochemical characterization and ligand-binding studies. Twenty units of outdated platelets obtained within 48–72 hr of collection is pooled and centrifuged at 3200 g for 30 min at 4°. The sedimented platelets are suspended in a platelet washing buffer consisting of 10 mM Tris, pH 7.6, 0.15 M NaCl, 0.6 mM EDTA, and 5 mM at 4° glucose after which they are centrifuged at 250 g for 30 min to remove contaminating erythrocytes. The platelets are then washed twice by centrifugation at 3200 g for 30 min, with the first resuspension performed in platelet washing buffer at 4° and the second in approximately 300 ml of warm (37°) 3 M KCl. After the second resuspension, the platelets are incubated for 30 min at 37°, then centrifuged at 8000 g for 10 min. The supernatant from this centrifugation is dialyzed for 18 hr against several changes of 10 mM Tris, pH 7.6, 0.15 M NaCl, 0.6 mM EDTA, 0.01% (w/v) NaN$_3$ (buffer A) and applied to a 10-ml column of wheat germ lectin-Sepharose 6-MB. The column is washed with buffer A until the absorbance of the eluate is equivalent to that of buffer A. The column is then developed with 2.5% N-acetylglucosamine in buffer A and the fractions with maximal absorbance at 280 nm are pooled and dialyzed against 10 mM Tris, pH 7.6, 1 M NaCl, 0.6 mM EDTA, 0.01% (w/v) NaN$_3$ (buffer B). After dialysis, the solution is concentrated to 15 ml by ultrafiltration (Diaflo PM30 membrane; Amicon,

[19] H. A. Cooper, W. P. Bennett, A. Kreger, D. Lyerly, and R. H. Wagner, *J. Lab. Clin. Med.* **97**, 379 (1981).

[20] B. Adelman, A. D. Michelson, J. Loscalzo, J. Greenberg, and R. I. Handin, *Blood* **65**, 32 (1985).

[21] M. Moroi, A. Goetze, A. Dubay, C. Wu, M. Hasitz, and G. A. Jamieson, *Thromb. Res.* **28**, 103 (1982).

[22] R. DeCristofaro, R. Landolfi, B. Bizzi, and M. Castagnola, *J. Chromatogr.* **426**, 376 (1988).

Danvers, MA) and applied to a 2.5 × 90-cm Sepharose 6B column. The column is developed with buffer B and the purified glycoprotein elutes in 25–35 ml. The glycocalicin may then either be concentrated by ultrafiltration or dialyzed against water and lyophilized. The yield of purified glycoprotein from a typical preparation is 0.5 mg, appears as a single M_r 110,000 band on a 7.5% polyacrylamide gel in the presence of sodium dodecyl sulfate and 0.6 M 2-mercaptoethanol, and stains with both Coomassie blue and periodic acid–Schiff's reagent.

Once isolated and purified, the characterization of glycocalicin by one of several possible biochemical or immunologic methods is essential to document unequivocally the identity of this protein. The compositional analysis of glycocalicin has been determined by two groups of investigators[1,23]; the results of their analyses are listed in Table I as mole percentages, including carbohydrate constituents, and as residues per molecule. The most abundant amino acids are threonine, leucine, and proline. The presence of four tryptophan residues per molecule has been demonstrated by using the m-iodosobenzoic acid method and is supported by the substantial intrinsic fluorescence of the molecule. These direct compositional data are supported by the derived sequence of the molecule determined from an analysis of the cDNA of glycoprotein Ib_α.[12] Inspection of the derived amino acid sequence also reveals the additional interesting features of 7 tandem, 24-amino acid leucine-rich glycopeptide (LRG) repeats located near the amino terminus of the polypeptide believed to contribute indirectly to function by participating in shear-dependent conformational changes within the molecule that expose the von Willebrand factor-binding site,[24] as well as a charged "hinge" domain comprising the von Willebrand factor-binding site.

The abundance of carbohydrate residues in the molecule and on the macroglycopeptide fragment produced by trypsin or chymotrypsin[25,26] is of interest in that it contains 64% of the total labile sialic acid of the platelet surface.[15] To date the only functional analysis of the carbohydrate portion of glycocalicin indicates that although removal of sialic acid and galactose residues does not affect ristocetin-dependent binding to von Willebrand factor, further removal of N-acetylglucosamine markedly reduces this binding.[27]

By electron microscopy, glycocalicin appears to be a semiflexible rod

[23] G. E. Carnahan and L. W. Cunningham, *Biochemistry* **22**, 5384 (1983).
[24] G. J. Roth, *Blood* **77**, 6 (1991).
[25] D. S. Pepper and G. A. Jamieson, *Biochemistry* **9**, 3706 (1970).
[26] A. J. Barber and G. A. Jamieson, *Biochemistry* **10**, 4711 (1971).
[27] A. D. Michelson, J. Loscalzo, B. Melnick, B. S. Coller, and R. I. Handin, *Blood* **67**, 19 (1986).

TABLE I
COMPOSITIONAL ANALYSIS OF GLYCOCALICIN[a]

Residues	Mol%	Mol% (including CHO)
Lys	6.3	3.3
His	2.1	1.1
Arg	2.3	1.1
Asp	8.8	4.7
Thr	14.0	7.6
Ser	8.9	4.9
Glu	10.5	5.8
Pro	12.6	6.8
Gly	5.1	2.7
Ala	4.4	2.3
1/2-Cys	ND[b]	ND
Val	4.6	2.5
Met	1.4	0.8
Ile	2.3	1.3
Leu	12.9	6.7
Tyr	2.1	1.2
Phe	3.0	1.6
NANA		12.5
Mannose		1.2
Fucose		1.9
Galactose		15.1
Glucose		2.3
GlcNAc		7.2
GalNAc		5.9

[a] From Ref. 23. CHO, carbohydrate; NANA, N-acetylneuraminic acid; GlcNAc, N-acetylglucosamine; GalNAc, N-acetylgalactosamine.
[b] ND, None detected.

of 56.5 × 5.4 nm, with a thickening at one end of the molecule.[28] It contains approximately 15% α helix,[17] and the native molecule contains sufficient tryptophan and tyrosine residues to generate a fluorescence spectrum; denaturation of the native molecule with 6 M guanidine hydrochloride markedly enhances intrinsic tryptophan fluorescence, suggesting that in the native structure these hydrophobic moieties are exposed to solvent or transfer fluorescence energy to other chromophores, thereby quenching fluorescence emission.

The most sensitive, accurate, and convenient method for identifying glycocalicin in our hands is an enzyme-linked immunosorbent assay

[28] J. W. Lawler, S. Margossian, and H. S. Slayter, *Fed. Proc., Fed. Am. Soc. Exp. Biol.* **39**, 1895 (1980).

(ELISA) method,[27] in which isolated glycocalicin competes with platelet surface glycoprotein Ib for a monoclonal antibody (see below) that recognizes glycocalicin epitopes. Microtiter wells (Immulon II) are incubated with 10 μg/ml poly(L-lysine) for 30 min at room temperature, after which the uncomplexed polymer is removed by flicking and aspiration. One hundred microliters of washed platelets at 10^8/ml in 10 mM Tris, pH 7.4, 0.15 M NaCl, and 4 mM EDTA is added to each well and the plates are centrifuged for 5 min at 800 g. Fifty microliters of 0.5% (w/v) formaldehyde in 10 mM Tris, pH 7.4, 0.15 M NaCl, and 4 mM EDTA are added to each well and incubated for 15 min at room temperature, after which the wells are washed twice with 10 mM Tris, pH 7.4, 0.15 M NaCl. The wells are then washed three times with 10 mM Tris, pH 8.0, 0.05% (w/v) Tween 20 (washing buffer). Each well is filled with 5 mg/ml bovine γ-globulin in 10 mM Tris, pH 7.4, 0.15 M NaCl and incubated for 30 min at 37°. The assay is performed using a mouse monoclonal antibody (6D1) directed against an epitope of glycocalicin[29] and sheep anti-mouse F(ab')$_2$ conjugated with horseradish peroxidase. After three washes of the wells with washing buffer, 25 μl of varying concentrations of glycocalicin in 10 mM Tris, pH 7.4, 0.15 M NaCl and 25 μl of 175 ng/ml 6D1 are added to each well and the wells incubated for 60 min at 37°. After three more washes with washing buffer, 50 μl of a 1 : 250 dilution of sheep anti-mouse (F(ab')$_2$ is added to each well and incubated for 60 min at 37°. The wells are then washed six times with washing buffer, 50 μl of a 1 : 1 : 18 (v/v) solution of 4% o-phenylenediamine, 0.3% (v/v) H$_2$O$_2$, and 17 mM citric acid in 65 mM phosphate, pH 6.3, is added to each well, and the wells incubated for 30 min at 37°. Color development is stopped by addition of 50 μl of 4.5 mM H$_2$SO$_4$. Optical density is read at 492 nm and expressed as a percentage of the optical density in assays with 6D1 without glycocalicin (i.e., 100% binding of monoclonal antibody to the fixed platelet surface). The optical density of wells without either 6D1 or glycocalicin is defined as 0% binding.

[29] B. S. Coller, E. I. Peerschke, L. E. Scuder, and C. A. Sullivan, *Blood* **61**, 99 (1983).

[26] Preparation and Functional Characterization of Monoclonal Antibodies against Glycoprotein Ib

By LESLEY E. SCUDDER, EFSTATHIA L. KALOMIRIS, and BARRY S. COLLER

Introduction

The glycoprotein Ib complex (GPIb complex) constitutes a nonintegrin platelet receptor for von Willebrand factor (see [24] in this volume). Monoclonal antibodies directed against the GPIb complex are useful tools for identifying and quantifying the GPIb complex, mapping functional domains of the receptor, and isolating the complex by immunoaffinity techniques.

Immunizations of Mice[1]

Materials

Trisodium citrate solution, 40%
Freund's complete adjuvant (Calbiochem, La Jolla, CA)
Tris-EDTA buffer: 10 mM Tris-HCl, 0.15 M NaCl, 10 mM NaEDTA, pH 7.4
Tris-NaCl buffer: 10 mM Tris-HCl, 0.15 M NaCl, pH 7.4
Mice: BALB/cBYJ (Jackson Laboratory, Bar Harbor, ME) 4–6 weeks old, either sex

Method

Whole blood is collected from normal donors by venipuncture and placed into plastic tubes (Falcon 2059, Oxnard, CA) containing 1/100 vol 40% citrate. The blood is centrifuged at 700 g for 3.5 min at 22° and the supernatant platelet-rich plasma (PRP) transferred to another tube with a plastic pipette (13-711-5; Fisherbrand, Pittsburgh, PA). The PRP is then washed with a fivefold volume of Tris-EDTA buffer and sedimented at ~1500 g for 8–10 min at 22°. The supernatant platelet-poor plasma is decanted and the platelet pellet resuspended in a combination of the residual ~0.1 ml of plasma and 0.2 ml Tris-EDTA buffer by gently aspirating the suspension into a plastic pipette tip (223-9000; Bio-Rad, Richmond,

---

[1] B. S. Coller, E. I. Peerschke, L. E. Scudder, and C. A. Sullivan, *Blood* **61**(1), 99 (1983).

CA) and dispensing it back into the tube several times until the suspension appears homogeneous. After the second wash, the platelets are resuspended in 0.1 ml of Tris-EDTA buffer, and mixed 1:1 with the adjuvant with the aid of a device made by cutting two 18-gauge, Luer-lok metal needles close to their hubs and welding them together tip to tip. The platelet suspension and an equal volume of well-mixed complete adjuvant (making sure to resuspend the pellet of mycobacteria, which tends to settle) are each drawn up into a 1-ml plastic syringe. The syringes are joined at opposite ends of the Luer-lok fittings and the pistons simultaneously pushed to and fro until the suspension becomes thick and milky white. The suspension is withdrawn into one of the syringes, and the latter is then disconnected from the device and fitted with a 23-gauge needle for injection into the mice.

In our study,[1] a BALB/c mouse receives six weekly intraperitoneal ~0.2-ml injections containing ~3×10^8 fresh, washed platelets. The last injection is of ~5×10^8 platelets in Tris-NaCl buffer without either EDTA or adjuvant and is given intravenously into the tail vein. Each of the seven weekly platelet suspensions is obtained from a different donor. Three to 4 days after the tail vein injection, the mouse is sacrificed for the fusion.

It should be noted that the immunization procedure detailed above is somewhat unconventional in that all six weekly injections of platelets include Freund's complete adjuvant. Most immunization schedules either use no adjuvant for cellular antigens or limit the complete adjuvant to only the first injection.[2-5] We have no evidence that our technique is superior to the others. It is important to limit the volume administered in the intravenous injection given at the end of the schedule. We injected approximately 0.1 ml and found that this was not well tolerated; some animals, including the one from which we ultimately obtained an antibody, suffered what appeared to be a cardiorespiratory arrest, either on a temporary or permanent basis. Because we were interested in obtaining an antibody that inhibited platelet function, we tried to minimize the processing procedures so that the antigen would be as close to the native conformation as possible. The search for antibodies for different purposes may influence the choice of preparation technique. For example, if one wants to develop an antibody that reacts well in immunoblotting antigens separated by

---

[2] J. W. Goding, "Monoclonal Antibodies: Principles and Practice." Academic Press, London, 1983.
[3] G. Galfre and C. Milstein, *Methods Immunol.* **73**, 3 (1981).
[4] S. F. de St. Groth and D. Scheidegger, *J. Immunol. Methods* **35**, 1 (1980).
[5] H. Zola and D. Brooks, in "Monoclonal Hybridoma Antibodies: Techniques and Applications" (J. G. Hurrell, ed.). CRC Press, Boca Raton, Florida, 1982.

SDS–polyacrylamide gel electrophoresis, it would probably be desirable to treat the antigen with sodium dodecyl sulfate (SDS) before immunizing.

Screening Assay

Our initial goal was to identify the ristocetin-dependent platelet receptor for von Willebrand factor (vWF) by preparing an antibody that would block the binding of vWF to ristocetin-treated platelets. We reasoned that such an antibody would also block the agglutination of platelets induced by ristocetin in the presence of vWF. We therefore devised a quick, easy, high-volume, and inexpensive screening assay employing formaldehyde-fixed platelets, normal plasma, and ristocetin.[1]

Materials

Ristocetin: 50 mg/ml in Tris-NaCl buffer, pH 7.4 (Lenau, Denmark)
Pooled normal plasma (PNP) (George King, Inc., Overland Park, KS)
Formaldehyde fixing solution: Just before use, add 2 ml of 37% formaldehyde (F-79; Fisher, Pittsburgh, PA) to 98 ml of Tris-NaCl buffer
Formaldehyde-fixed platelets: One unit of fresh platelet concentrate is mixed 1:1 (v/v) with the fixing solution and stored at 4° for up to several months. For use: 8–10 ml of fixed platelets is washed three times in Tris-NaCl buffer and resuspended to a count of $3-3.5 \times 10^8$/ml for platelet-rich buffer (PRB)

Method

All steps are carried out at 22°.

Add 50 μl of PRB to each well of a round-bottom microtiter plate (Linbro, Hamden, CT), and then 50 μl of test culture supernatant, test serum, test ascites, or buffer. After 2 min, add 10 μl of PNP to each well and gently shake by sliding the plate with the right hand against a stationary left thumb in three or four smooth, short strokes while keeping the plate flat on the bench top. Add 3 μl of ristocetin to each well and shake gently again until uniformly cloudy (ristocetin tends to form a white precipitate at the bottom of the well). Attach the plate with rubber bands to a platform rotator (Tektator V; American Scientific Products, Edison, NJ) and rotate for 8 min at ~280 rpm at 22°. The wells are then observed from the bottom with a magnifying mirror viewer (Cooke Microtiter Systems, Dynatech Laboratories, Alexandria, VA). Wells containing buffer or culture medium have marked agglutination (4+ rating), while those containing the supernatant of positive clones show less agglutination (0–3+). A positive control can be obtained by omitting the ristocetin or using plasma deficient in vWF (available from George King, Inc.).

Monoclonal Antibody Preparation[1-5]

Materials

RPMI: RPMI 1640 culture medium supplemented with penicillin (1000 U/ml) and streptomycin (100 µg/ml) (GIBCO, Grand Island, NY; or Flow Laboratories McLean, VA)

RPMI–FBS: RPMI 1640 supplemented with 10 or 20% fetal bovine serum (FBS) (200-6140; GIBCO)

NCTC 109 medium (M. A. Bioproducts, Walkersville, MD)

P3X63-Ag8.653 nonsecretory BALB/c mouse myeloma cell line[6]

NH_4Cl (0.83%): Filter sterilize through 0.22-µm filter (450-0020; Nalgene, Rochester, NY) and keep on ice

Polyethylene glycol (PEG): Originally we used PEG from Koch-Light (Coinbrook, Bucks, England; M_r 1000) as a 35% PEG solution in RPMI. Autoclave pure PEG to sterilize and put in 3.5-ml aliquots. Store at 22° in the dark. To prepare the 35% solution, heat one aliquot to 56° and add to 6.5 ml warmed RPMI. Currently, we use 40% PEG (M_r 1500) from Boehringer Mannheim Corp., GmbH (Indianapolis, IN; #783641), which comes as a 50% (w/v) solution in 75 mM HEPES buffer. Mix one 4-ml vial of PEG with 1 ml RPMI just before use

Note. There is evidence that many factors affect the fusion efficiency of different PEG preparations, and there may even be batch-to-batch differences with the same product. The pH of the solution may be important as well as contamination with aldehydes[2,7,8]; sterilization by pressure filtration rather than autoclaving may minimize the latter.

A (50× aminopterin): Dissolve 4.4 mg aminopterin in 500 ml distilled water. Filter sterilize and freeze 1-ml aliquots at −20°

HT (50× hypoxanthine and thymidine): Dissolve 68.1 mg hypoxanthine and 19.4 mg thymidine in 100 ml distilled water. Adjust to pH 7.0, filter sterilize, and freeze 1-ml aliquots at −20°

HAT medium: Add 1 ml A and 1 ml HT to 50 ml RPMI–20% (v/v) FBS–10% (v/v) NCTC 109. Make fresh

2,6,10,14-Tetramethylpentadecane (Pristane) (Sigma Chemical Corp., St. Louis, MO)

phosphate-buffered saline (PBS): 0.15 M NaCl, 0.01 M sodium phosphate, pH 7.4

Freezing solution: 10% (v/v) Dimethyl sulfoxide (DMSO) in 90% (v/v) FBS. Make fresh and filter sterilize

---

[6] J. F. Kearney, A. Radbruch, B. Liesgang, and K. Rajewsky, *J. Immunol.* **123,** 1548 (1979).
[7] R. D. Lane, R. S. Crissman, and M. F. Lachman, *J. Immunol. Methods* **72,** 71 (1984).
[8] J. L. Kadish and K. M. Wenc, *Hybridoma* **2,** 87 (1983).

Immunized mouse
Ketamine, 100 mg/ml (Aveco, Fort Dodge, IA)
Ethanol, 70%
Two sterile 23-gauge needles attached to 1-ml syringe barrels
Sterile flat-bottom, 96-well, microtiter plates with lids (Corning 25860, Corning, NY or Costar 3596, Cambridge, MA)
Sterile, 24-well microtiter plates with lids (Costar 3524)
Petri dishes, plastic pipettes (Falcon)
Flasks, 75-cm^2 (Falcon 3024 or Corning 25110-75)
Flasks, 25-cm^2 (25100-25; Corning)
Hood (VBM-400 Sterigard hood; Baker Co., Inc., Sanford, ME)
Dual-chamber, CO_2 water-jacketed incubator (3325; Forma Scientific, Marietta, OH)
Plastic tubes, 15 and 50 ml (Corning 25311 and Falcon 2070, respectively)

Method

Myeloma Cells. Two to 3 weeks before the fusion, the myeloma cells are thawed and maintained in RPMI with 10% FBS. Most investigators believe it is vital that the cells be in the log phase of growth at the time of the fusion and this is achieved by diluting two or three 75-cm^2 flasks to ~10^5 cells/ml 1 or 2 days before the fusion. On the day of the fusion, the cells are counted.

Spleen Cells. If possible, use a separate room and hood for the mouse surgery. Keep scissors and forceps immersed in 70% ethanol. The immunized mouse is anesthetized with ketamine (10 mg ip) and then sacrificed by cervical dislocation. First, secure the mouse by the tail and let it stand on the bench top. Pinch the nape of the neck from behind with the thumb and forefinger. Quickly, depress the nape of the neck to the bench top as the tail is snapped horizontally away from the body. The neck can be felt to snap. Wash the abdominal area with 70% ethanol, pick up the skin to avoid penetrating the viscera, and cut into the abdominal cavity. Aseptically remove the spleen and rinse in a 50-ml tube containing 20–30 ml RPMI. This tube can then be brought to the tissue culture hood. Empty the tube contents into a sterile petri dish and aspirate the medium. Add 10 ml of fresh RPMI to rinse again and aspirate the supernatant medium again. Use the two fine needles to tease the spleen apart with a "knife and fork" action. Tease the pieces as small as possible, but try to work rapidly (~5 min). Pipette 8–10 ml RPMI into the dish, withdraw all the cells and tissue chunks, transfer to a 50-ml tube, and fill to 35 ml with RPMI. After 10 min the large clumps will settle, and then transfer the top ~32 ml of smooth cell suspension to a clean 50-ml tube, centrifuge at 500 g for 10 min at 22°, and aspirate the supernatant. To lyse the red blood cells, add 5 ml of cold 0.83% (w/v) NH_4Cl to the pellet, place on ice for 10 min, add 5 ml of RPMI, and centrifuge at 500 g for 5 min at 22°. The red pellet should become white. If not, the 0.83% NH_4Cl step can be repeated. The final pellet should be gently resuspended in ~10 ml RPMI and a cell count obtained with a hemocytometer.

Fusion. All steps are conducted at 22° and the steps involving cell manipulation are conducted in a hood.

Add 4×10^8 spleen cells to 1×10^8 myeloma cells (4:1) in a 50-ml tube and then centrifuge at 500 g for 5 min. Aspirate the supernatant and gently tap the conical tube on the bench top to loosen the pellet. Add 3 ml of the 35% PEG or 1 ml of the 40% PEG using a wide-bore pipette (to avoid clogging by PEG and excess shear that would disrupt fused cells), mix gently by pipetting up and down two or three times, and immediately centrifuge at 500 g for 6 min. Do not aspirate supernatant. Add an additional 5 ml of RPMI, mix gently with the pipette, centrifuge again, but at 230 g for 6 min, and aspirate the supernatant. The pellet is then resuspended in 50 ml of fresh RPMI–20% FBS–10% NCTC 109. The cells can receive HAT medium now or after an overnight incubation in a 75-cm^2 flask. There may be a slight advantage to the overnight incubation, since contaminating fibroblasts and other cells adhere to the flask.

The next day the cells are mixed and decanted into a 50-ml tube, leaving unwanted cells attached to the flask. Add 1 ml of solution A and 1 ml of solution HT, mix, and aliquot 0.1-ml samples into five or six 96-well plates. Incubate for 4 days in a 37°, 5% CO_2 incubator and then add 0.1 ml of HAT medium to each well. Clones should start appearing as "fuzzy spots" at the periphery of the well ~2 weeks after the fusion; at that time, 0.2 ml of medium is withdrawn from each well showing cell growth and 0.2 ml of RPMI–20% FBS–HT medium is added. After another 8–9 days, 0.2 ml of medium is withdrawn again and replaced with 0.2 ml of RPMI–20% FBS–HT. About 4 weeks after the fusion, the cells can be transferred to the wells of a 24-well plate. Pipette the well contents up and down several times to remove most of the cells and place in a well of a 24-well plate. Add an additional 0.5 ml of RPMI–20% FBS–HT and, in about 4 days, add 1 ml of RPMI–20% FBS. Expand the cells to 25-cm^2 flasks (10 ml) and then to 75-cm^2 flasks (50 ml) by adding increasing amounts of medium. We routinely grow cells in RPMI–10% FBS, but one can decrease the cost by switching to horse serum or lower concentrations of FBS. Because mold periodically and unpredictably appears in individual flasks, we keep duplicate flasks growing at all times. The cells need to be "split," i.e., diluted 1:5 or 1:10, every 3–4 days, thus permitting twice weekly attention. To ensure monoclonality, the cells must be subcloned either by the limiting dilution technique[2-5,9] or by growth in soft agar.[2-5] Additional evidence for monoclonality can be obtained from subclass and light chain typing, by immunodiffusion, or by isoelectric focusing.

Freezing Clones

To avoid losing valuable clones, they should be frozen in several different aliquots as soon as their density reaches $\sim 10^5$/ml (usually 5–10 ml in flask). The cells are pelleted at 1000 g for 8 min at 22° and the culture supernatant aspirated and saved. The cells are resuspended in the freezing solution at $\sim 5 \times 10^5$ cells/ml and frozen at a rate of $\sim 1°$/min in a standardized biological freezing tray (Ro36-8c15, Union Carbide, Indianapolis, IN). They are stored indefinitely in the fluid phase of liquid nitrogen. For cells that are likely to be in storage for a number of years, it is important to thaw aliquots of cells every year or so to prepare fresh cells for refreezing.

Ascites Production

To obtain higher titers of antibody than can be achieved in culture medium, the hybrid cells can be grown as an ascites tumor in the abdominal cavity of a mouse. BALB/c mice are primed intraperitoneally with 0.5 ml of Pristane approximately 10–14 days before receiving the cells, which seems to be the best pretreatment period.[10,11] The cells are washed twice in PBS and injected intraperitoneally into 6-week-old BALB/c mice. Each mouse gets 0.1–0.2 ml usually containing $1-5 \times 10^6$ cells. In 2–4 weeks, the animal is anesthetized with ketamine (0.5–1 mg), the abdominal area is cleansed with alcohol, and the ascites drained through a 20-gauge needle into a 17×100-mm plastic tube (2059; Falcon). The fluid is heated to 56° for 10 min to defibrinate the ascites and then centrifuged at 1100 g for 10 min at 22°. The supernatant is saved frozen at $-20°$ until ready for purification. If desired, the procedure can be performed in a sterile hood.

Purification of Antibodies

Materials

Phosphate buffer: 0.1 M NaPO$_4$, 0.05% (w/v) NaN$_3$, pH 8.0
Citrate buffers: 0.1 M sodium citrate, 0.05% (w/v) NaN$_3$, pH 6.0, 4.5, 3.5, and 3.0
Note: Filter buffers through 0.45-μm membrane and deaerate

---

[9] H. A. Coller and B. S. Coller, *Hybridoma* **2**, 91 (1983).
[10] N. Hoogenraad, T. Helman, and J. Hoogenraad, *J. Immunol. Methods* **61**, 317 (1983).
[11] B. R. Brodeur, P. Tsang, and Y. Larose, *J. Immunol. Methods* **71**, 265 (1984).

Protein A-Sepharose Cl 4-B, 5–10 ml (Pharmacia, Piscataway, NJ), or
5–10 ml protein A-agarose (Boehringer-Mannheim, Indianapolis, IN)
Glass column: 0.8 × 20 cm (Bio-Rad)
Crystalline ammonium sulfate (Schwarz/Mann, Cambridge, MA)

Method

Sample Preparation from Culture Supernatant. One or 2 liters of culture supernatant (containing ~5–20 mg of monoclonal antibody) is stirred on ice while ammonium sulfate is slowly sprinkled in until 50% saturation (29.1 g/100 ml solution) is achieved. Stirring is continued on ice for 1 hr and then the solution is centrifuged for 20 min at 3000 g at 4°. After decanting the supernatant, the pellet is resuspended in as small a volume as possible of the phosphate buffer and dialyzed against an excess of the same phosphate buffer. If desired, one can check for the completeness of dialysis by adding a drop of the dialysate to a solution of saturated $BaCl_2$; a white precipitate indicates the presence of residual ammonium sulfate. If the antibody solution contains precipitate after dialysis, the suspension should be clarified by centrifugation before applying to the protein A column.

Sample Preparation from Ascites. Heat-defibrinated ascitic fluid is thawed at 37°, centrifuged at 1100 g for 10 min if any particulate matter remains, and diluted 1 : 1 with 0.1 M $NaPO_4$–0.05% NaN_3, pH 8.0 buffer.

Chromatographic Purification.[12] Protein A-agarose or protein A-Sepharose Cl 4B (5–10 ml) is washed in the phosphate buffer and the pH 3.0 citrate buffer before being poured into a 0.8 × 20 cm column. The column is eluted at 4° at rates up to 30–60 ml/hr depending on the manufacturer's recommendations. After the column is reequilibrated with the phosphate buffer, the sample is applied and the column eluted with the phosphate buffer until the optical density at 280 nm of the eluate returns to near baseline. The antibody, which is bound to the protein A, can then be eluted from the column by stepwise elutions with the citrate buffers at pH 6.0, 4.5, 3.5, and finally at pH 3.0. The IgG subclass will determine the pH at which the antibody elutes. Because IgG_1 binds least well to protein A, it is important that the pH of the sample and the buffer in the column be 8.0 before applying the sample. Even then, we have had protein A columns that did not completely bind IgG_1 antibodies, but did delay their elution. Thus, it is important to assay all of the column fractions for antibody until the behavior of the antibody on that particular column is determined. There may be a benefit in using a 1.5 M NaCl, 0.01 M $NaPO_4$, pH 8.0 buffer in the binding step, but we have not made direct comparisons. In

[12] P. L. Ey, S. J. Prowse, and C. R. Jenkin, *Immunochemistry* **15**, 429 (1978).

addition, a 3 M NaCl, 1.5 M glycine, pH 9.0 buffer is also said to improve the binding of IgG_1 to protein A columns. A propriety buffer system is also available that is said to improve the yield of IgG_1 antibodies (Bio-Rad). Antibodies of IgG_{2a} subclass usually do not elute until pH 4.5. Because we have observed that most of the IgG in fetal bovine serum elutes at pH 6, stepwise elutions with pH 6.0 and 4.5 buffers may yield a purer preparation than that obtained with just a single pH 4.5 elution. The column can be reequilibrated with the phosphate buffer and used again. The protein elution pattern is monitored by optical density at 280 nm and each of the citrate buffer fractions is adjusted to pH 7.0 ± 0.2 with 1 M Tris-HCl, pH 8.5. The fractions with protein are pooled and dialyzed against Tris-NaCl buffer containing 0.05% NaN_3, pH 7.4. Antibody concentration is estimated by absorption at 280 nm, assuming $A^{1\%}_{280\,nm}$ = 14. The purified antibody can be assayed for functional activity in the vWF screening assay. To establish the titer, serial dilutions are tested until no activity is present. We store antibody at 4° and have found that it retains activity for months to years under these conditions.

Iodination Procedure

Materials

Sephadex G-25 (Pharmacia) and disposable glass column (737-2240); alternatively, 10-ml columns of BioGel P-6 can be obtained prepacked (Econo-Pac 100G; Bio-Rad)
^{125}I (carrier-free NaI in 0.1 N NaOH; New England Nuclear, Boston, MA)
Iodogen (Pierce, Rockford, IL)
Glass scintillation vials with caps (2.5 × 5.5 cm)
Tris buffer, 0.05% NaN_3, pH 7.4
Tris buffer, 0.05% NaN_3, pH 7.4 with 0.2% (w/v) bovine serum albumin (BSA; Sigma, St. Louis, MO)
PBS, pH 7.4
PBS, pH 7.4 with 1% BSA (Sigma)
Saturated solution of sodium thiosulfate
TCA solution, 100% (w/v) (trichloroacetic acid, A-322; Fisher Scientific)

Method

A Sephadex G-25 or BioGel P-6 column with a bed volume of 9–12 ml is prerun with 50 ml of the Tris-NaCl–NaN_3 buffer containing 0.2% BSA at 22°. The column is equilibrated and eluted with the Tris-NaCl–NaN_3 buffer without the albumin until the optical density nears baseline.

Iodogen is dissolved in dichloromethane at 0.5 mg/ml and then 50 μl is used to coat the bottom of an uncapped glass scintillation vial (2.5 × 5.5 cm). The Iodogen is very gently swirled to ensure even coating as the dichloromethane evaporates. After evaporation, a stream of nitrogen is passed over the bottom of the vial and the vial is capped and stored at $-20°$. Although some vials retain their activity for several years under these conditions, others do not, and so we currently make freshly coated vials for each iodination. One millicurie of ^{125}I (in 10 μl) is then added to 1 ml of antibody solution (usually 1 mg/ml in 0.15 M NaCl, 0.01 M Tris/HCl, 0.05% sodium azide, pH 7.4), and after brief mixing the mixture is added to the Iodogen-coated vial. The vial is then capped, swirled intermittently for ~10 min, and then the reaction is terminated by applying the contents of the reaction vial to the equilibrated Sephadex G-25 column. After the first peak of ^{125}I-labeled protein elutes, the column is stopped to avoid eluting the free ^{125}I peak. The free ^{125}I peak is then eluted into saturated sodium thiosulfate and saturated sodium thiosulfate is passed through the column to eliminate any volatile ^{125}I. The individual fractions are characterized further by their optical density at 280 nm, TCA precipitability, and specific activities. In addition, each sample is tested by polyacrylamide gel electrophoresis to be certain that the radioactivity migrates with the antibody.

Trichloroacetic acid precipitability is performed by adding 1 μl of ^{125}I-labeled protein to 0.9 ml PBS–1% BSA. After mixing, 0.1 ml of 100% TCA is added, the solution is incubated on ice for 1 hr, and then centrifuged at 12,000 g at 22° for 3 min. Finally, 0.1 ml of the supernatant is removed and the radioactivity counted in a γ counter.

Binding Studies

Materials

Sucrose solution, 30% (w/v) (Fisher)
Microcentrifuge tubes, 0.4 ml (223-9502; Bio-Rad)
Microcentrifuge tubes, 1.5 ml (223-9501; Bio-Rad)
Microcentrifuge (Beckman Instruments, Irvine, CA; or 5413, Eppendorf, Hamburg, Germany)
^{125}I-labeled antibody
Platelet-rich plasma (PRP)

Method

Binding studies are performed at 22° using PRP prepared from whole blood collected into 1/100 vol of 40% sodium citrate. Platelet-rich plasma

(0.2 ml) is placed in a 1.5-ml microtube and ^{125}I-labeled antibody or a labeled/unlabeled mixture is added. After incubating the samples long enough for the reaction to reach equilibrium, duplicate 0.1-ml aliquots of PRP are gently layered over 0.1-ml samples of 30% sucrose contained in 0.4-ml microtubes. Layering of PRP is facilitated by fitting the pipette tip (223-9000; Bio-Rad) with a ~3-cm piece of microbore tubing (i.d. 0.050 in., 14-170-15E; Fisher Scientific). The duplicate microtubes are centrifuged at 22° for 3 min at 12,000 g, removed immediately from the centrifuge, and placed upright. The tip of each tube, containing the platelet pellet, is amputated with a dog nail cutter into a glass test tube (12 × 75 mm; Fisherbrand, Pittsburgh, PA) and the remaining portion of the microtube is rapidly placed into a second glass tube. Both tubes are counted in a γ counter. The fraction of radioactivity bound is calculated by dividing the counts in the pellet by the sum of the counts in the pellet and the supernatant.

Although the performance of binding studies is remarkably simple, their interpretation is not nearly as straightforward. In general, one wants to know the affinity of the antibody for the antigen and/or the total number of sites to which the antibody can bind. The mathematical rationales for using different equations and graphic formats to determine these parameters has been dealt with at great length by many authors and will not be reviewed here, but a few practical points will be emphasized.

1. As pointed out by Klotz,[13] if one chooses to use the method of Scatchard, it is imperative that data points be obtained at concentrations of the antibody high enough so that the free antibody concentration is considerably greater than the K_D.

2. Values for the number of antibody molecules bound at intermediate ratios of bound/free antibody concentrations are considerably more reliable than values obtained at either very high or very low bound/free ratios. Thus, preliminary experiments may be required to determine where this range is.

3. It should be fully appreciated that not all of the radioactivity added to a sample represents homogeneously labeled and active antibody molecules. Depending on the precise nature of the experiments, corrections may have to be made for (a) iodine that is not conjugated to antibody (or any other protein) as determined by the precipitability of the preparation with 10% TCA (usually <5% of total radioactivity is not precipitable), (b) trace contamination of the antibody preparations with nonimmunoglobulin molecules as judged by polyacrylamide gel electrophoresis (usually ~5–10%), (c) trace contamination with nonhybridoma immunoglobulin,

[13] I. M. Klotz, *Science* **217**, 1247 (1982).

especially if the antibody is purified from culture medium containing FBS with a significant immunoglobulin content, (d) hybridoma-derived antibody that was abnormally synthesized and so is inactive, (e) hybridoma-derived antibody that was inactivated by the purification procedure, (f) hybridoma-derived antibody that was inactivated during radiolabeling, and (g) hybridoma-derived antibody that was inactivated by radiation damage over time. Moreover, at the specific activities usually obtained with Iodogen or iodination ($\sim$1000 cpm/ng), only a small fraction of the antibody molecules have iodine atoms attached to them. Therefore, these preparations are, in fact, made up of at least two different species (unlabeled antibody and antibody labeled with only one iodine per antibody molecule) that may differ in their functional activity.

4. We have chosen to determine the radioactivity of both the platelet pellet and the supernatant because this permits the easy detection of errors in pipetting. In addition, it allows for the direct analysis of the free antibody rather than calculating it from the amount presumably added and the amount bound. Moreover, if all other factors are kept constant (amount of antibody added, platelet count, trichloroacetic acid precipitability of the antibody), then one can use the fraction of added antibody bound to normalize values obtained over a period of time without having to recalculate the specific activity of the antibody. The formula used is

$$\frac{(\text{Fraction bound})(\text{micrograms antibody per sample})(3.76 \times 10^{12})}{\text{Number of platelets per sample}}$$

$$= \text{molecules of antibody per platelet}$$

5. We have found the trichloroacetic acid precipitability of iodinated antibody preparations decreases significantly with time (from 95 to $\sim$60% in 3 months in some samples). We thus periodically check preparations that are stored for a prolonged period and discard those with significantly diminished precipitabilities.

6. With high-affinity antibodies, one method for determining the fraction of radioactivity not attached to active antibody [items (a–g) described above under point 3, above] is to add trace amounts of radiolabeled antibody to samples containing increasing numbers of platelets as recommended by Trucco and de Petris.[14] The fraction of antibody bound increases with increasing numbers of platelets and reaches a plateau at high platelet counts. One can then operationally define that plateau value as the fraction of radioactivity that is bound to active antibody. There are more elaborate alternative methods for

---

[14] M. Trucco and S. de Petris, in "Immunologic Methods" (I. Lefkovits and B. Pernis, eds.), p. 1. Academic Press, New York, 1981.

calculating this value,[15,16] but the above method provides a reasonable estimate for high-affinity antibodies. Use of this correction is crucial for determining the concentration of free antibody, since at or below the K_D, only a small fraction of antibody may be free, and so the radioactivity that is not attached to active antibody may equal or even exceed the true value of free antibody. Thus, errors of greater than 100% can be introduced. With our antibodies, the maximum fraction that can bind to platelets has varied from 60% to greater than 95%, depending on the preparation, even when the trichloroacetic acid-precipitability has been greater than 95%. It is difficult for us to be certain which of the factors detailed in point 3 (above) account for this variability. Our experience is, however, similar to that reported by others.[14]

7. When large numbers of samples need to be analyzed, it is cumbersome to have to perform multiple analyses on each sample to determine the maximum number of molecules that can bind per platelet by the conventional graphic techniques. An alternative that is widely used is to add an amount of antibody greatly in excess of the K_D (at least 20-fold) and simply measure the amount bound. Several precautions must be taken when using this approach: (a) As indicated above, this set of conditions puts the analysis in an area where the Scatchard plot is least reliable (for example, at very high concentrations, less than 5% of the added radioactivity may be bound and this not only becomes difficult to measure accurately, but also introduces a large mutliplier based on the high antibody concentration, which will then exaggerate any error). (b) To preserve radiolabeled antibody, it is common practice to mix unlabeled and radiolabeled antibody. The true specific activity of the resulting mixture is not easily determined, however, because it is difficult to assess the integrity of the unlabeled antibody with regard to items (b–d) in point 3 discussed above. Thus, it is best to try to perform studies involving direct comparisons using a single radiolabeled antibody preparation over the shortest time interval possible. Studies using different radiolabeled preparations over a prolonged period of time require more elaborate procedures to normalize data from batch to batch. Using the above procedure, we obtained results indicating that the number of antibody molecules bound per platelet was 26,000 ± 10,000 (mean ± SD; $n = 13$).

---

[15] S. J. Kennel, L. J. Foote, P. K. Lankford, M. Johnson, T. Mitchell, and G. R. Braslawsky, *Hybridoma* **2**(3), 297 (1983).

[16] T. Lindmo, E. Boven, F. Cuttitta, J. Fedorko, and P. A. Bunn, Jr., *J. Immunol. Methods* **72**, 77 (1984).

Radioimmunoelectrophoresis[17]

Materials

Buffers
  Tris–EDTA–glucose: 0.01 M Tris, 0.15 M NaCl, 1 mM EDTA, 5 mM glucose, pH 7.4
  Tris-glycine buffer: 0.038 M Tris, 0.1 M glycine, pH 8.7
  Triton X-100, 10% (w/v) (Sigma) in Tris-glycine buffer
Airfuge ultraspeed centrifuge (350624; Beckman)
Immunoelectrofilm Kit (Kallestad Laboratories, Chaska, MN)
Rabbit antiserum to human serum (518301; Calbiochem)
Rabbit antiserum to human glycocalicin
Rabbit antiserum to human thrombocytes (A225; Dako, Santa Barbara, CA)

Method

Platelet-rich plasma is prepared from whole blood collected into 1/100 vol of 40% sodium citrate and washed three times in Tris–EDTA–glucose buffer. The platelets are resuspended in the Tris-glycine buffer at a count of 6×10^9/ml, 1/9 vol of 10% Triton X-100 solution is then added, and the solution rocked at 4° for 30 min. After centrifuging at 80,000–100,000 g for 1 hr at 4°, the supernatant is removed, aliquotted, and stored frozen at −20°.

Immunoelectrophoresis is performed at 22° employing the 1.5% (w/v) agarose film and 0.065 M barbital/acetate buffer provided with the immunoelectrofilm kit. A 3-μl sample of solubilized platelet membranes is electrophoresed at 100 V for 50–75 min (using a sample of human serum in another well with bromphenol blue as a tracking dye as an indicator of when the electrophoresis is complete), and precipitin arcs are developed by overnight incubation at 22° with 0.1 ml of one of the rabbit antisera, to which is added ~100,000 cpm of ^{125}I-labeled monoclonal antibody. The agarose gel is washed extensively in 0.15 M NaCl, rinsed in distilled water, and dried with the aid of a hot air dryer. The dried gel is stained for 10 min in 0.05% (v/v) Amido black in 38% ethanol–10% acetic acid and destained in 92% methanol–8% acetic acid. The gel is autoradiographed in a cassette (Cronex Lightning Plus; Du Pont, Wilmington, DE) with X-ray film (XAR-5; Kodak, Rochester, NY) for variable periods of time.

[17] N. O. Solum, I. Hagen, C. Filion-Mykleburst, and T. Stabaek, *Biochim. Biophys. Acta* **597**, 235 (1980).

Preparation of the 6D1 Affinity Column

Materials

3-[Morpholino]propanesulfonic acid (MOPS) buffer (0.1 M), pH 7.0
Affi-Gel 10 (Bio-Rad)
Medium sintered glass funnel, 15 ml, connected to a vacuum line: Wash funnel with chromic–sulfuric acid cleaning solution and rinse extensively with distilled water
Glass rod (3 × 130 mm)
Distilled H_2O
6D1 monoclonal antibody (antibody directed at GPIb–glycocalicin)[1]
Ethanolamine (1 M), pH 8.0
Buffer 1: 0.15 M NaCl, 0.01 M Tris, 10 mM EDTA, 1% Triton X-100, pH 7.4
Buffer 2: 0.15 M NaCl, 0.01 M Tris, 10 mM EDTA, 0.05% Triton X-100, pH 7.4
Buffer 3: 1.5 M NaCl, 10 mM EDTA, 0.05% Triton X-100, pH 7.4
Buffer 4: 0.05 M diethylamine, 10 mM EDTA, 0.05% Triton X-100, pH 11.5
Disposable glass column, 0.8 × 7 cm (Bio-Rad)

Method

Purified 6D1 monoclonal antibody (10 mg in 15 ml) is dialyzed extensively against the MOPS buffer and stored at 4°.

Approximately 8 ml of Affi-Gel 10 slurry is placed in a sintered glass funnel, drained with vacuum, and the gel washed sequentially with three volumes of 4° water and 2 vol of 4° MOPS buffer while stirring vigorously with the glass rod. The suction is then broken gently so as to avoid drying the gel, and the latter transferred to a 50-ml test tube containing 15 ml of 0.67 mg/ml antibody solution. After rocking overnight at 4°, any unreacted sites on the gel are blocked by adding 1/10 vol 1 M ethanolamine and rocking for an additional 2 hr at 4°. The gel (3.5-ml packed volume) is packed in a 0.8 × 7-cm disposable glass column and washed sequentially with 50 ml of buffer 1, buffer 2, buffer 3, buffer 4, and buffer 1 again. The extent of coupling is assayed by protein determination of the antibody solution before and after reaction with the Affi-Gel. Unfortunately, this cannot be done simply by measuring the absorbance at 280 nm since the reaction releases a compound that absorbs at this wavelength.

Affinity Purification of Glycocalicin and/or Glycoprotein Ib

Materials

Outdated platelet concentrates
KCl, 3 M
Na$_2$EDTA, 269 mM (10% solution, pH 7.4)
Buffer A: 96.5 mM NaCl, 85.7 mM glucose, 8.5 M Tris, 10 mM EDTA, pH 7.4
Buffer B: 0.15 M NaCl, 0.01 M Tris, 1 mM EDTA, pH 7.4
Buffer C: 0.2 mM EDTA, 15.3 mM sodium azide, pH 7.0
Buffers 1, 2, 3, and 4 (see above)

Method

Glycocalicin and GPIb are extracted from platelets that are outdated but not older than 40 hr.[1,17,18] For purification of GPIb, 5–10 platelet concentrates are pooled and made 10 mM in EDTA. The red blood cells are removed by centrifuging repetitively at 400 g for 2 min, and the platelets are washed three times in buffer A. The platelet pellet is lysed in 0.15 M NaCl, 0.01 M Tris, 20 mM EDTA, 20 mM N-ethylmaleimide (NEM), 0.05% NaN$_3$, 1% Triton X-100, pH 7.4 by rocking at 4° for 1 hr. The sample is then centrifuged at 12,000 g for 15 min at 4° to remove the cell debris and the supernatant is recentrifuged at 100,000 g for 90 min at 4°. The extract is then applied to the 6D1 affinity column at approximately 5 ml/hr at 4°. The column is washed sequentially with a minimum of 10 bed volumes of buffer 1, buffer 2, and buffer 3. Glycoprotein Ib is eluted with buffer 4, dialyzed against buffer 2, and stored at −80°. Analysis by SDS–PAGE and periodic acid–Schiff staining of nonreduced samples reveals one major band of $M_r \sim 155,000$ corresponding to GPIb; on reduction with 5% mercaptoethanol, this band shifts to $M_r \sim 130,000$. A minor band of $M_r \sim 125,000$ corresponding to glycocalicin and a band at $M_r \sim 17,000$ are also present[1,18]; the latter band is GPIX, which is complexed with GPIb.

Because glycocalicin is cleaved from platelets by the action of the calcium-dependent protease that can be released from platelets,[17,18] glycocalicin-rich extracts are prepared by lysing outdated platelets in 3 M KCl. Approximately 10 units of outdated platelets are pooled, made 1 mM in EDTA, and centrifuged at 3000 g for 15 min at 4°. The pellet is resuspended in buffer B with care so as to avoid contamination with red blood cells. Any remaining red blood cells are then removed by repetitive centrif-

---

[18] B. S. Coller, E. L. Kalomiris, M. Steinberg, and L. E. Scudder, *J. Clin. Invest.* **73**, 794 (1984).

ugation at 400 g for 2 min at 4°. The platelets are then pelleted by centrifuging at 3000 g for 15 min at 4° and washed three times in buffer B. The platelet pellet is then resuspended in 3 M KCl (approximately 7 ml/unit used), incubated at 37° for 15 min with occasional swirling, and centrifuged at 3000 g for 15 min at 4°. The glycocalicin-containing supernatant is removed and recentrifuged twice at 12,000 g for 20 min at 4°. It is then dialyzed overnight at 4° against large volumes of buffer C and then against large volumes of distilled H_2O for at least 24 hr with several changes. During dialysis, a glycocalicin-rich precipitate forms that is isolated by centrifugation at 12,000 g for 45 min at 4°. The sediment is resuspended in a solution containing 154 mM NaCl, 0.2 mM EDTA, 15.3 mM NaN$_3$, pH 7.0 (using approximately 0.5 ml/unit used), stirred at 37° for 10 min, and centrifuged at 12,000 g for 45 min at 4° to remove the undissolved material. The glycocalicin is then purified from contaminants on a 6D1 affinity column as described for GPIb.

[27] Fibronectin Binding to Platelets

By JANE FORSYTH, EDWARD F. PLOW, and MARK H. GINSBERG

Introduction

Fibronectin is a large circulating glycoprotein that has been implicated in cell migration, opsonization, tissue development, and platelet adhesion.[1] Although qualitative studies have suggested the existence of cellular fibronectin receptors,[2,3] the first direct demonstration of fibronectin binding to a saturable cellular receptor was performed with platelets.[4] The performance of such assays has permitted the identification of a congenital deficiency of fibronectin receptors,[5] and establishment of two types of fibronectin interaction with platelets.[6] They have also provided important confirmation to the concept[7] that the Arg-Gly-Asp sequence in fibronectin mediates its binding to cell surfaces by the direct demonstration of inhibi-

---

[1] R. O. Hynes and K. M. Yamada, *J. Cell Biol.* **95**, 369 (1982).
[2] F. Grinnell, *J. Cell Biol.* **86**, 104 (1980).
[3] M. P. Bevilacqua, D. Amrani, M. W. Mosesson, and C. Bianco, *J. Exp. Med.* **153**, 42 (1981).
[4] E. F. Plow and M. H. Ginsberg, *J. Biol. Chem.* **256**, 9477 (1981).
[5] M. H. Ginsberg, J. Forsyth, A. Lightsey, J. Chediak, and E. F. Plow, *J. Clin. Invest.* **71**, 619 (1983).
[6] E. F. Plow, G. A. Marguerie, and M. H. Ginsberg, *Blood* **66**, 26 (1985).
[7] M. D. Pierschbacher and E. Ruoslahti, *Nature (London)* **309**, 30 (1984).

tion of fibronectin binding to a cell by such peptides.[8] In addition, assay of fibronectin binding to platelets provided[5] evidence that GPIIb–IIIa ($\alpha_{IIb}\beta_3$ integrin) was a multifunctional receptor, i.e., that fibrinogen and von Willebrand factor shared this receptor with fibronectin.[9–12] This multifunctional receptor exists in addition to the fibrinogen receptor belonging to the β_1 family of integrins ($\alpha_5\beta_1$ integrins).[13] Thus, the quantitative assessment of fibronectin binding to the platelet surface has proved of value in the general study of cellular adhesion, as well as in the specific analysis of platelet adhesive function.

The assays are performed with purified radiolabeled fibronectin and suspensions of washed platelets. The specific details for preparing and radiolabeling of the fibronectin and performance of the binding assays are described below.

Preparation of Gelatin-Sepharose

Fibronectin is isolated from human plasma by affinity chromatography on a gelatin-Sepharose column. Gelatin-Sepharose is available commercially, but is costly.

1. Wash 750 ml Sepharose 4B (Pharmacia, Piscataway, NJ) with distilled water and slowly stir the mixture in an ice bath in a fume hood.

2. Add 1.5 ml of 2 M Na_2CO_3 and stir on ice until the temperature of the mixture is less than 4°.

3. Dissolve 100 g cold CNBr (Sigma, St. Louis, MO) in 50 ml cold acetonitrile; it will take considerable time to dissolve the CNBr, but do not try to rush the process by heating it.

4. Slowly add the CNBr solution to the stirring Sepharose 4B, adding ice crystals occasionally to keep the solution cold. Stir the mixture for 10 min.

5. Filter the beads in a glass sintered funnel, then wash with 10 liters cold 0.2 M $NaHCO_3$, pH 9.5, and add ice crystals to cool.

[8] M. Ginsberg, M. D. Pierschbacher, E. Ruoslahti, G. Marguerie, and E. Plow, *J. Biol. Chem.* **260**, 3931 (1985).
[9] E. F. Plow, A. H. Srouji, D. Meyer, G. Marguerie, and M. H. Ginsberg, *J. Biol. Chem.* **259**, 5388 (1984).
[10] E. F. Plow, R. P. McEver, B. S. Coller, V. L. Woods, Jr., G. A. Marguerie, and M. H. Ginsberg, *Blood* **66**, 724 (1985).
[11] R. Pytela, M. D. Pierschbacher, M. H. Ginsberg, E. F. Plow, and E. Ruoslahti, *Science* **231**, 1559 (1986).
[12] E. F. Plow, M. D. Pierschbacher, E. Ruoslahti, G. A. Marguerie, and M. H. Ginsberg, *Proc. Natl. Acad. Sci. U.S.A.* **82**, 8057 (1985).
[13] R. O. Hynes, *Cell (Cambridge, Mass.)* **48**, 549 (1987).

6. Transfer the cake to a large beaker containing a chilled solution of 16 g of gelatin in 800 ml of 0.2 M NaHCO$_3$, pH 9.5. Incubate overnight at 4°, stirring slowly.

7. After the overnight incubation, the solution will be gelid and difficult to filter. At room temperature, filter the Sepharose and wash it with 10 liters of phosphate-buffered saline (PBS; 0.01 M Tris base, 0.15 M NaCl), pH 8.6; 10 liters of 0.01 M Tris, 2 M NaCl, pH 8.6; and 10 liters of PBS, pH 8.6. Store at 4° until used.

8. Pour the gelatin-Sepharose column at room temperature in a siliconized glass column. Before use, equilibrate the column with PBS (0.01 M sodium phosphate, 0.15 M NaCl) plus 5 mM EDTA, 0.05% (w/v) sodium azide, pH 7.0.

Preparation of Fibronectin

Fibronectin is prepared from human plasma drawn into 1/6 vol ACD [0.065 M citric acid, 0.085 M sodium citrate, 2% (w/v) dextrose]. The plasma should be fresh (not more than 2 days old) and kept at room temperature until used; freezing decreases the yield. The entire preparation is done at room temperature.

1. Centrifuge the plasma at 5000 rpm for 30 min, then filter it through Whatman (Clifton, NJ) filter paper (#2) by gravity. Add 0.05% sodium azide, 5 mM EDTA, 1 mM benzamidine to the plasma.

2. Load the plasma onto the preequilibrated column, then wash the column with successive washes of PBS, 1 mM benzamidine, pH 7.0; 1 M urea, 1 mM benzamidine, pH 7.0; and PBS, pH 7.0. The volume of each wash step should be two to three times the column bed volume.

3. Elute the column with 1 M NaBr, 0.02 M acetic acid, pH 5.0. As the fibronectin elutes, it appears as viscous strands. Therefore it is important to collect fractions by time rather than drop counting.

4. Pool the fractions with peak absorption at 280 nm and immediately dialyze the pool against PBS with three changes of buffer.

5. Store the fibronectin at $-70°$ in aliquots. Thaw only once.

6. Fibronectin is characterized by sodium dodecyl sulfate–polyacrylamide gel electrophoresis (SDS–PAGE), in which it yields a closely spaced doublet at M_r 225K reduced and a single M_r 450K band nonreduced. In addition, the concentration of fibronectin in solution can be estimated from its absorption coefficient $a = A_{1\,cm}^{0.1\%} = 1.28$.[14]

---

[14] M. Mosesson and C. A. Umfleet, *J. Biol. Chem.* **245**, 5728 (1970).

Radiolabeling of Fibronectin

1. To a tube containing 800 µl of fibronectin at approximately 1 mg/ml add 800 µCi ^{125}I and 20 µl of chloramine-T at 2 mg/ml in PBS, pH 7.0.
2. Mix and incubate at room temperature for 5 min.
3. Add 20 µl sodium metabisulfite at 2 mg/ml in PBS, pH 7.0, and 20 µl 1% (w/v) potassium iodide in distilled water.
4. Add 300 µl of a 1% BSA, 1 mM phenylmethylsulfonyl fluoride (PMSF) solution in PBS. Dialyze to remove free iodine.
5. Before dialysis the label is generally greater than 90% TCA precipitable and after dialysis it is greater than 99%.

Binding Assays

Suspension of washed platelets can be prepared by gel filtration as described in this series, Volume 169, p. 11. Specific techniques for doing this are described by Ginsberg et al.[15] Platelets are adjusted to a final cell concentration of 8×10^8 cells/ml.

1. Binding assays are performed in a 1.5-ml polypropylene Eppendorf centrifuge tube.
2. Precentrifuge a solution containing 10 µM ^{125}I-labeled fibronectin at 11,750 rpm for 5 min in a Beckman (Palo Alto, CA) microfuge B to remove potential aggregate.
3. Add 110 µl buffer or competing ligand to the Eppendorf tube followed by 50 µl of the ^{125}I-labeled fibronectin.
4. Add 20 µl platelet suspension and equilibrate at 37° in a water bath for 10 min.
5. Add 20 µl of the platelet stimulus (e.g., 10 units/µl human α-thrombin).
6. Incubate at 37° for desired time.
7. At selected time points triplicate 50-µl aliquots are layered onto 300 µl of 20% sucrose (ultrapure; Schwarz/Mann, Orangeburg, NY) in a 400-µl microfuge tube (West Coast Scientific, Emoryville, CA).
8. Centrifuge at 11,750 g for 5 min at room temperature.
9. The tube tips can then be amputated by slicing them with a razor blade, and counted in a γ scintillation spectrometer.

Several points are worthy of note. First, it is important to use the platelets within 2 hr of their preparation, as their capacity to respond to thrombin stimulation declines rapidly once the cells are washed. In con-

---

[15] M. H. Ginsberg, L. Taylor, and R. Painter, *Blood* **55**, 661 (1980).

trast, the platelets are stable in platelet-rich plasma, prior to washing, for at least 12 hr. In addition, if there is variability in replicate determinations, several potential errors should be explored. First, it is possible that platelets may be aggregating in the tubes. This problem can be readily addressed by reducing the platelet concentration, or avoiding agitation during incubations. A second potential problem is failure to preclear all the aggregates from the radiolabeled fibronectin solution. This should become apparent if platelet-free controls are included in the binding assay. Another source of variability may be a problem with amputating the tube tips. To avoid this, fresh razor blades should be used, and frequently replaced. In addition, an effort should be made to slice the tube tip just above the visible platelet pellet.

In addition to the platelet-free control alluded to above, other controls routinely included are those in which no stimulus is added, and those in which a 100-fold molar excess of unlabeled fibronectin is added as competing ligand.

Data Analysis. Binding data can be expressed as counts per minute bound, or picograms bound per specified number of platelets. This is readily calculated from the known specific activity of the ligand. Another popular method of expressing such data is in molecules per platelet, which is readily calculated from the approximate molecular weight for fibronectin of 4.5×10^5. In each case we routinely subtract the nonsaturable binding estimated by addition of greater than 100-fold molar excess of unlabeled fibronectin to the reaction mixture.

Analysis of the binding parameters generated from binding isotherms can utilize any of the popular means of expressing this data. It is important to recognize that fibronectin binding to platelets has not been established to obey simple mass action and, therefore, estimated parameters such as K_d and number of sites are only descriptive approximations. In the past, we have employed Scatchard plots to analyze such isotherms. In this case, it is advisable to manually subtract the nonsaturable binding from each data point. An alternative to this is to analyze the data utilizing nonlinear least-squares curve fitting (LIGAND program of Munson and Rodbard).[16] This program assumes simple mass action, binding to a finite number of independent binding sites, and that the nonsaturable binding component is a constant fraction of the free ligand. Fibronectin binding isotherms readily fit a single-site binding model, and estimates of K_d and numbers of sites confirming our published values with Scatchard plots have been obtained. This particular method is advantageous in the analysis of treatments thought to affect affinity or number of binding sites, because statisti-

[16] P. J. Munson and D. Rodbard, *Anal. Biochem.* **107**, 220 (1980).

cal estimation of the likelihood that a particular manuever affects either parameter can readily be obtained. Whatever the method of analysis of such binding isotherms, it is essential to present the raw binding data in addition to the transformed data.

Acknowledgment

Supported by Grants #HL28235, HL16411, and AM27214 from the NIH. This is publication #4376-IMM from the Research Institute of Scripps Clinic.

[28] Mathematical Simulation of Prothrombinase

By MICHAEL E. NESHEIM, RUSSELL P. TRACY, PAULA B. TRACY, DANILO S. BOSKOVIC, and KENNETH G. MANN

Introduction

Prothrombinase is a multicomponent enzyme complex involved in blood coagulation. It comprises a serine protease (factor Xa), a cofactor (factor Va), Ca^{2+}, and a catalytic surface.[1,2] In model systems *in vitro* the surface consists of negatively charged phospholipid vesicles, whereas the presumed surface *in vivo* consists of a component or components of the platelet. In model systems, as well as on platelets, the complex consists of a noncovalent but tightly bound complex of factor Xa and factor Va in which factor Va provides the equivalent of a surface receptor for factor Xa. These components together efficiently catalyze the proteolytic activation of the zymogen prothrombin to the blood-clotting enzyme thrombin. Because prothrombin is also a surface (phospholipid) binding protein, the reaction can be best described as a surface-dependent event, even though factor Xa alone will slowly catalyze prothrombin activation in solution in the absence of either factor Va, Ca^{2+}, or the catalytic surface. However, it does so at a rate about five orders of magnitude slower than the complete complex.[3-5]

[1] K. G. Mann, B. H. Odegaard, S. Krishnaswamy, P. B. Tracy, and M. E. Nesheim, in "Proteases in Biological Control and Biotechnology," p. 235. Alan R. Liss, New York, 1987.
[2] M. E. Nesheim, *Surv. Synth. Pathol. Res.* **31,** 219 (1984).
[3] M. E. Nesheim, J. B. Taswell, and K. G. Mann, *J. Biol. Chem.* **254,** 10952 (1979).
[4] N. H. Kane, M. J. Lindhout, C. M. Jackson, and P. W. Majerus, *J. Biol. Chem.* **255,** 1170 (1980).

The rate of enhancement that occurs by virtue of the surface-dependent reaction has been rationalized quantitatively and has been attributed to both surface condensation of components to achieve high location concentrations of substrate, and a 3000-fold enhancement of k_{cat} evoked in some manner by factor Va.[6,7]

The binding interactions between phospholipid vesicles and prothrombin, factor X (Xa), or factor Va have been characterized,[8] as have been the interactions of factor V and factor Va with bovine platelets.[9] In addition, coordinated interactions of factor Va and factor Xa with platelets have been determined.[10] The K_m and k_{cat} values for various combinations of Ca^{2+}, factor Va, factor Xa, and phospholipid vesicles as catalysts of prothrombin activation also have been determined.[3,5]

The binding parameters have allowed proposal of a structure for prothrombinase, and the kinetic parameters have provided insight into the functional consequences of the assembly of prothrombinase components. The binding and kinetic parameters together have allowed the generation of a first-generation computer model that simulates well the properties of prothrombinase (as assembled on phospholipid vesicles) over a wide range of concentrations of substrate, enzymatic components, and phospholipid.

The application of this model to prothrombinase assembled on phospholipid vesicles has successfully reproduced the observed behavior of the enzymatic complex. It has quantitatively predicted, for example, apparent inhibition by excess phospholipid or enzyme, the response of the reaction to various levels of Ca^{2+}, and a variable apparent K_m that depends on the concentration of phospholipid.[7,11]

The model has not been applied and tested with prothrombinase assembled on platelets because much of the binding and kinetic data required for its application have not been reported. The concepts of the model, however, are presumed applicable to prothrombinase assembled on platelets as well as phospholipid vesicles.

This chapter will provide an exposition of the approach to constructing a model of prothrombinase and, by inference, analogous clotting reactions.

---

[5] J. Rosing, G. Tans, J. Govers-Riemslag, R. Zwaal, and H. Hemker, *J. Biol. Chem.* **255**, 274 (1980).
[6] M. E. Nesheim, S. Eid, and K. G. Mann, *J. Biol. Chem.* **256**, 9874 (1981).
[7] M. E. Nesheim, R. P. Tracy, and K. G. Mann, *J. Biol. Chem.* **259**, 1447 (1984).
[8] J. W. Bloom, M. E. Nesheim, and K. G. Mann, *Biochemistry* **18**, 4419 (1979).
[9] P. B. Tracy, J. Peterson, F. W. McDuffie, and K. G. Mann, *J. Biol. Chem.* **254**, 10354 (1979).
[10] P. B. Tracy, M. E. Nesheim, and K. G. Mann, *J. Biol. Chem.* **256**, 743 (1981).
[11] M. M. Tucker, M. E. Nesheim, and K. G. Mann, *Biochemistry* **22**, 4540 (1983).

It will be constructed not only for systems in which the substrate and enzyme share common binding sites (e.g., phospholipid vesicles), but also for systems in which distinct sites exist on the surface for the substrate and enzyme (e.g., platelets). The development is similar for both systems, but the expected behavior exhibits differences by which the two systems may be distinguished experimentally.

Concepts of Model

For the purposes of modeling, the enzyme, E, refers to either factor Xa alone or a complex of factor Xa and factor Va. The substrate prothrombin is denoted S. Both E and S interact with a negatively charged phospholipid vesicle or a cellular surface and thereby concentrate in the vicinity of the surface. Both E and S are considered distributed between the bulk fluid and the surface, and reactions to produce the product of the reaction, P, occur both in bulk fluid and on the surface. Surface-bound E and S are considered to be concentrated within a "shell" around the surface (phospholipid vesicle in model systems), which has a thickness of about 100–300 Å, and constitutes an element of volume in which E and S have formal, and relatively high, local concentrations.

The shell is a mental construct used to assign volume-based concentrations to surface-bound components. Although hypothetical, its existence is based on the observations made by quasi-elastic light scattering that the hydrodynamic radii of complexes of phospholipid vesicles saturated with prothrombin or factor V exceed those of vesicles alone by about 100 Å.[12,13]

The tenets of the model are depicted in Fig. 1. The catalytic surface (e.g., phospholipid vesicle) is depicted by the central sphere, surrounded by the hypothetic shell. E and S distribute between bulk fluid and the vesicle (shell) according to their equilibrium binding constants (K_1 and K_2) and binding stoichiometries (n_1, n_2; i.e., moles phospholipid monomers per mole protein). E and S both interact in typical enzyme–substrate fashion in both bulk fluid and within the shell to produce the product, according to respective K_m and k_{cat} values. At equilibrium the fluid phase concentrations of E and S are [E] and [S], whereas the local concentrations within the shell are [E'] and [S']. Rates within the two regions, per unit volume, are given by typical Michaelis–Menten equations. The rate of the combined reactions is simply the sum of the individual reactions in the bulk fluid and the "shells." Because factor Va binds phospholipid much more tightly than factor Xa,[8] the enzyme in solution is implicitly free factor

[12] T. K. Lim, V. A. Bloomfield, and G. L. Nelsestuen, *Biochemistry* **16**, 4177 (1977).
[13] D. L. Higgins and K. G. Mann, *J. Biol. Chem.* **258**, 6503 (1983).

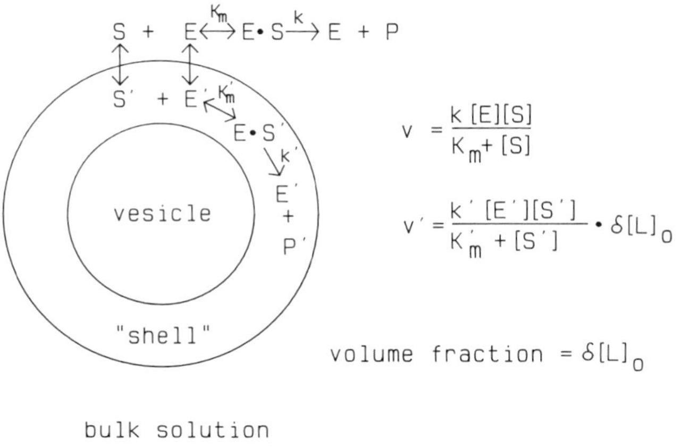

$$S + E \underset{K_m}{\leftrightarrow} E \cdot S \xrightarrow{k} E + P$$

$$S' + E' \underset{K_m'}{\leftrightarrow} E \cdot S' \xrightarrow{k'} E' + P'$$

$$v = \frac{k[E][S]}{K_m + [S]}$$

$$v' = \frac{k'[E'][S']}{K_m' + [S']} \cdot \delta[L]_0$$

volume fraction $= \delta[L]_0$

FIG. 1. Model of prothrombinase.

Xa, whereas that in the shell is a complex of factor Va and factor Xa. Because the latter is considerably more efficient than the former in prothrombin activation,[3,5] only the surface-dependent enzyme will contribute appreciably to the observed reaction under most conditions. In essence, the model to be developed in detail below is designed to determine the equilibrium distributions of E and S between bulk fluid and the region defined by the shell for any set of total concentrations of E, S, and L (surface, i.e., phospholipid concentration). From the equilibrium distributions, local and bulk concentrations of E and S then are calculated, and from these, local and bulk initial rates of conversion of S to P are calculated from the appropriate Michaelis–Menten-like rate equations.

The model will be developed for two kinds of surface. The first has indiscriminate affinities for both E and S, such that E and S can compete for available binding sites. Behavior of this kind is typified by phospholipid vesicles. Because the second kind of surface expresses distinct and exclusive sites or "receptors" for E and S, competition between E and S for the surface does not occur. Behavior of this kind is considered, as it might typify a cellular surface (such as that of the platelet) as the locus for interactions of E and S and subsequent production of the product of the reaction.

Equations of Binding Kinetics

According to the concepts presented in Fig. 1, the reaction between E and S occurs both in solution and in the shell, with initial rates given by Eqs. (1) and (2). Equation (1) applies in solution, whereas Eq. (2) applies in the shell. [E'] and [S'] imply local concentrations of E and S in the shell. Equation (2) is multiplied by the term $\delta[L]_0$ (the fraction of total volume occupied by the "second phase" determined by the shells) to express the rate with respect to the total volume of the system, rather than with respect to the volume of the shells. [E] and [E'] represent total enzyme in the two respective phases.

$$v = k[E][S]/(K_m + [S]) \tag{1}$$

$$v' = [k'[E'][S']/(K'_m + [S'])] \cdot \delta[L]_0 \tag{2}$$

If a fraction B of the substrate is bound, and a fraction β of the enzyme is bound, then bulk and local concentrations of these are given by Eqs. (3)–(6). In these equations $(1 - B)$ and $(1 - \beta)$ represent the fractions of substrate and enzyme not bound. $[E]_0$ and $[S]_0$ are the total concentrations of E and S, expressed with respect to the total volume of the system. Division of Eqs. (5) and (6) by $\delta[L]_0$ expresses the concentrations of bound enzyme and substrate, not with respect to the total volume, but rather in terms of local concentrations within the shell.

$$[S] = (1 - B)[S]_0 \tag{3}$$

$$[E] = (1 - \beta)[E]_0 \tag{4}$$

$$[S'] = B[S]_0/\delta[L]_0 \tag{5}$$

$$[E'] = \beta[E]_0/\delta[L]_0 \tag{6}$$

The values of B and β are needed to calculate formal concentrations of E and S. They are found through the following considerations.

Equations (7) and (8) represent the equilibrium interactions of S and E with receptors (or elements of the catalytic surface). Equations (9) and (10) are the mass action expressions for the interactions, where $[l_1]$ and $[l_2]$ represent the concentrations of unoccupied substrate and enzyme "receptors," respectively. K_1 and K_2 are the dissociation constants of the respective interactions.

$$S + l_1 \rightleftharpoons S \cdot l_1 \tag{7}$$

$$E + l_2 \rightleftharpoons E \cdot l_2 \tag{8}$$

$$[S][l_1] = K_1[S \cdot l_1] \tag{9}$$

$$[E][l_2] = K_2[E \cdot l_2] \tag{10}$$

Equations (9) and (10) may be divided, respectively, on both sides by $[S]_0$

and $[E]_0$, and thus can be expressed in terms of B and β, as in Eqs. (11) and (12).

$$(1 - B)[l_1] = K_1 B \tag{11}$$

$$(1 - \beta)[l_2] = K_2 \beta \tag{12}$$

The conservation equations for total receptor can now be used to eliminate $[l_1]$ and $[l_2]$ and thus solve for B and β. If the receptors for E and S are separate and distinct, two separate conservation equations exist, given by Eqs. (13) and (14).

$$[l_1]_0 = [l_1] + [S \cdot l_1] = [l_1] + B[S]_0 \tag{13}$$

$$[l_2]_0 = [l_2] + [E \cdot l_2] = [l_2] + \beta[E]_0 \tag{14}$$

In the terms on the far right-hand side of Eqs. (13) and (14), $B[S]_0$ and $\beta[E]_0$ have been used to express the concentrations of occupied receptors in terms of B, β, and total concentrations of S and E. The terms $[l_1]$ and $[l_2]$ are eliminated from Eqs. (13) and (14) through Eqs. (11) and (12), respectively, yielding the quadratic equations in B and β given by Eqs. (15) and (16).

$$0 = [S]_0 B^2 - ([S]_0 + [l_1]_0 + K_1)B + [l_1]_0 \tag{15}$$

$$0 = [E]_0 \beta^2 - ([E]_0 + [l_2]_0 + K_2)\beta + [l_2]_0 \tag{16}$$

The respective solutions for B and β are given in Eqs. (17) and (18) in terms of $[S]_0$, $[E]_0$, $[l_1]_0$, $[l_2]_0$, K_1, and K_2, i.e., in terms of readily measured parameters.

$$B = \frac{([S]_0 + [l_1]_0 + K_1) - \sqrt{([S]_0 + [l_1]_0 + K_1)^2 - 4[S]_0[l_1]_0}}{2[S]_0} \tag{17}$$

$$\beta = \frac{([E]_0 + [l_2]_0 + K_2) - \sqrt{([E]_0 + [l_2]_0 + K_2)^2 - 4[E]_0[l_2]_0}}{2[S]_0} \tag{18}$$

For any given choice of total concentrations of S, E, and the receptors l_1 and l_2, B and β are thus calculated from Eqs. (17) and (18). These values are then appropriately substituted into Eqs. (3)–(6) to determine local and bulk concentrations of E and S. The calculated concentrations are then inserted into Eqs. (1) and (2) to calculate simulated reaction rates from the reactions in solution and in the interface shell.

If E and S share the same surface, the binding of each is affected by the binding of the other. This situation is exemplified by prothrombinase assembled on phospholipid vesicles.[7] Under these conditions, the solution to the binding problem is more complex than it is in the case of separate receptors for E and S. In this case, a change in perspective from "sites" for S and E to phospholipid monomers is useful. If n_1 and n_2 monomers are nominally involved in creating sites for S and E (n_1 and n_2 are readily measured experimentally; see Ref. 8), then the concentration of sites l_1, expressed in units of the molar concentration of unoccupied monomers, [L], is $[l_1] = [L]/n_1$. Similarly, the concentration of unoccupied sites for E is given by $[l_2] = [L]/n_2$. The conservation of total phospholipid monomers then is expressed by Eq. (19). The terms n_1 and n_2 in Eq. (19) imply that $S \cdot l_1$ and $E \cdot l_2$ occupy n_1 and n_2 monomers, respectively.

$$[L]_0 = [L] + n_1[S \cdot l_1] + n_2[E \cdot l_2] \tag{19}$$

Equation (19) can be expressed further in terms of B, β, $[S]_0$, and $[E]_0$ as in Eq. (20).

$$[L]_0 = [L] + n_1 B[S]_0 + n_2 \beta [E]_0 \tag{20}$$

The mass action equations [Eqs. (11) and (12)], when rewritten in terms of [L], are given by Eqs. (21) and (22).

$$(1 - B)[L] = n_1 K_1 B \tag{21}$$

$$(1 - \beta)[L] = n_2 K_2 \beta \tag{22}$$

From these, β can be solved for in terms of B as in Eq. (23).

$$\beta = n_1 K_1 B / [n_2 K_2 + B(n_1 K_1 - n_2 K_2)] \tag{23}$$

In addition, [L] can be solved for in terms of B from Eq. (21), as in Eq. (24).

$$[L] = n_1 K_1 B / (1 - B) \tag{24}$$

The expressions for β and [L] of Eqs. (23) and (24) are then substituted into Eq. (20), resulting in Eq. (25), which expresses a relationship between B and the measurable variables $[S]_0$, $[E]_0$, $[L]_0$, K_1, K_2, n_1, and n_2.

$$0 = \frac{n_1 K_1 B}{(1 - B)} + n_1 [S]_0 B + \frac{n_1 n_2 K_1 [E]_0 B}{n_2 K_2 + B(n_1 K_1 - n_2 K_2)} - [L]_0 \tag{25}$$

Elimination of the terms in the denominator of Eq. (25) and rearrangement yields the cubic equation in B given by Eq. (26), with coefficients R, S, T, U, given by Eqs. (27)–(30). Note that R, S, T, and U contain only terms reflecting binding parameters and the initial concentrations of S, E, and L. Thus, each coefficient is experimentally assessable, and is constant for a particular choice of parameters.

$$0 = RB^3 + SB^2 + TB + U \tag{26}$$

$$R = n_1(n_2K_2 - n_1K_1)[S]_0 \tag{27}$$

$$S = n_1(n_1K_1 - 2n_2K_2)[S]_0 - n_1n_2K_1[E]_0$$
$$+ (n_1K_1 - n_2K_2)[L]_0 + n_1K_1(n_1K_1 - n_2K_2) \tag{28}$$

$$T = n_1n_2K_2[S]_0 + n_1n_2K_1[E]_0 + (2n_2K_2 - n_1K_1)[L]_0$$
$$+ n_1n_2K_1K_2 \tag{29}$$

$$U = -n_2K_2[L]_0 \tag{30}$$

Calculation of Binding Distributions of Substrate and Enzyme, Bulk and Local Concentrations of Substrate and Enzyme, and Initial Rate of Prothrombin Activation

For any given set of values of the initial concentrations of substrate, enzyme, receptors (or phospholipid), and values of K_1, K_2, k, k', K_m, K_m' and δ (plus n_1 and n_2 in the case of prothrombinase on phospholipid vesicles), the strategy to calculate binding distributions, bulk and local concentrations, and the initial rate of prothrombin activation involves (1) calculation of B and β [Eqs. (17) and (18) for discrete receptors, Eqs. (26) and (23) on vesicles], (2) calculation of bulk and local concentrations of E and S from values of B and β [Eqs. (3)–(6)], and (3) calculation of the rates of reactions in the bulk solution and the shell [Eqs. (1) and (2)].

If prothrombin activation on vesicles is modeled, 12 parameters are required to complete the calculations. They include binding constants and stoichiometries for the enzyme and substrate interactions with the receptors or phospholipid vesicles; initial concentrations of enzyme, substrate and phospholipid; k_{cat} and K_m values for the reactions both in solution and in the shells; and the conversion factor δ, which relates volume fraction of the shells to the phospholipid concentration. Published values of these parameters are available and are included in Table I. When discrete receptors for substrate and enzyme exist so that competitive binding does not occur, calculation of B and β (fractions of substrate and enzyme bound, respectively) is straightforward and is accomplished by insertion of the appropriate values into the right-hand sides of Eqs. (15) and (16).

When competition between S and E for the surface exists, such as occurs with phospholipid vesicles, a solution to the cubic equation [Eq. (26)] must be found. A simple procedure to approximate the solution of Eq. (26) to any arbitrary degree of precision involves binary search itera-

TABLE I
PARAMETERS OF BINDING AND KINETICS REQUIRED FOR SIMULATION OF
PROTHROMBINASE ON VESICLES[a]

Parameter	Value	Description	Ref.
K_1	2.3×10^{-6}	K_D for II, PCPS interactions, M	8
K_2	7.0×10^{-10}	K_D for Xa, Va, PCPS interaction, M	8
k	0.61	k_{cat} Xa, in solution, min^{-1}	5
k'	2100	k_{cat} Xa · Va · PCPS (shells), min^{-1}	5, 6
$[S]_0$	1.39×10^{-6}	Total initial concentration of II, M	—
$[E]_0$	1.0×10^{-8}	Total concentration of enzyme, M	—
$[L]_0$	1.5×10^{-5}	Initial PCPS concentration, M	—
n_1	104	Stoichiometry of PCPS, II interaction (mol PCPS/mol II)	8
n_2	91	Stoichiometry of PCPS, Va interaction (mol PCPS/mol Va)	8
K_m	1.31×10^{-4}	K_m in solution	5
K_m'	1.31×10^{-4}	K_m in "shells"	—
δ	30^b	Conversion factor for calculating volume fraction of shells from $[L]_0$, M^{-1}	—

[a] II, Prothrombin; PCPS, phosphatidylcholine/phosphatidylserine vesicles (3 : 1); Va · Xa, complex of factor Va and factor Xa. The choices of $[S]_0$, $[E]_0$, and $[L]_0$ are arbitrary. Those indicated are useful default values as the value of $[S]_0$ approximates plasma prothrombin concentrations, the value of $[E]_0$ is small compared to $[S]_0$, and $[L]_0$ is sufficient to maximize the rate with respect to phospholipid at these values of $[S]_0$ and $[E]_0$.

[b] The value of 30 is considerably greater than the value of 3.34 inferred by consideration of the difference between hydrodynamic radii of vesicles alone and vesicles saturated with protein (see text). A value for δ of 3.34 implies an increase of 100 Å, whereas δ = 30 implies an increase of about 330 Å. The effect of using 30 instead of 3.34 is to decrease values of calculated local concentrations by a factor of 3.34/30 = 0.11. Empirical observation[7] indicates a seemingly better fit of simulation to observation using the larger value of δ. This may reflect that the almost infinitely sharp gradient of concentration implied by the "shell" model unrealistically overestimates the gradient that likely exists in real systems.

tion with a desktop computer. The entries in Table II represent a bisection algorithm,[14] written in Microsoft BASIC for the Macintosh computer (Apple Computer Company, Cupertino, CA and included in the "Clotspeed" program, Ref. 7), that quickly performs the iteration. Because the expression on the right-hand side of Eq. (26) evaluates to zero only for the "correct" value of B, this value is approached by evaluating

[14] R. L. Burden, J. D. Faires, and A. C. Reynolds, "Numerical Analysis," 2nd ed., p. 21. Prindle, Weber and Schmidt, Boston, 1981.

TABLE II
Subroutine in BASIC for Estimation of B by Bisection Algorithm[a]

```
FindB:                  'Subroutine to estimate value of B
  Precision = .000001
  Border1 = 0: f1 = U
  Border2 = 1: f2 = R+S+T+U
  IF f1 = 0 THEN root = Border1: GOTO FinalRoot
  IF f2 = 0 THEN root = Border2: GOTO FinalRoot
  IF f1*f2 > 0 THEN PRINT "PROBLEM: NO roots or EVEN # of roots!": RETURN

Border:
  Border3 = (Border1+Border2)/2
  f3 = R*Border3*Border3*Border3 + S*Border3*Border3 + T*Border3 + U
  IF f3 = 0 THEN root = Border3: GOTO FinalRoot
  Change = ABS(Border3-Border1)
  If Change < Precision THEN root = Border3: GOTO FinalRoot
  If f3*f2 < 0 THEN Border1 = Border3: f1 = f3: GOTO Border
  Border2 = Border3: f2 = f3: GOTO Border

FinalRoot:
  B = root
RETURN
```

[a] The values of R, S, T, and U are calculated from the binding parameters of Table I according to Eqs. (27)–(30).

the right-hand side of Eq. (26) at two different selected values of B, and then determining whether a change in sign occurs in the value of the right-hand side of Eq. (26) for the two choices. If so, the correct value lies between the two choices. If not, the correct value lies outside the interval. Because the correct value ranges between zero and one, the search starts on the intervals 0 to 0.5 and 0.5 to 1.0. The interval on which a change in sign of the expression on the right-hand side of Eq. (26) occurs is then subdivided and the search is continued. This procedure is continued until the interval on which the correct value of B lies is arbitrarily small. In the algorithm of Table II the limit of precision is set at 1 part in 10^6.

Alternatively, it is possible to calculate the value(s) of B directly, by an application of the solution formulas for a general cubic equation.[15] The advantages of this direct analytic method are exact value(s) of B as well as faster execution, which is most noticeable when doing repeated evaluations for series of starting parameters. The advantage of the bisection algorithm, however, is that it is simpler to implement.

[15] S. M. Selby, "CRC Standard Mathematical Tables," 20th ed., p. 103, Chem. Rubber Publ. Co., Cleveland, Ohio, 1972.

TABLE III
Parameters Calculated from Variables of Table I and Solution to Binding Equations on Simulation of Prothrombinase on Vesicles

Parameter	Description	Equation
$p(1)$	Fraction of substrate bound	(26)
$p(2)$	Fraction of enzyme bound	(23)
$p(3)$	Bulk concentration of substrate	(3)
$p(4)$	Bulk concentration of enzyme	(4)
$p(5)$	Local concentration of substrate	(5)
$p(6)$	Local concentration of enzyme	(6)
$p(7)$	Rate of solution reaction	(1)
$p(8)$	Rate of local reaction	(2)
$p(9)$	Sum of $p(7)$ and $p(8)$	(1) + (2)

Once B is found, all other desired parameters are calculated as described in the first paragraph of this section. Useful parameters that can be calculated from the solution to Eq. (26) and subsequent calculations are indicated in Table III. These include fractions of enzyme and substrate bound, bulk and local concentrations of the enzyme and substrate, and initial rates in the bulk solution and in the local shells.

Magnitude of δ

The term δ of Eqs. (5), (6), and Table I, when multiplied by the concentration of the components of the catalytic surface, yields the fraction of total volume of the system occupied by the second phase, or shell (in the case of vesicles). It thus, in effect, determines the magnitude of the increase in formal concentration of E and S that occurs on their binding to the catalytic surface. For example, if the volume of the second phase is 0.01% of the total volume, and 1.0% of the available substrate is bound, its formal concentration in the shell will be approximately 100-fold greater than the bulk concentration. That is, $[S'] = 0.01[S]_0/0.0001 = 100[S]_0$ from Eq. (5).

Because the value of δ is based on a hypothetical construct (albeit one motivated by experimental observation), it depends on the assumptions and observations on which the construct is based. If the vesicles are composed of phosphatidylcholine and phosphatidylserine (3:1) of 160-Å radius, the value of δ is determined by the magnitude of the increase in hydrodynamic radius ($\sim$100 Å) that occurs when the vesicles are saturated with either prothrombin or factor V, plus the observation that vesicles of this size are composed of about 10,000 phospholipid monomers.[12] A shell defined by two concentric spheres of 160 and 260 Å has a formal volume $V = 5.6 \times 10^{-20}$ liters/vesicle. The volume fraction of the shells is thus given by: $\delta[L]_0 = V[\text{vesicles}]6.02 \times 10^{23} = V[L]_0 \times 6.02 \times 10^{23}/10,000$. Here [vesicles] and $[L]_0$ are molar concentrations of vesicles and phospholipid monomers, respectively. Thus, δ is the proportionality constant relating the concentration of phospholipid to the volume fraction comprising the second phase or shells. In this example, δ has a value of $3.34\ M^{-1}$ when the concentration of phospholipid is expressed as moles phospholipid monomer per liter.

Attributes, Uses, and Limitations of Model

One of the strengths of the model is that it is constructed around parameters that are determined independently of the model. The first 10 parameters of Table I, for example, are either initial concentrations of components E, S, and L, or parameters of binding and kinetics determined by independent measurements. Therefore, the model and its ability to simulate with reasonable fidelity experimental observations relies on the validity of its concepts and assumptions, and not on the arbitrary adjustment of parameters to reconcile the model with experimental data.

Another strength is that the model is based on correlations of binding interactions with expressed functions and straightforward notions of local and bulk concentrations. It thereby offers rationalizations of observed rates in terms of binding and concentrations. Because parameters can be varied mathematically at will, the influences on the surface of binding properties of the individual components can be investigated and rationalized theoretically, and then tested experimentally.

Another strength of the model is that its concepts are not unduly abstract, thereby providing a good theoretical construct for the conceptualization of prothrombinase. Thus, it serves as a useful tool with which to generate hypotheses for further experimentation and intuitive explanations for observed functional properties. If prothrombin activation by a system comprising five components (prothrombin, Ca^{2+}, phospholipid, factor Va, and factor Xa) were to be characterized experimentally by systematic changes in the levels of each of the components between arbitrary limits in 10 increments, 100,000 observations minimally would be required. The model therefore can introduce a substantial degree of efficiency in choosing from among numerous experimental conditions those likely to be most illuminating.

The model has proven useful to date in rationalizing numerous, seemingly unusual, properties of prothrombinase. For example, the reaction rate is profoundly retarded by excess levels of either phosphatidylcholine/ phosphatidylserine (PCPS) or factor Xa · Va. The apparent inhibition by excess PCPS can be rationalized simply as a manifestation of "dilution" of surface-bound substrate in excess volume of the shells. Similarly, inhibition by excess enzyme can be rationalized on the basis of competition between enzyme and substrate for limited catalytic surface. Excess enzyme, in effect, lowers the local concentration of substrate as the surface approaches saturation with enzyme. The net result of this is decreased rates of reaction. The model also rationalizes the dependence of apparent K_m on the concentrations of PCPS.[5,16] As the phospholipid concentration increases, the volume fraction of the second phase, or shells, increases; thus, higher levels of total substrate are required to obtain a given local concentration (which determines the rate) and higher apparent K_m values are measured experimentally.

Generally, the model has proven very useful in rationalizing numerous properties of the prothrombinase-catalyzed activation of prothrombin. It nonetheless is clearly only an approximation of the real system. The concept of the shell, for example, implies a rectangular concentration gradient of substrate and enzyme with respect to distance from the catalytic surface. This clearly is only a crude approximation of the gradients that might exist in reality. In addition, the catalyst, E, is considered a single entity, whereas in the real system it consists of two components, factor Xa and factor Va. Prothrombin activation is considered a single event, whereas in reality two proteolytic cleavages are required to convert the zymogen to thrombin. The role of Ca^{2+} in the reactions is only implicit, although it could be modeled conceivably by changes in the affinity constants K_1 and K_2. The model implies equilibrium between substrate and enzyme with the surface. Under the real conditions in which substrate turnover occurs, however, transport of substrate to, and release of product from, the catalytic surface may contribute rate-determining features of the reaction. In addition, the model does not include feedback reactions (activation of factor V, or catalysis of cleavage of fragment 1 by thrombin), and it does not include potential effects of products on the reactions.

Nonetheless, in spite of these limitations, the model has proven useful and should continue to be a useful tool for further investigation not only of the properties of prothrombinase and other complex enzymes involved in coagulation, but also of other surface-dependent enzymatic systems.

[16] O. Malhotra, M. E. Nesheim, and K. G. Mann, *J. Biol. Chem.* **260,** 279 (1985).

[29] Platelet Factor Xa Receptor

By PAULA B. TRACY, MICHAEL E. NESHEIM, and
KENNETH G. MANN

Introduction

The serine protease factor Xa is the enzymatic constituent of the multicomponent coagulation complex, prothrombinase. Factor Xa, the protein cofactor factor Va, Ca^{2+}, and an appropriate "surface membrane" or "receptor" are required for proper assembly of the protein components and comprise the functional catalytic unit.[1,2] *In vivo,* the platelet is considered the most likely source of the "surface receptor" and constitutes the site at which the catalyst assembles and prothrombin activation occurs in the normal hemostatic mechanism.[3-7] Initial studies aimed at the elucidation of the functional interaction of the prothrombinase complex components with platelets were performed by Miletich and colleagues[4,5] working with human proteins and platelets, and by Dahlbäck and Stenflo[8] working in a homologous bovine system. Both groups of investigators demonstrated that platelet activation was absolutely required to support a platelet–factor Xa interaction; however, it was not clear whether activation was required to expose a "cryptic" factor Xa membrane receptor or, perhaps, release a sequestered platelet component. Miletich and colleagues provided evidence suggesting that factor V or Va formed a major component of the platelet factor Xa receptor. Their studies showed reduced binding of factor Xa to platelets in the presence of alloantibodies to factor V[5] and in patients congenitally deficient in factor V.[9] Additional support was provided by Stenflo and Dahlbäck,[10] who reported that the platelet factor Xa receptor was destroyed by activated protein C, a potent

---

[1] K. G. Mann, this series, Vol. 45, p. 123.
[2] J. W. Suttie and C. M. Jackson, *Physiol. Rev.* **57**, 1 (1977).
[3] P. N. Walsh, *Blood* **43**, 597 (1974).
[4] J. P. Miletich, C. M. Jackson, and P. W. Majerus, *Proc. Natl. Acad. Sci. U.S.A.* **74**, 4033 (1977).
[5] J. P. Miletich, C. M. Jackson, and P. W. Majerus, *J. Biol. Chem.* **253**, 6908 (1978).
[6] P. B. Tracy, M. E. Nesheim, and K. G. Mann, *J. Biol. Chem.* **256**, 743 (1981).
[7] W. H. Kane and P. W. Majerus, *J. Biol. Chem.* **257**, 3963 (1982).
[8] B. Dahlbäck and J. Stenflo, *Biochemistry* **17**, 4938 (1978).
[9] J. P. Miletich, D. W. Majerus, and P. W. Majerus, *J. Clin. Invest.* **62**, 824 (1978).
[10] J. Stenflo and B. Dahlbäck, *in* "The Regulation of Coagulation" (K. G. Mann and F. B. Taylor, eds.), p. 225. Elsevier-North-Holland, New York, 1980.

proteolytic inactivator of factor Va. Thus, the hypothesis was put forth that factor Va is the platelet receptor for factor Xa.

Because purified, homogeneous bovine factor V could be prepared in our laboratory,[11,12] we tested this hypothesis by direct measurements of the binding interactions of factors V, Va, and Xa with bovine platelets and established a direct relationship between factor Va and factor Xa binding to the platelet surface. Initially, the binding parameters governing the interactions of factors V and Va with unstimulated or thrombin-stimulated platelets were assessed using radioiodinated proteins and analyses of equilibrium binding isotherms.[13] The combined interactions of factors Va and Xa at the surface of unstimulated platelets were then investigated by direct, simultaneous, equilibrium binding measurements of both ligands to platelets using a dual-radiolabel technique.[6] Results obtained from the equilibrium binding experiments were verified using analyses of the kinetics of prothrombin activation. The latter studies allowed interpretation of the binding parameters that govern the "functional" factor Va–factor Xa–platelet binding sites involved in prothrombin activation.[6] The combined data indicate that factor Va forms the receptor for factor Xa on the surface of unstimulated platelets and results in a 1:1 factor Va–factor Xa, Ca^{2+}-dependent, stoichiometric complex of high affinity (apparent $K_D \approx 10^{-10}$ M).

In additional studies we demonstrated that factor Va bound to platelets is cleaved by a platelet-associated protease, as well as by factor Xa during complex formation.[14] Neither of these proteolytic events appears to have an effect on the cofactor activity of factor Va in prothrombin activation. Finally, we were able to demonstrate that only one peptide, the COOH terminal-derived component E of the two-subunit factor Va molecule (component D, 94,000 Da, and component E, 74,000 Da), is required to mediate both the binding of factor Va to the platelet surface and the subsequent interaction of factor Xa with platelet-bound factor Va.[15]

We have employed analyses of the kinetics of prothrombin activation to infer the functional parameters governing *human* factor Va and factor Xa binding.[16] Because normal human platelets contain a large intracellular

---

[11] M. E. Nesheim, K. H. Myrmel, L. Hibbard, and K. G. Mann, *J. Biol. Chem.* **254**, 508 (1979).
[12] M. E. Nesheim and K. G. Mann, *J. Biol. Chem.* **254**, 1326 (1979).
[13] P. B. Tracy, M. E. Nesheim, J. M. Peterson, F. C. McDuffie, and K. G. Mann, *J. Biol. Chem.* **254**, 10354 (1979).
[14] P. B. Tracy, M. E. Nesheim, and K. G. Mann, *J. Biol. Chem.* **258**, 662 (1983).
[15] P. B. Tracy and K. G. Mann, *Proc. Natl. Acad. Sci. U.S.A.* **80**, 2380 (1983).
[16] P. B. Tracy, L. L. Eide, and K. G. Mann, *J. Biol. Chem.* **260**, 2119 (1985).

store of factor V,[17] kinetic studies were performed with platelets obtained from a factor V-deficient individual totally devoid of both plasma and platelet factor V activity and antigen.[17] The factor Xa binding parameters obtained were verified using platelets from normal persons. These data parallel studies in the bovine system and indicate that the formation of a functional prothrombinase complex on the human platelet surface is mediated by a membrane-bound, stoichiometric complex of factor Va and factor Xa (1 : 1) of high affinity (apparent $K_D \approx 10^{-10} M$). The methodology and protocols developed and used to define these binding events are the subject of this report.

Binding Interactions of Bovine Factors Va and Xa with Bovine Platelets

Materials

Homogeneous, single-chain bovine factor V (M_r 330,000; $E^{1\%}_{280\,nm} = 9.6$) is isolated and assayed for procoagulant activity as described by Nesheim et al.[11,12] Activation to factor Va is accomplished by addition of bovine α-thrombin to a final concentration of 2 NIH units/ml ($1.9 \times 10^{-8} M$) and incubation at 37° for 3 min.[12] Bovine α-thrombin (M_r 37,400; $E^{1\%}_{280\,nm} = 19.5$) is prepared as described by Lundblad et al.[18] Bovine factor X (M_r 55,000; $E^{1\%}_{280\,nm} = 12.4$) and bovine prothrombin (M_r 72,000; $E^{1\%}_{280\,nm} = 14.4$) are prepared and assayed for procoagulant activity as described by Bajaj and Mann.[19] Factor Xa is prepared using electrophoretically homogeneous factor X-activator from Russell's viper venom[20] immobilized on activated CH-Sepharose (Sigma Chemical Co., St. Louis, MO) and equilibrated in 0.02 M Tris-HCl, 0.15 M NaCl, 2.5 mM Ca^{2+}, pH 7.4. Bovine platelets are isolated, washed free of plasma, and characterized as described previously.[13]

Apiezon oil can be obtained from J. B. Biddle Co. (Blue Bell, PA). Tris base, bovine serum albumin, fat-free bovine serum albumin, and Tagit reagent [N-succinimidyl 3-(4-hydroxyphenyl)propionate] are obtained from Sigma. Carrier-free sodium [^{125}I]- and [^{131}I]iodide are obtained from New England Nuclear Co. (Boston, MA). Bolton–Hunter reagent for radioiodination of proteins is prepared as described.[21] The potent, reversible thrombin inhibitor, dansylarginine N-(3-ethyl-1,5-pentanediyl)amide

---

[17] P. B. Tracy, L. L. Eide, E. J. W. Bowie, and K. G. Mann, *Blood* **60**, 59 (1982).
[18] R. L. Lundblad, R. C. Uhteg, C. N. Vogel, H. S. Kingdon, and K. G. Mann, *Biochem. Biophys. Res. Commun.* **66**, 482 (1975).
[19] S. P. Bajaj and K. G. Mann, *J. Biol. Chem.* **248**, 7729 (1973).
[20] W. Kisiel, M. A. Hermodson, and E. W. Davie, *Biochemistry* **15**, 4901 (1976).
[21] A. E. Bolton and W. M. Hunter, *Biochem. J.* **133**, 529 (1973).

[(DAPA), $K_D = 1.43 \times 10^{-8}$ M] is prepared as described by Nesheim et al.[22]

Radiolabeling of Proteins

Bovine factor V is radiolabeled[6,23] with ^{125}I or ^{131}I using freshly prepared Bolton–Hunter reagent at a ratio of 2 mCi iodine/mg protein. Factor V (1.0 ml, 0.5 mg/ml) is dialyzed against 0.1 M sodium borate (pH 8.5) for 3 hr and then treated with 1 mCi of Bolton–Hunter reagent for 20 min at 0°. The iodination reaction is quenched with 50 μl of 2 M glycine in 0.1 M sodium borate, pH 8.5. The radioiodinated factor V is diluted with 3 vol of Tris–borate buffer (0.01 M Tris–borate, 1 mM $CaCl_2$, pH 8.3) and applied to a 1.0-ml Cibacron Blue Sepharose column (Pharmacia, Piscataway, NJ) equilibrated with the same buffer to separate the iodinated factor V from other products of the conjugation reaction. The column is washed with at least 30 ml of Tris–borate buffer, followed by 30 ml of the same buffer containing 0.1 M NaCl, until the majority of free iodide has been washed from the column. The radioiodinated factor V is eluted with the Tris-borate buffer containing 0.3 M NaCl. Approximately 50–60% of the protein is recovered in three 2.0-ml fractions, which are then pooled, dialyzed against 50% glycerol (v/v), and stored at $-20°$. Following iodination and rechromatography, the factor V preparations have specific radioactivities ranging from 250 to 1000 cpm/ng (0.03–0.12 mol iodine/mol protein), with greater than 98% of the radioactivity being trichloroacetic acid (TCA)-precipitable.

Factor V has the same electrophoretic mobility on sodium dodecyl sulfate (SDS) gel electrophoresis prior to and after labeling with at least 85% of the total isotope corresponding to a protein band of homogeneous, single-chain factor V. Even though the Bolton–Hunter method is a relatively mild iodination procedure, some factor Va peptides are always produced during the process of labeling. They can be visualized by autoradiography of the dried gels as detailed below. These peptides are removed by immunoadsorption chromatography of the labeled preparations using an immobilized anti-bovine factor V monoclonal antibody[24] that does not bind the single-chain molecule in the presence of Ca^{2+} (20 mM) but rather

---

[22] M. E. Nesheim, F. G. Prendergast, and K. G. Mann, *Biochemistry* **18**, 996 (1979).
[23] M. E. Nesheim, J. A. Katzmann, P. B. Tracy, and K. G. Mann, this series, Vol. 80, p. 249.
[24] W. B. Foster, M. M. Tucker, J. A. Katzmann, and K. G. Mann, *J. Biol. Chem.* **258**, 5608 (1983).

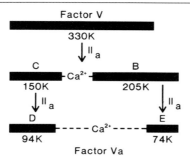

FIG. 1. Schematic representation of the thrombin-catalyzed cleavage of bovine factor V. The individual components shown represent the precursors of as well as the two-subunit factor Va molecule. The letters above the corresponding bars identify each of the components, and the apparent molecular weights are indicated below each bar. Those components that remain associated due to an integral metal ion are connected by dotted lines. Components D and E (94K and 74K, respectively) comprise the two-subunit factor Va molecule. (From Ref. 25.)

binds the factor Va peptides of 205,000, 94,000, and 74,000 Da (Fig. 1).[25] This additional chromatography step results in radioiodinated factor V preparations that retain 90–100% of their cofactor activity and in which greater than 95% of the radioactivity is associated with the single-chain protein.

^{125}I-Labeled factor Va is prepared by thrombin activation of the radioiodinated factor V preparations. ^{125}I-Labeled factor V (6×10^{-8} M) is incubated with α-thrombin [2 NIH units (U)/ml] for 3 min at 37° and then assayed to determine whether activation is complete or if additional incubation time is required. Following complete activation, DAPA (30 μM) is added to inhibit additional thrombin proteolytic activity. Factor Va solutions are stored on ice and used within 3 hr of preparation. Autoradiographic visualization and densitometric scanning of the radioiodinated factor Va preparations following electrophoresis in SDS indicate that at least 90% of the isotope resides in the two peptides comprising the two-subunit factor Va molecule, component D (94,000 Da) and component E (74,000 Da), as shown in Fig. 1 and indicating that the molar ratios of iodine present in both the factor V and factor Va preparations are nearly identical. Consequently, 1 mol of radioiodinated factor V is presumed to yield 1 mol of radioiodinated factor Va, and no correction for radioactivity not associated with factor Va is required.

^{125}I-Labeled factor X is also prepared using Bolton–Hunter reagent as

[25] D. L. Higgins and K. G. Mann, *J. Biol. Chem.* **258**, 6503 (1983).

detailed above for factor V; however, a ratio of 4 mCi of ^{125}I per milligram of protein is used.[6] ^{125}I-Labeled factor X (0.5 mg/ml; 1.0 ml) is separated from other labeled products of the conjugation reaction by gel filtration on Sephadex G-25 (10 ml) equilibrated in 0.02 M Tris-HCl, 0.15 M NaCl, pH 7.4. Normally, at least 90% of the protein is recovered, dialyzed against 50% glycerol (v/v), and stored at $-20°$. The ^{125}I-labeled factor X preparations have specific radioactivities of 470–1700 cpm/ng (0.01–0.04 mol ^{125}I/mol of protein), and retain 85–100% of their coagulant activity. ^{125}I-Labeled factor Xa (3.6×10^{-7} M) is prepared using immobilized RVV-X[26] equilibrated in 0.02 M Tris-HCl, 0.15 M NaCl, 2.5 mM Ca^{2+}, stored on ice, and used within 3 hr of preparation.

^{125}I-Labeled Factor V and Factor Va Platelet Binding Measurements

The binding of ^{125}I-labeled factor V and factor Va to bovine platelets[13] is measured by centrifugation through oil as described by Miletich and co-workers for factor Xa binding.[4,5,9] Experiments were designed to answer the following questions: (1) Do factor V and factor Va bind to platelets, and if so, what are the equilibrium binding parameters governing these interactions? (2) Do factor V and factor Va interact similarly with platelets? (3) Is platelet activation required to mediate these binding interactions? To address these questions, multiple binding experiments are performed at ambient temperature, 22–24°, using three different platelet preparations[26]: (1) washed, unstimulated platelets, (2) platelets intentionally activated with α-thrombin in the absence of stirring to prevent platelet aggregation (1 NIH U thrombin/1 $\times$ 10^8 platelets for 10 min, 22°, followed by the addition of 30 μM DAPA), and (3) platelets collected, washed, and assayed in the presence of prostaglandin E (PGE$_1$, 5 μM) to reduce the likelihood of any platelet activation. When factor Va binding is measured with unstimulated platelets, DAPA (30 μM) is included in the reaction mixture to ensure that the catalytic amount of thrombin present from the stock solution of added factor Va does not activate the platelets.[13]

Reaction mixtures (2 ml) contain 1 $\times$ 10^8 platelets/ml and varying amounts of the radiolabeled protein in 0.02 M Tris-HCl, 0.14 M NaCl, 0.1% (w/v) glucose, 0.5% (v/v) fat-free bovine serum albumin, 2.5 mM Ca^{2+}, pH 7.4 (binding study buffer). Binding measurements are made following a 30-min incubation, as initial measurements of the kinetics of factor Va binding indicated that, even though initial binding is very rapid, steady state binding is not reached until 30 min has elapsed. Duplicate 0.5-ml aliquots of the reaction mixture are carefully layered onto 0.5 ml

---

[26] W. Kisiel, M. A. Hermodson, and E. W. Davie, *Biochemistry* **15**, 4901 (1976).

of an oil mixture consisting of 1 part Apiezon oil and 9 parts n-butyl phthalate in 1.5-ml Eppendorf conical centrifuge tubes. Following centrifugation at 12,000 rpm for 2 min in a Brinkmann (Westbury, NY) microcentrifuge, the supernatant is removed for determination of unbound radioiodide activity in a Beckman (Palo Alto, CA) γ counter. The remaining oil layer is carefully removed and the bottom of the tube, containing the platelet pellet and bound radioiodinated protein, is cut off and the radioactivity measured. The sum of the radioactivity recovered in the supernatant and pellet should be equal to the amount of radioactivity present in a 0.5-ml aliquot of the unseparated reaction mixture. With knowledge of the specific radioactivity (cpm/mol) of the radiolabeled protein, an accurate calculation of the actual ligand concentration in the reaction mixture is made by determining the radioactivity present in multiple aliquots of the unseparated reaction mixture.

Nonspecific binding due to isotope entrapment is estimated in parallel reaction mixtures containing a 100-fold molar excess of unlabeled protein. Specific binding is obtained by subtraction of nonspecifically bound isotope from the total isotope for each reaction mixture.

Initial experiments are done to determine the time dependence and reversibility of the factor V and factor Va interactions with platelets. Because reversible and saturable binding can be achieved with both ligands, steady state factor V and factor Va binding is measured by employing a concentration range from 3×10^{-11} to 6×10^{-9} M factor V or factor Va. Specific factor V and factor Va platelet binding data are subjected to Scatchard analysis[27] using Eq. (1):

$$b/f = (NP_0/K_D) - (1/K_D) b[V]_0 \qquad (1)$$

in which b and f are fractions of factor V (Va) bound and free, respectively, N is the number of factor V (Va) binding sites/platelet, P_0 is the molar concentration of platelets, K_D is the dissociation constant for the factor V (Va) binding site interaction, and $[V]_0$ is the nominal concentration of factor V (Va). The dissociation constant and the number of binding sites are calculated from the slope and vertical intercept, respectively.

The binding of factor Va to platelets was saturable, exchangeable, of high affinity, and independent of the state of platelet activation. As shown in Table I, Scatchard analyses of the individual binding isotherms have been interpreted to indicate the existence of two classes of binding sites. The binding data were also analyzed and fitted for two classes of binding sites as described by Klotz and Hunston.[28] Thrombin pretreatment of the

---

[27] I. M. Klotz and D. L. Hunston, *Biochemistry* **10**, 3065 (1971).
[28] P. B. Tracy and K. G. Mann, *Fed. Proc., Fed. Am. Soc. Exp. Biol.* **42**, 31 (1983).

TABLE I
BINDING PARAMETERS OF FACTOR V AND Va INTERACTION WITH PLATELETS

	Factor Va[a]				Factor V[b]	
	High-affinity sites		Low-affinity sites			
Platelet	K_D ($M \times 10^{10}$)	n^c	K_D ($M \times 10^9$)	n^c	K_D ($M \times 10^9$)	n^c
Unstimulated	4.0 ± 2.1	837 ± 48	3.6 ± 0.8	3403 ± 30	2.8 ± 1.5	816 ± 92
Thrombin activated	3.4 ± 1.7	827 ± 230	2.9 ± 1.3	3496 ± 104	3.6 ± 2.4	919 ± 283

[a] Average of four separate Scatchard analyses of binding studies done on different days and shown ± SD, indicates two classes of factor Va platelet-binding sites.
[b] Average of five separate Scatchard analyses of binding studies done on different days and shown ± SD, indicates a single class of factor V platelet-binding sites.
[c] n, Number of ligand-binding sites per platelet.

platelets was not required for, or had any effect on, factor Va binding, because thrombin-activated platelets (as well as platelets collected and assayed in the presence of PGE_1), showed factor Va binding characteristics identical to those of unstimulated platelets (Table I).

The binding of factor V to unstimulated and thrombin-activated platelets, as well as PGE_1-treated platelets, indicated that platelets most likely possess a single class of factor V binding sites (Table I). Factor Va can displace all bound factor V, whereas factor V will displace only the less tightly bound factor Va. Our data indicate that both factor V and factor Va will bind to unstimulated platelets; however, platelets express a binding domain selective for factor Va.[13]

^{125}I-Labeled Factor Xa Binding to Unstimulated Bovine Platelets

Using the ligand binding methodology detailed above, experiments were designed to determine whether platelet-bound factor V and factor Va facilitate the binding of factor Xa to unstimulated platelets, and if platelet activation is required as well. The concentration-dependent binding of ^{125}I-labeled factor Xa to unstimulated platelets is measured in the presence and absence of added, unlabeled factor V or Va. Reaction conditions are as described previously.

Our initial experiments included determining the time dependence and reversibility of steady state factor Xa binding. Because equilibrium and reversible binding was achieved by 15 min, all subsequent binding measurements were made following a 15-min incubation of the ligands with platelets. Initially, three separate binding experiments were done in which the ^{125}I-labeled factor Xa was varied from 2.7×10^{-11} to 4.2×10^{-9} M in

(1) the absence of added factor V or Va, (2) 1.2×10^{-8} M factor V, or (3) 1.2×10^{-8} M factor Va. Nonspecific binding due to isotope entrapment was again determined in parallel reaction mixtures containing a 100-fold molar excess of unlabeled factor Xa. The specific ^{125}I-labeled factor Xa binding data were plotted as molecules of factor Xa bound per platelet vs the nominal ^{125}I-labeled factor Xa concentration present in each reaction mixture, as based on the specific activity of the radiolabeled factor Xa (cpm/ng). These initial experiments indicated that both factor V and factor Va, at concentrations well below their physiological concentrations, will promote the binding of factor Xa to unstimulated platelets. The data were consistent with earlier factor V and Va binding data, because three times more factor Xa was platelet associated in the presence of factor Va as compared to factor V. The addition of factor V resulted in approximately 800 molecules of factor Xa bound, whereas with the addition of factor Va, 2300 factor Xa molecules were platelet associated.

This postulated factor V–factor Xa–platelet interaction may be of physiological significance, as one would predict that at plasma concentrations, factor V is always associated with unstimulated platelets.[13] Furthermore, complexation of factor Xa with either platelet-bound[28] or lipid-bound[29] factor V results in the cleavage of factor V to yield the active cofactor.[29] Even though the rate of the factor Xa-catalyzed activation of factor V is two orders of magnitude slower than the rate achieved with thrombin, the potential for factor V and factor Xa to associate in the prothrombinase complex on the platelet surface suggests a mechanism by which thrombin is initially produced. Thrombin then provides positive feedback by activating additional factor V to factor Va.

Because factor Va is the more efficient cofactor in the prothrombinase complex, the next series of experiments were done to determine the dependence of factor Xa binding on added factor Va. Two binding experiments were performed in which the ^{125}I-labeled factor Xa concentration was varied from 2.7×10^{-11} to 3.1×10^{-9} M in the presence of either 1.5 nM factor Va or 6.1 nM factor Va, and unstimulated or thrombin-activated platelets. The data indicated that factor Xa binding to platelets is dependent on the factor Va concentration because the factor Xa binding obtained in the presence of 6.1 nM factor Va was greater than that observed with 1.5 nM factor Va, and because thrombin activation of the platelets prior to the binding studies had little or no effect on the amount of factor Xa bound. These data suggest that factor Xa binding is directly related to the amount of factor Va bound to the platelet, and is independent of platelet activation. Consequently, double-label binding experiments were per-

[29] W. B. Foster, M. E. Nesheim, and K. G. Mann, *J. Biol. Chem.* **258,** 13970 (1983).

formed using ^{125}I-labeled factor Xa and ^{131}I-labeled factor Va in order to make simultaneous measurements of the binding of both ligands to unstimulated platelets.

Coordinate Binding of ^{125}I-Labeled Factor Xa and ^{131}-I-Labeled Factor Va to Unstimulated Bovine Platelets[6]

Reaction conditions are as described previously with the exception that prothrombin (1.39 μM) is also included. The addition of prothrombin and subsequent production of thrombin had no effect on the factor Xa binding data described above. Prothrombin is included in order to compare the binding data obtained using radiolabeled proteins with binding data obtained using analyses of kinetics of prothrombin activation, which is described in detail below (Prothrombin Activation Experiments Monitored with DAPA). The binding of ^{125}I-labeled factor Xa and ^{131}I-labeled factor Va to unstimulated platelets was studied as a function of the concentrations of both ligands. The concentration dependence of ^{125}I-labeled factor Xa binding in the absence or presence of eight different concentrations of ^{131}I-labeled factor Va (0.1 to 1.43 nM) was studied at five different factor Xa concentrations ranging from 0.7 to 4.2 nM. The highest concentration of factor Va present was sufficient to saturate the high-affinity factor Va binding sites previously described. All binding measurements were made following a 15-min incubation and nonspecific binding corrections were made by including a 100 M excess of each of the unlabeled proteins in parallel reaction mixtures. All radioactivity measurements were made in a three-channel Beckman Biogamma γ counter (equivalent to the Beckman 4000 γ counter available currently). Double-isotope counting conditions were established using the channels ratio method described in the instruction manual accompanying the instrument. Knowledge of the specific radioactivity of each ligand (cpm/ng of protein) allowed calculation of the molecules of each ligand bound per platelet. When the binding data obtained at each factor Xa concentration are analyzed in terms of molecules of factor Xa bound per platelet vs molecules of factor Va bound per platelet, the data indicate that factor Xa binding is dependent on both the factor Xa and factor Va concentrations. The data are linear and are depicted in Fig. 2. The nine data points shown in each panel represent the amount of platelet-bound factor Xa (vertical axis) and platelet-bound factor Va (horizontal axis) calculated for each of the concentrations of factor Va added to the reaction mixtures; added factor Xa was held constant at the concentration shown in each panel. These data indicate that at a fixed factor Xa concentration, the amount of factor Xa bound to platelets increases linearly with increased factor Va binding as the factor Va concen-

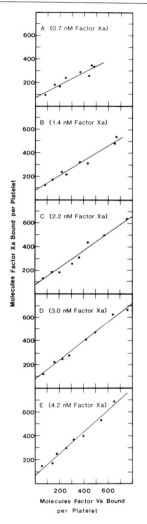

FIG. 2. Concentration-dependent binding of ^{125}I-labeled factor Xa to platelet-associated ^{131}I-labeled factor Va. (A–E) Binding experiments in which the amount of both ligands bound to platelets was determined from reaction mixtures containing a fixed amount of ^{125}I-labeled factor Xa at the concentration shown, as the ^{131}I-labeled factor Va concentration was varied from 0.1 to 1.43 nM. The molecules of factor Xa bound per platelet (vertical axis) and factor Va bound per platelet (horizontal axis) are plotted a a function of the eight different factor Va concentrations employed. (From Ref. 6.)

TABLE II
RATIO OF MOLECULES OF FACTOR Xa AND FACTOR Va BOUND
TO UNSTIMULATED PLATELETS AS FUNCTION OF FACTOR Xa
CONCENTRATION[a]

Figure 2 (part)	$[Xa]_0 \times 10^9 \, M$	Molecules of factor Xa bound / Molecules of factor Va bound
A	0.7	0.54
B	1.4	0.64
C	2.2	0.73
D	3.0	0.84
E	4.2	0.90

[a] The data shown here represent the numerical values obtained for the slopes of the lines shown in Fig. 2.

tration is increased. As the factor Xa concentration is increased, more factor Xa is bound to the platelets.

In Fig. 2A–E, the slopes of the lines are obtained by linear regression analysis and represent the molar ratio of platelet-bound factor Xa to platelet-bound factor Va (Table II). The data indicate that at lower, nonsaturating concentrations of factor Xa, more factor Va is bound to the platelet than factor Xa. As the factor Xa concentration is increased, more factor Xa binding sites (i.e., platelet-bound factor Va) are filled. When factor Xa is present at a saturating concentration with respect to its platelet binding sites, then factor Xa and factor Va are bound in a 1:1 molar ratio.

Derivation of Expression for Interaction of Factor Xa with Platelets

If the interaction of factor Xa with the platelet surface is described as factor Xa binding to the platelet-bound factor Va (pVa), Eq. (2) can be

$$pVa + Xa \rightarrow pVa \cdot Xa \quad (2)$$

applied. The dissociation constant for this process can be expressed as in Eq. (3).

$$K_D = [pVa][Xa]/[pVa \cdot Xa] \quad (3)$$

or,

$$K_D = [pVa](f/b) \quad (4)$$

where f and b are equal to the fraction of free and bound factor Xa in the system. [pVa], the concentration of platelet-bound factor Va not complexed with factor Xa, can be expressed in terms of total factor Va bound to platelets, $[pVa_t]$, and nominal factor Xa concentration, $[Xa]_0$. This relationship is shown in Eq. (5).

$$[pVa] = [pVa_t] - nb[Xa]_0 \quad (5)$$

in which n represents the binding stoichiometry of factor Va to factor Xa. Substitution of Eq. (5) into Eq. (4) gives

$$K_D = ([pVa_t] - nb[Xa]_0)(f/b) \quad (6)$$

or, on rearrangement,

$$[pVa_t]/b[Xa]_0 = (1/f)(K_D/[Xa]_0) + n \quad (7)$$

The term $1/f$ was approximately 1.0 at all times since experimental conditions were chosen such that the fraction of free factor Xa varied from 0.9 to 0.97. Under experimental conditions, Eq. (7) can be replaced, without significant error, by Eq. (8), in which $1/f = 1.0$.

$$[pVa_t]/b[Xa]_0 = (K_D/[Xa]_0) + n \quad (8)$$

Equation (8) expresses a linear relationship between the reciprocals of the nominal concentrations of factor Xa and the ratio of total factor Va and factor Xa bound to platelets. The slope is equal to the apparent dissociation constant for the binding of factor Xa to its platelet receptor, factor Va, and the vertical axis intercept is equal to the stoichiometry of binding. (This derivation appears as in the appendix of Ref. 6.)

Thus, a double-reciprocal plot of the data shown in Table II is linear, and is shown in Fig. 3. The apparent K_D obtained is 6×10^{-10} M, with a binding stoichiometry of 1.04 mol of factor Xa bound per mole of factor Va bound.

Kinetic Determination of Binding Interaction of Factor Va with Platelets

The coordinate binding data strongly suggest that factor Va is the receptor for factor Xa on the surface of unstimulated bovine platelets and that they form a stoichiometric complex of high affinity. The factor Va binding data indicate that this protein interacts with two distinct platelet sites with apparent K_D values of 4×10^{-10} M ($n \approx 800$ to 900) and 3×10^{-9} M ($n \approx 3500$), respectively.[13] To determine whether these factor Va binding sites are "functional" sites and involved in the factor Xa-catalyzed activation of prothrombin, analyses of the kinetics of prothrombin activation were used to construct a functional factor Va platelet binding isotherm. Experiments were performed in which the influence of factor Va on the rate of prothrombin activation was taken to reflect factor Va binding to a site on the bovine platelet surface, thereby forming the receptor for

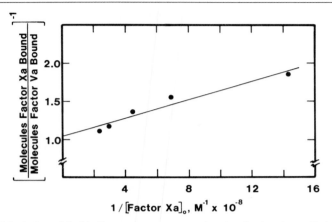

FIG. 3. Calculation of the binding parameters governing the platelet–factor Xa interaction when "modeled" as factor Xa binding to platelet-bound factor Va. The binding of factor Xa to platelets can be modeled as factor Xa binding to platelet-bound factor Va as described by Eq. (8) in the text. Accordingly, a double-reciprocal plot of the data presented in Table II is shown in which the ratio of molecules of platelet-bound factor Va to platelet-bound factor Xa is plotted as a function of the factor Xa concentration. An apparent K_D of 6×10^{-10} M and a binding stoichiometry of 1.04 molecules of factor Xa bound per molecule of factor Va bound are calculated from the slope and intercept, respectively. (From Ref. 6.)

factor Xa. The rationale for this approach is based on the observation that deletion of the cofactor, factor Va, from the complete catalyst complex results in a 10,000-fold decrease in the reaction rate,[30] and thus an ineffective enzyme complex.[31] Furthermore, a platelet-independent, kinetically discernible factor Va–factor Xa complex occurs at only 0.1% of the rate of the complete catalyst. Consequently, there is no thrombin formed unless factor Va binds to the platelet surface to perform its function.

Prothrombin Activation Experiments Monitored with DAPA[6]

The factor Xa-catalyzed conversion of prothrombin to thrombin is monitored continuously through the enhanced fluorescence of the DAPA–thrombin complex.[23,29] Prothrombin activation experiments are done at ambient temperature (22–24°) in 0.02 M Tris-HCl, 0.15 M NaCl, pH 7.4. Reaction mixtures (1.5 ml) contain 1.39 μM prothrombin, 2.5 mM Ca^{2+}, 3×10^8 platelets/ml, 3 μM DAPA, 5 nM factor Xa, and varying concentrations of factor Va. Experiments are initiated with factor Xa following a 2-min preincubation of the platelet reaction mixture with factor

---

[30] Data were obtained using potential physiological concentrations of proteins.
[31] M. E. Nesheim, J. B. Taswell, and K. G. Mann, *J. Biol. Chem.* **254**, 10952 (1979).

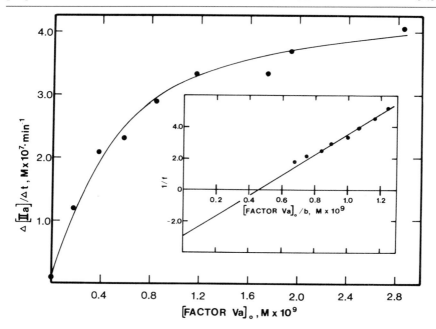

FIG. 4. Kinetic determination of the "functional" factor Va binding sites on platelets. The conversion of prothrombin to thrombin was monitored as a function of the factor Va concentration under conditions in which it was the rate-limiting component. Platelet number (3×10^8) was held constant; prothrombin was present at a physiological concentration (1.39 μM); and factor Xa (5 nM) and Ca^{2+} (2.5 mM) were present at saturating concentrations with respect to the rate at which thrombin was generated. Thrombin generation was monitored continuously through the enhanced fluorescence of the DAPA–thrombin complex as described in the text. Results are expressed as the amount of thrombin generated per minute as a function of the factor Va concentration. The double-reciprocal plot obtained from the saturation curve is shown as an inset, as described by Eq. (9) in the text. An apparent K_D of 1.6×10^{-10} M is calculated from the slope, and the presence of 915 factor Va platelet-binding sites is determined from the intercept. (From Ref. 6.)

Va. The time course of thrombin formation (i.e., DAPA fluorescence) is continuously recorded as described by Nesheim et al.[23,31] The reaction velocity is determined from the initial slope of the recorded data.

For all activation mixtures, saturating concentrations of factor Xa and Ca^{2+} are held constant, and the factor Va concentration is varied from 0.17 to 2.85 nM. The dependence of the rate of prothrombin activation on the factor Va concentration is presumed to reflect the factor Va-dependent binding interactions contributing to the assembly of the entire catalyst (Fig. 4). When the further addition of factor Va produced no additional increase in the rate of thrombin generation, it was presumed that the functional factor Va–platelet binding sites are occupied. Values corresponding to the fraction of sites bound and the fraction of sites free are calculated as the ratio of observed velocity to limit velocity, because velocities less than saturating are considered directly proportional to the extent that the available sites are filled. Equation (9) demonstrates a linear relationship between the fraction of free sites ($1/f$) and the factor Va concentration ($[Va]_0$) divided by the fraction of bound sites (b), as derived previously by Nesheim et al.[31]

$$1/f = (1/K_D)([Va]_0/b) - (nP_0/K_D) \tag{9}$$

In Eq. (9), P_0 is the molar concentration of platelets; K_D is the *apparent dissociation constant* for the factor Va–platelet interaction; and n is the apparent number of factor Va binding sites per platelet. Equation (9) is used to construct a double-reciprocal plot shown as an inset in Fig. 4. The apparent K_D calculated from the slope is equal to 1.6×10^{-10} M; the number of functional factor Va binding sites per platelet, calculated from the intercept, is 915. A second experiment, using different platelets, yielded similar results ($K_D = 2.2 \times 10^{-10}$ M, $n = 927$). It should be noted that analyses of equilibrium binding experiments, using ^{125}I-labeled factor Va, yielded similar results for the high-affinity factor Va binding sites ($K_D = 4 \times 10^{-10}$ M; $n = 800$ to 900). Thus these high-affinity binding sites, once occupied by factor Va, most likely represent the factor Xa receptor sites involved in the formation of a *functional* prothrombinase complex at the platelet surface.

Visualization of the Platelet–Factor Xa Receptor: Platelet-Bound Factor Va

We typically visualize and assess the interaction of bovine ^{125}I-labeled factor Va with bovine platelets by coupling binding studies with solubilization of platelet pellets containing the bound radiolabeled protein followed by electrophoresis in SDS and subsequent autoradiography.[14]

Electrophoresis and Autoradiography of ^{125}I-Labeled Factor Va Bound to Platelets

The interaction of ^{125}I-labeled factor Va with platelets is measured by centrifugation through oil as described previously. ^{125}I-labeled factor Va (0.76 nM) is incubated with washed bovine platelets (2×10^8 platelets/ml) in the presence of 30 μM DAPA. At various times, duplicate aliquots are removed from the reaction mixture and centrifuged through oil for 1 min.

One aliquot is processed for determination of bound radioactivity as described above. The supernatant and oil layer are rapidly removed from the duplicate aliquot, and the platelet pellet is solubilized in 10% (v/v) acetic acid, frozen, and lyophilized. Lyophilized samples are solubilized in 2% (w/v) SDS, 2% (v/v) 2-mercaptoethanol and subjected to electrophoresis in polyacrylamide gradient slab gels (5–15%) according to Neville.[32] The ^{125}I-labeled factor Va peptides associated with platelets are visualized by subjecting dried gels to autoradiography at $-70°$ using Kodak (Rochester, NY) XR-1 film and Cronex Lightning Plus intensifying screens (Du Pont, Wilmington, DL).

The ^{125}I-labeled factor Va used in these studies is prepared by thrombin activation of single-chain ^{125}I-labeled factor V, as detailed above, which results in three radioiodinated factor Va peptides, components D, E, and F. Component D (M_r 94,000) and component E (M_r 74,000) comprise the two-subunit factor Va molecule (Fig. 1) and require Ca^{2+} to maintain factor Va integrity and activity, whereas component F (M_r 71,000) is one of three activation peptides that do not appear to be essential for factor Va activity.[33] An autoradiograph, depicting the factor Va peptides bound to platelets with time, is shown in Fig. 5 and indicates that components D and E rapidly associate with platelets whereas component F does not bind to platelets. An additional component of approximately 90,000 Da that appears with time has been labeled component D' because it appeared to be arising from component D, through the action of a platelet-associated protease. This proteolytic activity is inhibited by protease inhibitors such as leupeptin (0.5 mM) and pepstatin A (0.05 mM), and also by platelet inhibitors such as PGE (5 μM). Furthermore, the proteolytic activity expressed by platelets appears to be membrane related, because Triton X-100 platelet lysates do not express this activity. Thus, factor Va bound to platelets consists of three peptides, components D, D', and E.[14]

Proteolysis of Platelet-Bound Factor Va Induced by Complex Formation with Activated Protein C and Factor Xa

The ability of the serine protease, activated protein C, to cleave the platelet-associated factor Va peptides was studied, as was the ability of factor Xa to protect against this activity. To determine the effect of activated protein C on platelet-bound factor Va, ^{125}I-labeled factor Va (0.76 nM) is incubated with platelets for 15 min. Activated protein C (3.2 ×

---

[32] D. M. Neville, Jr., *J. Biol. Chem.* **246**, 6328 (1971).
[33] M. E. Nesheim, L. S. Hibbard, P. B. Tracy, J. W. Bloom, K. W. Myrmel, and K. G. Mann, in "The Regulation of Coagulation" (K. G. Mann and F. B. Taylor, eds.), p. 145. Elsevier/North-Holland, New York, 1980.

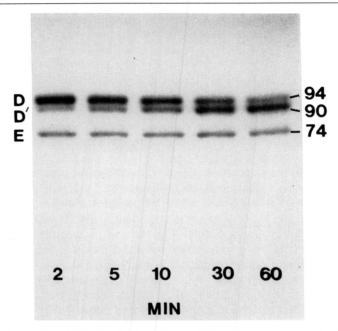

FIG. 5. Autoradiographic visualization of factor Va bound to platelets. At the time points indicated, aliquots were removed from a platelet–^{125}I-labeled factor Va reaction mixture, centrifuged through oil, and processed for SDS electrophoresis and autoradiography. The apparent molecular weight of each peptide is shown ($\times$ 10^{-3}). The time-dependent appearance of component D' is apparent. (From Ref. 14.)

10^{-8} M) is then added and aliquots removed from the reaction mixture at 0.5, 1, 2, 5, and 10 min and processed for autoradiography as described above. The protection afforded by prior complex formation of platelet-bound factor Va with factor Xa is examined by adding factor Xa (1.52 $\times$ 10^{-8} M) to the platelet–factor Va reaction mixture immediately prior to the addition of activated protein C.

The effect of factor Xa alone on platelet-bound factor Va is also studied by adding varying concentrations of factor Xa (0.76–15.2 nM) to a reaction mixture containing platelets (2 $\times$ 10^8 platelets/ml) and factor Va (0.76 nM) that had been incubated for 15 min. Examination of the autoradiographs indicated that platelet-bound factor Va is cleaved by both activated protein C and factor Xa to yield the cleavage products depicted schematically in Fig. 6. Activated protein C rapidly cleaves both components D and D' to yield three peptide products, with secondary time-dependent cleavages occurring in component E. Factor Xa predominantly cleaves component

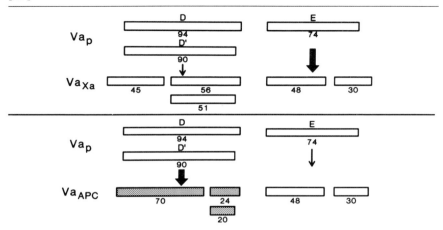

FIG. 6. Schematic representation of the activated protein C (APC) and factor Xa-induced cleavages in platelet-bound factor Va. (From Ref. 14.)

E, with secondary cleavages occurring in components D and D'. The factor Va cleavage products resulting from activated protein C and factor Xa proteolysis of components D and D' are indeed different, while the cleavages in component E induced by both proteases appear to be the same, or very similar. Because the activated protein C cleavages are not inhibited by antithrombin III and heparin, these cleavages are not due to contamination with factor Xa. While the cleavages induced by activated protein C inactivate platelet-bound factor Va and are of regulatory significance, the factor Xa-induced cleavages appear to have no obvious effect on the ability of factor Va to function in the prothrombinase complex.[14]

Subunit of Factor Va, Component E, Mediating Prothrombinase Complex Assembly at Platelet Surface

As discussed above and shown in Figs. 5 and 6, factor Va bound to platelets consists of three peptides. Component D' appears to arise as a result of proteolysis of component D, possibly mediated by the Ca^{2+}-dependent sulfhydryl platelet protease described in both the bovine and human systems.[34–36] Because the activity of this protease is dependent on millimolar concentrations of Ca^{2+}, experiments were performed initially to determine if platelets isolated in the presence of Na_2EDTA would no

[34] J. A. Truglia and A. Stracher, *Biochem. Biophys. Res. Commun.* **100**, 814 (1981).
[35] D. R. Phillips and M. Jakábová, *J. Biol. Chem.* **252**, 5602 (1977).
[36] M. Sakon, J. Kambayashi, H. Ohno, and G. Kosaki, *Thromb. Res.* **24**, 207 (1981).

longer exhibit proteolytic activity toward factor Va. Observations made during these experiments enabled us to hypothesize that component E of the two-subunit factor Va molecules mediates the binding of factor Va to the platelet surface.

Binding of Factor Va to Platelets Mediated through Component E

Bovine platelets are isolated from blood in which acid–citrate–dextrose (ACD) solution [0.15 M sodium citrate/citric acid/2% (w/v) glucose, pH 4.5] is used as the anticoagulant (ACD platelets).[26] Alternatively, blood is drawn into 5 mM Na$_2$EDTA (final concentration) and platelets are isolated and washed as indicated above except that 5 mM Na$_2$EDTA is included in the first two washes; the final wash and resuspension contains no Na$_2$EDTA (EDTA platelets). In addition, EDTA platelets are recalcified with 2.5 mM CaCl$_2$ for 10 min (recalcified EDTA platelets) prior to the initiation of studies described below. The assessment of binding of ^{125}I-labeled factor Va to these platelets, as well as electrophoresis and autoradiography of the radioiodinated peptides associated with these platelets, is done as described previously. The time course of factor Va binding to these three different platelet preparations is shown in Fig. 7A; an autoradiography depicting the factor Va peptides associated with these same platelets is shown in Fig. 7B. The ACD platelets bound the greater amount of factor Va and exhibited proteolytic activity toward bound factor Va, as shown by the appearance of component D' with the concomitant loss of component D. The autoradiography depicting the factor Va peptides associated with the EDTA platelets indicates that at the time of maximal factor Va binding (2 min), components D and E were platelet associated; however, component D slowly dissociated from the platelet surface and only component E remained bound. Because component D' is not visible at any time, these results can be interpreted as indicating that the residual Na$_2$EDTA in the platelet preparation inhibits the activity of the platelet-associated protease normally associated with our other platelet preparations. In addition, it appears that residual, platelet-associated Na$_2$EDTA chelates the Ca^{2+} required to maintain factor Va subunit integrity and activity, thus resulting in the loss of component D from the platelet surface. From this observation, we postulate that component E primarily mediates the binding of factor Va to the platelet surface, and that component D interacts with the platelet through its association with component E. Recalcification of EDTA platelets partially restored their ability to bind both components of the factor Va molecule. The appearance of component D' was not apparent until 10 min after the addition of factor

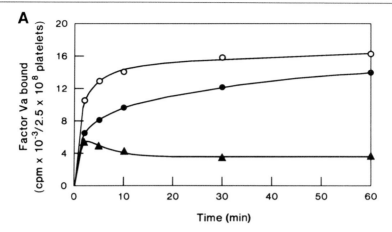
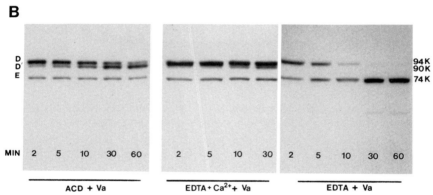

Fig. 7. Interaction of factor Va with platelets determined by direct binding measurements (A) and electrophoresis and autoradiography of factor Va bound to platelets (B). (A) The time-dependent binding of ^{125}I-labeled factor Va to ACD platelets (○), recalcified EDTA platelets (●), or EDTA platelets (▲). (B) Autoradiographic visualization of the factor Va peptides associated with either ACD, recalcified EDTA, or EDTA platelets at time points that correspond to the binding isotherms in (A). (From Ref. 15.)

Va to these platelets, indicating that recalcification did not completely restore the platelet Ca^{2+}-dependent proteolytic activity.

Because these experiments suggested that the binding of factor Va to the platelet surface is mediated primarily through component E, we tested this hypothesis by comparing the interaction of factor Va with ACD platelets when (1) factor Va was present as the two-subunit functional molecule, or (2) as an inactive mixture of both components D and E whose subunit

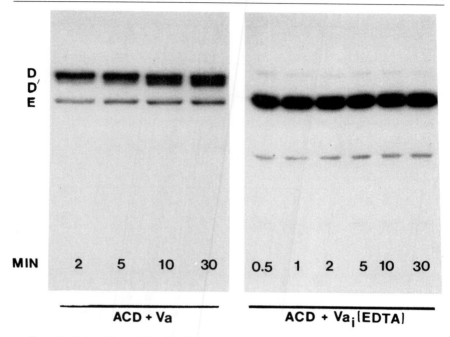

FIG. 8. Autoradiographic visualization of the interaction of factor Va and EDTA-inactivated factor Va [factor Va$_i$(EDTA)] with platelets (Modified from Ref. 15.)

interaction had been destroyed by preincubation with Na$_2$EDTA. ^{125}I-Labeled factor Va (61 nM) is functionally inactivated and subunit dissociation accomplished by addition of 5 mM Na$_2$EDTA (final concentration). Incubation at ambient temperature (approximately 10 min) is allowed until no factor Va activity remains as determined by bioassay. When Na$_2$EDTA-inactivated factor Va [factor Va$_i$(EDTA)] is used, additional Ca^{2+} was not included in the reaction mixtures. (Our buffers routinely contain approximately 0.1 μM Ca^{2+} from our water source.) Either ^{125}I-labeled factor Va$_i$(EDTA) or active ^{125}I-labeled factor Va (0.4 nM) is incubated with ACD platelets (2 × 10^8 platelets/ml). Aliquots are removed at various times and processed for autoradiography as described previously (Fig. 8). The resulting autoradiograph depicting the factor Va peptides associated with platelets incubated with factor Va$_i$(EDTA) indicated that only component E bound to the platelet surface even though both components D and E were present in the incubation mixture. There was no indication of component D being bound even after a 30-min incubation. Thus, component E mediates factor Va binding to the platelet surface and component D associates with platelets only through its metal ion-dependent association with component E.

Component E Forms Part of Receptor for Factor Xa at Bovine Platelet Surface

On binding to platelet-bound factor Va, factor Xa rapidly cleaves component E into two peptides having a molecular weight of approximately 48,000 and 30,000 (Fig. 6). This observation was used to design experiments to determine if factor Xa would interact with component E bound to platelets in the absence of component D. ^{125}I-labeled factor Va (0.4 nM) is incubated for 30 min with EDTA platelets (2 × 10^8/ml). Under these conditions, only component E is bound to the platelet membrane as shown in Fig. 9A (left) by autoradiography. After the 30-min incubation, the platelets are washed twice to remove any unbound components D and E and resuspended in binding study buffer containing 2.5 mM Ca^{2+} (because Ca^{2+} is required for factor Xa binding to platelets[4,5]). Factor Xa (1.5 nM) is then added, and aliquots removed at the time points indicated and processed for electrophoresis and autoradiography (Fig. 9A, center). A standard showing the factor Xa-induced cleavages of platelet-bound factor Va in which component E is rapidly cleaved to peptides of approximately 48,000 and 30,000 Da is shown at the extreme right of Fig. 9, with secondary cleavages in components D and D'. The autoradiograph indicates that addition of factor Xa to a platelet–component E complex results in the factor Xa-induced cleavage of component E to yield the typical factor Xa-derived products. In a parallel experiment utilizing unlabeled factor Va and ^{125}I-labeled factor Xa, binding measurements indicated that radiolabeled factor Xa became associated with the platelet surface in a time-dependent manner and was displaced by unlabeled factor Xa (Fig. 9B). Thus, component E, in the absence of component D, will form a receptor for factor Xa at the platelet surface. Because component E alone mediates both factor Va binding to the platelet surface and binds factor Xa, we have identified component E as the factor Va peptide that facilitates the assembly of the prothrombinase complex on the bovine platelet surface.

Human Platelet Factor Xa Receptor

The use of equilibrium binding techniques has not been successful in yielding stoichiometric and quantitative data in the human platelet system. Equilibrium binding measurements of human factor Va with unstimulated platelets have been reported by Kane and Majerus[7] as nonsaturable; thus the data preclude quantitative interpretation of binding affinity and stoichi-

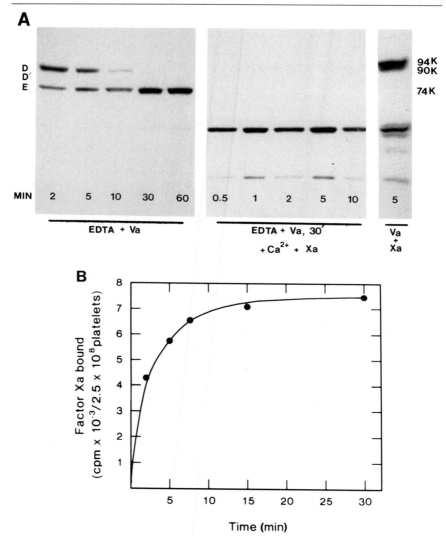

FIG. 9. Interaction of factor Xa with component E bound to platelets determined by autoradiographic visualization of the factor Xa-induced cleavages of component E (A) and direct binding measurements (B). (A) Autoradiography of the ^{125}I-labeled factor Va peptides associated with platelets indicates that only the 74K component E is bound to EDTA platelets after a 30-min incubation (left). After 30 min, these platelets were washed twice, recalcified, and factor Xa was then added. Autoradiographic visualization of the factor Va peptides bound to platelets following the addition of factor Xa indicates that component E bound to platelets is rapidly cleaved by factor Xa to yield peptides of 48K and 30K (center). A standard showing the factor Xa-induced cleavages in platelet-bound factor Va in which component E

ometry. We also have been unable to achieve saturable factor Va-platelet binding to human platelets even when amounts of factor Va approaching its physiological concentration (7 μg/ml) have been employed. In addition, human platelets contain a significant reserve of factor V,[17] which may egress from the platelet and cause problems in determining the specific activity of the radiolabeled protein. This factor precludes any radiolabeled ligand binding measurements to normal thrombin-activated platelets. Analyses of the kinetics of prothrombin activation provided quantitative evaluation of the functional prothrombinase platelet binding sites in the bovine system equivalent to the values obtained for the high-affinity factor Va binding sites determined from equilibrium binding studies. The equivalence of the data obtained from kinetic measurements to those determined by equilibrium binding measurements suggests that the kinetic approach would provide valid estimates of the binding parameters. We have therefore used the kinetic approach to calculate the stoichiometry and binding parameters governing the functional interactions of human factor Va and human factor Xa with isolated, normal human platelets and platelets obtained from a factor V-deficient individual who was complete devoid of both plasma and platelet factor V.[17]

Materials

Human factor V (M_r 330,000; $E^{1\%}_{280\,nm}$ = 9.6) is purified using a hybridoma antibody, prepared in this laboratory, as described by Katzmann et al.,[37] and converted to factor Va by incubation of factor V with catalytic amounts of thrombin.[23] The bioactivity of factor V is assessed as described previously.[23,37] Human factor X (M_r 50,000; $E^{1\%}_{280\,nm}$ = 11.6) and human prothrombin (M_r 72,000; $E^{1\%}_{280\,nm}$ = 14.2) are isolated as described by Bajaj et al.[38] Factor Xa is prepared using factor X activator purified from Russell's viper venom[20] immobilized on activated CH-Sepharose (Sigma). N-2-Hydroxyethylpiperazine-N'-2-ethanesulfonic acid (HEPES) was obtained

[37] J. A. Katzmann, M. E. Nesheim, L. S. Hibbard, and K. G. Mann, *Proc. Natl. Acad. Sci. U.S.A.* **78**, 162 (1981).
[38] S. P. Bajaj, S. I. Rapaport, and C. Prodanos, *Prep. Biochem.* **11**, 397 (1981).

---

is rapidly cleaved to yield peptides of 48K and 30K, with secondary cleavages occurring in components D and D', is shown at the right. (B) The time course of binding of ^{125}I-labeled factor Xa to component E bound to platelets. The experiment described in (A) above was modified by incubating unlabeled factor Va with EDTA platelets for 30 min. The ability of ^{125}I-labeled factor Xa to bind to these washed and recalcified platelets was then determined. (From Ref. 14.)

from Sigma. Apyrase was the generous gift of Dr. Ralene Kinlough-Rathbone (McMaster University Medical Center, Hamilton, Ontario, Canada). The preparation of all other reagents, with the exception of platelets, has been detailed above.

Platelets are isolated and washed to remove plasma by strict adherence to the method of Mustard et al.[39] Blood is obtained by venipuncture from healthy, nonsmoking, adult males who have not ingested aspirin in the last 3 weeks. Following the collection of blood (six parts) into ACD anticoagulant (one part), all subsequent manipulations are performed at 37°. Platelet-rich plasma (PRP) is obtained by centrifugation of the anticoagulated blood at 190 g for 15 min. Washed platelets are obtained by repeated resuspension and centrifugation (1000 g) using an albumin (0.35%, w/v)/Tyrode's solution (0.137 M NaCl, 2.7 mM KCl, 12 mM NaHCO$_3$, 0.36 mM NaH$_2$PO$_4$, 1 mM MgCl$_2$, 2 mM CaCl$_2$, 5 mM dextrose) containing apyrase (30 μg/ml) throughout and heparin (50 NIH U/ml) in the initial wash. Following the third and final wash and centrifugation, the platelets are suspended in Tyrode's wash medium containing 5 mM HEPES and 3 μg/ml apyrase, pH 7.35. The platelet count in the final suspension varied from 4 to 8 × 10^8 platelets/ml as determined by cell counts using a Coulter counter (Coulter Electronics, Hialeah, FL).

Kinetic Determination of Binding Interaction of Factor Xa with Normal, Thrombin-Activated Human Platelets

Prothrombin activation mixtures are monitored to determine the influence of factor Xa concentrations on the rate of thrombin generation in the presence of thrombin-activated normal platelets, Ca^{2+}, and factor Va. The velocity changes accompanying increases in factor Xa concentration are interpreted as reflecting the interaction of factor Xa with its receptor, platelet-bound factor Va. Platelets are activated intentionally with thrombin both to release and activate platelet-associated factor V, thereby eliminating the potential influence of platelet factor V activation on the reaction rate. Exogenous, excess factor Va was also added to ensure that factor Va was not a limiting component because, in contrast to the data of Miletich et al.,[9,40] our initial kinetic experiments indicated that thrombin-activated platelets do not release sufficient factor Va to saturate all the platelet factor Va binding sites capable of participating in prothrombin activation. Prothrombin activation studies (1.5 ml) are performed at 22–24°

[39] J. F. Mustard, D. W. Perrie, N. G. Ardlie, and M. A. Packham, *Br. J. Haematol.* **22**, 193 (1972).
[40] J. P. Miletich, W. H. Kane, S. L. Hofmann, N. Stanford, and P. W. Majerus, *Blood* **54**, 1015 (1979).

in a HEPES/Tyrode's buffer and contained 1.39 μM prothrombin, 3 μM DAPA, 1 × 10^8 thrombin-activated platelets, 5 nM factor Va, and varying concentrations of factor Xa. The experimental protocol includes thrombin activation of the platelets suspended in HEPES/Tyrode's buffer (0.5 NIH U/ml, 2 min, 22°) *without stirring* followed by the addition of DAPA and prothrombin. The contents of the cuvette are mixed gently by inversion. Factor Va is added and allowed to incubate with the platelets for 2 min to facilitate the required platelet–factor Va interaction. Prothrombin activation is initiated by the addition of factor Xa and monitored as described in detail above. This protocol results in reaction conditions in which a physiological concentration of prothrombin, saturating concentrations of factor Va and Ca^{2+}, and a defined number of platelet receptor sites are held constant and factor Xa is varied from 0.05 to 5.0 nM. When the further addition of factor Xa produces no additional increase in the velocity of thrombin formation, it is presumed that the functional factor Xa platelet binding sites are saturated. A typical saturation curve is shown in Fig. 10 and the double-reciprocal plot is shown as an inset. The apparent K_D is calculated from the slope and number of functional factor Xa platelet binding sites calculated from the intercept as detailed above in our studies using bovine platelets. These data are interpreted presuming a 1 : 1 stoichiometry of factor Xa with platelet-bound factor Va and indicate that factor Xa binds to 2500 sites with high affinity (apparent $K_D = 0.9 \times 10^{-10}$ M) and expressing a $k_{cat} = 13$ mol of thrombin/sec/mol of factor Xa bound. Several of these analyses performed with different platelet preparations ($n = 8$) indicate that normal human thrombin-activated platelets possess approximately 2700 ± 1000 factor Xa binding sites with an apparent $K_D = 1.2 \pm 0.6 \times 10^{-10}$ M, and a $k_{cat} = 19 \pm 7$ mol of thrombin/sec/mol of factor Xa bound.

Kinetic Determination of Stoichiometry of Platelet–Factor Va–Factor Xa Binding Interactions

The store of factor V in the granules of normal platelets[17] represents approximately 25% of the total blood factor V. Variable release of this store of factor V prevents a reciprocal determination of the functional factor Va–platelet binding sites. However, a kinetic assessment of factor Va binding can be performed using platelets from a factor V-deficient individual totally devoid of both plasma and platelet factor V,[16,17] but otherwise exhibiting normal platelet function in routine coagulation testing. Furthermore, the use of factor V-deficient platelets allows quantitation of both factor Va and factor Xa binding parameters and therefore accurate determination of the stoichiometries governing these interactions. Thus,

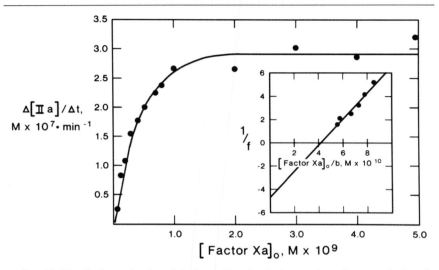

FIG. 10. Kinetic determination of the factor Xa–platelet interaction using normal, thrombin-activated human platelets. The conversion of prothrombin to thrombin was monitored as a function of the factor Xa concentration under conditions in which platelets (1×10^8) were held constant, prothrombin was present at a physiological concentration (1.39 μM), and factor Va (5 nM) and Ca^{2+} (2 mM) were present at concentrations that are saturating with respect to the rate of thrombin generation. Thus, velocity changes induced by varying the factor Xa concentration presumably reflect its binding to its receptor, platelet-bound factor Va. Prothrombin activation was initiated by the addition of factor Xa, and initial velocities of thrombin formation were obtained from recorded fluorescence tracings. The double-reciprocal plot obtained from the rate saturation curve is shown as an inset and indicated that factor Xa interacts with 2500 binding sites with an apparent K_D of 0.9×10^{-10} M. (From Ref. 16.)

prothrombin activation is monitored as a function of both the factor Va concentration [when factor Xa (5 nM) is present at a fixed and saturating amount] or varying the factor Xa concentration [when factor Va (5 nM) is present at a fixed and saturating amount] in the presence of thrombin-activated factor V-deficient platelets. The same experimental protocol is followed as detailed previously for platelets from a normal individual. The influence of factor Va concentration on thrombin generation produced the binding isotherm shown in Fig. 11 (open circles). Varying the factor Xa concentration produced the binding isotherm depicted by the filled circles. These data indicate that factor Va and factor Xa form a 1 : 1 stoichiometric complex on the surface of thrombin-activated human platelets. Graphic analysis of the individual ligand binding isotherms allowed calculation of 4700 factor Va binding sites (apparent $K_D = 2.1 \times 10^{-10}$ M) and 5100 factor Xa binding sites (apparent $K_D = 2.3 \times 10^{-10}$ M) expressing similar k_{cat} values per site (14–17 mol of thrombin/sec/mol ligand bound).

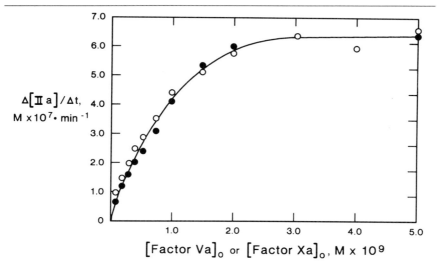

FIG. 11. Stoichiometry of the factor Xa–factor Va–platelet interaction determined kinetically using thrombin-activated factor V-deficient platelets. Platelets obtained from an individual totally devoid of both plasma and platelet factor V[17] were used in prothrombin activation reactions in order to assess the influence of both factor Va (○) and factor Xa (●) on the rate of thrombin generation. The data shown indicate that factor Va and factor Xa form a 1:1 stoichiometric complex on the platelet surface because the saturation curves obtained when either component is limiting are superimposable.

To substantiate the use of a kinetic approach for calculating the factor Va and factor Xa platelet binding parameters, we monitored prothrombin activation at three different concentrations of factor V-deficient platelets as a function of varying factor Va concentration. Factor Va titrations are performed in the presence of saturating amounts of factor Xa (5 nM) with (1) 0.5×10^8, (2) 1.0×10^8, or (3) 2.5×10^8 platelets/ml. These data are shown in Table III. As expected, the calculated apparent K_D and number of receptor sites expressed per platelet remain constant, as does the k_{cat}, when expressed per receptor site. However, the absolute rate of thrombin generation at saturating concentrations of both factor Va and factor Xa is directly related to the platelet (receptor site) concentration, because that concentration reflects the *total* number of available enzyme binding sites present in the reaction mixture.

Summary

The assembly and function of the prothrombinase complex on the bovine and human platelet membrane is mediated through binding interactions in which factor Va bound to the platelet surface forms at least part

TABLE III
KINETIC DETERMINATION OF FACTOR Va–PLATELET INTERACTION USING DIFFERENT
CONCENTRATIONS OF FACTOR V-DEFICIENT PLATELETS

Platelet number/ml ($\times\ 10^{-8}$)	Calculated binding parameters			Absolute rate of thrombin generation[c] ($M\ \times\ 10^7$/min)
	Apparent K_D ($M\ \times\ 10^{10}$)	n[a]	k_{cat}[b]	
0.5	2.3	5100	17	3.5
1.0	2.1	4700	14	6.8
2.5	2.5	4900	16	16.0

[a] n, Number of factor Va binding sites per platelet.
[b] Moles of thrombin per second per mole of factor Va bound.
[c] Factor Va and factor Xa were present at saturating amounts so as not to be limiting components.

of the "receptor" for factor Xa in a 1:1 stoichiometric complex.[6,16] A model depicting these binding interactions is shown in Fig. 12. Data from our laboratory indicate that the prothrombinase catalyst assembles in an analogous manner on the surface of monocytes,[16,41] lymphocytes,[16] neutrophils,[16] and well-defined phospholipid vesicles employed in model systems.[31,33] The 74,000-Da subunit of factor Va, component E, which mediates the binding of factor Va to either bovine platelets,[15] human monocytes,[42] or phospholipid vesicles,[25] is shown binding to the cell membrane through its putative "receptor." The 94,000-Da subunit of factor Va, component D, is associated with the membrane surface through its metal ion-dependent interaction with component E. Factor Va forms at least part of the receptor that mediates the binding of factor Xa to an appropriate membrane surface, because component E has been shown to contribute significantly to the interaction of factor Xa with either the platelet,[15] monocyte,[42] or vesicle[43] membrane surface.

Our data do not preclude the possibility that component D contributes to the binding of factor Xa and the function of the prothrombinase complex. Component D appears to be important for several reasons. Cleavage of component D by activated protein C results in the complete loss of factor Va cofactor activity.[44,45] An interaction between factor Xa and

[41] P. B. Tracy, M. S. Rohrbach, and K. G. Mann, *J. Biol. Chem.* **258,** 7264 (1983).
[42] P. B. Tracy, unpublished observations, 1983.
[43] M. M. Tucker, W. B. Foster, J. A. Katzmann, and K. G. Mann, *J. Biol. Chem.* **258,** 1210 (1983).
[44] F. J. Walker, P. W. Sexton, and C. T. Esmon, *Biochim. Biophys. Acta* **571,** 333 (1979).
[45] M. E. Nesheim, W. Canfield, W. Kisiel, and K. G. Mann, *J. Biol. Chem.* **257,** 1443 (1982).

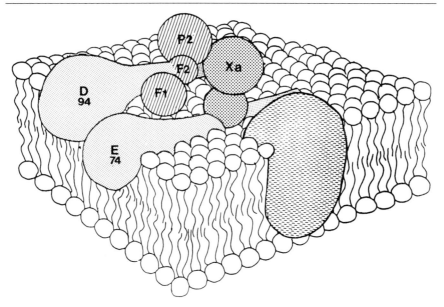

FIG. 12. Model of the activation of prothrombin as catalyzed by the prothrombinase complex at the platelet surface. Factor Va, a two-subunit molecule composed of a 74K subunit, component E, and a 94K subunit, component D, is shown bound to the cell membrane lipid bilayer through an interaction of component E with its putative cell membrane "receptor." Component D is associated with the membrane surface through its metal ion-dependent interaction with component E. Factor Va forms at least part of the factor Xa receptor through interactions in which factor Xa binds to the membrane-bound component E in a 1:1 molar stoichiometry. Prothrombin composed of fragment 1 (F1), fragment 2 (F2), and prothrombin 2 (P2) domains is shown in association with the membrane-bound complex of factor Va and factor Xa.

component D is implied from the observation that factor Xa protects factor Va from activated protein C inactivation.[14,44,45] Furthermore, the binding of factor Xa to platelet-bound factor Va results in the time-dependent cleavage of components D and D'.[14] Because component D is not required absolutely for prothrombinase complex assembly, we would speculate that it may be important in mediating prothrombin binding (depicted as a three-domain molecule) and increasing the catalytic efficiency of the enzymatic complex.

Although a substantial amount of data has been reported indicating that factor Va bound to an appropriate membrane surface forms the receptor facilitating the subsequent binding of factor Xa, no data have been reported regarding the component(s) that mediate(s) factor Va binding to the platelet surface. Striking differences between the cell surface and phospholipid membrane studies suggest that cells express factor Va receptors distinguishable from the hydrophobic and ionic interactions[25,46] that appear to mediate factor Va binding to the pure phospholipid mixtures studied to date. For instance, the binding affinity of factor Va to the platelet surface is three orders of magnitude tighter[13] than its affinity for phospholipid vesicles.[25,46] We have prepared a monoclonal antibody that has the capacity of blocking factor Va binding to platelets without influencing the binding of factor Va to phospholipids.[47] Finally, Miletich and co-workers studied a patient who had a congenital deficiency of factor Xa binding to her thrombin-activated platelets in the presence of exogenous factor Va,[40] suggesting the congenital absence of a factor Va "receptor" on the surface of these cells. However, because the ability of these platelets to support a factor Va interaction has not yet been studied, it is impossible to know if they are deficient in a factor Va "receptor," or deficient in another component apart from factor Va that is required to facilitate a platelet–factor Xa interaction. Even though these observations are circumstantial, they may suggest that a more complex prothrombinase receptor exists on the platelet surface than can be explained by lipid alone.

Studies have been initiated to monitor factor Va binding to the human platelet surface by using a noninhibitory monoclonal antibody that binds to the light chain of the factor Va molecule. Factor Xa binding was determined directly using radiolabeled ligand. Even though factor Va bound to the unstimulated platelet surface, platelet activation was required to achieve maximal binding. The concentrations of thrombin used to stimulate the platelets that effected maximal factor Va binding did not affect factor Xa binding. Significantly higher thrombin concentrations were required to achieve maximal factor Xa binding. Consequently, factor Va binding alone was not sufficient to support the subsequent binding of factor Xa. These data suggest that human platelets express an inducible receptor for factor Xa that, in addition to platelet-bound factor Va, is required to facilitate a platelet–factor Xa interaction.[48]

---

[46] J. W. Bloom, M. E. Nesheim, and K. G. Mann, *Biochemistry* **18,** 4419 (1979).
[47] M. M. Tucker, P. B. Tracy, and K. G. Mann, unpublished observations.
[48] C. H. Catcher, L. S. O'Rourke, and P. B. Tracy, *Blood* **74,** Suppl. 1, 1094 (1989).

[30] Binding of Coagulation Factor XIa to Receptor on Human Platelets

By DIPALI SINHA and PETER N. WALSH

General Methods

Principle

Human platelets have been shown to promote the proteolytic activation of coagulation factors XII and XI.[1] An apparent requirement for this contribution of platelets to contact activation is that first the protein cofactor noncovalently associates with factor XI in plasma, and then factor XI binds to activated platelets in the presence of high molecular weight kininogen.[2] The factor XIa formed as a consequence binds specifically to a site on platelets distinct from that for factor XI,[3] and retains its structural and functional properties.[3,4]

Materials

Proteins. Factor XI (specific activity = 270 U/mg)[5,6] and high molecular weight kininogen[7] (15 U/mg specific activity) are purified as previously described. Factor XIa is prepared by incubation with factor XIIa.[6] Prekallikrein (133 U/mg specific activity),[8] kallikrein,[9] human α-thrombin,[3] and prothrombin[6] are prepared as described. All proteins should be >98% pure as judged by sodium dodecyl sulfate (SDS) polyacrylamide slab gel electrophoresis.[10]

Other Reagents. Methyl silicone oil (1.0. DC200) and Hi Phenyl silicone oil (125 DC550) are obtained from William F. Nye, Inc. (Fairhaven, MA); carrier-free Na^{125}I-labeled Bolton–Hunter reagent and sodium

---

[1] P. N. Walsh and J. H. Griffin, *Blood* **57**, 106 (1981).
[2] J. S. Greengard, M. J. Heeb, E. Ersdal, P. N. Walsh, and J. H. Griffin, *Biochemistry* **25**, 3884 (1986).
[3] D. Sinha, F. S. Seaman, A. Koshy, L. C. Knight, and P. N. Walsh, *J. Clin. Invest.* **73**, 1550 (1984).
[4] P. N. Walsh, D. Sinha, F. Kueppers, F. S. Seaman, and K. B. Blanstein, *J. Clin. Invest.* **80**, 1578 (1987).
[5] B. N. Bouma and J. H. Griffin, *J. Biol. Chem.* **252**, 6432 (1977).
[6] P. N. Walsh, H. Bradford, D. Sinha, J. R. Piperno, and G. P. Tuszynski, *J. Clin. Invest.* **73**, 1392 (1984).
[7] D. M. Kerbiriou and J. H. Griffin, *J. Biol. Chem.* **254**, 12020 (1979).
[8] D. M. Kerbiriou, B. N. Bouma, and J. H. Griffin, *J. Biol. Chem.* **255**, 3952 (1980).
[9] B. N. Bouma, L. A. Miles, G. Beretta, and J. H. Griffin, *Biochemistry* **19**, 1151 (1980).
[10] U. K. Laemmli, *Nature (London)* **227**, 680 (1970).

[^{51}Cr]chromate are from New England Nuclear (Boston, MA); Factor XI-deficient plasma is from George King Biomedical (Overland Park, KS); Ultrol HEPES is from Calbiochem Corp. (San Diego, CA); and the chromogenic substrate Pyr-Glu-Pro-Arg-*p*-nitroanilide hydrochloride (S-2366) is from A. B. Kabi Peptide Research (Stockholm, Sweden).

Radiolabeling

For binding experiments, factor XI is radiolabeled with ^{125}I either by the procedure of Bolton and Hunter[11] or by the Iodogen method[12] as previously described.[3] Free ^{125}I is then separated from the labeled protein by passage through a 1-ml Sephadex G-25 column,[13] and the protein is then dialyzed in the presence of ovalbumin (1 mg/ml). ^{125}I-Labeled factor XIa is then prepared by incubating the zymogen with factor XIIa.[3]

Assays

Coagulation proteins are assayed as previously described,[3] utilizing appropriate congenitally deficient substrate plasmas. Factor XIa is measured by amidolytic assay[3,14] and by radioimmunoassay.[3,14] Protein assays are performed by the method of Lowry *et al.*[15]

Preparation of Cell Suspensions

Platelet-rich plasma is obtained from citrated human blood and gel filtered on a Sepharose 2B column into calcium-free HEPES-buffered Tyrode's solution, pH 7.4, containing bovine serum albumin (1 mg/ml), 138 mM NaCl, 2.7 mM KCl, 1.0 mM MgCl$_2$ · 6H$_2$O, 3.3 mM NaH$_2$PO$_4$ · H$_2$O, 15 mM N-2-hydroxyethylpiperazine-N'-2-ethanesulfonic acid (HEPES), and 5.5 mM dextrose as previously described.[3] Erythrocyte suspensions are prepared from fresh human blood washed five times in Hanks' balanced salt solution to remove plasma and other cells as previously described.[3] The red cells are finally suspended in calcium-free HEPES-buffered Tyrode's solution, pH 7.4.

---

[11] A. E. Bolton and W. M. Hunter, *Biochem. J.* **133**, 529 (1973).
[12] P. J. Fraker and J. C. Speck, *Biochem. Biophys. Res. Commun.* **80**, 849 (1978).
[13] G. P. Tuszynski, L. Knight, J. R. Piperno, and P. N. Walsh, *Anal. Biochem.* **106**, 118 (1980).
[14] C. F. Scott, D. Sinha, F. S. Seaman, P. N. Walsh, and R. W. Colman, *Blood* **63**, 42 (1984).
[15] O. H. Lowry, H. J. Rosebrough, A. L. Farr, and F. J. Randall, *J. Biol. Chem.* **193**, 365 (1951).

Binding Measurements

To measure the binding of factor XIa to platelets over varying incubation times, at variable concentrations of added factor XIa, or under differing conditions, gel-filtered platelets (2–4 × 10^8/ml) or control cells (e.g., erythrocytes) are incubated at 37° in 1.5-ml polypropylene centrifuge tubes (Sarstedt, Inc., Princeton, NJ) with ^{125}I-labeled factor XIa, in the presence or absence of high molecular weight kininogen or thrombin (or other platelet agonists). Determinations of specificity of binding and measurements of nonspecific binding are made in the presence of 50- to 100-fold molar excesses of unlabeled factor XIa or other proteins. At specified incubation times, 100-μl aliquots are layered over a mixture of silicone oils, (1 vol of DC200 to 5 vol DC550) contained in microsediment tubes with narrow-bore extended tips (Sarstedt, Inc.). After centrifugation for 2 min in a microfuge (model B; Beckman Instruments, Inc., Cedar Grove, NJ), the tips containing the sediments are cut off with wire cutters and the sediments and supernatants are counted separately in a γ counter.

Validation of Binding Assay

To validate the binding assay it is necessary to ascertain that it separates bound from free ligand. Therefore, it should initially be demonstrated that the fraction of cells sedimented is acceptably high and that the fraction of protein not specifically bound (i.e., the "trapped volume") is acceptably low. In experiments with platelets labeled with ^{51}Cr, >94% of the ^{51}Cr is recovered in the pellet, thus confirming that any ligand bound to the platelets would be expected to appear in the sediment. When platelets are incubated with ^{125}I-labeled bovine serum albumin, the fraction of radioactivity appearing in the pellet is routinely <0.005, whereas when ^{125}I-labeled factor XIa is centrifuged over silicone oils in the absence of platelets, the fraction of radioactivity in the pellet is <0.001. This procedure is thereby validated as a method to separate platelets from unbound ligand.

It is also necessary to demonstrate that the radiolabeled ligand behaves in all respects, especially its binding capacity, in a manner similar to the unlabeled protein. We utilize determinations of clotting activity and amidolytic activity to compare unlabeled factor XIa with ^{125}I-labeled factor XIa (0.5–1.1 × 10^6 cpm/μg), and to confirm that >85% of the functional activity of the ligand is retained after radiolabeling. To determine the binding characteristics of the labeled ligand compared with unlabeled protein, varying ratios of ^{125}I-labeled and unlabeled factor XIa are incubated for 5 min at 37° with gel-filtered platelets (2.45 × 10^8/ml), thrombin (0.1 U/ml), and high molecular weight kininogen (12 μg/ml), keeping the

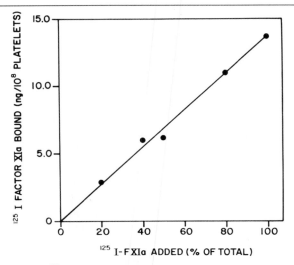

FIG. 1. Binding of ^{125}I-labeled factor XIa to thrombin-treated platelets in the presence of high molecular weight kininogen. Experimental details are given in the text. (Reprinted with permission from Ref. 3.)

total concentration of factor XIa constant (0.10 μg/ml). When binding of ^{125}I-labeled factor XIa is plotted as a function of the fraction of added ^{125}I-labeled factor XIa, a straight line is obtained (Fig. 1). This indicates that the affinities of the labeled and unlabeled factor XIa are similar.

Finally, it is necessary to confirm that the radioactivity recovered in the sediment in fact represents the ligand, that is, ^{125}I-labeled factor XIa, and not a contaminant. This is determined by incubating gel-filtered platelets (4.5 × 10^8/ml) with ^{125}I-labeled factor XIa (2.5 μg/ml) under conditions determined to be optimal for binding, i.e., an 8-min incubation in the presence of high molecular weight kininogen (20 μg/ml) and thrombin (0.1 U/ml). After centrifugation through 20% (w/v) sucrose, the platelet pellet is solubilized in 2.0% (w/v) sodium dodecyl sulfate containing 1 mM diisopropyl fluorophosphate, 1 mM benzamidine, and 1 mM EDTA and analyzed by autoradiography of 7.5% polyacrylamide gel electrophoretograms. The results of this experiment, shown in Fig. 2, indicate that the radioactivity recovered in the pellet (Fig. 2, lane 3) corresponds to polypeptides of M_r 48,000 and 32,000, which are indistinguishable from free factor XIa (Fig. 2, lane 1). This experiment confirms that the bound and free proteins are electrophoretically identical. Therefore, the ligand is factor XIa and not a contaminant. This result also provides no evidence for proteolytic degradation of the bound factor XIa or for the formation of high molecular weight covalent complexes.

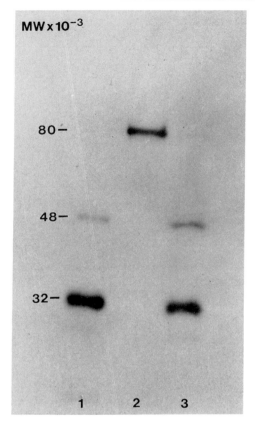

FIG. 2. Reduced SDS gel electrophoresis (7.5% polyacrylamide) and autoradiography of ^{125}I-labeled factor XIa (lane 1), ^{125}I-labeled factor XI (lane 2), and bound ^{125}I-labeled factor XIa after centrifugation through 20% sucrose. Experimental details are given in the text. (Reprinted with permission from Ref. 3.)

Binding Characteristics

Time Course and Requirements

Figure 3 shows the results of a progress curve of binding of factor XIa (added at a concentration of 0.6 μg/ml) to gel-filtered platelets (2.04 × 10^8/μl) incubated at 37° in the presence or absence of high molecular weight kininogen (12.5 μg/ml) and thrombin (0.25 U/ml). It is apparent that binding of factor XIa to platelets requires the presence of high molecular weight kininogen and that the rate and extent of binding are increased

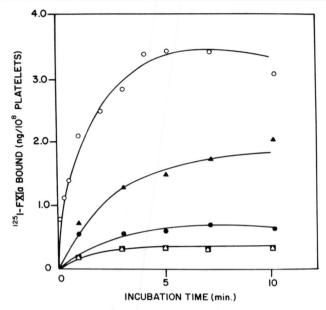

FIG. 3. Progress curves of binding of ^{125}I-labeled factor XIa to platelets or erythrocytes. Details are presented in the text. Incubation mixtures contained platelets + thrombin + high molecular weight kininogen (○); platelets + high molecular weight kininogen (▲); platelets + thrombin (●); platelets alone (△); or erythrocytes + thrombin + high molecular weight kininogen (■). (Reprinted with permission from Ref. 3.)

when platelets are stimulated with thrombin. The optimal conditions for binding have been previously published.[3]

Specificity

The specificity of binding of factor XIa to platelets is determined both with respect to the ligand bound and with respect to the cell that binds it. First, a variety of proteins in 50- to 100-fold molar excess are added with ^{125}I-labeled factor XIa to platelet suspensions incubated with factor XIa, high molecular weight kininogen, and thrombin to determine whether related or unrelated proteins compete with and prevent binding of factor XIa. Results previously published[3] indicate that factor XIa binding to platelets is unaffected by the presence of prothrombin, prekallikrein, factor XIIa, or even the zymogen, factor XI, which is structurally identical to the enzyme, except that factor XIa has been proteolytically activated by cleavage of a single peptide bond in each of the two identical M_r 80,000 subunits of factor XI. In contrast, unlabeled factor XIa, added in 50-fold

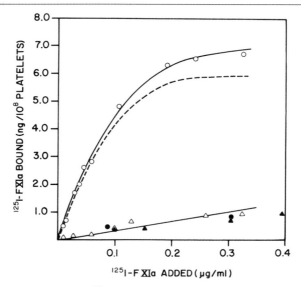

FIG. 4. Saturable binding of ^{125}I-labeled factor XIa to platelets in the presence of high molecular weight kininogen and thrombin (○). Experimental details are given in the text. Other data shown represent nonspecific binding of ^{125}I-labeled factor XIa to platelets in the presence of high molecular weight kininogen + thrombin + a 50-fold excess of unlabeled factor XIa (●); platelets alone (△); or washed erythrocytes + thrombin + high molecular weight kininogen (▲). (Reprinted with permission from Ref. 3.)

excess to ^{125}I-labeled factor XIa, prevents the binding of the labeled ligand ~90%. Factor XIa binding therefore can be shown to be exquisitely specific for the ligand because the binding site recognizes factor XIa but not factor XI.

Specificity of the binding of factor XIa to platelets is demonstrated for the cell as well as for the ligand by data presented in Figs. 3 and 4. Thus, when erythrocytes replace platelets in the binding assay after incubation of factor XIa with high molecular weight kininogen and thrombin, only a low level of nonspecific, nonsaturable binding, unaffected by the presence of high molecular weight kininogen or thrombin, is observed.

Saturability and Reversibility

To determine whether specific binding sites for factor XIa are saturated on activated platelets in the presence of high molecular weight kininogen, the incubation mixture is varied over about two orders of magnitude as shown in Fig. 4. Here, gel-filtered platelets ($2.2 \times 10^8/\mu l$) are incubated for 5 min at 37° with various concentrations of ^{125}I-labeled factor XIa in the presence of high molecular weight kininogen (12 µg/ml) and thrombin (0.1 U/ml) and binding is determined as described above. "Nonspecific binding," determined in the presence of excess unlabeled factor XIa, is low level and nonsaturable, and similar results are obtained when high molecular weight kininogen and thrombin are excluded from the incubation mixture or when red cells replace platelets in the presence of high molecular weight kininogen and thrombin. When "nonspecific binding" is subtracted from total binding, a hyperbolic curve, depicted by the dashed line in Fig. 4, describes specific binding.

Reversibility of factor XIa binding to platelets is examined by incubating reactants in a time course experiment similar to that depicted in Fig. 3 until equilibrium is achieved at 4–5 min, then adding a 50- to 100-fold excess of unlabeled factor XIa and determining the amount of factor XIa bound at subsequent time points. Results of such an experiment, previously published,[3] indicate that 35–40% of the bound ligand is displaced within 10 min.

Determination of Number of Binding Sites and Affinity of Binding

Because of the incompleteness of reversal of binding, we have estimated the characteristics of binding (number of sites and affinity of binding) from direct plots of bound vs added ligand as shown in Fig. 4. Such estimates indicate that the concentration of added factor XIa at which all factor XIa binding sites are occupied ranges between 0.2 and 0.4 µg/ml (1.25–2.5 nM) and that the number of specific binding sites ranges from 3.0 to 11.5 ng/10^8 platelets (mean of 6 ng/10^8 platelets). An assumption that must be made in order to derive number of binding sites and the affinity of binding from Scatchard analysis is that the ligand is freely dissociable from its binding site.[16] If one assumes equilibrium between free and bound ligand by carrying out binding measurements at early time points when the majority of factor XIa binding is dissociable, and determines binding constants by Scatchard analysis, the dissociation constant (K_D) obtained is 3.24 nM for the data shown in Fig. 4 and the amount bound is 8.24 ng/10^8 platelets, which is in good agreement with estimates obtained by direct plots. It can therefore be calculated that the number of specific binding sites on activated platelets is 110–450 per platelet.

Characterization of Bound Ligand

An essential component of the evaluation of a ligand–receptor interaction is the functional characterization of a bound ligand. We therefore have carried out detailed evaluations of the capacity of platelet-bound

[16] G. Scatchard, *Ann. N.Y. Acad. Sci.* **51**, 660 (1949).

factor XIa to catalyze the conversion of factor IX to factor IXa utilizing an assay in which the release of a tritium-labeled activation peptide from factor IX is monitored to determine rates of factor IX activation.[6] Using this assay, the activity of platelet-bound factor XIa as a factor IX activator is fully retained on the platelet surface.[17] In addition we have studied the complex interactions between platelets, factor XIa, α_1-protease inhibitor, and factor IX activation, because α_1-protease inhibitor is the major plasma inhibitor of factor XIa.[4] These studies show that platelets are capable not only of protecting factor XIa from inactivation in the presence of α_1-protease inhibitor, but that platelets also secrete an inhibitor of factor XIa.[4] These results support the view that platelets can regulate factor XIa-catalyzed factor IX activation by secreting an inhibitor of factor XIa that may act primarily outside the platelet microenvironment and by protecting factor XIa from inhibition, resulting in the localization of factor IX activation by factor XIa bound to its platelet receptor.

Acknowledgments

This work was supported by NIH Grants HL25661 and HL14217, by Grant CTR1389 from the Council for Tobacco Research, Inc., and by a grant from the American Heart Association Pennsylvania Affiliate. The authors are grateful to Terry Wyllner for typing the manuscript.

[17] P. N. Walsh, D. Sinha, A. Koshy, F. S. Seaman, and H. Bradford, *Blood* **68**, 225 (1986).

[31] High Molecular Weight Kininogen Receptor

By JUDITH S. GREENGARD and JOHN H. GRIFFIN

Introduction

High molecular weight kininogen is a nonenzymatic cofactor in the reactions of the contact phase of blood coagulation.[1-3] It appears to act as a surface receptor for factor XI and prekallikrein during their surface-dependent activation by factor XIIa.

Early investigations suggested that platelets could promote contact

[1] J. H. Griffin and C. G. Cochrane, *Proc. Natl. Acad. Sci. U.S.A.* **73**, 2554 (1976).
[2] H. L. Meier, J. V. Pierce, R. W. Colman, and A. P. Kaplan, *J. Clin. Invest.* **60**, 18 (1977).
[3] J. H. Griffin and C. G. Cochrane, *Semin. Thromb. Hemostasis* **5**, 254 (1979).

activation.[4,5] Later studies showed that stimulated platelets provide a surface on which the reactions of the contact phase of intrinsic coagulation could occur.[6] Specifically, factor XI was shown to become activated by factor XIIa, and factor XII to be activated by kallikrein in the presence of washed stimulated platelets. Surprisingly, factor XI was also activated in the absence of added factor XII when kallikrein was added to stimulated washed platelets.[6] Previously it had been assumed that kallikrein does not activate factor XI on negatively charged model surfaces.[7] Each of these reactions occurred more readily in the presence of high molecular weight kininogen.[6] This suggested that high molecular weight kininogen might act as a surface binding site for factor XI on the platelet surface, as it does on artificial, negatively charged surfaces.[8] Thus, it seemed probable that platelets would possess high-affinity specific binding sites for high molecular weight kininogen.

Many interactions between platelet receptors and their ligands are dependent on calcium ions. In some cases, for example the binding of fibrinogen to platelets, the receptor itself consists of a calcium-dependent complex of two membrane glycoproteins, GPIIb and GPIIIa.[9-11] On the other hand, the binding of the vitamin K-dependent protein, factor Xa, requires calcium ions, probably at least in part to maintain the ligand in native conformation and to bind the molecule to the membrane.[12]

High molecular weight kininogen has within its sequence a region that is unusually rich in histidine residues.[13,14] This histidine-rich region is important for the procoagulant activity of high molecular weight kininogen[15-17] and is necessary for the binding of the protein to negatively charged

[4] P. N. Walsh, *Br. J. Haematol.* **22,** 237 (1972).
[5] P. N. Walsh, *Br. J. Haematol.* **22,** 393 (1972).
[6] P. N. Walsh and J. H. Griffin, *Blood* **57,** 106 (1981).
[7] C. G. Cochrane and J. H. Griffin, *Adv. Immunol.* **33,** 241 (1982).
[8] R. C. Wiggins, B. N. Bouma, C. G. Cochrane, and J. H. Griffin, *Proc. Natl. Acad. Sci. U.S.A.* **74,** 4636 (1977).
[9] G. O. Gogstad, I. Hagen, M. B. Krutnes, and N. O. Solum, *Biochim. Biophys. Acta* **689,** 21 (1982).
[10] I. Hagen, O. J. Bjerrum, G. O. Gogstad, R. Korsmo, and N. O. Solum, *Biochim. Biophys. Acta* **701,** 1 (1982).
[11] T. J. Kunicki, D. Pidard, J. P. Rosa, and A. T. Nurden, *Blood* **58,** 268 (1981).
[12] J. P. Miletich, C. M. Jackson, and P. W. Majerus, *Proc. Natl. Acad. Sci. U.S.A.* **74,** 4033 (1977).
[13] Y. N. Han, H. Kato, S. Iwanaga, and M. Komiya, *J. Biochem. (Tokyo)* **83,** 223 (1978).
[14] Y. N. Han, M. Komiya, S. Iwanaga, and T. Suzuki, *J. Biochem. (Tokyo)* **77,** 55 (1975).
[15] H. Kato, Y. N. Han, S. Iwanaga, N. Hashimoto, T. Sugo, S. Fujii, and T. Suzuki, in "Kininogenases" (G. C. Haberland, J. W. Kohen, and T. Suzuki, eds), p. 66. Schattauer, Stuttgart, 1977.
[16] H. Kato, T. Sugo, N. Ikari, N. Hashimoto, S. Iwanaga, and S. Fujii, *Thromb. Haemostasis* **42,** 262 (1979).

surfaces.[18] It is known that zinc ions are commonly chelated to proteins via the imidazole side chains of histidine residues.[19-21] This prompted us to investigate the participation of zinc ions in the interaction of high molecular weight kininogen with the platelet surface.

We show here that platelets possess specific, high-affinity receptors for high molecular weight kininogen and that binding of the ligand requires zinc ions. The data suggest that under plasma conditions the high molecular weight kininogen receptors on stimulated platelets would be saturated, and indicate a role for zinc ions either in contact phase reactions or in platelet receptor–ligand interactions.

Methodology

Proteins

High molecular weight kininogen is purified according to Kerbiriou and Griffin.[22] Briefly, citrated plasma is chromatographed over DEAE-Sepharose in a Tris–succinate buffer, pH 8.2, and eluted with a salt gradient. Fractions containing high molecular weight kininogen are pooled and chromatographed over SP-Sephadex in a sodium acetate buffer, pH 5.3, and eluted with a salt gradient. A single polypeptide chain of M_r 110,000 with specific activity of 14 clotting units per milligram is obtained,[22] which is radiolabeled with ^{125}I using the chloramine-T technique.[23] The radiolabeled protein retains its procoagulant activity. In our experience, one preparation that had lost its procoagulant activity after radiolabeling also lost its platelet binding activity. Proteins are stored frozen at $-80°$ in 6 mM sodium acetate, 0.15 M NaCl, pH 5.3. Prior to each experiment, solutions of radiolabeled proteins are centrifuged 1.5 min at 10,000 g at 22° to remove sedimentable labeled particles. This minimizes background precipitation. Human α-thrombin is prepared as described.[24]

---

[17] T. Sugo, N. Ikari, H. Kato, S. Iwanaga, and S. Fujii, *Biochemistry* **19**, 3215 (1980).
[18] A. G. Scicli, G. R. Waldmann, J. A. Guimaraes, G. Scicli, O. A. Carretero, H. Kato, Y. N. Han, and S. Iwanaga, *J. Exp. Med.* **149**, 842 (1979).
[19] J. F. Chlebowski and J. G. Coleman, *Met. Ions Biol. Syst.* **6**, 2 (1976).
[20] J. A. Hartsuck and W. N. Lipscomb, *in* "The Enzymes" (P. D. Boyer, ed.), Vol. 3, p. 1. Academic Press, New York, 1971.
[21] B. W. Matthews, J. N. Jansonius, P. M. Colman, B. P. Schoenborn, and D. Dupourque, *Nature (London), New Biol.* **238**, 37 (1972).
[22] D. M. Kerbiriou and J. H. Griffin, *J. Biol. Chem.* **254**, 12020 (1979).
[23] P. McConahey and F. Dixon, *Int. Arch. Allergy Appl. Immunol.* **29**, 185 (1966).
[24] T. Morita, S. Iwanaga, and T. Suzuki, *J. Biochem. (Tokyo)* **9**, 1089 (1976).

Platelet Washing Procedure

Calcium-free HEPES–Tyrode's buffer consists of 138 mM NaCl, 2.7 mM KCl, 1.0 mM MgCl$_2$ · 6H$_2$O, 3.5 mM N-2-hydroxyethylpiperazine-N'-2-ethanesulfonic acid (HEPES), 5.5 mM dextrose, 3.3 mM NaH$_2$PO$_4$ · H$_2$O, titrated to pH 7.35 or 6.5.[25] This buffer is most conveniently prepared as a 10-fold concentrated solution and stored frozen. When diluted for use, the buffer pH is readjusted, 1.0 mg/ml bovine serum albumin (BSA) is added, and the solution is filtered through a 0.45-μm Millipore (Bedford, MA) filter. Blood is collected from healthy human donors. It is important that donors be medication free, because many common drugs contain aspirin or other antiplatelet compounds.[26] Plastic tubes and pipettes are used throughout. Blood is drawn through a 19-gauge butterfly needle and mixed with one-sixth volume acid citrate dextrose. The blood is centrifuged for 15 min at 160 g to obtain platelet-rich plasma. Centrifugations are performed at room temperature without braking. The platelet pellet is gently resuspended in HEPES–Tyrode's buffer, pH 6.5, and gel filtered at room temperature over Sepharose CL-2B that has been washed with acetone followed by saline containing 0.02% (w/v) sodium azide. It is then equilibrated in HEPES–Tyrode's buffer, pH 7.35.[27] A bed volume of 50 ml is sufficient to filter the platelets obtained from 170 ml blood. The platelet fraction is in the excluded volume of this column, while the plasma elutes in the included volume. The platelets may be counted by phase-contrast microscopy or electronically (Coulter Electronics, Hialeah, FL). Platelets are warmed to 37° in a constant-temperature water bath before use. When ^{125}I-labeled factor XI or ^{125}I-labeled BSA is added as a tracer to the platelet-rich plasma, less than 0.06% is retained in the platelet eluate.[6]

Methodology for Binding Assays

All experiments are performed at 37° without stirring. Platelets (2–3 × 10^8/ml) are incubated in HEPES–Tyrode's buffer, pH 7.35, in a 1.5-ml plastic Eppendorf conical centrifuge tube with mixtures of radiolabeled and unlabeled high molecular weight kininogen, divalent metal ions, and platelet agonists. Solutions of divalent metal ions are filtered through 0.1-μm Millipore filters before use. The platelet agonist (ADP or thrombin) is added last, at time zero. Aliquots are removed and centrifuged in microsediment tubes (Sarstedt, Princeton, NJ) containing a mixture of silicone oils (five parts

[25] S. Timmons and J. Hawiger, *Thromb. Res.* **12**, 297 (1978).
[26] D. A. Triplett, C. S. Harms, P. Newhouse, and C. Clark, *in* "Platelet Function; Laboratory Evaluation and Clinical Application" (D. A. Triplett, ed.), p. 292. Educ. Prod. Div., Am. Soc. Clin. Pathol., Chicago, 1978.
[27] O. Tangen, H. J. Berman, and P. Marfey, *Thromb. Diath. Haemorrh.* **25**, 268 (1971).

DC550 to one part DC200; William F. Nye Specialty Lubricants, New Bedford, MA).[28] The tips of the tubes containing the platelet sediments are amputated using wire cutter pliers and the upper portion and the tip are placed in separate tubes for counting. Radioactivity in both the sediments and the supernatants is determined in order to give independent measurements of bound and free ligand. The advantage of centrifugal techniques for the separation of ligand bound to receptors from free ligand has been discussed.[29] One critical point is that separation of bound and free ligand must be accomplished in a time less than one-seventh of the $t_{1/2}$ for dissociation of bound ligand.[30] This is indeed the case for high molecular weight kininogen binding to platelets, in which the $t_{1/2}$ for dissociation is approximately 3 min,[31] while separation occurs in less than 30 sec.

Controls for Binding Assays

To show that the density of the silicone oil mixture is such that the platelets sediment while the buffer remains suspended above, a number of control experiments are useful. ^{51}Cr-Labeled platelets are used to measure the ability of platelets to sediment through the oil mixture. Washed platelets are centrifuged at 800 g for 15 min and resuspended at 10^9/ml in HEPES–Tyrode's buffer, pH 6.5. These are incubated with 50 μCi of sodium [^{51}Cr]chromate (Amersham, Arlington Heights, IL) for 30 min at 37°. The radiolabeled platelets are gel filtered to remove unabsorbed radioactivity, centrifuged over the silicone oil mixture, and the percentage of sedimented radioactivity is determined. Unless efforts are made to demonstrate that no unabsorbed radioactivity remains with the platelets, this percentage gives only the lower limit of sedimentable platelets. In our hands more than 85% of the platelets sediment to the tube bottom.[31] To determine the "trapped volume" of buffer that is brought down through the oil by sedimenting platelets, unlabeled platelets are incubated with a small amount of ^{125}I-labeled BSA. As the buffer contains 1 mg/ml BSA, the percentage of radiolabel in the pellet is assumed to correspond to the percentage of aqueous volume that sediments. In a typical experiment, we find this to be 8 fl/platelet, or 0.2% of the total volume. It may be concluded from these control experiments that when binding assays are performed, the amount of radiolabeled ligand in the supernatant does not represent a large amount bound to unsedimented platelets, and the amount

---

[28] H. Feinberg, H. Michael, and G. V. R. Born, *J. Lab. Clin. Med.* **84**, 926 (1974).
[29] A. M. Siegl, this series, Vol. 68, p. 179.
[30] J. P. Bennett, Jr., *in* "Neurotransmural Receptor Binding" (H. I. Yamura, J. J. Enna, and M. J. Euhar, eds.), p. 57. Raven Press, New York, 1978.
[31] J. S. Greengard and J. H. Griffin, *Biochemistry* **23**, 6863 (1984).

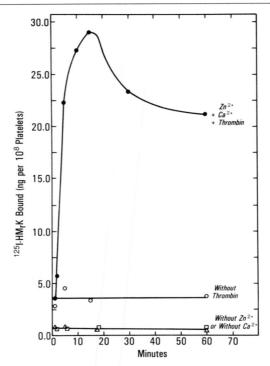

FIG. 1. Binding of ^{125}I-labeled high molecular weight kininogen to platelets. Platelets were incubated unstirred at 37° with 200 ng/ml ^{125}I-labeled high molecular weight kininogen, 0.1 U/ml α-thrombin, or buffer, with or without 25 μM ZnCl$_2$ and 2.0 mM CaCl$_2$. At the indicated times, aliquots were removed and centrifuged. (Reprinted with permission from Greengard and Griffin.[31])

in the platelet pellet does not correspond to ligand molecules in the "trapped volume."

Binding Data

A typical experiment showing binding of ^{125}I-labeled high molecular weight kininogen is illustrated in Fig. 1. Washed platelets were incubated at 37° with a mixture of unlabeled and ^{125}I-labeled ligand in the presence or absence of ZnCl$_2$, CaCl$_2$, and thrombin. Aliquots were removed and centrifuged at the indicated times. Binding required both zinc and calcium ions, as well as platelet stimulation. Maximum binding was reached in approximately 20 min and declined slightly thereafter. This may be due to proteolysis of receptors, platelet aggregation, or other causes. Experiments that study variations of other parameters, such as reagent concentrations, were performed as early as possible after maximal binding is reached. We used a 20-min time point for such experiments. This time course (Fig. 1) is consistent with those reported for other platelet receptors, such as those for fibrinogen, fibronectin, and factor V.[32-34] Control experiments showed that maximal binding occurred when platelets were stimulated with more than 0.25 NIH units (U)/ml thrombin or more than 5 μM ADP.[31] These data are also consistent with optimal levels of agonists giving stimulation for expression of other receptors.[35-37]

Further control experiments showed that when binding to erythrocytes was compared to binding to platelets, more than six times as much high molecular weight kininogen was bound to the platelets as to the erythrocytes.[31] Association of high molecular weight kininogen with the erythrocytes was not dependent on $ZnCl_2$, $CaCl_2$, or thrombin, and it did not increase with time of incubation, suggesting that the binding to platelets is not due to some generalized property of phospholipid membranes on blood cell surfaces. Binding of ^{125}I-labeled high molecular weight kininogen to thrombin-stimulated platelets was studied as a function of the ionic concentration of zinc and calcium (Fig. 2). At all concentrations of $CaCl_2$ tested, binding displayed a marked dependence on the concentration of $ZnCl_2$. At each concentration of $CaCl_2$, binding was maximal at 100–500 μM $ZnCl_2$, and declined at higher $ZnCl_2$ concentrations. Binding was also dependent on $CaCl_2$ concentration, with an optimum at 2 mM. The major effect of increasing the $CaCl_2$ concentration was to lower the threshold of $ZnCl_2$ concentration necessary to yield significant binding of high molecular weight kininogen from 100 μM $ZnCl_2$ in the absence of $CaCl_2$ to 10–25 μM at 2–5 mM $CaCl_2$. Calcium ions appear to potentiate the effect of low concentrations of zinc ions without substituting for them. The horizontal bar in Fig. 2 shows the approximate range of zinc ions in plasma as determined by atomic absorption.[38] At approximately plasma levels of calcium ions, namely 2 mM,[38] considerable high molecular weight kininogen binding was observed in this range of $ZnCl_2$ concentrations.

[32] P. B. Tracy, J. M. Peterson, M. E. Nesheim, F. C. McDuffie, and K. G. Mann, *J. Biol. Chem.* **254**, 10354 (1979).
[33] G. A. Marguerie, E. F. Plow, and T. S. Edgington, *J. Biol. Chem.* **254**, 5357 (1979).
[34] E. F. Plow and M. H. Ginsberg, *J. Biol. Chem.* **256**, 9477 (1981).
[35] E. F. Plow and G. A. Marguerie, *Blood* **56**, 557 (1980).
[36] T. Fujimoto, S. Ohara, and J. Hawiger, *J. Clin. Invest.* **69**, 1212 (1982).
[37] W. H. Kane, M. J. Lindhout, C. M. Jackson, and P. W. Majerus, *J. Biol. Chem.* **255**, 1170 (1980).
[38] J. Woo, J. Treuting, and D. C. Cannon, in "Clinical Diagnosis and Management by Laboratory Methods" (J. B. Henry, ed.), 16th ed., Vol. 1, p. 259. Saunders, Philadelphia, 1979.

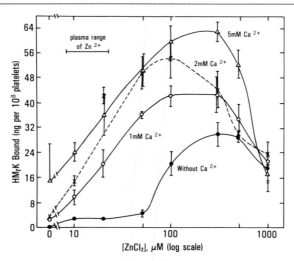

FIG. 2. Effect of calcium ions on binding of ^{125}I-labeled high molecular weight kininogen to platelets at various concentrations of zinc ions. Platelets were incubated at 37° with the indicated concentrations of $ZnCl_2$, ^{125}I-labeled high molecular weight kininogen (225 ng/ml), thrombin (0.1 U/ml), and $CaCl_2$ at 0 mM (●), 1 mM (○), 2 mM (×), or 5 mM (△). After 20 min, aliquots were removed and the amount of ^{125}I-labeled high molecular weight kininogen binding was determined. (Reprinted with permission from Greengard and Griffin.[31])

The possibility that the interaction observed between a protein and its putative receptor represents some nonspecific stickiness of the platelet surface must be excluded. One method of demonstrating specificity is to perform competition experiments, showing that although the unlabeled protein competes with the radiolabeled protein for binding sites, other unrelated proteins do not compete. The results of such an experiment are shown in Table I. A 120-fold molar excess of either one-chain or two-chain high molecular weight kininogen competed away more than 95% of the ^{125}I-labeled one-chain high molecular weight kininogen, but a variety of other proteins at similar concentrations failed to compete significantly. This implies that these binding sites are indeed specific for high molecular weight kininogen.

Analysis of Saturation Data

In principle, almost any two proteins could complex through ion-exchange interactions under sufficiently nonstringent conditions. To assess the physiological significance of an observed binding interaction, it is necessary to ask whether a significant amount of binding would occur under the conditions that resemble the physiological milieu. For receptor binding, two

TABLE I
COMPETITION FOR BINDING OF ^{125}I-LABELED HIGH MOLECULAR WEIGHT
KININOGEN WITH VARIOUS PROTEINS[a]

Competing protein	Percentage control binding
None	100
Single-chain high molecular weight kininogen	3.5
Kinin-free high molecular weight kininogen	7.5
Soybean trypsin inhibitor	79
Factor XII	87
Prekallikrein	93
Ovomucoid trypsin inhibitor	103
Ovalbumin	101

[a] Platelets were incubated with 25 μM ZnCl$_2$, 2.0 mM CaCl$_2$, 0.1 U/ml thrombin, and 200 ng/ml ^{125}I-labeled high molecular weight kininogen mixed with buffer or 60 μg/ml of various proteins. After 20 min, samples were centrifuged. Binding of ^{125}I-labeled high molecular weight kininogen was compared to the control binding in the absence of competing proteins. From Greengard and Griffin.[31]

parameters relating to this point may be measured: the apparent dissociation constant [K_D (app)], and the number of binding sites per cell. If the available concentration of the ligand is far below the K_D (app), then little binding will occur. For available concentrations that are comparable to the K_D (app), the number of binding sites per cell at saturation indicates how many molecules of ligand actually may associate with the cell.

To measure these parameters for the binding of high molecular weight kininogen to the platelet, a saturation experiment was performed. The amount of ligand bound was determined at different high molecular weight kininogen concentrations. The result is shown in Fig. 3 (closed circles). As with many such saturation curves, there are at least two components to the binding, a saturable and an apparently nonsaturable component. The nonsaturable component is usually measured as that binding of radiolabeled ligand that cannot be competed away by a large excess of unlabeled ligand at the concentrations tested. In the experiment illustrated in Fig. 3, this component was fitted as a parameter under the assumption of absolute nonsaturability by the data-fitting computer program LIGAND.[39] The resulting value was verified by the competition method. This value was subtracted from each point, yielding the saturation curve indicated by the open circles. The apparently nonsaturable component of the binding is often equated with nonspecific binding, but this may not be the case. Such

[39] P. J. Munson and D. Rodbard, *Anal. Biochem.* **107,** 220 (1980).

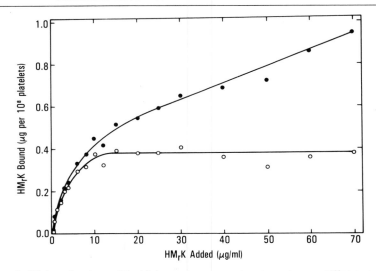

FIG. 3. High molecular weight kininogen concentration dependence of ^{125}I-labeled high molecular weight kininogen binding to platelets. Platelets were incubated with 25 μM ZnCl$_2$, 2.0 mM CaCl$_2$, 0.1 U/ml thrombin, and mixtures of ^{125}I-labeled and unlabeled high molecular weight kininogen at various concentrations. The amount bound was determined at 20 min. Nonsaturable binding was fitted as a parameter using the LIGAND computer program[39] and verified by competition analysis. (●) Total binding; (○) saturable binding. (Reprinted with permission from Greengard and Griffin.[31])

binding may represent specific binding to a large but finite number of low-affinity specific sites.

There are several methods available to obtain the binding parameters from a saturation curve such as Fig. 3. The simplest and least statistically reliable method is to draw the saturation plateau by eye and extrapolate to the y axis to obtain the number of binding sites, using half this value to extrapolate to the x axis to obtain the K_D (app). The subjectivity of this method leads to a large error and irreproducibility. Other graph methods are much less subjective. If binding is at equilibrium and there is a single class of independent binding sites, a Scatchard analysis may be performed.[40] An experiment is analyzed by this method in Fig. 4. Because high molecular weight kininogen binding meets the Scatchard criterion of reversible binding,[31,40] such an analysis will yield accurate results. The ratio of concentrations of bound ligand (B) and free ligand (F) is plotted against bound ligand (B). The slope on the line yields the negative of the inverse of K_D (app), while the x intercept yields the concentration of binding sites, from which the number of receptors per cell can be calcu-

[40] G. Scatchard, *Ann. N. Y. Acad. Sci.* **251**, 660 (1949).

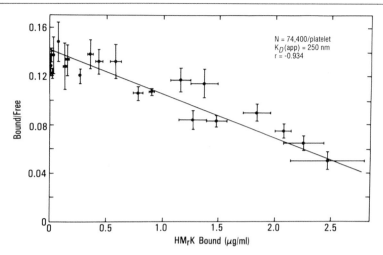

FIG. 4. Least-squares fit of Scatchard analysis of saturation binding data. Platelets were incubated with 500 μM $ZnCl_2$, 0.5 U/ml thrombin and mixtures of ^{125}I-labeled and unlabeled high molecular weight kininogen at various concentrations. The amount bound was determined at various concentrations after 20-min incubations.

lated. Variations on this method that can permit analysis of binding to two classes of sites have been described.[29] Due to the magnification of error inherent in calculating the ratio of B/F, it is important to measure F independently, rather than to measure only B and calculate F by subtraction. The error bars in Fig. 4 illustrate the problem associated with this method, namely, that small errors are magnified near either axis, making the extrapolations more difficult. A second graph method has been proposed by Klotz,[41] in which B is plotted against log F. The advantage of such a plot is that the approach to saturation can be evaluated by the location of the inflection point on the semilog plot relative to the highest B value plotted. However, since the size of the error increases with increasing magnitude of F in such a plot, it becomes difficult to determine either the precise point of inflection or the level of the plateau. As discussed by Munson and Rodbard,[42] the method of graph transformation chosen does not alter the information content of the data, but merely presents the distortion due to error in a different way. They suggest that computerized nonlinear least-squares curve fitting of raw data will yield statistically optimized results.[42] This avoids the distortions of the error distribution in the data due to graph transformations and allows the use of objective

[41] I. M. Klotz, *Science* **217**, 1247 (1982).
[42] P. J. Munson and D. Rodbard, *Science* **220**, 979 (1983).

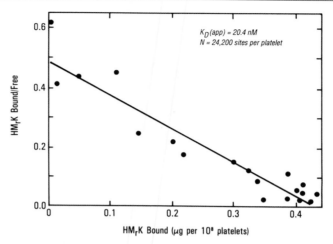

Fig. 5. Nonlinear curve fitting of saturation data in Fig. 3, displayed as a Scatchard plot. Nonsaturable binding, apparent dissociation constant, and number of binding sites were fitted as parameters using the LIGAND computer program.[39] Parameters are listed as ± standard deviation. (Reprinted with permission from Greengard and Griffin.[31])

criteria for goodness of fit.[42] Such a computerized fit[39] of data in the saturation curve (Fig. 3) is displayed as a Scatchard plot in Fig. 5.

The Scatchard plots shown in Figs. 4 and 5 are typical of those obtained under a variety of metal ion conditions using platelets from several donors (Table II). At plasma concentrations of zinc ions (25 μM) and calcium

TABLE II
LIGAND[a]-DERIVED BINDING CONSTANTS FOR HIGH MOLECULAR WEIGHT KININOGEN BINDING TO THROMBIN-STIMULATED PLATELETS

ZnCl$_2$ (μM)	CaCl$_2$ (mM)	Donor number	K_D (app) (nM)	Binding sites per platelet
500	2.0	1	78.1 ± 2.3	83,800 ± 23,400
	2.0	2	117.3 ± 50.1	45,100 ± 19,900
150	2.0	1	58.1 ± 9.3	158,900 ± 22,600
	2.0	3	46.9 ± 5.9	126,800 ± 8,400
	2.0	1	89.6 ± 8.9	142,700 ± 9,900
25	2.0	4	29.8 ± 5.6	21,000 ± 2,200
	2.0	5	20.4 ± 2.9	24,200 ± 1,900
	2.0	1	18.5 ± 2.5	32,200 ± 2,700
	2.0	6	20.5 ± 3.7	38,900 ± 4,100
	2.0	7	26.5 ± 3.6	15,300 ± 1,600
	1.0	7	57.8 ± 12.2	9,000 ± 1,700
	3.0	7	43.7 ± 7.0	27,700 ± 3,500

[a] Munson and Rodbard.[39]

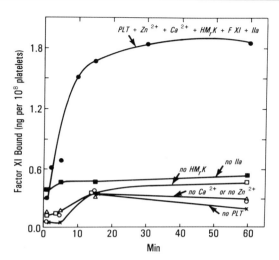

FIG. 6. Binding of ^{125}I-labeled factor XI to platelets. Platelets were incubated without stirring at 37° with ^{125}I-labeled factor XI (50 ng/ml), high molecular weight kininogen (5 μg/ml), or buffer, ZnCl$_2$ (25 μM) or buffer, CaCl$_2$ (2.0 mM) or buffer, and thrombin (0.1 U/ml) or buffer. At the indicated times, aliquots were removed and centrifuged. (From Greengard et al.[49])

ions (2 mM), the K_D (app) ranges from 18 to 30 nM, and the number of receptors per platelet from 15,000 to 39,000. This degree of variation is comparable to interdonor variations observed with fibrinogen binding to platelets.[43] Because the concentration of high molecular weight kininogen in plasma exceeds 600 nM, it may be predicted that these sites are saturated in plasma when they are available on stimulated platelets. At zinc ion concentrations much higher than those in plasma, both the K_D (app) and the number of sites per platelet increase. It is possible that the additional sites arise due to nonphysiologic binding to alternative locations on the platelet surface, which is permitted by the reduction in stringency afforded by the increased zinc ion concentration. In contrast, the major effect of changing the calcium ion concentration appears to be on the number of sites alone, but the effect is slight over the range tested.

Discussion

The data presented here show that high molecular weight kininogen binds to stimulated platelets with high affinity. The binding requires the presence of zinc ions. When physiologic concentrations of calcium ions

[43] G. A. Marguerie, T. S. Edgington, and E. F. Plow, *J. Biol. Chem.* **255**, 154 (1980).

are present, near optimal binding occurs at physiologic concentrations of zinc ions. At these metal ion concentrations, saturation experiments yield apparent dissociation constants far below the concentration of high molecular weight kininogen in plasma. The classic criteria for concluding that a binding interaction is receptor mediated are saturability, reversibility, specificity, and the demonstration that binding parallels physiologic function with respect to time course and concentration.[44] Data presented here and elsewhere show that high molecular weight kininogen binding to the platelet satisfies the first three criteria.[31] Because the ligand is neither a hormone that induces effects on the platelet, nor a zymogen with inducible enzymatic activity, it does not satisfy the last criterion directly. The major physiologic function of high molecular weight kininogen is to act as a surface receptor for factor XI and prekallikrein.[1-3] Presumably, therefore, its function in binding to the platelet is to provide a binding site for one or both of these proteins. The data in Fig. 6 indicate that high molecular weight kininogen supports factor XI binding to the platelet under the same ion concentrations of zinc and calcium that are near optimal for high molecular weight kininogen binding, and that the time course of factor XI binding is similar to that of high molecular weight kininogen binding (Fig. 1). It may be concluded that the effects of high molecular weight kininogen parallel its binding at least with respect to time course. Therefore, it may be inferred that the interaction of high molecular weight kininogen with stimulated platelets is indeed receptor mediated.

Recently other investigators have identified high molecular weight kininogen receptors on unstimulated platelets, neutrophils, and endothelial cells,[45-47] as well as low molecular weight kininogen receptors on unstimulated platelets[48] and the high molecular weight kininogen-dependent binding of factor XI to stimulated platelets.[49]

Acknowledgments

The skilled technical assistance of Eva Ersdal, Mary Ann Batard, and Jennifer Oldstone is gratefully acknowledged, as is the efficient secretarial assistance of Cheryl McLean and Jeanne Schoenwandt.

[44] O. Hechter, *Adv. Exp. Med. Biol.* **96**, 1 (1975).
[45] E. J. Gustafson, D. Schutsky, L. Knight, and A. M. Schmaier, *J. Clin. Invest.* **78**, 310 (1986).
[46] E. J. Gustafson, A. M. Schmaier, Y. T. Wachtfogel, N. Kaufman, U. Kucich, and R. W. Colman, *J. Clin. Invest.* **84**, 28 (1989).
[47] F. Van Iwaarden, P. G. de Groot, and B. N. Bouma, *J. Biol. Chem.* **263**, 4698 (1988).
[48] F. J. Meloni and A. M. Schmaier, *J. Biol. Chem.* **266**, 6786 (1991).
[49] J. S. Greengard, M. J. Heeb, E. Ersdal, P. N. Walsh, and J. Griffin, *Biochemistry* **25**, 3884 (1986).

[32] Binding Characteristics of Homologous Plasma Lipoproteins to Human Platelets

By ELISABETH KOLLER and FRANZ KOLLER

Assay Methods

Principle

Human plasma lipoproteins of the high-density class (HDL) and the low-density class (LDL) show fast and saturable binding to human platelets. The parameters of these interactions are determined applying lipoproteins labeled covalently in their protein moieties (either with ^{125}I or with fluorescent dyes).

Buffers

Buffer A: Tyrode's solution without Ca^{2+} (NaCl 137 mM, KCl 2.7 mM, NaHCO$_3$ 11.9 mM, MgCl$_2$ 1.0 mM, NaH$_2$PO$_4$ 0.42 mM, D-glucose 5.5 mM, human serum albumin 3.5 g/liter, pH 6.5). To improve the pH stability, 5.0 mM HEPES was added in all experiments requiring longer incubation periods

Buffer B: As above, with addition of 100 μg/ml apyrase (Sigma, St. Louis, MO) and 10 units heparin/ml

Buffer C: Buffer A, containing 2 mM CaCl$_2$, pH adjusted to 7.35

Buffer L: Buffer C, without albumin and glucose

Procedures

Preparation of Washed Human Platelets. Platelet-rich plasma (PRP) is obtained from ACD-blood of healthy donors immediately after drawing by sedimentation at 120 g at room temperature for 20 min. After addition of apyrase (50 μg/ml) platelets are isolated either by gel filtration on Sepharose 2B[1] or by repeated centrifugations.[2] The final suspension medium in both cases is buffer C.

[1] O. Tangen, H. J. Berman, and P. Marfey, *Thromb. Diath. Haemorrh.* **25**, 268 (1971).
[2] J. F. Mustard, D. W. Perry, N. G. Ardlie, and M. A. Packham, *Br. J. Haematol.* **22**, 193 (1972). Briefly, PRP is spun twice for 5 min at 800 g in small tubes (5 ml) at room temperature. The combined pellets are resuspended in buffer B at 37°. The same sequence of centrifugation is repeated three times, first in identical manner, then with resuspension in buffer A, and finally in buffer C.

Preparation of Lipoproteins. The supernatant of the initial 800 g sedimentation of PRP, platelet poor plasma (PPP), is spun sequentially at 100,000 g and 5° following stepwise adjustment of required density values by addition of solid KBr. Prior to each flotation step the residual plasma is overlayered with KBr solutions of appropriate density to remove plasma proteins adhering to the lipoprotein fraction. The mimimal times of separation are 16 hr for very low-density lipoprotein (VLDL) floated at $d = 1.006$, 20 hr for LDL isolated at $d = 1.006$ to 1.063, and at least 36 hr each for HDL_2 at $d = 1.063$ to 1.125 and for HDL_3 at $d = 1.125$ to 1.21. Finally, lipoproteins are dialyzed exhaustively against buffer L and passed through filters of 0.45-μm pore size. The composition of each preparation is examined by determination of all components and compared with average literature values.[3] Based on this known composition, determinations of protein concentration routinely are sufficient to calculate lipoprotein concentrations.

^{125}I Labeling of Lipoproteins. The method of Salacinski *et al.*,[4] using insoluble 1,3,4,6-tetrachloro-3α,6α-diphenylglycoluril (Iodogen), is applied. Iodogen (80 μg in 50 μl of dry CH_2Cl_2) is transferred to a Beckman (Palo Alto, CA) microfuge vial and the solvent removed by aspiration with dry nitrogen. Two microliters of $Na^{125}I$ (3.7 MBq/ml; 62.9 MBq/μg) is added, immediately followed by addition of 200 μl of lipoprotein solution containing about 1 mg protein/ml. The reaction is allowed to proceed for 10 to 15 min at 5°. The liquid phase is then transferred to another vial containing 20 μl of 2 M aqueous KI and 800 μl of buffer L. After mixing, the content of this vial is applied to a 10-ml column of Sephadex G-25 fine, equilibrated in the same buffer, and fractionated at room temperature, collecting 10 drops per fraction. Radioactively labeled lipoprotein appears as a very narrow peak without detectable contamination with lower molecular weight material. The radioactivity incorporated typically is about 40 kBq/μg of protein. More than 90% of the label introduced by this procedure is precipitable by trichloroacetic acid in all lipoprotein density classes. The portion of the total label associated with the lipid moiety was found to be $\leq 8\%$ (HDL_3) by delipidation following Folch *et al.*[5]

Fluorescence Labeling of Lipoproteins. Standard procedures are applied to achieve covalent modification of different lipoprotein fractions

---

[3] S. Eisenberg, *Ann. N. Y. Acad. Sci.* **348**, 30 (1980); J. C. Osborne, Jr., and H. B. Brewer, Jr., *Adv. Prot. Chem.* **31**, 253 (1977).

[4] P. R. P. Salacinski, C. H. McLean, J. E. C. Sykes, V. V. Clement-Jones, and P. J. Lowry, *Anal. Biochem.* **117**, 139 (1981).

[5] J. Folch, M. Lees, and G. H. S. Stanley, *J. Biol. Chem.* **226**, 497 (1957).

with fluorescein isothiocyanate (FITC) at pH 8.5, and with N-iodoacetyl-N'-(S-sulfo-1-naphthyl)ethylenediamine (1,5-I-AEDANS) at pH 6.5.[6] Purification of the labeled lipoprotein is achieved by centrifugation and repeated gel filtration at high and low salt concentrations, finally with buffer L. The amounts of covalently attached dyes are calculated spectroscopically using $\varepsilon_{495} = 37,500$ for FITC, and $\varepsilon_{350} = 6600$ for 1,5-I-AEDANS, respectively.[7]

Radioactive assay method: A series of incubations in conical microfuge tubes is carried out at 37° for 15 min. Each vial contains 0.2 ml of washed platelets suspended in buffer C (7–30 × 10^8 cells/ml), varying amounts of labeled lipoprotein, and buffer L up to a total volume of 0.5 ml. To determine the amount of unspecific, nonsaturable binding, otherwise identical incubations were performed with the addition of unlabeled lipoprotein. The amounts of added unlabeled lipoprotein were sufficient to yield total lipoprotein concentrations at least 100 times higher than those required for half-maximal saturation of the high-affinity sites. After the desired incubation times the incubation mixtures are layered on top of 0.5 ml 20% sucrose solution in buffer L and spun in a Beckman Microfuge at 8740 × g for 45 sec. The tips of the tubes containing the platelets are amputated after careful aspiration of the supernatant and radioactivity is determined in both tips and supernatants. Almost identical results were obtained with the following procedure formerly in use, which, however, is not fast enough to be applied to large series of determinations.[8] The platelets are sedimented by centrifugation of the incubation mixtures (8740 × g, 45 sec). The pellets are resuspended three times in 150 μl of ice-cold buffer C by repeated agitation of the medium against the platelet sediment by use of an appropriately sized plastic tip attached to a hand-operated automatic pipette and immediate centrifugation for 30 sec, the whole washing procedure being accomplished within less than 5 min. Finally the platelet pellet is resuspended in another 150 μl of buffer C and transferred to a new plastic tube. Radioactivity in the latter as well as in the original supernatant solution and in the three washing solutions is determined in a γ counter.

The concentration of the fraction of lipoprotein remaining unbound at equilibrium is derived from the sum of counts of all supernatant solutions;

---

[6] K. G. Mann and W. W. Fish, this series, Vol. 26, p. 28; E. N. Hudson and G. Weber, *Biochemistry* **12**, 4154 (1973).

[7] C. R. Cantor and P. R. Schimmel, "Biophysical Chemistry." Freeman, New York, 1980.

[8] E. Koller, F. Koller, and W. Doleschel, *Hoppe-Seyler's Z. Physiol. Chem.* **363**, 395 (1982).

the washed platelet suspension yields the bound fraction of radioactive ligand. At low total lipoprotein concentrations (final concentrations in the vials in the nanomolar range) at least triplicate determinations should be performed; at micromolar total concentrations of ligand two series of experiments will be sufficient.

Fluorescence method: To minimize disturbation by scattering, all determinations of fluorescence intensities are performed with plane polarized light. The excitation bandpass should be adjusted to less than 4 nm, that in the emitted light beam to about 5 nm. The cell holder should be thermostatted to about 25°. The fluorescence intensities (either indicated directly or calculated as $I = I_{\parallel} + 2I_{\perp}\sigma$) are determined at 350/475 nm when working with 1,5-I-AEDANS-labeled lipoproteins and at 495/520 nm for FITC-modified ligands.

Stock solutions of labeled lipoprotein are prepared by dilution with buffer L. Undiluted lipoprotein together with dilutions to one-third and one-fifth usually will suffice to obtain a complete binding isotherm. Platelet suspension (0.1 ml; 10 to 20 × 10^8 cells/ml) is then pipetted into a 1 × 1 cm standard fluorescence cuvette and 1.9 ml of buffer L is added. After mixing and determination of the background fluorescence intensity the chromophore-labeled lipoprotein is added stepwise (in portions of 10 or 20 μl) to this suspension. After each addition the system is allowed to equilibrate (with occasional mixing) for 15 min at 25°,[9] then the fluorescence intensities are determined as above. Five (or 10) consecutive additions of lipoprotein are recommended for each of the 3 prepared stock solutions. The same sequence of additions of ligand is repeated with mixtures containing (1) 50 μl of the original platelet suspension per 2 ml, (2) 200 μl/2 ml, and (3) no platelets at all, otherwise proceeding in an identical manner.

All measured fluorescence intensities are corrected for inner filter effect and scattering. The following procedure is both simple and reliable: A high-purity, reasonably water-soluble fluorescence standard with high quantum yield in aqueous solution and a sufficiently large absorption coefficient in the spectral range of interest is appropriately diluted with buffer L and fluorescence is determined under the conditions described above. The same is done for equally concentrated solutions of this standard chromophore, but in the presence of 0.05, 0.1, and 0.2 ml of the platelet suspension under investigation. Comparison of these

---

[9] Fifteen minutes might be too short to allow equilibration with very low concentrations of ligand. In these rare cases no further increase in fluorescence intensities was detectable after incubation periods of 30 min.

relative fluorescence intensities yields empirical factors correcting for optical effects related to the platelets in suspension. Effects associated with the added lipoprotein normally are negligible, since the latter will add very little to the total absorption at the wavelength of excitation. Possible complications due to interactions of the standard dye with platelets can be excluded by measuring at two temperatures separated by about 10 to 15°. In the absence of interactions the correction factors for optical effects should show no detectable temperature dependence.[10]

The (corrected) values of relative fluorescence intensity are increased in the presence of human platelets when compared with the series in plain buffer. The reciprocals of these changes of fluorescence intensity ($1/\Delta I$) are plotted against the reciprocal concentration of lipoprotein as shown in Fig. 1. Slightly curved binding isotherms are obtained, which approach linearity with decreasing concentration of platelets. The left part of the plot can be linearly extrapolated to infinite ligand concentration with proper accuracy. Based on the assumption of kinetically indistinguishable receptor sites the intercept on the vertical axis is $1/nPb$, that with the abscissa corresponds to K. P represents the concentration of platelets (in molarities, treating platelets as molecules), n is the average number of receptor sites per single platelet, K stands for the thermodynamic association constant for binding to the receptors, and b is the incremental change of fluorescence intensity of the average chromophore molecule on binding, depending on instrumental parameters and to some extent on the degree of lipoprotein labeling. This value must be known to get access to the number of sites/platelets, which can be achieved without any additional measurement according to Fig. 2. For different (preferentially low) concentrations of total lipoprotein (T) kept constant, the corresponding values of $T/\Delta I$ are plotted against the reciprocals of platelet concentration. Extrapolation of these more or less linear plots to the ordinate should yield a common intercept, its numerical value being $1/b$.

Again the accuracy of the individual determinations is crucial for

---

[10] The deviation of the apparent fluorescence intensities observed in the presence of platelets from the corrected values is less pronounced than might be anticipated. As typical examples from our experiments the fluorescence intensities measured at 520 nm had to be corrected by factors of 0.979 and 0.919 for 100 and 300 µl of original platelet suspension (16.2 × 10^8/ml), respectively. We used lysine derivatives of both 1,5-I-AEDANS and FITC as standard dyes to determine these values (adding 40 µl of a solution in buffer L, containing about 5 mg of dye/ml). Reaction of either fluorescent derivative with the platelet plasma membrane proceeds only very slowly (if at all) and can easily be corrected for by extrapolating back to mixing time.

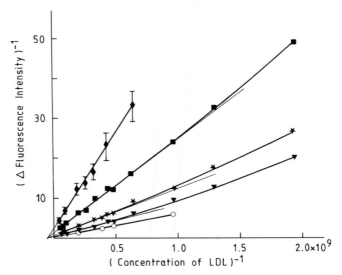

FIG. 1. Double-reciprocal plot illustrating LDL binding to blood platelets at 37°. Low-density lipoprotein was labeled with FITC (44 : 1), binding leading to increased fluorescence at 495/520 nm. The final concentrations of platelets were 0.425 (◆), 0.85 (■), 1.7 (★), 2.95 (▼), and 4.25×10^8 cells/ml (○). Error bars are shown for the most diluted series.

the reliability of the final results. Therefore the reference series (in the absence of platelets) should be repeated twice. The experiments in the presence of platelets should be performed in duplicate. Quite similar results have been obtained with the two types of chromophores described. The use of FITC is to be preferred, however, because of its markedly higher fluorescence yield and its absorption at longer wavelengths.

Unnecessary illumination of samples should be avoided. The serum albumin included in the platelet suspension buffer systems should be of the highest available degree of purity (delipidated if possible) to minimize background emission close to the wavelengths applied in this assay. Partial peroxidation of lipoprotein lipids can produce similar complications. Although several methods are known to inhibit lipid peroxidation, the best means to avoid this interference is to use only freshly prepared lipoproteins, since plasma lipoproteins in any case *in vitro* show a distinct tendency to change structure, size, and composition with time.

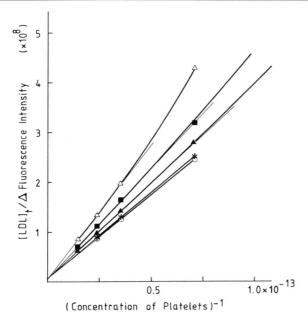

Fig. 2. Extrapolation to infinite platelet concentration of the reciprocal changes of fluorescence (495/520 nm) of FITC-labeled LDL. The individual series were determined at constant total concentrations of LDL (LDL_t): 0.35 (○), 0.52 (∗), 1.04 (▲), 3.1 (■), and 6.7 × 10^{-9} M (△). The data are obtained from the series of experiments illustrated in Fig. 1.

General Considerations and Problems in Receptor Binding Studies Involving Lipoproteins

Both size and surface structure make plasma lipoproteins (especially of the lower buoyant density classes) highly adherent to most common laboratory equipment materials, including glass and various types of plasticware. Our attempts to find any combination of material and pretreatment virtually excluding nonspecific adhesion were unsuccessful.

Binding data generally are best obtained by determination of the amount of free (unbound) ligand, calculating the bound fraction as the difference between this concentration and the known total ligand concentration. In the case of plasma lipoproteins (especially at low degrees of receptor saturation, i.e., at low ligand concentration), this difference represents not only ligand bound to the platelets but also

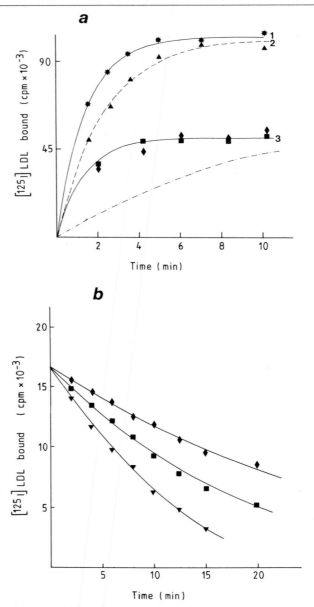

Fig. 3. (a) Time dependence of binding of ^{125}I-labeled LDL to blood platelets. 1. Incubation of 4.5×10^8 platelets/ml with 1.9×10^{-7} M LDL. The solid line is calculated with $k_{ass} = 6.3 \times 10^4$ M^{-1} sec^{-1}. 2. As above, with the addition of HDL$_3$ (4.4×10^{-7} M); the dotted line was obtained with $k_{ass} = 3.7 \times 10^4$ M^{-1} sec^{-1}. 3. Incubation of the same number of

adherent to the vessel wall. Therefore, the amount of platelet-bound lipoprotein must be determined directly. With radioactively labeled ligand this means separation of platelets and supernatants. As shown in Fig. 3, binding of lipoproteins to the platelet surface proceeds quite rapidly, and dissociation of this complex is still rather fast, the corresponding rate constants being $6.3 \times 10^4 \, M^{-1} \, \text{sec}^{-1}$ and $1.27 \times 10^{-3} \, \text{sec}^{-1}$, respectively, at 37°. Consequently, the separation must be accomplished in a rather short time. Separation by filtration was completely unsatisfactory again because of nonspecific binding of free lipoprotein to the filter matrix, irrespective of material, pretreatment, and washings. Separation by sedimentation as described obviously is fast enough not to deteriorate the results to any serious extent. The sucrose centrifugation procedure described above overcomes the problem of dissociation of bound ligand by avoiding time consuming washing steps in the separation of bound and free ligand. This method is, however, not applicable when investigating the kinetics of platelet–ligand interaction because of the delicate, rather long-lasting overlayering process. In those cases the centrifugation/resuspension procedure can be successfully applied. The platelet button obtained by the first centrifugation step may contain a rather large volume of occluded supernatant fraction, which has to be completely removed to achieve reliable results. To allow complete washing out of these contaminants, the pellet must be capable of being resuspended in rather short times. Resuspension by vortexing in our hands was far less satisfactory than the application of the method described above. Removal of the supernatant is again best done with an automatic pipette, leaving a layer of about 1 mm depth on top of the pellet. If carefully done, no loss of platelets will occur under these conditions. This volume of top layer after three washings will be sufficiently diluted to represent only negligible contami-

---

platelets with half the amount of [^{125}I]LDL as above. The same amount of (■) unlabeled LDL and (♦) FITC-labeled LDL (36 : 1), respectively, was added. The total concentration of LDL was thus kept unchanged as compared with experiments 1 and 2. The solid line is calculated based on the assumption of identical binding behavior of unmodified LDL and covalently modified ligands. The dashed–dotted line represents the calculated binding kinetics assuming binding of radiolabeled ligand only. (b) Kinetics of dissociation of the ^{125}I-labeled LDL–receptor complex. Data were obtained by resuspension of 0.96×10^8 platelets incubated with LDL in 1 ml buffer C and incubation for the indicated time periods. (♦) No addition; the solid line is the theoretical behavior calculated with $k_{\text{ass}} = 6.3 \times 10^4 \, M^{-1} \, \text{sec}^{-1}$ and $k_{\text{diss}} = 1.27 \times 10^3 \, \text{sec}^{-1}$; (■) addition of HDL$_3$ ($0.76 \times 10^{-6} \, M$); the kinetic parameters are changed to $k_{\text{ass}} = 3.7 \times 10^4 \, M^{-1} \, \text{sec}^{-1}$ and $k_{\text{diss}} = 2.2 \times 10^{-3} \, \text{sec}^{-1}$, respectively; (▼) in the presence of 1800 units of heparin/ml.

nation. An alternative method, originally successfully applied by us to check our results obtained by the latter method,[8] is the separation by gel filtration on a semimicroscale. Small (5–7 ml bed volume) columns of BioGel A-150m (Bio-Rad, Richmond, CA) or similar materials, equilibrated with buffer L at 5°, are "calibrated" with respect to the elution zones of platelets and lipoproteins and then loaded with the mixtures to be separated. The platelet fraction is eluted within 3–4 min, fast enough to give correct results. With the cited column material almost complete recovery of the applied radioactivity was achieved. Since the more slowly eluting lipoprotein fraction must be collected quantitatively as well, this method is certainly too slow to be applied routinely.

Aviram et al.[11] and Mazurov et al.[12] developed two principally similar assay procedures to study the binding of LDL. Both methods, however, include rather slow washing steps, leading to unnecessary dilution. Significant losses by dissociation therefore can be expected, leading to nonlinear binding kinetics and isotherms.

The method of radioiodination does to some extent have influence on the results. The ability of the labeled lipoprotein to bind to the platelet plasma membrane is not impaired by using either the Bolton–Hunter reagent (as originally done by us[8]) or the iodine monochloride method, as for the majority of lipoprotein receptor studies in other cell systems.[13] The latter method, however, leads to some labeling of the lipid moiety (up to 25% of the total labeling), which increases the possibility of artifacts by lipid-exchange reactions between cells and ligands. The Bolton–Hunter method almost completely avoids this source of error, but the degree of labeling is markedly lower than when applying Iodogen, leading to less accuracy in the most sensitive part of the binding isotherm.

Covalent labeling of the ligand with a fluorescent chromophore involves several disadvantages and two major benefits. The method principally must be expected to be less accurate than the former one, basically because of the necessity to work with solutions diluted to a higher degree. Furthermore, changes in intensity are determined rather than intensities themselves. Taking into account the slightly lower sensitivity of the method per se, this means a considerably higher demand for experimental skill and reproducibility. On the other hand this method works without separation of bound and unbound ligand and therefore allows direct obser-

[11] M. Aviram, J. G. Brook, A. M. Lees, and R. S. Lees, *Biochem. Biophys. Res. Commun.* **99**, 308 (1981).
[12] A. V. Mazurov, S. N. Predbrazhensky, V. L. Leytin, V. S. Repin, and V. N. Smirnov, *FEBS Lett.* **137**, 319 (1982).
[13] D. W. Bilheimer, S. Eisenberg, and R. I. Levy, *Biochim. Biophys. Acta* **260**, 212 (1972).

vation of unstable (fast dissociating) complexes. In addition, changes with time can easily be observed, and it has at least the potential to provide some structural information on the receptor–ligand complex concomitantly.

Covalent modification of lysine residues of lipoprotein (LDL) does not seriously affect their ability to bind to blood platelets[11,12,13a] (cf. Fig. 3). This behavior differs markedly from the results obtained with human fibroblasts. Accordingly, the interaction with the plasma membrane of platelets remains intact after incorporation of up to 50 molecules per molecule of lipoprotein (LDL) of either of the two fluorescent dyes cited.

The interaction of platelets with rhodamine isothiocyanate (RITC)-labeled LDL has been studied by Mazurov et al.[12] The authors applied the technique of flow cytofluorimetry and found this method superior to an assay based on ^{125}I labeling.

Characteristics of Binding

Specificity

As mentioned above, the major problem associated with the physicochemical characteristics of plasma lipoproteins is their tendency to associate to both polar and nonpolar surfaces. The specific nature of the observed binding to blood platelets has been demonstrated (mostly for LDL) by various methods, including replacement of bound labeled ligand by unlabeled lipoprotein, release induced by heparin of specifically bound LDL, and inhibition of binding by polyclonal and monoclonal antibodies against glycoprotein IIb and IIIa, which have been identified as lipoprotein-binding proteins,[14] and by polyclonal antibodies against apo B,[15] respectively.

Binding studies with lipoprotein receptors of various types of mammalian tissue have sometimes revealed an additional nonspecific (nonsaturable, low affinity) binding. Under the experimental conditions described above, this is practically absent for lipoproteins with density > 1.006. Very low-density lipoproteins, however, show nonspecific, nonsaturable binding to the platelet surface to an extent making its elimination impossi-

[13a] The authors report a serious inhibition of binding to platelets of LDL following peracetylation. Careful examination of their results, however, indicates that the capability of LDL modified in this way to compete with ^{125}I-labeled LDL for binding declines only after prolonged incubation at 37° (more than 60 min).
[14] E. Koller, F. Koller, and B. R. Binder, *J. Biol. Chem.* **248,** 12412 (1989).
[15] E. Koller, *FEBS Lett.* **200,** 97 (1986).

ble. Binding to specific sites may still occur, but under these conditions it escapes detection.

Number and Affinity of Sites

The binding behavior of the different classes of lipoproteins can be interpreted by assuming one single class of macroscopically homogeneous receptor sites each. Cooperativity between sites and/or clustering following binding cannot be detected (they are ruled out by the observation that "fixation" of the platelet surface by treatment with formaldehyde prior to the addition of ligand has little influence on the binding isotherms). Quantitative description is summarized in Table I, including the data given by other authors. The conclusions drawn from preliminary investigations with respect to the assumed correlation between binding and apoprotein composition are included. As far as tested by us, every type of lipoprotein interferes with binding of any other class of lipoprotein in a mixed-type, noncompetitive way. Thus, the binding of one class of lipoprotein alters the characteristics of the remaining free binding sites for nonidentical lipoproteins, but already bound lipoproteins are also affected by subsequent binding of some different class of lipoprotein (Table II).

Temperature Dependence

Within the range from 4 to 37° the association constant for the receptor binding of LDL shows only a slight increase, indicating an entropy-driven reaction; the rate constant of association exhibits a more pronounced temperature dependence (Table III).

Uptake of Lipoproteins

The probability of incorporation of surface-bound ligand into the platelet interior is rather small, taking into consideration (1) the almost complete replacement of bound labeled ligand by unlabeled ligand applied in proper excess, (2) the lack of influence of fixation with formaldehyde, and (3) (at least as far as energy-consuming processes of uptake are concerned) the almost complete lack of temperature dependence. On the other hand, since phospholipid exchange between the plasma membrane of intact platelets and HDL has been demonstrated,[16] the binding isotherms obtained by application of radiolabeled ligands could be affected by partial exchange

[16] D. G. Hassall, K. Desai, J. S. Owen, and K. R. Bruckdorfer, *Platelets* **1**, 29 (1990).

TABLE I
BINDING OF PLASMA LIPOPROTEINS TO HUMAN BLOOD PLATELETS

Lipoprotein class	Authors	Number of sites per platelet (± SD)	K^a ($\times 10^{-7}$, M^{-1}) (± SD)	Approximate half-maximal saturation (μg protein/ml)	k_{assoc}^b ($\times 10^{-4}$) M^{-1} sec^{-1}	k_{dissoc}^c ($\times 10^3$) sec^{-1}	Time to attain equilibrium (min)	Apolipoproteins involved
LDL	Koller et al.[d]	1470 ± 640	6.2 ± 2.2	10	3.5	1.3	5–10	n.d.
	Koller et al.[e]	1840 ± 750	5.7 ± 2.4	10	6.3	1.27	5–10	apo B/E
	Hassall et al.[17]	1965 ± 177	5	10	n.d.	n.d.	1–10	apo B
	Mazurov et al.[12]	1900[f]	n.d.	40[f]	n.d.	n.d.	5–15	n.d.
	Curtiss et al.[g]	7075 ± 4800	2.5	20	n.d.	n.d.	120	apo B
	Aviram et al.[20]	n.d.	n.d.	300	n.d.	n.d.	n.d.	n.d.
HDL$_2$	Koller et al.[e]	1520 ± 450	6.1 ± 3.0	10	3.5	n.d.	5–10	n.d.
HDL-E	Desai et al.[21]	4193 ± 1337	0.12	470	n.d.	n.d.	15	apo E
HDL$_3$	Koller et al.[d]	3200 ± 410	9.0 ± 1.7	1	6.7	1.08	5–10	apo E
	Koller et al.[e]	3200 ± 1100	9.0 ± 2.3	1	n.d.	n.d.	5–10	
HDL$_2$ + HDL$_3$	Curtiss et al.[g]	1585 ± 390	3.1 ± 2.8	8	n.d.	n.d.	120	apo AI/AII
	Aviram et al.[20]	n.d.	n.d.	400	n.d.	n.d.	>180	n.d.

Note. n.d., not determined.
[a] Binding constant.
[b] Rate constant for association.
[c] Rate constant for dissociation.
[d] As determined with ^{125}I-labeled LDL, cf. ref. 8.
[e] As determined with FITC-labeled LDL, unpublished.
[f] As estimated from Fig. 2 in ref. 12.
[g] L. K. Curtiss and E. F. Plow, *Blood* **64**, 365 (1986).

TABLE II
BINDING OF INDIVIDUAL CLASSES OF PLASMA LIPOPROTEINS IN PRESENCE OF
INHIBITORY LIPOPROTEIN SUBCLASSES[a]

Inhibitor	Association constants for binding of indicated class of lipoprotein[b] (M^{-1}, × 10^{-7})			Number of LDL-binding sites[c]
	K_{HDL_3}	K_{HDL_2}	K_{LDL}	
HDL$_3$	(9.0)	ND[d]	0.8	910
HDL$_2$	2.3	(6.1)	5.5	1860
LDL	1.6	5.4	(5.7)	(1840)
VLDL	7.5	ND[d]	5.0	1290

[a] The pattern of inhibition in most cases is more complex than predicted by simple competition for common sites. Binding data in these cases were analyzed in analogy to noncompetitive (allotopic) enzyme inhibition schemes [cf. H. B. Halsal, *Trends Biochem. Sci.* **5**, IX–X, and references herein (1980)].
[b] Binding constants as indicated represent the calculated limiting affinity at saturating concentrations of inhibitor.
[c] The number of binding sites as a rule is also reduced in the presence of competing lipoproteins. The values as presented in the table are lower limits observed at saturating concentration of inhibitor.
[d] ND, Not determined.

of the (small) fraction of labeled lipids. As outlined elsewhere in detail,[8] the results, however, are barely affected by stoichiometric exchange. Slightly hyperbolic curves should be expected in Scatchard analysis, extrapolation leading to the correct value of n (number of sites per platelet) and to only slightly incorrect values for K. Assuming nonstoichiometric exchange (i.e., net transfer of lipids from lipoproteins to platelets) similar plots would be obtained. The apparent binding constants, however, would then represent the affinity of the lipid-deficient, rather than of the native lipoprotein to the receptor.

Identification of the Lipoprotein Binding Proteins

As reported recently,[14] the two major lipoprotein-binding membrane proteins were purified to apparent homogeneity and identified as glycoprotein IIb (GPIIb) and IIIa (GPIIIa), respectively, by their polypeptide size and by specific antibodies against these glycoproteins. The possible exis-

TABLE III
TEMPERATURE DEPENDENCE OF LDL BINDING

Temperature (°C)	K (M^{-1}, $\times$ 10^{-7})	k_{ass} (M^{-1} sec^{-1}, $\times$ 10^{-4})
37	5.7	6.3
25	5.2	5.1
22	4.9	3.9
12	4.4	3.7
4	3.9	2.9

tence of a further receptor protein(s) cannot be ruled out. In some of our membrane preparations a third lipoprotein binding protein was detectable, and Hassal et al.[17] reported a single lipoprotein binding protein with molecular weight of 140 kDa.

Physiological Function of Blood Platelet Lipoprotein Receptors

The most important role of lipoprotein receptors in a wide variety of cells is to supply them with cholesterol and to regulate the biosynthesis of cholesterol (apoB/E-receptor-mediated LDL pathway).[18] Platelets lack the ability to synthesize cholesterol and so the physiological role of lipoprotein-binding proteins can be expected to be somewhat different from that of the apoB/E receptor. This assumption is further supported by the identification of GPIIb and GPIIIa as binding proteins which are distinct from the apoB/E receptor, and by the presence of saturable LDL binding sites on platelets of patients with homozygous familial hypercholesterolaemia, lacking the classical LDL receptor.[17] There is, however, evidence for lipid exchange between different types of lipoprotein molecules and the platelet plasma membrane. As a consequence of this exchange the functional state of the platelets may be altered.[16,19] Most importantly, the binding of lipoproteins by platelets may be directly related to platelet activation. The long-known enhancement of aggregation by LDL and, though less unambiguous, the opposing effect

[17] D. G. Hassall, K. Desai, J. S. Owen, and K. R. Bruckdorfer, *Platelets* **1**, 29 (1990).
[18] R. W. Mahley and T. L. Innerarity, *Biochim. Biophys. Acta* **737**, 197 (1983).
[19] F. Martin-Nizard, B. Richard, G. Tropier, A. Nouvelot, J. C. Fruchart, P. Duthilleul, and C. Delbart, *Thromb. Res.* **46**, 811 (1987).

of HDL support this hypothesis.[20-24] Furthermore, it has been demonstrated in the last few years that LDL causes inhibition of adenylate cyclase,[25] evokes protein phosphorylation and mobilization of thromboxane A_2,[26] and induces the release of inositol phosphates.[27] These effects definitely bring LDL binding in close relationship with platelet reactivity, although they become significant only at high LDL concentrations, far beyond that leading to half-saturation of binding sites.[28]

GPIIb-IIIa fulfills a key role in the course of platelet activation. The agonist-induced binding of fibrinogen to GPIIb-IIIa is a necessary prerequisite for aggregation. Consequently, the binding of lipoproteins to this membrane protein complex might have major effects on platelet function *in vivo*. In fact, we could show that fibrinogen binding to ADP or thrombin-stimulated platelets is significantly enhanced in the presence of LDL.[29]

[20] M. Aviram and J. G. Brook, *Atherosclerosis* **46**, 259 (1983).
[21] K. Desai, K. R. Bruckdorfer, R. A. Hutton, and J. S. Owen, *J. Lipid Res.* **30**, 831 (1989).
[22] E. Koller, Th. Vukovich, W. Doleschel, and W. Auerswald, *Atherogenese 4* (Suppl. IV), 53 (1979).
[23] D. G. Hassall, J. S. Owen, and K. R. Bruckdorfer, *Biochem. J.* **216**, 43 (1983).
[24] R. Farbiszewski, Z. Skrzydlewski, and K. Worowski, *Thromb. Diath. Haemostas.* **21**, 89 (1963).
[25] K. R. Bruckdorfer, S. Buckley, and D. G. Hassall, *Biochem. J.* **223**, 189 (1984).
[26] H. E. Andrews, J. W. Aitken, D. G. Hassall, V. O. Skinner, and K. R. Bruckdorfer, *Biochem. J.* **242**, 559 (1987).
[27] M. Knorr, R. Locher, E. Vogt, W. Vetter, L. H. Block, F. Ferracin, H. Lefkovits, and A. Pletscher, *Eur. J. Biochem.* **172**, 753 (1988).
[28] Calculations based on the *in vitro* interaction between platelets and LDL might, however, be misleading. The presence of additional classes of lipoproteins obviously markedly reduces the strength of this binding. The degree of saturation of sites *in vivo* may therefore be well below 100%.
[29] E. Koller, F. Koller, and B. R. Binder, *Thromb. Haemostas.* **62**, (abstr 830) 261 (1989).

[33] Platelet Insulin Receptor

By Anthony S. Hajek and J. Heinrich Joist

Introduction

Platelets contain insulin receptors with characteristics that are similar to the insulin receptors found in other types of cells. These include an alkaline pH binding optimum, site–site interactions between receptors (i.e., negative cooperativity), numbers of binding sites per cell surface area, and a high-affinity binding constant in the nanomolar range.[1,2] Insulin binding to platelets can be demonstrated and quantitated by conventional hormone binding techniques,[3,4] including experiments where the association of ^{125}I-labeled insulin with the platelet insulin receptor is competitively inhibited by native insulin. The resulting data can be used for Scatchard analysis[5] to calculate the number of binding sites on the platelet plasma membrane and to obtain a value for the dissociation constant. Fluorescein isothiocyanate (FITC)-labeled insulin has also been used successfully to study platelet insulin receptors.[2] Presented here is the more traditional method, which utilizes radioiodinated insulin as the ligand.

Methods

Binding Assay Buffer

N-2-Hydroxyethylpiperazine-N'-2-ethanesulfonic acid (HEPES; 100 mM), NaCl (120 mM), MgSO$_4$ (1.2 mM), KCl (2.5 mM), glucose (10 mM), EDTA (1 mM), and 10 g/liter bovine serum albumin, pH 8.0.[6] The buffer should be freshly prepared on the day of usage.

Platelet Isolation

Blood is collected by clean venipuncture, using a two-syringe technique, into 1/7 vol of acid–citrate–dextrose (ACD) solution, pH 4.5. Platelets are isolated by differential centrifugation according to the method of Mustard et al.[7] Platelets prepared by this method appear to be morphologically (disk shape, normal number and distribution of subcellular organelles) and functionally (normal in vitro aggregation to ADP in the presence of added fibrinogen) intact, i.e., similar to platelets in citrated platelet-rich plasma. All steps in this procedure are carried out at 37° to guarantee maximum effectiveness of apyrase in removing extracellular ADP, and thus prevent platelet activation.[1] The ACD blood is centrifuged at 180 g for 15 min. The platelet-rich plasma is removed and centrifuged at 2000 g

---

[1] A. S. Hajek, J. H. Joist, R. K. Baker, L. Jarett, and W. H. Daughaday, *J. Clin. Invest.* **63,** 1060 (1979).
[2] R. Shimoyama, *J. Clin. Endocrinol. Metab.* **53,** 502 (1981).
[3] J. Roth, this series, Vol. 37, p. 66.
[4] T. Kono, this series, Vol. 37, p. 193.
[5] G. Scatchard, *Ann. N.Y. Acad. Sci.* **51,** 660, 1949.
[6] M. M. Rechler and J. M. Podskalny, *Diabetes* **25,** 250 (1976).
[7] J. F. Mustard, D. W. Perry, N. G. Ardlie, and M. A. Packham, *Br. J. Haematol.* **22,** 193 (1972).

for 10 min. The platelet sediment is immediately resuspended in Tyrode's solution (8.0 g/liter NaCl, 0.2 g/liter KCl, 1.0 g/liter $NaHCO_3$, 0.05 g/liter $NaH_2PO_4 \cdot H_2O$, 0.438 g/liter $CaCl_2 \cdot 6H_2O$, pH 7.35) containing 25 units/ml heparin, 3.5 g/liter bovine serum albumin, and 3 units/ml of potato apyrase prepared as described elsewhere.[8] Approximately 0.5 ml of packed platelets is added to 10 ml of solution. Following centrifugation of the mixture at 1000 g for 10 min, the platelets are resuspended in 10 ml of the above washing solution (except that heparin is omitted), the mixture is again centrifuged at 1000 g for 10 min, and the platelets are resuspended in HEPES binding assay buffer at pH 8.0. It is important to determine the white blood cell (WBC) concentration in the cell suspension at this point microscopically (hemocytometer) since blood mononuclear cells (monocytes, lymphocytes) also bind insulin.[9,10] White blood cell contamination of 1 WBC/100,000 platelets or less can be readily achieved by this method and appears acceptable.

Insulin Iodination

Crystalline, porcine, monocomponent insulin may be iodinated by a variety of methods[11-13] or obtained commercially. The preparation should contain 0.4–0.5 iodine atoms per molecule of insulin. Because iodinated insulin deteriorates over time to yield products that are biologically inactive, it is advisable that each preparation be used within 2 weeks after iodination. The insulin preparation can be purified by means of gel-filtration chromatography (Sephadex G-50)[2] or other suitable gel-filtration methods.

Insulin Binding Assay

All materials (tubes, pipettes) that will come in contact with the labeled insulin should be made of polypropylene. This material does not adsorb insulin as readily as glass or polystyrene so that background radioactivity is minimized. Washed platelets (0.5–2.0 × $10^6/\mu l$ in HEPES binding assay buffer, pH 8.0) are incubated with a mixture of ^{125}I-labeled insulin and unlabeled insulin (0–10 μg/ml). The mixture is briefly

[8] J. Molnar and L. Lorand, *Arch. Biochem. Biophys.* **93**, 353 (1961).
[9] R. A. Schwartz, B. S. Bianco, B. S. Handwerger, and C. R. Kahn, *Proc. Natl. Acad. Sci. U.S.A.* **72**, 474 (1975).
[10] J. R. Gavin, P. Gorden, J. Roth, J. A. Archer, and D. N. Buell, *J. Biol. Chem.* **248**, 2202 (1973).
[11] J. J. Conahey and F. J. Dixon, this series, Vol. 70, p. 210.
[12] M. Morrison, this series, Vol. 70, p. 214.
[13] J. J. Langone, this series, Vol. 70, p. 221.

vortexed and incubated for 3 hr at 24°. Since the platelets tend to settle to the bottom of the tubes with time it is advisable to vortex the tubes gently every 20 min. At the end of the incubation period the mixture is again vortexed and duplicate 0.5-ml aliquots are removed and layered over 0.5 ml of ice-cold HEPES binding buffer in a 1.5-ml microfuge tube and immediately centrifuged in a cold room for 5 min in a microfuge (Beckman Instruments, Inc., Spinco Div., Palo Alto, CA). The supernatant buffer is removed by a two-step method, i.e., the bulk of the supernatant fluid is aspirated, and the remaining traces of buffer are allowed to drain down the inside walls of the microfuge tubes and are removed by a second aspiration. The tips of the tubes containing the platelet pellet are cut off and counted for radioactivity in a γ counter. Specific binding of ^{125}I-labeled insulin is determined by subtracting the amount of radioactivity (counts per minute, cpm) associated with the platelet pellet in the presence of 10 μg/ml of unlabeled insulin from the total cpm determined in the platelet sediment.

Comments

Under the conditions of the assay described here, hormone degradation [as determined by trichloroacetic acid (TCA) precipitation and rebinding experiments] and receptor degradation were not observed.[1]

Specificity

At a platelet concentration of 800,000 μl in the incubation mixture, binding reaches a steady state at 2 to 3 hr (half-maximal binding is reached by 30 min), at which time approximately 2% of the labeled hormone is bound to the platelets. Binding of approximately 90% of the labeled insulin is competitively inhibited by 10 μg/ml of native insulin (specific binding). As little as 3 μg/ml of labeled insulin (a physiological blood concentration) may cause a 40–50% decrease in specific binding of the iodoinsulin.[1] Further evidence for the specificity of the binding of insulin to the platelet surface was obtained from experiments showing that with catfish insulin and porcine proinsulin (which in humans are biologically less potent than native porcine insulin) substantially higher doses are required to produce a decrease in the specific binding of ^{125}I-labeled insulin as compared to porcine insulin. Furthermore, diisopropyl fluorophosphate (DFP)-treated thrombin, human prolactin, human growth hormone, and glucagon did not inhibit the binding of radiolabeled insulin to platelets.[1]

Receptor Site Number and Binding Affinity

Scatchard plots[5] of the competitive binding assay data show a curvilinear pattern from which one can derive a high-affinity dissociation constant of approximately $3 \times 10^9 \, M^{-1}$. The curvilinear pattern of the plot is consistent with either the presence of two classes of binding sites with different affinities or a single population of insulin receptors that display enhanced dissociation of labeled insulin in the presence of excess unlabeled hormone (negative cooperativity). The number of 500–600 insulin binding sites per platelet, derived from Scatchard plot analysis, is appreciably less than that reported for human lymphocytes $(12,500)^{10}$ and erythrocytes (2000).[14] However, when expressed as number of binding sites per platelet surface area, the value of approximately $25/\mu m^2$ is very similar to the insulin receptor density observed in other insulin binding cell types.[1,9,10] It is important to note that the estimation of the density of platelet insulin receptors is based on the assumption of a smooth, continuous outer cell membrane. Since the platelets appear to have an extensive spongelike, internal surface-connecting canalicular system,[15] the actual platelet surface area potentially accessible for insulin binding is likely to be greatly in excess of that estimated on the basis of the measurement of platelet size. This means that the value of 25 insulin receptors/μm^2 for human platelets[1] is probably an overestimate.

The method outlined about is straightforward and yields consistent results. The values for the high-affinity dissociation constant and platelet receptor number obtained with this method are lower than those obtained with FITC-labeled insulin binding experiments,[2] but are in agreement with values reported for ^{125}I-labeled insulin binding in other systems, as well as with alternate methods of analysis (i.e., Lineweaver–Burke-type plots).[4]

The issue of the biologic significance of the platelet insulin receptor remains unsettled. Whereas earlier reports indicated that insulin may increase lactate formation in human platelets[16] and inhibit platelet aggregation in vitro,[17] an extensive study in our laboratory[18] failed to show any effect of insulin on in vitro secretion, glucose transport and metabolism, or protein phosphorylation in human platelets in response to stimulation

---

[14] P. DeMeyts, J. Roth, D. M. Neville, Jr., J. R. Gavin, and M. A. Lesniak, *Biochem. Biophys. Res. Commun.* **55**, 154 (1973).
[15] J. G. White, *Am. J. Pathol.* **66**, 295 (1972).
[16] S. Karpatkin, *J. Clin. Invest.* **46**, 409 (1967).
[17] A. A. Hassanein, T. A. El-Garf, and Z. El-Baz, *Thromb. Diath. Haemorrh.* **27**, 114 (1972).
[18] J. H. Joist, L. Babb, and R. K. Baker, unpublished observations.

by different agonists. Others reported a lack of correlation between ADP-induced platelet aggregation and serum insulin levels in patients with type I and type II diabetes[19] and a lack of effect of continuous subcutaneous insulin infusion on platelet aggregation to ADP and epinephrine or thromboxane B_2 generation.[20]

[19] D. B. Jones, T. M. Davis, E. Brown, R. D. Carter, J. I. Mann, and R. J. Prescott, *Diabetologia* **29**, 291 (1986).
[20] L. H. Monnier, M. Rodier, A. Gancel, P. Crastes De Paulet, C. Colette, M. Piperno, and J. Crastes DePaulet, *Diabete Metab.* **13**, 210 (1987).

[34] Membrane-Impermeant Cross-Linking Reagents for Structural and Functional Analyses of Platelet Membrane Glycoproteins

By JAMES V. STAROS, NICOLAS J. KOTITE, and LEON W. CUNNINGHAM

The interactions among platelet membrane glycoproteins and between these molecules and macromolecular components of the subendothelial matrix and of the plasma have been areas of great and growing interest. Chemical cross-linking is a technique that has proven useful in many studies of protein–protein interactions.[1-4] Indeed, cross-linking has provided useful information concerning platelet supramolecular structure[5-8] and the interactions of specific macromolecules[7,9-14] or synthetic pep-

[1] F. Wold, this series, Vol. 25, p. 623.
[2] K. Peters and F. M. Richards, *Annu. Rev. Biochem.* **46**, 523 (1977).
[3] T. H. Ji, this series, Vol. 91, p. 580.
[4] J. V. Staros and P. S. R. Anjaneyulu, this series, Vol. 172, p. 609.
[5] G. E. Davies and J. Palek, *Blood* **59**, 502 (1982).
[6] S. M. Jung and M. Moroi, *Biochim. Biophys. Acta* **761**, 152 (1983).
[7] N. J. Kotite, J. V. Staros, and L. W. Cunningham, *Biochemistry* **23**, 3099 (1984).
[8] A. Sonnenberg, H. Janssen, F. Hogervorst, J. Calafat, and J. Hilgers, *J. Biol. Chem.* **262**, 10376 (1987).
[9] N. E. Larsen and E. R. Simons, *Biochemistry* **20**, 4141 (1981).
[10] J. Lahav, M. A. Schwartz, and R. O. Hynes, *Cell (Cambridge, Mass.)* **31**, 253 (1982).
[11] J. Takamatsu, M. K. Horne, III, and H. R. Gralnick, *J. Clin. Invest.* **77**, 362 (1986).
[12] F. C. Molinas, J. Wietzerbin, and E. Falcoff, *J. Immunol.* **138**, 802 (1987).
[13] M. Jandrot-Perrus, D. Didry, M.-C. Guillin, and A. T. Nurden, *Eur. J. Biochem.* **174**, 359 (1988).
[14] R. K. Andrews, J. J. Gorman, W. J. Booth, G. L. Corino, P. A. Castaldi, and M. C. Berndt, *Biochemistry* **28**, 8326 (1989).

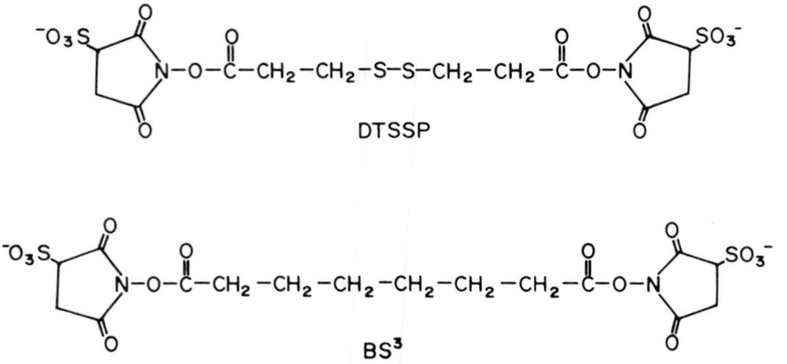

FIG. 1. The structures of 3,3'-dithiobis(sulfosuccinimidyl propionate) (DTSSP) and bis(sulfosuccinimidyl) suberate (BS³). (Reprinted from Ref. 20 with permission. Copyright 1982 American Chemical Society.)

tides related to such macromolecules[15–18] with platelet surface components.

One complication encountered in cross-linking studies of cell surfaces has been that the reagents available prior to the last decade were all membrane permeant, so that internal cellular components as well as surface components reacted with them. For example, when intact platelets were treated with the membrane-permeant reagent 3,3'-dithiobis(succinimidyl propionate), cytoskeletal components were the most prominent cross-linked complexes detected.[5] For many studies that focus on macromolecular interactions at the membrane surface, it would be desirable to restrict reactions to the extracytoplasmic surface of the cell. The introduction of membrane-impermeant cross-linkers[4,19–21] provides a class of reagents that meets this criterion. This chapter will focus on the preparation of bifunctional N-hydroxysulfosuccinimide active esters of dicarboxylic acids (Fig. 1) as high-yield, membrane-impermeant cross-linking reagents,[20] and their application to human platelet structure and function.[7]

[15] S. A. Santoro and W. J. Lawing, Jr., *Cell (Cambridge, Mass.)* **48**, 867 (1987).
[16] S. E. D'Souza, M. H. Ginsberg, S. C.-T. Lam, and E. F. Plow, *J. Biol. Chem.* **263**, 3943 (1988).
[17] S. E. D'Souza, M. H. Ginsberg, T. A. Burke, S. C.-T. Lam, and E. F. Plow, *Science* **242**, 91 (1988).
[18] S. E. D'Souza, M. H. Ginsberg, T. A. Burke, and E. F. Plow, *J. Biol. Chem.* **265**, 3440 (1990).
[19] J. V. Staros, D. G. Morgan, and D. R. Appling, *J. Biol. Chem.* **256**, 5890 (1981).
[20] J. V. Staros, *Biochemistry* **21**, 3950 (1982).
[21] J. V. Staros, *Acc. Chem. Res.* **21**, 435 (1988).

FIG. 2. The reaction of a sulfosuccinimidyl active ester with a primary amino group to form an amide linkage with release of N-hydroxysulfosuccinimide.

Sulfosuccinimidyl active esters react with amino groups to form stable amide bonds, with loss of N-hydroxysulfosuccinimide (Fig. 2). The sulfonate groups on the sulfosuccinimide rings render these reagents highly water soluble and membrane impermeant.[20,21] In the absence of nucleophiles, sulfosuccinimidyl active esters hydrolyze very slowly as compared with their rate of reaction with amino groups,[22] resulting in a very high yield of covalent linkage.

Beside the chemical properties of a cross-linking reagent, another important criterion in reagent design is span, i.e., the distance between the reacted groups on the protein spanned by the cross-linker. For the all-trans configurations of the two reagents shown in Fig. 1, this distance is 11–12 Å, as measured with CPK models. What does this distance mean in molecular terms? Within myoglobin, which is essentially a bundle of closely packed α helices, the center-to-center distance of α helices in pairwise contact averages 9 Å.[23] Thus, the two cross-linkers shown in Fig. 1 span a distance comparable to the center-to-center distance between two closely packed α helices. It is important to remember when interpreting experiments employing these reagents that the cross-links formed are, on the scale of proteins, intramolecular in terms of distance. When intermolecular cross-links are formed, it is implied that the two participating proteins were in very close proximity at the time of cross-linking.

Experimental Procedures

Reagents[24]

Synthesis of N-Hydroxysulfosuccinimide Sodium Salt.[20] It is very important to keep N-hydroxymaleimide, the immediate precursor of N-hydroxysulfosuccinimide, in an inert atmosphere. Therefore, solutions

[22] P. S. R. Anjaneyulu and J. V. Staros, *Int. J. Pept. Protein Res.* **30**, 117 (1987).
[23] T. J. Richmond and F. M. Richards, *J. Mol. Biol.* **119**, 537 (1978).
[24] Commercial sources are now available for N-hydroxysulfosuccinimide sodium salt (Fluka Chemical Corp., Ronkonkoma, NY, Pierce Chemical Co., Rockford, Il, Aldrich Chemical Co., Milwaukee, WI) and for cross-linking reagents incorporating this compound (Pierce Chemical Co.).

of this compound are prepared in a glove bag filled with N_2 and are transferred to a purged and N_2-filled reaction vessel in a syringe. In this manner, N-hydroxymaleimide (Fluka Chemical Corp.) (1.45 g, 12.8 mmol) is dissolved in absolute ethanol (15 ml) and the resulting solution is transferred to the reaction vessel under N_2. An aqueous solution (10 ml) of $Na_2S_2O_5$ (1.22 g, 6.4 mmol) is then added with stirring. The reaction mixture is stirred at room temperature for 2 hr under N_2, and the vessel is then opened to the atmosphere. After the solvent is removed by rotary evaporation (bath temperature ~40°), the product, a thick yellow oil, is dissolved in 50 ml of H_2O, filtered, and lyophilized. The resulting light yellow solid is triturated overnight with anhydrous ether. An off-white powder is then recovered by filtration [yield, 2.66 g (96%)]. This product forms a single spot when subjected to thin-layer chromatography on silica gel plates (0.20 mm with fluorescent indicator on aluminum backing from EM Industries, Hawthorn, NY) developed in 5:2:3 1-butanol–acetic acid–water. It gives a single peak when subjected to ion-pair reversed-phase high-performance liquid chromatography (HPLC) using a C_{18} column (Alltech, Deerfield, IL; Cat. No. 600RP) and a mobile phase of aqueous 10 mM tetrabutylammonium formate, pH 4.0 and methanol (60:40). If necessary, the product may be recrystallized from 95% ethanol.

Synthesis of 3,3'-Dithiobis(sulfosuccinimidyl propionate) Disodium Salt (DTSSP).[20] N-Hydroxysulfosuccinimide sodium salt (0.44 g, 2.0 mmol), 3,3'-dithiodipropionic acid (Aldrich Chemical Co., Milwaukee, WI) (0.21 g, 1.0 mmol), and N,N'-dicyclohexylcarbodiimide (Aldrich) (0.46 g, 2.2 mmol) are dissolved in 5.0 ml of anhydrous dimethylformamide. The reaction vessel is capped and the reaction mixture stirred overnight at room temperature. The reaction mixture is then cooled to 3° and stirred for 2–3 hr. The precipitated dicyclohexylurea is removed by filtration and washed with a small quantity of dry dimethylformamide. The product is then precipitated from the pooled filtrate by addition of ~20 vol of ethyl acetate, recovered by filtration, and stored in a vacuum dessicator [yield, 0.40 g (65%)], assuming a pure anhydrous product. The noncleavable cross-linker bis(sulfosuccinimidyl) suberate (BS[3]) is synthesized by the same method by substituting suberic acid (0.175 g, 1.0 mmol) for 3,3'-dithiodipropionic acid. DTSSP, BS[3], and other homologous cross-linkers are routinely assayed in our laboratory by testing their ability to cross-link rabbit muscle aldolase, as described below.

Methods

Cross-Linking of Rabbit Muscle Aldolase.[20] A suspension of crystalline rabbit muscle aldolase in 2.5 M $(NH_4)_2SO_4$ (type IV, Sigma Chemical Co., St. Louis, MO) is exhaustively dialyzed against 50 mM sodium phosphate, pH 7.4. The final concentration of aldolase is determined by absorbance at 280 nm ($E_{280}^{1\%}$ = 9.38).[25] Equal aliquots are diluted with 50 mM sodium phosphate, pH 7.4, and are treated with various concentrations of cross-linker that have been prepared immediately before use as a 10 mM stock solution in the same buffer. After incubation for 30 min at room temperature, the reactions are quenched by addition of one-sixth volume of 50 mM ethanolamine, 20 mM N-ethylmaleimide, 50 mM sodium phosphate, pH 7.4. To each sample is then added sodium dodecyl sulfate (SDS) gel-solubilizing solution, with or without reductant, as required. Samples are incubated at 50° for 3 min, and then stored at −65°.

Treatment of Platelets with DTSSP.[7] Platelets are isolated[26,27] from freshly drawn blood and resuspended in platelet buffer (137 mM NaCl, 2.7 mM KCl, 4.25 mM Na$_2$HPO$_4$, 1.5 mM KH$_2$PO$_4$, 5 mM glucose 2 mM EDTA, pH 7.4) at a concentration of 0.5–1 × 10^9/ml. For some experiments, platelets are radiolabeled by the periodate–boro[^{3}H]hydride method.[7] A 10 mM stock of DTSSP is prepared in the same buffer immediately before use. Appropriate aliquots of the DTSSP stock are added to samples of the platelet suspension (5 × 10^8/ml), and the reaction is allowed to proceed for 20 min at room temperature. The reaction is quenched by addition of Tris-HCl, pH 7.4, to a final concentration of 0.2 M, and by incubation for 5 min. (Subsequent studies have suggested that other primary amines are better quenching agents than Tris. In other studies, such as the cross-linking of rabbit muscle aldolase described above, we have employed ethanolamine. We prepare a quench buffer by adding ethanolamine, typically to a final concentration of 20–50 mM, to a sample of the same buffer used for the reaction, and adjusting the pH with HCl back to its previous value.) Six volumes of platelet buffer are then added to the samples, which are then pelleted by centrifugation at 1200 g for 10 min at 4°.

For aggregation assays, the platelets are washed once in aggregation buffer (platelet buffer without EDTA and with the addition of bovine serum albumin to 0.35%) by resuspension and pelleting as above, and are finally resuspended in aggregation buffer at a concentration of 2.5 × 10^8/ml. Immediately prior to assay, samples are adjusted to 2 mM CaCl$_2$, 1 mM MgCl$_2$, and 0.05% fibrinogen and are subjected to an aggregation assay[28,29] in a Chronolog aggregometer (Chronolog Corp., Haverstown, PA).

[25] J. W. Donovan, *Biochemistry* **3**, 67 (1964).
[26] S. A. Santoro and L. W. Cunningham, *Proc. Natl. Acad. Sci. U.S.A.* **76**, 2644 (1979).
[27] See also J. F. Mustard, R. L. Kinlough-Rathbone, and M. A. Packham, this series, Vol. 169, p. 3; S. Timmons and J. Hawiger, *ibid.* p. 11.
[28] S. A. Santoro and L. W. Cunningham, *J. Clin. Invest.* **60**, 1054 (1977).
[29] See also M. Zucker, this series, Vol. 169, p. 117.

For analysis of the effects of cross-linking on the profile of platelet surface glycoproteins, the platelets are washed once as above, but in platelet buffer containing 20 mM N-ethylmaleimide. This treatment alkylates free thiols to prevent thiol–disulfide exchange from scrambling the cross-linked products when the cells are disrupted.[19,20,30] Finally, the platelets are washed once more in platelet buffer without N-ethylmaleimide, and the resulting pellets are stored at $-70°$ until dissolved for analysis by SDS–polyacrylamide gel electrophoresis.

Results

Cross-Linking of Rabbit Muscle Aldolase. Once a cross-linking reagent has been prepared or purchased, it is important to assay its cross-linking activity in a well-defined system. We routinely assay all new reagents as well as all new batches of our well-characterized reagents by testing their ability to cross-link rabbit muscle aldolase, a tetrameric protein, under standard conditions. An example of a quality control assay testing a new batch of DTSSP against the current stock of the same reagent is shown in Fig. 3. In the assay the two batches of DTSSP were reacted with aldolase under identical conditions. This assay is also very useful for testing whether a new reagent is as efficient a cross-linker as a given standard or whether a proposed reaction buffer will support efficient cross-linking. Thus it is often useful to react aldolase with the cross-linker of choice in a new buffer and to compare the results to the reaction of identical concentrations of aldolase in the standard buffer with identical concentrations of the same reagent. Control samples, reacted with DTSSP but dissolved in SDS gel-solubilizing solution containing dithiothreitol and subjected to SDS–polyacrylamide gel electrophoresis under reducing conditions, ran essentially quantitatively as monomers, i.e., like the sample in Fig. 3, lane 1.[20]

Effect of DTSSP or BS^3 on Platelet Aggregation.[7] Figure 4 shows the results of treating platelets with various concentrations of DTSSP or BS^3 and subjecting them to an assay for aggregation induced by collagen. The response to collagen of platelets treated with either cross-linker was sharply reduced. Under the conditions of the experiment depicted in Fig. 3, the EC_{50} of DTSSP for inhibition of collagen-induced aggregation was ~ 2 μM and that of BS^3 was ~ 8 μM.

To distinguish whether the inhibition of collagen-induced platelet aggregation was the result of cross-linking of surface components or simply of the acylation of surface groups, platelets were treated with sulfosuccinimidyl propionate, a monofunctional analog of DTSSP, and were then

[30] K. Wang and F. M. Richards, *J. Biol. Chem.* **249**, 8005 (1974).

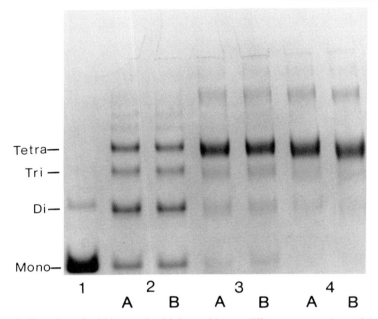

FIG. 3. Reaction of rabbit muscle aldolase with two different preparations of DTSSP. Rabbit muscle aldolase, 1 mg/ml, was reacted with the following concentrations of DTSSP, as described in the test: lane 1, 0; lanes 2, 0.05 mM; lanes 3, 0.20 mM; lanes 4, 1.0 mM. After quenching the reaction, samples were taken up in SDS gel-solubilizing solution without reductant and were subjected to SDS-polyacrylamide gel electrophoresis under nonreducing conditions. A and B denote two different synthetic preparations of DTSSP. Mono, Di, Tri, and Tetra refer to the positions of monomers, dimers, trimers, and tetramers of aldolase subunits, respectively.

subjected to the aggregation assay. No inhibition of collagen-induced platelet aggregation was observed up to a concentration of 200 μM sulfosuccinimidyl propionate, suggesting that it is cross-linking and not simply acylation of surface components that results in the observed inhibition.

The inhibition of aggregation appears to be specific to the fibrillar collagen induction pathway. Treatment of platelets with DTSSP was found to have no effect on aggregation in response to thrombin or on platelet–collagen adhesion.[7]

Effect of BS^3 on Sodium Dodecyl Sulfate-Polyacrylamide Gel Profile of Platelet Surface Glycoproteins.[7] To assess whether intermolecular cross-links formed between platelet surface glycoproteins might correlate with the inhibition of collagen-induced aggregation, radiolabeled platelets were treated with BS^3, dissolved in SDS gel buffer, and subjected to SDS–polya-

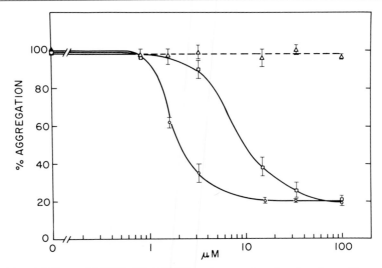

FIG. 4. Effects of DTSSP (○), BS^3 (□), and a monofunctional analog, sulfosuccinimidyl propionate (SSP) (△), on collagen-induced platelet aggregation. The concentration of SSP used was twice that indicated on the abscissa, so that the concentration of reactive groups would be the same as in the samples with the cross-linkers. Percentage aggregation was calculated from the slope of the aggregation profiles. The values shown represent the average of three determination ± SD. (Reprinted from Ref. 7 with permission. Copyright 1984 American Chemical Society.)

crylamide gel electrophoresis. The resulting gel was subjected to fluorography,[31] and the fluorograph was scanned with a fiber optic densitometer (model 800; Kontes). The results of this analysis are shown in Fig. 5.

At low concentrations of BS^3, which effectively inhibit collagen-induced platelet aggregation, radiolabeled bands corresponding to glycoproteins IIb, IIIa, and IV were significantly reduced in the SDS gel profile. In addition, the bands corresponding to glycoproteins IIb and IIIa disappeared in parallel, suggesting that they were cross-linked to one another. This observation is consistent with independent evidence that these proteins exist as a 1 : 1 noncovalent complex[32] that has been recognized as an important platelet integrin.[33]

---

[31] W. M. Bonner and R. A. Laskey, *Eur. J. Biochem.* **46**, 83 (1974).
[32] L. K. Jennings and D. R. Phillips, *J. Biol. Chem.* **257**, 10458 (1982).
[33] R. O. Hynes, *Cell (Cambridge, Mass.)* **48**, 549 (1987).

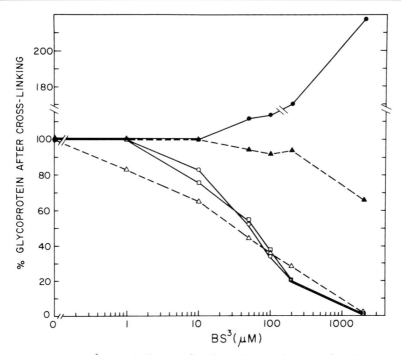

FIG. 5. Effect of BS^3 on the SDS gel profile of radiolabeled platelet surface glycoproteins. Radiolabeled proteins were treated with varying concentrations of BS^3. The reactions were quenched, and the samples were dissolved in SDS gel-solubilizing solution and were subjected to SDS–polyacrylamide gel electrophoresis. The resulting gel was subjected to fluorography, and the resulting fluorograph was analyzed densitometrically. The percentage of each glycoprotein at various concentrations of BS^3 was calculated with reference to the uncross-linked control. (●), 150K; (▲), GPIb; (○), GPIIb; (□), GPIIIa; (△), GPIV. (Reprinted from Ref. 7 with permission. Copyright 1984 American Chemical Society.)

Discussion

The advent of high-yield, membrane-impermeant cross-linking reagents has allowed the experimenter to probe protein–protein interactions at one face of a membrane. Platelets appear to present an especially rich variety of questions that can be addressed with these reagents. These include the identification of platelet surface proteins that specifically interact with protein components of the subendothelial matrix and of the plasma, as well as possible changes in tertiary or quaternary structure of surface proteins that accompany platelet activation.

Initial studies in which these reagents have been used to cross-link platelet surfaces have resulted in the intriguing observation that these reagents specifically inhibit collagen-induced platelet aggregation but not adhesion to collagen or thrombin-induced platelet aggregation. The mechanism by which this specific inhibition occurs is not known. Control experiments with a monofunctional sulfosuccinimidyl ester have discounted the possibility that it is simply acylation of specific residues that gives rise to the observed inhibition. Perhaps a tertiary or quaternary structural change in a specific surface protein is a required step in platelet activation by collagen, and cross-linking by BS^3 or DTSSP locks this protein in the unactivated state.

The selective reduction in intensity of bands corresponding to several major glycoproteins in an SDS gel profile of platelets treated with one of these reagents suggests that candidates for surface proteins involved in collagen-induced platelet activation might be explored by this approach. However, much additional work needs to be done before this question is clarified. Radioisotopically labeled BS^3 or DTSSP may be useful in this endeavor.

Acknowledgment

Work in this laboratory was supported by grants from the National Institutes of Health, DK25489 and DK31880.

[35] Surface Labeling of Platelet Membrane Glycoproteins

By DAVID R. PHILLIPS

Introduction

Many reactions related to the hemostatic effectiveness of the platelet (e.g., binding of platelet agonists, platelet adhesion, platelet aggregation, and platelet procoagulant activity) occur on specific glycoproteins on the outer surface of the platelet plasma membrane.[1] Identification of the membrane glycoproteins involved in these reactions has been facilitated by procedures that specifically label platelet surface proteins. These procedures attempt to introduce specifically and exclusively a radioactive label only into macromolecules on the outer surface of the membrane. The basic premise of these procedures is that the labeling agent does not penetrate

[1] N. Kieffer and D. R. Phillips, *Annu. Rev. Cell Biol.* **6**, 329 (1990).

TABLE I
RADIOLABELING PLATELET MEMBRANE GLYCOPROTEINS

Labeling method	Functional groups labeled	Major glycoproteins labeled[a]
Lactoperoxidase-catalyzed iodination (^{125}I or ^{131}I)	Tyrosine (histidine)	GPIIb (α_{IIb}), GPIIIa (β_3), GPIV, [GPIIa (β_1)]
Periodate/sodium boro[^{3}H]hydride labeling	Sialic acid	GPIb, [GPIIb, GPIIIa, GPIV, GPV, GPIX]

[a] Glycoproteins shown in brackets are labeled less intensely.

the plasma membrane, and hence only membrane surface components are radiolabeled.

Two methods of labeling membrane surface glycoproteins will be described in this report: lactoperoxidase-catalyzed iodination[2] and periodate/sodium boro[^{3}H]hydride labeling.[3] These procedures were selected for presentation because they are sensitive, introduce minimal alterations into platelet glycoproteins, together label all known surface glycoproteins, and are widely used. The sites of labeling and the major glycoproteins identified by these techniques are extensive and up to 40 glycoproteins can be identified. The major ones are summarized in Table I. Additional agents have been identified that also label cell surface proteins on platelets, and the reader is referred to the original publications for a description of the procedures using them: neuraminidase/galactose oxidase/sodium boro[^{3}H]hydride[4], transglutaminase[5], diazotized diiodosulfanylic acid[6], and Iodogen.[7]

Protein labeling procedures are characteristically more facile with proteins in solution than with proteins in membranes. Consequently, platelets must be washed to remove plasma proteins prior to labeling, which results in two limitations of these techniques. First, buffers that include protein to "stabilize" platelets during isolation cannot be used for washing. Second, because radiolabeling of platelets involves numerous washing steps (first to eliminate plasma proteins and second to reduce the concentration of unincorporated isotope), the labeled platelets are usually less reactive than

[2] D. R. Phillips, *Biochemistry* **11**, 4582 (1972).
[3] T. L. Steck and G. Dawson, *J. Biol. Chem.* **249**, 2135 (1974).
[4] D. R. Phillips and P. P. Agin, *J. Clin. Invest.* **60**, 535 (1977).
[5] T. Okumura and G. A. Jamieson, *J. Biol. Chem.* **251**, 5944 (1976).
[6] J. N. George, R. D. Potterf, D. C. Lewis, and D. A. Sears, *J. Lab. Clin. Med.* **88**, 232 (1976).
[7] G. P. Tuszynski, L. C. Knight, E. Kornecki, and S. Srivastava, *Anal. Biochem.* **130**, 166 (1983).

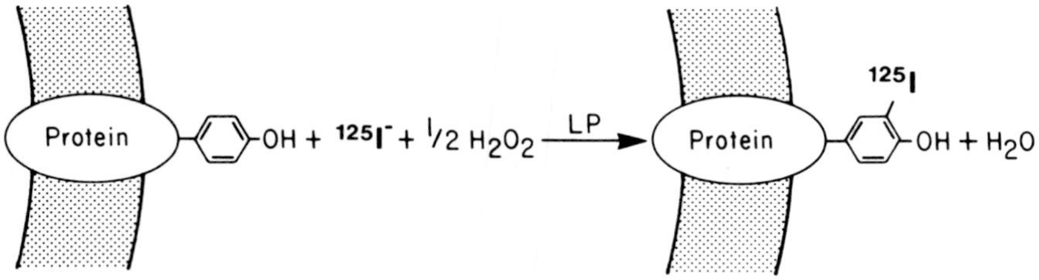

FIG. 1. The lactoperoxidase-catalyzed iodination reaction.

unlabeled platelets and have lost their characteristic discoid morphology. These limitations can be minimized by judicious selection of platelet isolation conditions.[8]

Lactoperoxidase-Catalyzed Iodination

Lactoperoxidase is used to oxidize ^{125}I so that it rapidly iodinates tyrosine (and to a lesser extent histidine). The reaction catalyzed is diagrammed in Fig. 1. Because the platelet membrane is impermeable to lactoperoxidase (M_r 78,000), the iodination reaction occurs primarily with proteins on the platelet surface.[2,4] The reader is referred to Ref. 9 for a complete discussion of the iodination reaction and a description of the products produced.

Platelets from freshly drawn blood are washed and suspended at ambient temperature in normal Tyrode's buffer (138 mM sodium chloride, 2.7 mM potassium chloride, 12 mM sodium bicarbonate, 0.36 mM sodium phosphate, 1.8 mM calcium chloride, 0.49 mM magnesium chloride, and 5.5 mM glucose, pH 7.4) to a platelet count of 10^9/ml. Other platelet washing buffers have been used, and all have proven suitable, providing they are free of protein, do not contain inhibitors of lactoperoxidase, and do not disproportionate hydrogen peroxide. To 1 ml of the platelet suspension, 1 mCi of carrier-free Na^{125}I is added, with gentle stirring at ambient temperature, followed by 0.25 nmol of lactoperoxidase (Sigma, St. Louis, MO) and five 10-μl aliquots of freshly prepared 3 mM hydrogen peroxide, added at 10-sec intervals. The use of fresh isotope (within 1 month of preparation) is desirable to minimize the presence of I$_2$ and other oxidized species of iodide. Lactoperoxidase is stable when stored frozen in solution. The 3 mM hydrogen peroxide solution should be freshly prepared at 4° in 1 mM EDTA to avoid any disproportionation reaction.

[8] J. E. B. Fox, C. C. Reynolds, and J. K. Boyles, this volume [6].
[9] M. Morrison, this series, Vol. 70, p. 214.

The labeled platelets are diluted 10-fold with Tyrode's buffer, and are sedimented by centrifugation at 2000 g for 15 min at 4°. The labeled platelets are washed twice by resuspension in 10 ml of the Tyrode's buffer followed by centrifugation. Less than 1% of the isotope is covalently bound to protein under these conditions. Higher yields can be achieved through more additions of hydrogen peroxide but should be avoided because platelets are activated by high concentrations of this oxidizing agent. Most of the ^{125}I in the washed platelets (>90%) is intracellular, not bound to protein, and difficult to remove by these procedures. Platelets labeled by lactoperoxidase-catalyzed iodination are functional in that they secrete serotonin and aggregate when treated with thrombin.

A modification of the lactoperoxidase-catalyzed iodination procedure can be used to monitor changes of the platelet surface that occur during physiological responses of the platelet.[10] In this modification, the iodination reaction is initiated by hydrogen peroxide and is terminated 15 sec later by the addition of catalase, which disproportionates all remaining hydrogen peroxide. An example of this reaction is illustrated in Fig. 2, in which platelet surface proteins were iodinated during aggregation. In this instance, platelet aggregation is monitored by light scattering in an aggregometer. Na^{125}I and lactoperoxidase are added before the agonist (thrombin). The iodination reaction is initiated 1 min later by the single addition of hydrogen peroxide. After 15 sec, the iodination reaction is terminated by adding catalase. To separate platelets and associated proteins from the nonassociated secreted proteins that are also labeled, the samples are carefully layered on 15% sucrose in Tyrode's solution and centrifuged for 3 min at 8500 g (microfuge; Beckman Instruments, Palo Alto, CA). The resulting pellet is processed for electrophoresis. The proteins that have become labeled during platelet stimulation are identified by comparing the proteins labeled in stimulated platelets to those labeled in unstimulated platelets.

Periodate/Sodium Borohydride Labeling

The periodate/sodium boro[^{3}H]hydride procedure labels sialic acid residues on membrane glycoproteins and glycolipids[3] (see Fig. 3). Periodate is used to cleave oxidatively carbon–carbon bonds between the 7–8 or 8–9 positions of sialic acid residues. The resulting aldehyde is then reduced with sodium boro[^{3}H]hydride, producing the stable, radiolabeled alcohol derivative. Labeling can be restricted to the outer membrane surface by equilibrating platelets to 4° (which renders the membrane impermeable to the oxidizing agent) before exposing them to periodate.

[10] D. R. Phillips, L. K. Jennings, and H. R. Prasanna, *J. Biol. Chem.* **255**, 11629 (1980).

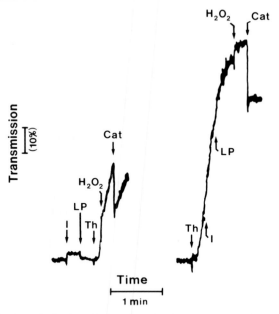

FIG. 2. Iodination of platelets during thrombin-induced aggregation. Washed platelets (5×10^8) were suspended in 0.5 ml Tyrode's solution and iodinated either 10 or 60 sec after the induction of aggregation. The reaction was performed with stirring: aggregation was monitored by the decrease in light transmittance. Additions at the arrows were as follows: I, 0.25 mCi carrier-free ^{125}I; LP, 1.5×10^{-10} mol lactoperoxidase; Th, 0.05 unit thrombin; H_2O_2, 3.5×10^{-9} mol hydrogen peroxide; Cat, 3.5×10^{-9} mol catalase.

Platelets from freshly drawn blood are washed and suspended in 1 ml of a buffer containing 150 mM sodium chloride and 10 mM N-2-hydroxyethylpiperazine-N'-2-ethanesulfonic acid (HEPES), pH 7.6. Tris buffers are to be avoided as they quench the periodate. The suspension is equilibrated to 4° and, after the addition of sodium periodate (1 mM final concentration, freshly prepared), is incubated an additional 10 min at 4° in the

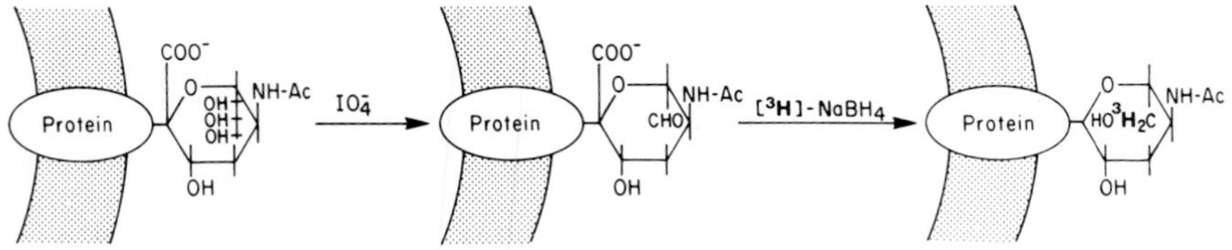

FIG. 3. The periodate/sodium borohydride labeling procedure.

dark. All remaining steps are performed at ambient temperature. The periodate-treated platelets are removed from solution by centrifugation at 800 g for 10 min at 4°, washed once with the HEPES buffer, and resuspended in 1 ml of this buffer. Sodium boro[^{3}H]hydride (0.5 mCi) is added, and the suspension is incubated for 5 min. The sodium boro[^{3}H]hydride should be stored in one-use-size aliquots at $-80°$ in 0.1 M sodium hydroxide and thawed immediately before use. The volume added should not affect the pH of the platelet suspension. The labeled platelets are centrifuged, washed once with the HEPES buffer, and resuspended in the original volume. Higher specific activities of labeled platelets can be achieved by (1) using more sodium boro[^{3}H]hydride, (2) decreasing the pH of the periodate oxidation to 6, and (3) increasing to 8 the pH of the borohydride reduction.[11]

Identification of Labeled Membrane Glycoproteins

Membrane glycoproteins labeled by either the lactoperoxidase or periodate procedures are readily identifiable by sodium dodecyl sulfate (SDS)–polyacrylamide gel electrophoresis, as is illustrated in Fig. 4. In platelets labeled by lactoperoxidate-catalyzed iodination, the labeled proteins are detected by autoradiography of the dried gel.[4] As shown in lanes 2 and 4 of Fig. 4, glycoprotein (GP) IIb and GPIIIa are the most prominently labeled bands, reflecting their abundance in platelets ($\sim$50,000 copies per platelet). Glycoprotein IIIa (M_r 114,000, reduced) labels approximately three times more intensely by this method than does GPIIb (M_r 132,000, reduced), even though the two glycoproteins are present in equal concentrations; both of the disulfide-linked subunits of GPIIb, GPIIb$_\alpha$, and GPIIb$_\beta$ are labeled. The identity of the GPIIb and GPIIIa bands can be confirmed by analyzing various parameters of these glycoproteins: (1) characteristic shifts in molecular weight on disulfide reduction (M_r 95,000 for nonreduced GPIIIa and M_r 142,000 for nonreduced GPIIb[12]); (2) their absence in platelets from patients with Glanzmann's thrombasthenia[4], (3) staining by the periodic acid–Schiff reagent[2], (4) binding to *Lens culinaris* lectin[4], and (5) coimmunoprecipitation from Ca^{2+}-containing buffers by monoclonal antibodies[13] or antibodies specific for one of the two glycoproteins in the complex. Electrophoresis of nonreduced samples of SDS-solubilized platelets should be performed immediately after solubilization to avoid protein polymerization. Alternatively, the

---

[11] B. Steiner, K. J. Clemetson, and E. F. Lüscher, *Thromb. Res.* **29**, 43 (1983).
[12] D. R. Phillips and P. P. Agin, *J. Biol. Chem.* **252**, 2121 (1977).
[13] R. P. McEver, E. M. Bennett, and M. N. Martin, *J. Biol. Chem.* **258**, 5269 (1983).

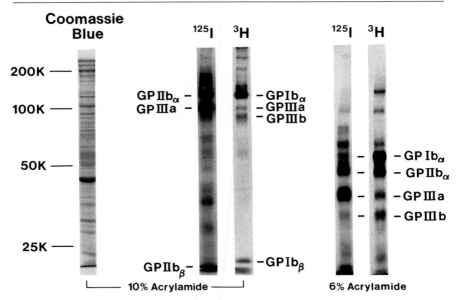

FIG. 4. Detection of cell surface glycoproteins by lactoperoxidase-catalyzed iodination and periodate/sodium boro[^{3}H]hydride labeling. Proteins in lanes 1–3 were separated by electrophoresis in SDS through 10% polyacrylamide gels; 6% polyacrylamide gels were used in lanes 4 and 5. Lane 1 shows the Coomassie Brilliant blue-stained proteins in whole platelets; lanes 2 and 4 are autoradiograms of SDS gels showing the cell surface glycoproteins labeled by lactoperoxidase-catalyzed iodination; lanes 3 and 5 are fluorograms of SDS gels showing the cell surface glycoproteins labeled by the periodate/sodium boro[^{3}H]hydride technique. The molecular weight scale is indicated.

solubilized samples can be stored in an anaerobic solution or treated with N-ethylmaleimide.[12] The GPIIIb band (M_r 97,000, also termed GPIV) is the primary labeled band just below reduced GPIIIa, and is recognizable because, unlike GPIIIa, it does not change electrophoretic mobility on disulfide bond reduction[4] and is resistant to chymotrypsin hydrolysis on intact platelets.[5] There are several labeled bands above GPIIb$_\alpha$ that can be visualized by nonreduced–reduced two-dimensional electrophoresis.[12] Identity of these and other bands is also confirmed by electrophoresis according to O'Farrell (see Ref. 14) and immunoprecipitation with monospecific antibodies.[15]

In platelets labeled by the periodate/sodium boro[^{3}H]hydride procedure, the labeled glycoproteins are detected by autofluorography of the

[14] K. J. Clemetson, A. Capitanio, and E. F. Lüscher, *Biochim. Biophys. Acta* **553**, 11 (1979).
[15] R. P. McEver and M. N. Martin, *J. Biol. Chem.* **259**, 9799 (1984).

dried SDS–polyacrylamide gel.[16] Lanes 3 and 5 of Fig. 4 illustrate the glycoproteins labeled by this procedure; increased exposure of the film to the gel will permit detection of more than 30 labeled glycoproteins. Glycoprotein Ib is the most prominent of the labeled glycoproteins, reflecting its abundance in platelets (~30,000 copies per platelet) and its high sialic acid content. Both of the disulfide-linked subunits of GPIb are labeled by this procedure; GPIb$_\alpha$, M_r 128,000–141,000; and GPIb$_\beta$, M_r 22,000. The heterogeneity in the molecular weight of GPIb$_\alpha$ is due to the presence of allelic variants, which are usually quite rare.[17] The identity of GPIb can be confirmed by several of its properties: (1) characteristic shift in molecular weight on reduction of disulfide bonds (M_r 170,000, nonreduced[12]), (2) absence in platelets from patients with Bernard–Soulier syndrome[18], (3) intense staining by the periodic acid–Schiff reagent[2], (4) binding to wheat germ agglutinin[19], (5) immunoprecipitation with monoclonal antibodies[20], and (6) selective hydrolysis by Ca^{2+}-dependent protease treatment of intact platelets.[21] Glycoprotein IX (M_r 17,000), also labeled by the periodate procedure, exists as a complex with GPIb[22] and therefore coimmunoprecipitates with GPIb.

Fibrinogen, thrombospondin, and other α granule proteins and glycoproteins are not normally labeled by either of the surface-labeling methods described here. They do become prominently labeled components, however, if platelets have undergone the release reaction prior to labeling.[10] Accordingly, care must be taken to maintain platelets in an unactivated state during isolation and to complete the labeling procedure without delay after isolation.

Acknowledgments

The work was supported by Grants HL28947 and HL 32254 from the National Institutes of Health.

The author wishes to thank James X. Warger and Norma Jean Gargasz for graphics, Barbara Allen and Sally Gullatt Seehafer for editorial assistance, and Michele Prator and Linda Harris Odumade for manuscript preparation.

---

[16] W. A. Bonner and R. A. Laskey, *Eur. J. Biochem.* **46,** 83 (1974).
[17] M. Moroi, S. M. Jung, and N. Yoshida, *Blood* **64,** 622 (1984).
[18] A. T. Nurden and J. P. Caen, *Nature (London)* **255,** 720 (1975).
[19] K. J. Clemetson, S. L. Pfueller, E. F. Lüscher, and C. S. P. Jenkins, *Biochim. Biophys. Acta* **464,** 493 (1977).
[20] A. J. McMichael, N. A. Rust, J. R. Pilch, R. Sochyinsky, J. Morton, D. Y. Mason, C. Ruan, G. Tobelem, and J. Caen, *Br. J. Haematol.* **49,** 501 (1981).
[21] N. Yoshida, B. Weksler, and R. Nachman, *J. Biol. Chem.* **258,** 7168 (1983).
[22] M. C. Berndt, C. Gregory, A. Kabral, H. Zola, D. Fournier, and P. A. Castaldi, *Eur. J. Biochem.* **151,** 637 (1985).

[36] Evaluation of Platelet Surface Antigens by Fluorescence Flow Cytometry

By BURT ADELMAN, PATRICIA CARLSON, and ROBERT I. HANDIN

Introduction

Fluorescence flow cytometry has proven to be a useful technique for identifying cell surface-associated antigens. Large numbers of cells can be examined simultaneously and more than one property of each cell analyzed, thereby facilitating the identification of subpopulations. Both intrinsic constitutive membrane proteins and molecules adsorbed or bound to the cell surface can be identified. For example, by using antibodies directed against unique surface markers this method has been used to identify T lymphocyte subtypes within heterogeneous leukocyte preparations.[1]

This chapter will focus on methods for the analysis of platelets by fluorescence flow cytometry. As an example, we describe the identification of glycoprotein Ib (GPIb), the platelet receptor for von Willebrand factor (vWF),[2] by incubation of platelets with anti-GPIb antibody. Other investigators have used similar methods to study the glycoprotein IIb–IIIa complex[3] and platelet-associated immunoglobulin,[4-6] complement,[7] blood group antigens,[8] and decay-accelerating factor.[9] The technique of fluorescence flow cytometry can also be used to study the binding of proteins like vWF, fibrinogen, and fibronectin to the platelet surface. In addition, the method can be used to screen monoclonal and polyclonal antibodies directed against platelet antigens.

For a complete discussion of the theory and instrument design underlying the development of fluorescence flow cytometry equipment and

---

[1] E. L. Reinherz and S. F. Schlossman, *Cell (Cambridge, Mass.)* **19,** 821 (1980).
[2] B. Adelman, A. D. Michelson, R. I. Handin, and K. A. Ault, *Blood* **66,** 423 (1985).
[3] L. K. Jennings, R. A. Ashmun, W. C. Wang, and M. E. Dockter, *Blood* **68,** 173 (1986).
[4] J. Lazarchick and S. A. Hall, *J. Immunol. Methods* **87,** 257 (1986).
[5] J. Lazarchick, P. V. Genco, S. A. Hall, A. D. Ponzio, and N. M. Burdash, *Diagn. Immunol.* **2,** 238 (1984).
[6] C. S. Rosenfeld and D. C. Bodensteiner, *Am. J. Clin. Pathol.* **85,** 207 (1986).
[7] V. Martin, K. A. Ault, and R. I. Handin, *Blood* **54,** Suppl. 1, 112 (1979).
[8] R. A. Dunstan and M. B. Simpson, *Br. J. Haematol.* **61,** 603 (1985).
[9] A. Nicholson-Weller, J. P. March, C. E. Rosen, D. S. Spicer, and K. F. Austen, *Blood* **65,** 1237 (1985).

TABLE I
PREPARATION OF PLATELETS FOR ANALYSIS BY
FLUORESCENCE FLOW CYTOMETRY

1. Prepare platelet-rich plasma
2. Fix platelets in 2% (v/v) formaldehyde
3. Wash fixed platelets
4. Incubate platelets with primary or control antibody
5. Wash platelets free of excess antibody
6. Incubate platelets with fluorescein-labeled second antibody
7. Wash platelets free of excess antibody
8. Analyze by flow cytometry

methods the reader is referred to a review[10] and two texts[11,12] on this subject.

Preparation of Platelets and Fluorescent Staining

We routinely treat platelets with formaldehyde[13] prior to incubation with antibody (see Table I). A major advantage of formaldehyde fixation is inhibition of degrading enzymes present within platelets that might remove surface antigens. This is a particular problem with GPIb, as it is rapidly released from the platelet surface by a calcium-dependent protease.[14] Fixation also prevents platelet aggregation during washing steps and enhances the stability of the stained cells. We find that formaldehyde-treated platelets can be stored at 4° for up to 5 days prior to analysis with minimal loss of the fluorescent signal. The effect of fixation on the immunoreactivity, platelet content, and distribution of any antigen studied must be determined by the investigator. If fixation alters the immunoreactivity of a particular antigen it is possible to treat platelets with formaldehyde after immunostaining.[8]

Whole blood is drawn through a 21-gauge butterfly-type needle into a plastic syringe containing 3.8% (w/v) sodium citrate. Nine parts of blood are mixed with one part anticoagulant using a double-syringe technique. First draw 3 ml of blood into a syringe that is then discarded. A second syringe containing sodium citrate is used to draw the blood. The anticoagu-

[10] D. R. Parks and L. A. Herzenberg, this series, Vol. 108, p. 197.
[11] M. A. Van Dilla, P. N. Dean, O. D. Laerum, and M. R. Melamed, "Flow Cytometry: Instrumentation and Data Analysis." Academic Press, London, 1985.
[12] H. M. Shapiro, "Practical Flow Cytometry." Alan R. Liss, New York, 1985.
[13] J. P. Allain, H. A. Cooper, R. H. Wagner, and K. M. Brinkhous, *J. Lab. Clin. Med.* **85**, 318 (1975).
[14] D. R. Phillips and M. Jakábová, *J. Biol. Chem.* **253**, 3435 (1978).

lated blood is mixed and then centrifuged in polypropylene test tubes at 200 g for 10 min at room temperature. If blood is obtained from a normal donor, 15–30 ml will provide an adequate number of platelets for study although studies have been performed on as little as 1 ml of whole blood. The upper two-thirds of the platelet-rich plasma (PRP) is collected and mixed with an equal volume of buffered 2% (v/v) formaldehyde that has been warmed to 37°. Buffered formaldehyde (2% formaldehyde in TBS: 10 mM Tris, 0.01 M NaCl, pH 7.4) is kept refrigerated and made fresh every week from a stock solution of 37% formaldehyde (Fisher Scientific Co., Fair Lawn, NJ). After incubation for 30 min at 37° the platelets are centrifuged at 3200 g for 10 min at 4°, resuspended in washing buffer (TBS containing 2 mM EDTA), and washed twice by centrifugation at 3200 g for 10 min at 4°. After washing, the platelets are suspended in TBS without EDTA. If staining and analysis are not done immediately the platelets should be stored at 4° in TBS containing 0.02% sodium azide. Prior to use, previously stored platelets should be centrifuged and resuspended in fresh buffer. Although we have used TBS in this procedure other physiologic buffers should be equally effective.

Immunofluorescent Staining

Glycoprotein Ib was detected on formaldehyde-fixed platelets by indirect immunofluorescent staining (Fig. 1). In our studies we have used both monoclonal and polyclonal antibodies directed against the α chain of GPIb. Monoclonal antibodies were produced by immunizing mice with intact platelets or with purified glycocalicin, a proteolytic fragment of the α chain of GPIb (see [25] in this volume). Polyclonal antibodies were produced by immunizing rabbits with glycocalicin. The second agent is always a fluorescein isothiocyanate (FITC)-conjugated, affinity-purified, F(ab')$_2$ fragment of appropriate specificity (Cooper Biomedical, Inc., Malvern, PA).

Formaldehyde-fixed platelets in TBS were diluted to a concentration of 50,000 platelets/μl. They were then incubated for 30 min at 4° with primary antibody (antibody may be in purified form or used as ascites or serum). After incubation the platelets were washed three times by centrifugation (3200 g for 15 min at 4°) and resuspended in TBS at the original volume. The platelets are then incubated with the appropriate FITC-labeled second antibody for 30 min at 4° and then again washed three times as mentioned above. Each antibody should be used at a saturating concentration, which must be determined by trial and error.

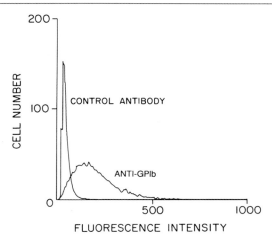

FIG. 1. Fluorescence flow cytometry analysis of GPIb on formaldehyde-treated platelets. Platelets were prepared as described in text and stained with 3G6, a monoclonal antibody directed against the α chain of GPIb or with a control, nonspecific antibody. Fluorescence analysis was performed on an Ortho 50H flow cytometer equipped with a 100-mW argon ion laser operating at 488 nm and 50 mW. Fluorescence intensity is displayed on a linear scale in which the scale units are channel numbers. Each curve represents 10,000 platelets. Both the mean and peak fluorescence signal from the specifically stained platelets is greater than that of the nonspecifically stained platelets.

Fluorescence Flow Cytometry

All flow cytometers currently marketed utilize similar technology for sample presentation, illumination, and raw data capture. Differences exist in the number, power, and placement of light sources and in capabilities for data analysis and presentation. The methods described here should be applicable to machines with even the most basic configuration. Some flow cytometers use a mercury arc lamp rather than a laser as the source of illumination. We have not had adequate experience with such machines to comment on their use, so that this discussion is based on experience only with cytometers equipped with a laser.

Identification of specific cell populations within a sample prior to fluorescence analysis is usually determined by light scatter measurements. Forward-angle light scatter is proportional to overall cell size, while 90° scatter is related to internal cell structure. The two scatter signals taken together can be used to distinguish different cell types in mixed samples (such as whole blood), and to identify dead cells in homogeneous cell preparations. Once a cell population or subpopulation is identified by

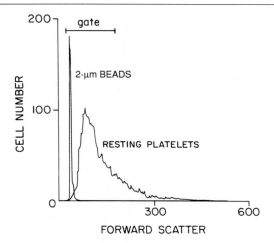

Fig. 2. Determination of the appropriate sizing gate for analysis of platelets by flow cytometry. This composite picture includes the forward-angle scatter analysis of 2-μm beads and formaldehyde-treated resting platelets. The gate is set to include all particles falling within the region indicated. This region will include approximately 80% of the resting platelets that are 2 μm and greater in diameter. Each curve represents 10,000 platelets and is displayed on a linear scale in which the scatter units are channel numbers.

its light scatter signal (forward scatter alone or forward- and right-angle scatter), a gate can be drawn that will segregate the population of interest and the machine instructed to collect fluorescent signals from all particles whose scatter characteristics fall within the defined gate. Gate boundaries are delimited by user-defined upper and lower channels.

Light scatter from nonspherical cells, such as platelets, is affected by the orientation of each target cell flowing past the laser beam. Nonspherical cells tend to orient with their long axes parallel to the direction of flow. Even aligned in this manner, platelets will pass through the laser beam in various orientations and produce a range of scatter signals because of their discoid shape. In addition, aggregated platelets will produce varying scatter signals depending on aggregate size and orientation.

We use forward light scatter measurements derived from 2-μm beads (Polysciences, Warrington, PA) and formaldehyde-treated resting platelets to develop a sizing gate prior to analysis of fluorescent labeled platelets (Fig. 2). We have not found that concurrent 90° scatter adds significantly to this process when analyzing a homogeneous sample derived from platelet-rich plasma or washed platelets. On the other hand, use of both signals is helpful when analysis is being performed on a heterogeneous sample containing all blood cell elements. The forward scatter gate is

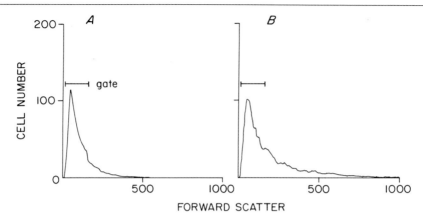

FIG. 3. Forward-scatter analysis of resting (A) and ADP-stimulated (B) platelets demonstrating platelet aggregates. Platelets in platelet-rich plasma were stimulated by incubation with 5 μM ADP for 20 min prior to formaldehyde treatment. Control platelets were not stimulated. The gate shown in (A) includes 80% of the resting platelets that are 2 μm and greater in diameter. As seen in (B), after ADP stimulation and aggregate formation this same gate includes only 55% of the total number of particles analyzed. The aggregated platelets are represented by that portion of curve (B) that extends beyond the gate and up to channel 1000.

selected to include single platelets and exclude aggregates and platelet fragments. The lower channel is determined by displaying the forward-angle scatter histogram generated by 2-μm beads, thus excluding platelet fragments. The upper channel, chosen from a point on the forward-scatter histogram of fixed, resting platelets is set so that 80% of the analyzed cells will be contained within the gate. The placement of this channel is based on our analysis of forward-angle scatter histograms generated by platelet preparations that intentionally contain platelet aggregates. These samples are produced by adding 5 μM ADP to PRP prior to formaldehyde treatment (Fig. 3).

Fluorescence analysis of platelets that fall within the sizing gate is similar to that used for any other FITC-labeled cell. Excitation of the dye is achieved with an argon ion laser emitting a 488-nm laser beam. We have successfully used lasers with power ratings of 100 mW and 5 W. For these lasers, the corresponding power output at 488 nm is 50 mW and 280–300 mW, respectively. The power output of the laser determines, in part, the sensitivity of the system. For most targets, the power of the exciting laser beam will determine the strength of the excitation signal emitted. If the copy number of the antigen of interest is very low its detection may be enhanced by using a more powerful laser. Direct immunofluorescent staining of antigens with FITC-labeled primary antibodies may also be facilitated by higher power lasers. With the platelets diluted to 50,000 μl we adjust the sample flow rate so that the cells are analyzed at a rate of 300 to 500/sec.

The fluorescence signal can be subjected to either linear or logarithmic amplification. Linear amplification is best suited for signals that vary over only a small range. Logarithmic amplifiers are better suited for analysis of data that vary over a wide range of fluorescence intensity. Detection of subpopulations, particularly if they are very much brighter or dimmer than the majority of cells, will be aided by logarithmic amplification.[10]

Data Analysis and System Calibration

Because current methods in flow cytometry do not permit direct quantitation of cell surface antigens, results are usually expressed in arbitrary units. Often these units refer to the peak or mean channel number describing the fluorescence histogram of a cell population. Fluorescence density calculations based on individual cell fluorescence intensity (channel number) divided by its corresponding 90° scatter channel number may also be utilized.

To compare results from one experiment to another and from day to day, it is necessary to have a method for instrument calibration. We have found that commercially available beads are too bright and prefer using glutaraldehyde-fixed chicken erythrocytes. Using the fluorescence photomultiplier gain control we locate the erythrocyte fluorescence histogram so that peak fluorescence is located midway along the fluorescence intensity axis. Similarly, we have also utilized Fluorotrol GF (a research reagent prepared by Ortho Diagnostic Systems, Westwood, MA) for machine calibration. Fluorotrol GF is a mixture of thymocyte nuclei in which approximately 20% of the nuclei are unstained, 40% are stained with 50,000 FITC molecules per nucleus, and 40% are stained with 220,000 FITC molecules per nucleus. Depending on the laser power of a specific instrument, fluorescence analysis of Fluorotrol GF will produce a two- or three-peak histogram (in low-powered systems autofluorescence from the unstained nuclei may not be detected). The fluorescence peaks can be aligned over the same channels each day and thus the machine repeatedly calibrated.

Analysis of Platelets without Fixation and in Whole Blood

As mentioned previously, the effect of formaldehyde fixation on a specific antigen must be determined experimentally. Saunders *et al.* reported that the distribution of the Zw^a antigen is altered by fixation.[15] Other

[15] P. W. G. Saunders, B. E. Durack, and H. K. Narang, *Br. J. Haematol.* **62,** 631 (1986).

investigators have used flow cytometry techniques to examine intrinsic or adsorbed platelet antigens without prior platelet fixation. Nicholson-Weller et al. have described the presence of decay-accelerating factor, a complement regulatory substance, on the surface of platelets.[9] They utilized washed platelets for their studies and analyzed them shortly after preparation. Dunstan and Simpson have used flow cytometry to analyze platelet content of ABH, Ii, Lewis, P, PL^{A1}, Bak, and HLA class I antigens.[8] In their studies platelets were immunostained after washing and then formaldehyde fixed. Jennings and others reported on the analysis of the platelet glycoprotein IIb–IIIa complex in dilute PRP or whole blood.[3] To keep the platelets from aggregating during antibody staining and analysis they mixed previously anticoagulated PRP or whole blood with an equal volume of a buffer containing prostacyclin (0.154 M NaCl, 0.01 M Tris, 0.0005 M Ca_2Cl, 50 μM prostacyclin). The applicability of this method for analysis of GPIb has not been evaluated. Moake et al. reported that prostacyclin can inhibit vWF-dependent platelet agglutination by altering the platelet surface.[16] Although prostacyclin did not block vWF binding to platelets, it is not known whether GPIb immunoreactivity is altered.

Other studies have focused on identification of circulating activated platelets using flow cytometry techniques. Specific surface antigens have been identified that predict the activated state.[17,18] Similarly, the presence of platelet-derived microparticles has been associated with intravascular platelet activation.[19,20]

Acknowledgments

B. Adelman is the recipient of NHLBI Clinical Investigator Award HL01053. Additional support was provided by a grant from the Council for Tobacco Research U.S.A., Inc.

---

[16] J. L. Moake, S. S. Tang, J. D. Olson, J. H. Troll, P. L. Cimo, and P. J. A. Davies, *Am. J. Physiol.* **241,** H54 (1981).
[17] C. L. Berman, E. L. Yeo, J. D. Wencel-Drake, B. C. Furie, M. H. Ginsberg, and B. Furie, *J. Clin. Invest.* **78,** 130 (1986).
[18] S. J. Shattil, M. Cunningham, and J. A. Hoxie, *Blood* **70,** 307 (1987).
[19] C. S. Abrams, N. Ellison, A. Z. Budzynski, and S. J. Shattil, *Blood* **75,** 128 (1990).
[20] J. N. George, E. B. Pickett, S. Saucerman, R. P. McEver, T. J. Kunicki, N. Kieffer, and P. J. Newman, *J. Clin. Invest.* **78,** 340 (1986).

[37] Identification of Platelet Membrane Target Antigens for Human Antibodies by Immunoblotting

By DIANA S. BEARDSLEY

Introduction

The destruction of blood platelets by antibodies plays an important role in immune-mediated thrombocytopenia. Autoantibodies are involved in autoimmune thrombocytopenia (ITP), and alloantibodies are involved in neonatal alloimmune thrombocytopenia (NATP), posttransfusion purpura (PTP), and resistance to platelet transfusion. For many years, detection and study of these anti-platelet antibodies depended on complement fixation, chromium release, and antiglobulin consumption assays[1] or quantitation of platelet-associated immunoglobulin.[2] Although these approaches can indicate the presence of anti-platelet antibodies, none allows further examination of the antigens involved in the reaction. It is these antigens, however, that mark the platelet for destruction.

It is possible by using immunoblotting techniques to identify on platelets the antigenic proteins that are targets for many of these antibodies. That approach is the subject of this chapter. First, the thrombocytopenic syndromes in which anti-platelet antibodies have been implicated will be briefly reviewed. The current state of our understanding of the composition of the platelet surface will be mentioned, and then the technique of immunoblotting will be detailed. Finally, results obtained by this technique will be discussed. Although the clinical value of platelet antigen identification has not yet been fully assessed, a number of important observations have already been made toward understanding the molecular details of these thrombocytopenic syndromes.

Syndromes of Immune Platelet Destruction

Autoantibodies

Idiopathic or autoimmune thrombocytopenic purpura (ITP) is an acquired disorder in which platelets are rapidly destroyed in the reticuloendothelial system.[2-4] In children, ITP frequently follows a viral illness and

---
[1] N. R. Shulman, V. J. Marder, M. C. Hiller, and E. I. Collier, *Prog. Hematol.* **4**, 222 (1964).
[2] J. G. Kelton and S. Gibbons, *Thromb. Haemostasis* **8**, 83 (1982).
[3] S. Karpatkin, *Blood* **56**, 329 (1980).
[4] A. L. Lightsey, *Pediatr. Clin. North Am.* **27**, 293 (1980).

usually resolves spontaneously.[4] In adults, however, ITP more commonly has an insidious onset and becomes a chronic problem.[2,3,5] More than 30 years ago, Harrington demonstrated that the globulin fraction of plasma contains the platelet-destructive agent responsible for most cases of ITP.[6] The various forms of therapy commonly employed (steroids, splenectomy, immunosuppressives, and intravenous γ-globulin) are all aimed at interfering with antibody-mediated platelet destruction. However, the responsible antibodies have not been studied in detail. The target antigens are thought to be "public" antigens, because they are present on most normal platelets, but until recently they had not been further characterized.

Immune platelet destruction is a common feature of acquired immunodeficiency syndrome (AIDS) and of human immunodeficiency virus (HIV) infection.[7,8] The thrombocytopenia is generally thought of as ITP. Because circulating immune complexes, often a part of AIDS, are usually cleared by platelets, it has been proposed that AIDS-related ITP may represent an immune complex disease rather than autoimmune platelet destruction.[7] Autoantibodies have also been implicated.[8] Using immunoblotting techniques, a 25-kDa antigen was suggested as a target for HIV-associated autoantibodies.[9] However, other investigators have not confirmed this finding. Immune thrombocytopenia and autoantibodies can also accompany systemic lupus erythematosus,[10] although different, intracellular targets have been reported in this disease.[11,12]

Some drug-induced thrombocytopenias may be caused by induced autoantibodies or by another type of autoimmune platelet destruction in which the antigen–antibody unit is complete only in the presence of the drug, acting as a hapten. The most completely studied examples have been quinidine or quinine-dependent antibodies.[13,14] The antibodies responsible

---

[5] R. McMillan, *N. Engl. J. Med.* **304,** 1135 (1981).
[6] W. J. Harrington, V. Minnich, J. W. Hollingworth, and C. V. Moore, *J. Lab. Clin. Med.* **38,** 1 (1951).
[7] C. M. Walsh, M. A. Nardi, and S. Karpatkin, *N. Engl. J. Med.* **311,** 625 (1984).
[8] S. Savona, M. A. Nardi, E. T. Lennette, and S. Karpatkin, *Ann. Intern. Med.* **102,** 737 (1985).
[9] R. B. Stricker, D. I. Abrams, L. Corash, and M. A. Shuman, *N. Engl. J. Med.* **313,** 1375 (1985).
[10] T. Asano, B. C. Furie, and B. Furie, *Blood* **66,** 1254 (1985).
[11] C. Kaplan, P. Champeix, D. Blanchard, J. Y. Muller, and J. P. Cartron, *Br. J. Haematol.* **67,** 89 (1987).
[12] M.-N. Guilly, F. Danon, J. C. Brouet, M. Bormens, and J.-C. Courvalin, *Eur. J. Cell Biol.* **43,** 266 (1987).
[13] W. Lerner, R. Caruso, D. Faig, and S. Karpatkin, *Blood* **66,** 306 (1985).
[14] R. B. Stricker and M. A. Shuman, *Blood* **67,** 1377 (1986).

for heparin-associated thrombocytopenia and platelet agglutination have been studied by immunoprecipitation and immunoblotting.[15]

Alloantibodies

The existence of platelet-specific alloantigens has been accepted for more than 25 years since Shulman reported studies of two rare types of alloimmunization.[1,16,17] Posttransfusion purpura (PTP) is an enigmatic syndrome characterized by sudden severe thrombocytopenia with complement-fixing cytolytic antibodies occurring 7–10 days after a previously immunized antigen-negative individual receives blood containing antigen-positive platelets.[16] The basis for destruction of the antigen-*negative* platelets of the patient after immune stimulation by antigen-*positive* platelets remains a mystery. Alloimmunization can also occur during pregnancy; maternal anti-platelet IgG antibodies that cross the placenta can lead to a severe thrombocytopenia in the fetus in the syndrome of neonatal alloimmune thrombocytopenic purpura (NATP).[17,18] For affected infants, the risk of central nervous system hemorrhage or death is significant (10–20%).[19] Most cases of PTP and NATP are due to incompatibility of the platelet-specific antigen Pl^{A1} (or Zw^a).[16,20] Immunoblotting has been used to demonstrate anti-Pl^{A1} antibodies in PTP and NATP sera and to confirm that this antigen is a part of GPIIIa.[21-23] Other platelet-specific antigens that have been reported are listed in Table I.[1,16,20,24-27]

Platelet-specific alloantigens can influence the response to platelet transfusion,[28] although HLA alloantigens are certainly important for the

---

[15] D. M. Lynch and S. E. Howe, *Blood* **66**, 1176 (1985).
[16] N. R. Shulman, R. H. Aster, A. Leitner, and M. C. Hiller, *J. Clin. Invest.* **40**, 1597 (1961).
[17] N. R. Shulman, R. H. Aster, H. A. Pearson, and M. C. Hiller, *J. Clin. Invest.* **41**, 1059 (1962).
[18] J. G. Kelton, V. S. Blanchette, W. E. Wilson, P. Powers, K. R. Mohan Pai, S. Beffer, and R. D. Barr, *N. Engl. J. Med.* **302**, 1401 (1980).
[19] H. A. Pearson, N. R. Shulman, V. J. Marder, and T. E. Cone, Jr., *Blood* **23**, 154 (1964).
[20] J. J. van Loghem, H. Dorfmeijer, and M. van der Hart, *Vox Sang.* **4**, 161 (1959).
[21] T. J. Kunicki and R. H. Aster, *Mol. Immunol.* **16**, 353 (1979).
[22] D. J. S. Beardsley, J. E. Spiegel, R. I. Handin, M. M. Jacobs, and S. E. Lux, *J. Clin. Invest.* **74**, 1701 (1984).
[23] R. McMillan, D. Mason, P. Tani, and G. M. S. Schmidt, *Br. J. Haematol.* **51**, 297 (1982).
[24] J. Moulinier, *Proc. Congr. Eur. Soc. Haematol., 6th*. p. 817. Karger, New York, Basel (1958).
[25] C. M. van der Weerdt, *Histocompat. Test.*, **2**, 161 (1965).
[26] A. E. G. Kr. von dem Borne, E. von Riesz, F. W. A. Verheught, J. W. ten Cate, J. G. Koppe, C. P. Engelfriet, and L. E. Nijenhuisle, *Vox Sang.* **39**, 113 (1980).
[27] B. Boizard and J. L. Wautier, *Vox Sang.* **46**, 47 (1984).
[28] J. H. Hermann, T. S. Kickler, and P. M. Ness, *Blood* **66**, 279a (1985).

TABLE I
PLATELET-SPECIFIC ALLOANTIGENS

Antigenic systems	Frequency	Associated with[a]	Ref.
Pl^A (Zw)	Pl^{A1} (Zw^a) = 98%	PTP, NATP	16, 20
	Pl^{A2} (Zw^b) = 28%	NATP	
DUZO	23% positive	NATP	24
Pl^E	Pl^{E1} = 99%	Antibody developed in transfused patient	1
	Pl^{E1} = 4%	NATP	
Ko	Ko^a = 15%	Transfusion alloimmunization	25
	Ko^b = 99%		
Bak	Bak^a = 91%	NATP, PTP	26
(Probably = Lek)	Lek^a = 98%		27

[a] PTP, Posttransfusion purpura; NATP, neonatal alloimmune thrombocytopenia.

survival of transfused platelets.[29] The HLA system, however, is not amenable to study by the immunoblotting technique.[22]

Platelet Surface Proteins

The composition of the platelet membrane has been discussed in detail in [35] in this volume. Identification of the surface-accessible proteins has been done by using reagents that do not enter the cytoplasm and thus radiolabel only the surface proteins. Phillips[30] devised a system of naming the platelet surface proteins in groups (I, II, III, etc.) according to decreasing apparent molecular weight on polyacrylamide gel electrophoresis, with identification of individual proteins depending on their mobilities before and after reduction of disulfide bonds. Table II summarizes the current state of our understanding of the major platelet surface proteins[30–32] (also see Refs. 33, 34).

Immunoblotting Technique

The general approach to antigen identification by immunoblotting is illustrated in Fig. 1 and technical specifics of the method are given in the figure caption.[22] Electrophoretically separated platelet proteins (step A)

[29] R. A. Yankee, F. G. Grumet, and G. Rogentine, *N. Engl. J. Med.* **281**, 1208 (1969).
[30] D. R. Phillips and P. P. Agin, *J. Biol. Chem.* **252**, 2121 (1977).
[31] D. R. Phillips, *Thromb. Haemostasis* **42**, 1638 (1979).
[32] J. P. Caen, A. T. Nurden, and T. J. Kunicki, *Philos. Trans. R. Soc. London, Ser. B* **294**, 281 (1981).
[33] D. R. Phillips and J. Jakábová, *J. Biol. Chem.* **252**, 5602 (1977).
[34] D. F. Mosher, A. Vaheri, J. J. Choate, and C. G. Gahmberg, *Blood* **53**, 3 (1979).

TABLE II
Properties of Glycoproteins

Glycoprotein	Apparent molecular weight[a]		Functional importance	Ref.
	Nonreduced	Reduced		
Ia	153,000	167,000	Absent from platelets that fail to aggregate with collagen	33
Ib	170,000	143,000 ( ) 22,000 ( )	Absent from Bernard–Soulier platelets	32
Ic	148,000	134,000 ( ) 27,000 ( )		
IIa	138,000	157,000		
IIb	142,000	132,000 ( ) 23,000 ( )	Absent from Glanzmann's thrombasthenic platelets	32
IIIa	99,000	110,000		
IV	97,000	97,000		
V	68,000–85,000	85,000	Cleaved by thrombin	34
IX		44,000		

[a] Approximate, may vary with electrophoresis conditions.

immobilized on nitrocellulose (step B) are incubated with the antibody in question (step C), followed by a labeled antiglobulin (step D) for detection of anti-platelet antibodies. It is occasionally helpful to screen for anti-platelet antibodies using intact or solubilized platelets bound to microtiter wells or to nitrocellulose dot blots without prior electrophoretic separation. However, for human antibodies, this is usually not very informative because all platelets contain IgG. Furthermore, IgG and immune complexes can bind to platelets in a nonspecific manner, i.e., independent of Fab–antigen binding. Therefore, antigen identification requires that the nonimmune (or "nonspecific") and immune (or "specific") immunoglobulin binding be distinguished. The initial electrophoresis separates nonspecifically bound IgG from platelet proteins and allows this distinction to be made.

Platelet Isolation

The pertinent antigenic proteins for antibody-mediated platelet destruction are surface membrane proteins. However, most investigators use solubilized whole platelets as the protein targets for immunoblotting.

Because a calcium-activated protease is the most important proteolytic enzyme present in platelets,[33] calcium chelation is essential during the solubilization process. Although white blood cell contamination is usually less than 1/1000 platelets, some workers include inhibitors of leukocyte serine proteases such as phenylmethylsulfonyl fluoride (PMSF) or diisopropyl fluorophosphate (DFP).

Platelets are isolated from whole blood containing citrate and 10 mM EDTA by differential centrifugation after gel filtration. The gel filtration can be omitted with some increase in the amount of IgG present with the platelets. The addition of EDTA to inhibit the calcium-dependent protease and N-ethylmaleimide to block sulfhydryls groups results in minimal, reproducible nonspecific binding of IgG to a band at approximately 200K apparent molecular weight. The solubilized samples are electrophoresed immediately after solubilization, as stored samples show evidence of proteolysis.

The platelets used as targets are usually normal platelets. However, it is often helpful to study platelets congenitally deficient in one or more of the major surface glycoproteins to identify further an antigenic protein.[32,35] For example, the 100-kDa antigenic target protein for some ITP autoantibodies was shown to be GPIIIa because it was present on platelets from normal donors but not on platelets from patients with Glanzmann's thrombasthenia, in which the platelets lack GPIIb–IIIa.[22] In such cases, it is important to use only platelets whose precise surface protein composition has been determined by sodium dodecyl sulfate-polyacrylamide gel electrophoresis (SDS–PAGE) of surface-labeled platelets because the clinical phenotypes of Glanzmann's thrombasthenia and Bernard–Soulier syndrome can be caused by different degrees of biochemical defect in protein composition.[36,37]

Electrophoretic Separation (Step A)

Polyacrylamide gel electrophoresis of proteins solubilized in SDS using systems similar to those described in [35] in this volume is used to separate platelet antigens and other immunoglobulin binding proteins according to apparent molecular weight. The Laemmli system[38] is used with a 3% (w/v)

---

[35] H. K. Nieuwenhuis, J. W. N. Akkerman, K. S. Sakariassen, W. P. M. Houdijk, P. F. E. M. Nievelstein, and J. J. Sixma, *Thromb. Haemostasis* **54,** 124a (1985).
[36] I. Hagen, A. T. Nurden, O. J. Bjerrum, N. O. Solum, and J. P. Caen, *Protides Biol. Fluids* **27,** 875 (1979).
[37] K. Tenoue, S. Hasegunos, N. Yamamoto, K. Yamamoto, A. Yamaguchi, and H. Yamazaki, *Thromb. Haemostasis* **54,** 182a (1985).
[38] U. K. Laemmli, *Nature (London)* **222,** 688 (1970).

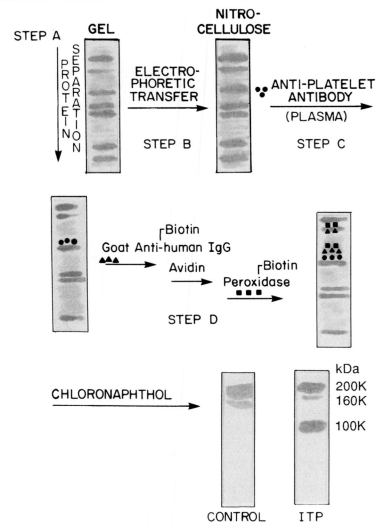

FIG. 1. Immunoblotting technique. For platelet preparation, platelet-rich plasma is obtained from blood collected in 10 mM EDTA/ACD-A (1:9, volume of EDTA/ACD-A to blood volume by centrifugation at 230 g for 20 min at room temperature. The supernatant is gel filtered over Sepharose 2B (Pharmacia, Uppsala, Sweden) and then centrifuged at 800 g for 20 min at room temperature. The pelleted platelets are then washed three times in a TES–Tyrode's buffer containing 136 mM NaCl, 10 mM N-tris(hydroxymethyl)-2-aminoethanesulfonic acid (TES) (Calbiochem, Los Angeles, CA), 2.6 mM KCl, 0.5 mM NaH_2PO_4, 2 mM EDTA, 2 mM $MgCl_2$, and 5.5 mM glucose, pH 7.4. Protein concentration is determined. For SDS–PAGE, 10^9 platelets in 0.5 ml of TES–Tyrode's buffer are solubilized by addition

polyacrylamide stacking gel and a separating slab gel of between 5 and 12% polyacrylamide, depending on the apparent molecular weight of suspected antigens. Parallel separation of proteins of known molecular weight allows calculation of the apparent molecular weight of antigenic proteins. Although a semilog plot of molecular weight versus relative mobility can be derived from a gel lane of standard proteins stained with silver or Coomassie Brilliant blue, it is also possible to use standards transferred to the nitrocellulose paper. Prestained marker proteins are most convenient (Bethesda Research Laboratories, Gaithersburg, MD), but proteins can also be stained directly on the nitrocellulose by using India ink[39] or colloidal metal techniques[40] that are sensitive to less than 10 ng of protein. The precise location of the gel lanes can be marked by adding methyl green to

[39] K. Hancock and V. C. W. Tsang, *Anal. Biochem.* **133,** 157 (1983).
[40] N. Moeremans, G. Daneels, and J. DeMey, *Anal. Biochem.* **145,** 315 (1985).

---

of 0.5 ml of 60 mM Tris-HCl, pH 6.8, 0.002% (v/v) bromphenol blue, 2.5% (v/v) glycerol, 3% (w/v) SDS with or without 5% (v/v) 2-mercaptoethanol, and then boiled for 5 min. Step A (Electrophoresis): Using the Laemmli discontinuous buffer system,[38] a 1.5-mm slab gel is prepared, consisting of a 3% stacking gel and a 5, 7.5, or 12% separating gel. Platelet proteins (100–200 μg) are applied to each lane. To the first and last lanes 15 μl of prestained molecular weight markers (BRL, Gaithersburg, MD) are applied. Electrophoresis at 15 V continues until a bromphenol blue tracking dye reaches the bottom of the gel. Near the end of the electrophoresis, a marker dye of 10 μl of 0.01% methyl green in sample buffer is added to each well. This will transfer to the nitrocellulose, indicating each lane of proteins. Step B (Transfer to nitrocellulose): The polyacrylamide gel is soaked in blotting buffer (0.042 M Tris, 0.19 M glycine, 10% methanol) for 30 min at room temperature and then placed on a piece of nitrocellulose paper (0.45-μm pore size) (Schleicher & Schuell, Keene, NH) and wetted with blotting buffer to remove all air bubbles. The paper is placed toward the anode of the blotting apparatus (LKB). Close contact is maintained with filter paper and Scotch Brite (3M Co., Minneapolis, MN) pads for upholstering. Transfer occurs in 5 liters of buffer for 90 min at 80 V (0.5 A). The nitrocellulose is dried between filter paper and cut at the methyl green lane markers. Step C (Antibody binding): Nitrocellulose strips containing a single lane of platelet proteins are incubated first with 1.5% (v/v) goat serum in phosphate-buffered saline (PBS) for 1 hr to saturate protein-binding sites and then with a 1:50 dilution of plasma or serum in PBS with 1% (v/v) bovine serum albumin (PBS/1% BSA) for 1 hr. During incubation the tubes are mixed on a rocking mixer. The strips are washed three to five times for 5 min with 2.5 ml of PBS/1% BSA or PBS between subsequent incubation steps. Step D (Antibody localization): A 1:200 dilution of biotinylated affinity-purified goat anti-human IgG (Vector Laboratories, Burlingame, CA) in PBS/1% BSA is added for 15 min. After exposure for 30 min to a 1% (v/v) solution of avidin and biotinylated horseradish peroxidase (Vector Laboratories, Burlingame, CA) in PBS containing 0.1% (v/v) Tween 20 detergent, 2.5 ml of 0.02% (w/v) chloronaphthol, 0.006% (v/v) hydrogen peroxide in PBS is added. Purple bands develop within 2 min. All solutions are prepared fresh each day; azide is not added because it inhibits the peroxidase reaction.

the sample in the well prior to electrophoresis and again near the end of electrophoresis. The dye will transfer to the nitrocellulose paper, indicating the location of the lanes containing platelet proteins.

Transfer to Nitrocellulose (Step B)

The separated proteins are electrophoretically transferred from the polyacrylamide gel to nitrocellulose paper using the Western blotting technique.[41] Most laboratories use a commercial apparatus (e.g., Bio-rad, Richmond, CA; E-C Corp., St. Petersburg, FL; or LKB, Rockville, MD). Nitrocellulose paper with a 0.45-μm pore size is commonly used, although Stricker and Shuman[14] found it necessary to use a 0.1-μm pore size to immobilize a 25K antigenic protein. The blotting buffer can be reused twice. To transfer strongly basic proteins, it may be necessary to alter the buffer.[42] The gel is soaked in blotting buffer to allow diffusion of SDS and size equilibration of the gel prior to the electrophoretic transfer. Some investigators add 0.1% (w/v) nonionic detergent to the blotting buffer. Transfer of proteins up to 200K apparent molecular weight from a 1.5-mm thick 7.5% gel is usually complete in 90 min at 80 V or 0.5 A but should be assayed for each apparatus and type of gel by staining the gel after transfer with Coomassie Brilliant blue.

Immune Staining (Steps C and D)

After transfer of platelet proteins to the nitrocellulose paper, the lanes containing platelet proteins are cut out and stored for up to 6 months in individual tubes. Prior to antibody incubation, remaining nonspecific protein binding sites are blocked by incubation with irrelevant protein. Bovine serum albumin and bovine γ-globulin in phosphate-buffered saline or 1% (v/v) normal goat serum can be used,[22] although some investigators prefer gelatin or nonfat dry milk. The nitrocellulose is then incubated with the antibody being studied, usually as a 1 : 100 dilution of serum or plasma. Plasma gives a stronger reaction in some cases; but usually either is adequate. Unbound IgG is removed by multiple washes, and bound immunoglobulin is localized by incubation with an antiglobulin reagent. The preferred method (Fig. 1) uses biotin-conjugated antiglobulin and horseradish peroxidase with avidin. Alternatively, affinity-purified goat anti-human IgG that has been radioiodinated by the chloramine-T reaction[43] or conjugated to peroxidase may be used.

[41] H. Towbin, T. Staehelin, and J. Gordon, *Proc. Natl. Acad. Sci. U.S.A.* **76**, 4350 (1979).
[42] B. Szewczyk and L. M. Kozloff, *Anal. Biochem.* **150**, 403 (1985).
[43] M. W. Hunter and F. C. Greenwood, *Nature (London)* **194**, 495 (1962).

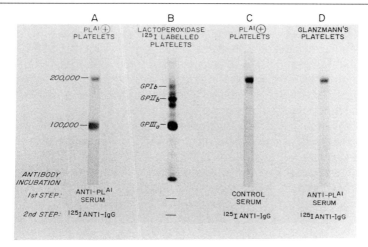

FIG. 2. Pl^{A1}. Autoradiograms of platelet protein preparations separated by SDS–PAGE, transferred to nitrocellulose, and subjected to antibody incubations as indicated at the bottom of the figure. Note the M_r 100,000 band that comigrates with GPIIIa.

Immunoblotting Results

"Nonspecific" Binding of Immunoglobulin to Platelets

When separated platelet proteins immobilized on nitrocellulose are incubated with control serum and then exposed to labeled antiglobulin as detailed above, IgG binding is noted at approximately 200K (nonreduced) and 45K (reduced) apparent molecular weight. This represents immunoglobulin binding in a nonimmune or "nonspecific" manner, probably via the Fc binding protein of similar apparent molecular weight isolated by Cheng and Hawiger.[44] An important feature of the immunoblotting technique is that this nonimmune IgG binding is physically separated from many important platelet antigens. Occasionally a band is seen at 160K, probably due to IgG isolated with the platelets.

Platelet Alloantigens

Incubation of Pl^{A1}-positive platelet proteins with an anti-P1^{A1} serum is illustrated in Fig. 2 (lane A). Note that, in addition to the nonspecific 200K band, IgG binds to a protein at 100K that comigrates with GPIIIa. The antigen is absent from type I Glanzmann's thrombasthenic platelets known

[44] C. M. Cheng and J. Hawiger, *J. Biol. Chem.* **254**, 2165 (1979).

to lack GPIIb–IIIa (lane D, Fig. 2) and is destroyed by disulfide reduction. Similar findings have been reported by other investigators[23,27,45] and are consistent with the isolation of the Pl[A1] antigen with GPIIIa.[21] Bak[a] [26] and the probably identical Lek[46] antigen have been localized to GPIIb using immunoblotting. Immunoblotting may prove to be particularly valuable in studying infants with NATP. In these situations, it is possible to assess maternal and neonatal serum for the presence of antibodies reactive against paternal, but not against maternal, platelet proteins. In such cases even new antigens are amenable to study by immunoblotting because antigen-positive and antigen-negative platelets are available from the father and mother, respectively. In preliminary results of studies of patients who were refractory to platelet transfusions, Hermann *et al.* reported that no serum alloantibodies were detectable by immunoblotting, although they were able to detect anti-Pl[A1] posttransfusion antibodies.[28]

Drug-Induced Thrombocytopenia

For about 10 years, heparin-associated thrombocytopenia has been known to have an immune etiology.[47] However, the precise mechanism for destruction of platelets has not been clear. This problem was studied using an immunoblotting technique.[15] Immunoglobulin binding to proteins of 182K, 124K, and 82K was reported to be enhanced when heparin was added to the incubating plasma; however, only the 82K protein was detected by immunoprecipitation. Work by Adelman, however, suggests that some heparin-dependent antibodies bind to an antigen on GPIb.[48]

Circulating autoantibodies from five patients with quinidine purpura have been shown by Stricker and Shuman,[14] using immunoblotting techniques, to be directed against an antigen on GPV. Because GPV exists in the membrane as part of a complex that includes GPIb, immunoprecipitation studies have previously suggested that GPIb might contain the antigenic quinidine antibodies.[49]

Autoantibodies in Autoimmune Thrombocytopenia

Immunoblotting studies of sera from patients with ITP have shown that some of these patients have circulating autoantibodies against GPIIIa.[22] Figure 3 illustrates results with one such patient. Although IgG in control

[45] D. V. Devine and W. F. Rosse, *Blood* **64**, 1240 (1984).
[46] N. Kieffer, B. Boizard, D. Didry, J.-L. Wautier, and A. T. Nurden, *Blood* **64**, 1212 (1984).
[47] D. B. Cines, P. Kaywin, M. Bina, A. Tomaski, and A. D. Schreiber, *N. Engl. J. Med.* **303**, 788 (1980).
[48] B. Adelman, *Clin. Res.* **34**, 449a (1986).
[49] L. Degos, G. Tobelem, P. Lethielleux, S. Levy-Toledano, J. Caen, and J. Colombani, *Blood* **50**, 899 (1977).

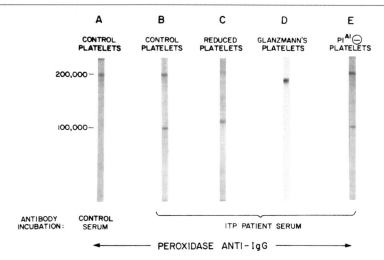

FIG. 3. Chronic ITP. Photographs of nitrocellulose transfers of platelet proteins incubated with ITP serum or control serum as indicated, followed by biotinylated goat anti-human IgG, avidin, and biotinylated horseradish peroxidase. Colored bands appeared after addition of chloronaphthol. Note the specific M_r 100,000 protein that binds IgG. This protein is absent from Glanzmann's thrombasthenic platelets and is converted to an apparent molecular weight of 110,000 on reduction of disulfide bonds.

serum bound only the 200K band (Fig. 3, lane A), IgG from ITP patient serum bound to a platelet protein of apparent molecular weight 100K (nonreduced) and 110K (reduced). This protein was absent from Glanzmann's thrombasthenic platelets, suggesting that the target antigen is on GPIIIa. Using different techniques, other investigators have implicated the GPIIb–IIIa complex as a target for ITP antibodies.[50,51] However, the ITP target antigen is not the same as Pl[A1] because it survives disulfide reduction and is present on Pl[A1]-negative platelets (lane E, Fig. 3). To demonstrate that this antibody was a true autoantibody, these studies were repeated using platelets from the patient as the source of target proteins from immunoblotting and the results were identical. Anti-GPIIIa antibodies occur less frequently in children with acute ITP than in adults with chronic ITP.[22,52]

Not all ITP antibodies that bind to a protein at 100K are directed against

[50] V. L. Woods, E. H. Oh, D. Mason, and R. McMillan, *Blood* **63**, 368 (1984).
[51] E. F. van Leeuwen, J. T. M. van der Ven, C. P. Engelfriet, and A. E. G. Kr. von dem Borne, *Blood* **59**, 23 (1982).
[52] R. McMillan, P. Tani, F. Millard, P. Berchtold, L. Renshaw, and V. L. Woods, Jr., *Blood* **70**, 1040 (1987).

GPIIIa. In some patients antibodies were detected against a protein at 100K that is present on thrombasthenic platelets and that does not change its apparent molecular weight after disulfide reduction.[53] The clinical importance of antibodies against this unidentified protein is not yet known.

In acute ITP following primary varicella infection, antibodies against an 85K protein were detected in all five cases studied.[54] In two separate immunoblotting studies of thrombocytopenic patients with systemic lupus erythematosus, autoantibodies against a 66K intracellular protein were noted.[11,12] Thus, evidence is developing that the presence of circulating autoantibodies against different target antigens may correlate with different specific clinical settings in ITP. It will be especially helpful if particular target antigens could be used to predict the clinical course. There is not yet enough experience to know whether such prognosis will be possible, but several prospective studies are in progress to test this hypothesis.

To summarize, identification of platelet allo- and autoantigens by immunoblotting has been helpful in studying immune platelet destruction. The techniques outlined are currently in use in many laboratories and further investigations are likely to yield a better understanding of a number of important clinical syndromes and to help in characterizing the molecular details of platelet antigens.

[53] D. J. S. Beardsley, H. Taatjes, and S. E. Lux, *Clin. Res.* **32,** 494a (1984).
[54] D. J. S. Beardsley, J. S. Ho, and E. C. Beyer, *Blood* **66,** (1985).

[38] Crossed Immunoelectrophoresis of Human Platelet Membranes

By SIMON KARPATKIN, SABRA SHULMAN, and LESLIE HOWARD

The technique of crossed immunoelectrophoresis (CIE) of platelet membranes was introduced for the study of platelet membrane cell surface antigens[1,2] because of several advantages over sodium dodecyl sulfate–polyacrylamide gel electrophoresis (SDS–PAGE): (1) It does not completely denature membrane proteins, so that intrinsic biologic activity can often be assayed[3-11]; (2) it is 10 times more sensitive than SDS–PAGE

[1] I. Hagen, O. J. Bjerrum, and N. O. Solum, *Eur. J. Biochem.* **99,** 9 (1979).
[2] S. Shulman and S. Karpatkin, *J. Biol. Chem.* **256,** 4320 (1980).
[3] T. J. Kunicki, D. Pidard, J.-P. Rosa, and A. T. Nurden, *Blood* **58,** 268 (1981).

stained with Coomassie Brilliant blue[2]; (3) it may be used quantitatively because the peak areas of individual immunoprecipitate arcs are proportional to the antigen–antibody ratios[12]; (4) various lectins, antibodies, or other ligands can be employed in intermediate spacer gels to provide an immunoaffinoelectrophoresis pattern[1,2,13,14]; (5) comparative CIE studies of anti-membrane antiserum absorbed with whole cells as well as isotopic labeling of the intact platelet surface enable conclusions to be drawn regarding the relative surface location of various membrane antigens[1,2]; and (6) amphiphilic proteins can be recognized by charge-shift CIE and crossed hydrophobic interaction immunoelectrophoresis.[1] Examples of these advantages will be detailed below.

Platelet membranes or intact washed platelets are solubilized in a nonionic detergent, Triton X-100, and separated by charge and size by electrophoresis on a slab of 1% (w/v) agarose poured onto a glass plate. Following this procedure, a strip of the electrophoresed proteins and glycoproteins is retained (approximately 20% of the total area), and fresh agar, containing rabbit anti-platelet or anti-platelet membrane antibody, is poured onto the glass, replacing the 80% discarded agarose. The separated proteins are then electrophoresed into the antibody-containing agarose in the second dimension, 90° from the first dimension, at pH 8.6, the isoelectric point of rabbit antibody (making this antibody immobile during electrophoresis). As the antigens move into the antibody-containing agarose, they complex with the antibody in the soluble phase and immunoprecipitate at antigen–antibody equivalence. Thus, a low concentration of antigen with respect to antibody will form a relatively smaller immunoprecipitate arc than that formed by a higher concentration of antigen that would have to travel further into the

---

[4] L. Howard, S. Shulman, S. Sadanandan, and S. Karpatkin, *J. Biol. Chem.* **257**, 8331 (1982).
[5] I. Hagen, O. J. Bjerrum, G. Gogstad, R. Korsmo, and N. O. Solum, *Biochim. Biophys. Acta* **701**, 1 (1982).
[6] G. O. Gogstad, I. Hagen, M.-B. Krutnes, and N. O. Solum, *Biochim. Biophys. Acta* **689**, 21 (1982).
[7] G. O. Gogstad, F. Brosstad, M.-B. Krutnes, I. Hagen, and N. O. Solum, *Blood* **60**, 663 (1982).
[8] G. O. Gogstad and F. Brosstad, *Thromb. Res.* **29**, 237 (1983).
[9] I. Hagen, F. Brosstad, N. O. Solum, and K. Korsmo, *J. Lab. Clin. Med.* **97**, 213 (1981).
[10] G. O. Gogstad, N. O. Solum, and M.-B. Krutnes, *Br. J. Haematol.* **53**, 563 (1984).
[11] S. Karpatkin, R. Ferziger, and D. Dorfman. *J. Biol. Chem.* **261**, 14266 (1986).
[12] M. Karpatkin, L. Howard, and S. Karpatkin, *J. Lab. Clin. Med.* **104**, 223 (1984).
[13] I. Hagen, A. Nurden, O. J. Bjerrum, and N. O. Solum, *J. Clin. Invest.* **65**, 722 (1980).
[14] D. Varon and S. Karpatkin, *Proc. Natl. Acad. Sci. U.S.A.* **80**, 6992 (1983).

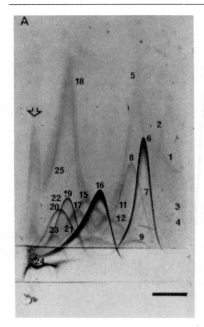
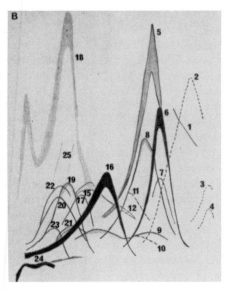

FIG. 1. Crossed immunoelectrophoresis of immunoprecipitates of 50 μg of human platelets solubilized in Tris–glycine buffer, pH 8.7, containing 1% (v/v) Triton X-100, 0.04 M Tris, 0.1 M glycine, pH 8.7, using antibodies raised against whole platelets, stained with Coomassie Brilliant blue. The cathodal side is at the left, the anodal at the right. (A) Photograph; (B) drawing. In order to obtain better resolution, an antibody-free intermediate gel was inserted. Dashed lines of drawing represent precipitate observed irregularly. Bar (A): 1-cm scale. (Taken from Hagen et al.[1])

antibody-containing agarose before reaching equivalence. The duration of electrophoresis time in the second dimension is not so important because immunoprecipitate arcs formed in the agarose are relatively immobile after reaching equivalence.

Approximately 20 immunoprecipitate arcs have been observed with intact platelet extracts[1] employed as antigen (Fig. 1), and 10 immunoprecipitate arcs observed in purified platelet membranes[2] (Fig. 2). These have been identified with specific antibodies against known antigens (Fig. 3) employed in intermediate spacer gels[13] as well as coelectrophoresis experiments. In the latter situation, the purported purified antigen is run independently as well as together with the platelet extract. The purported purified antigen as well as the unknown antigen should have immunoprecipitate arcs of identity and immunoprecipitate areas that are additive[2] (Fig. 4).

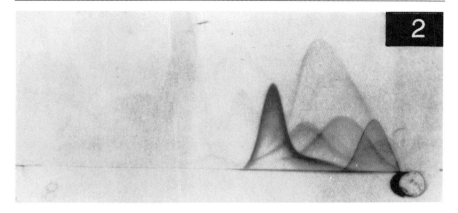

FIG. 2. Crossed immunoelectrophoresis of a human platelet membrane preparation. Fifty micrograms of membrane protein in 1% (v/v) Triton X-100, 0.07 M Tris, 0.02 M barbital buffer, pH 8.6, was applied to an 8 × 10-cm gel. The cathodal side is at the right, the anodal side is at the left. (Shulman and Karpatkin.[2])

Eight immunoprecipitate arcs have been identified in solubilized platelet extracts[15] (Fig. 5).

Assay for Intrinsic Biologic Activity. Crossed immunoelectrophoresis of human platelet membranes reveals a major antigen that is absent on platelets in Glanzmann's thrombasthenia[2,13] (Fig. 6). It was originally designated antigen 10 by one group[2] and 16 by another.[1] The former group postulated that this major antigen might represent a complex, because its immunoprecipitate arc was variable from one preparation to another and often had lines of identity with other immunoprecipitate arcs. For example, whenever the cathodal limb of the major immunoprecipitate arc 10 was absent, another arc appeared with lines of identity merging with 10 (Fig. 7). It was subsequently shown that this major antigen is composed of two separate glycoproteins, GPIIb and GPIIIa, which are held together by Ca^{2+} and dissociated by divalent cation chelating agents[3,4] (Figs. 8 and 9). This complex could not have been demonstrated with SDS–PAGE, as this procedure denatures the complex and reveals the separated components GPIIb and GPIIIa with molecular weights of 93,000 and 125,000, respectively. The GPIIb–GPIIIa complex has subsequently been shown to be the fibrinogen receptor,[7,16–18] which is necessary for platelet aggregation

[15] J. N. George, A. T. Nurden, and D. R. Phillips, *N. Engl. J. Med.* **311,** 1084 (1984).
[16] R. L. Nachman and L. L. K. Leung, *J. Clin. Invest.* **69,** 263 (1982).
[17] R. P. McEver, E. M. Bennett, and M. N. Martin, *J. Biol. Chem.* **258,** 5269 (1983).
[18] R. L. Nachman, L. L. K. Leung, M. Kloczewiak, and J. Hawiger, *J. Biol. Chem.* **259,** 8584 (1984).

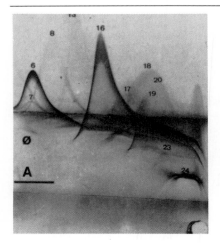
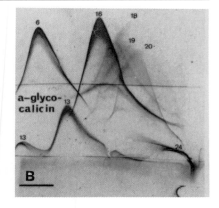
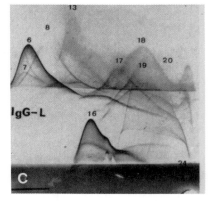

Fig. 3. Cross immunoelectrophoresis of 100 μg of 1% (v/v) Triton X-100-solubilized platelet preparation employing an intermediate spacer gel with (A) buffer, (B) anti-glycocalicin antibody, and (C) antibody IgG-L obtained from a multiply transfused patient with Glanzmann's thrombasthenia, which induces thrombasthenia-like reactivity with normal platelets *in vitro*. (Hagen et al.[13])

with physiologic agonists. Indeed, with CIE, ^{125}I-labeled fibrinogen binding could be demonstrated with the associated GPIIb–GPIIIa–Ca^{2+} complex,[7] but not with the EGTA-dissociated complex (Fig. 10), indicating that receptors for fibrinogen are accessible within the immunoprecipitate arc despite the presence of antibody. Platelet aggregation does not take place in Glanzmann's thrombasthenia, since the fibrinogen receptor is absent.

Other examples of retention of biologic activity have been demonstrated, wherein platelet proteins have been shown to bind to immobilized

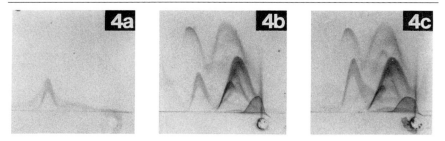

FIG. 4. Crossed immunoelectrophoresis of coelectrophoresis of human albumin with platelet membranes. (a) Human albumin (100 ng); (b) platelet membranes (50 μg); (c) coelectrophoresis of both. (Shulman and Karpatkin.[2])

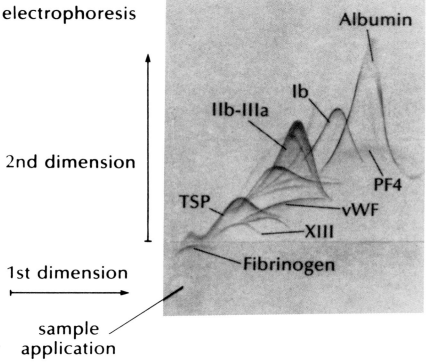

FIG. 5. Composite drawing of immunoprecipitate arcs identified on CIE of Triton X-100-solubilized intact platelets. Fibrinogen, thrombospondin (TSP), von Willebrand factor (vWF), and platelet factor 4 (PF4) are derived from platelet α granules. Factor XIII α chain is located in the soluble cytoplasm. (George et al.[15])

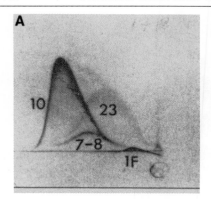 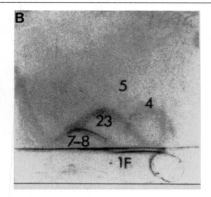

Fig. 6. Crossed immunoelectrophoresis of platelet membrane preparations obtained from (A) a normal subject, and (B) a patient with severe Glanzmann's thrombasthenia. Membranes were solubilized in 1% (v/v) Triton X-100 without EDTA. (Howard et al.[4])

thrombin[9] and immobilized heparin.[10] In addition, $^{45}Ca^{2+}$ has been shown to bind to GPIIb as well as the GPIIb–GPIIIa complex[11] of Triton X-100-solubilized membranes (Fig. 12), and platelet factor XIII enzymatic activity[8] has been demonstrated on CIE in Triton X-100-solubilized platelets (Fig. 11).

Sensitivity of Antigen–Antibody Immunoprecipitate Arcs. The second dimension on CIE is equivalent to the Laurell immunoelectrophoresis procedure[19] and is capable of detecting nanogram quantities of antigen present in platelet membranes (Fig. 4).

Quantitation of Immunoprecipitate Arcs. The peak areas of individual immunoprecipitate arcs are proportional to the antigen–antibody ratios,[12]

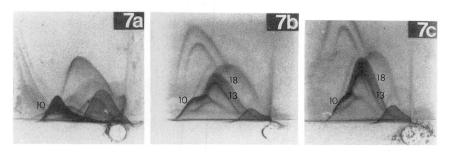

Fig. 7. Relationship of antigen 10 with antigens 13 and 18 on CIE of platelet membrane preparations from normal subjects. (a) Antigen 10 has a complete cathodal tail with antigens 13 and 18 not visible; (b) absent cathodal tail with antigens 13 and 18 visible; (c) relatively more of antigens 13 and 18 visible. (Shulman and Karpatkin.[2])

[19] C. B. Laurell, *Anal. Biochem.* **10**, 358 (1965).

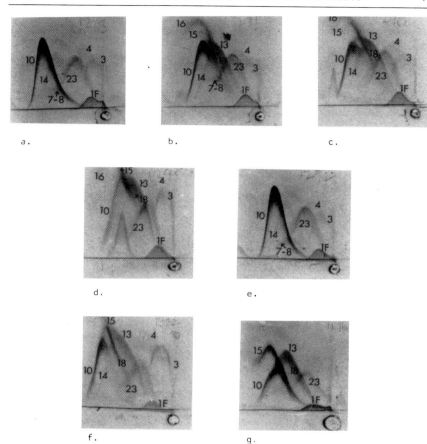

FIG. 8. Effect of EDTA, EGTA, sodium citrate, and Ca^{2+} on the CIE pattern obtained from Triton X-100-solubilized platelet membranes. (a) No EDTA in extraction buffer; (b) 1 mM EDTA; (c) 2 mM EDTA; (d) 5 mM EDTA; (e) 1 mM EDTA plus 4 mM $CaCl_2$; (f) 1 mM EGTA; (g) 25 mM sodium citrate in electrophoresis well. (Howard et al.[4])

as in the Laurell technique[19] (Fig. 12). Crossed immunoelectrophoresis has been employed to measure heterozygosity of Glanzmann's thrombasthenia by quantifying the major antigen complex.[20]

Affinity Crossed Immunoelectrophoresis with Antigen Ligands. The use of an intermediate spacer gel has been helpful in identifying specific carbohydrates as well as other antigenic determinants. An intermediate

[20] F. H. Hermann, M. Meyer, G. O. Gogstad, and N. O. Solum, *Thromb. Res.* **32**, 615 (1983).

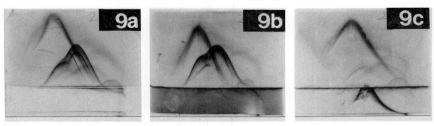

FIG. 9. Effect of antibodies against GPIIb and GPIIIa on the EGTA-dissociated complex, when placed in the intermediate spacer gel. (a) Membranes extracted in Triton X-100 electrophoresed with 10 mM EGTA in the electrophoresis well and buffer in the intermediate spacer gel; (b) same as (a), with rabbit anti-GPIIb in the spacer gel; (c) same as (a), with anti-GPIIIa in the spacer gel. (Howard et al.[4])

spacer gel containing carbohydrate lectins or specific antibodies is poured between the rabbit anti-membrane antibody and the strip of agarose containing the electrophoretically separated antigens. As the antigen is electrophoresed upward into the agarose, it will be retarded by prior binding to ligands in the intermediate spacer gel if these antigens contain specific carbohydrate moieties or antigenic determinants. Figure 9 demonstrates the use of specific antibodies against GPIIb and GPIIIa in the intermediate spacer. Figure 13 demonstrates the use of immobilized concanavalin A in the intermediate spacer, indicating the presence of α-methylmannoside residues. Concanavalin A, as well as wheat germ agglutinin, *Rinus communis,* and *Lens culinaris* have also been used in the first dimension to retard electrophoretic separation of antigens.[1] Monoclonal antibodies against specific antigenic epitopes have also been employed in the intermediate spacer gel.[14,17,21,22] However, because these are generally nonprecipitating, it was first necessary to label the monoclonal antibody with ^{125}I. Thus, Fig. 14 demonstrates the Coomassie Brilliant blue stain and autoradiograph of a CIE experiment in which an ^{125}I-labeled monoclonal antibody against GPIIb as well as the GPIIb–GPIIIa complex was employed in the spacer intermediate gel.[14] Note the comigration of the isotopically labeled monoclonal antibody with antigen and coprecipitation with nonradioactive rabbit anti-platelet membrane antibody at antigen–antibody equivalence.

Relative Surface Location of Various Membrane Antigens. The surface location of membrane antigens, as well as their relative surface location in relation to other surface antigens, can be determined by graded adsorption of the rabbit anti-platelet membrane antibody with increasing

---

[21] R. R. Montgomery, T. J. Kunicki, C. Taves, D. Pidard, and M. Corcoran, *J. Clin. Invest.* **71,** 385 (1983).

[22] P. J. Newman, R. W. Allen, R. A. Kahn, and T. J. Kunicki, *Blood* **65,** 227 (1985).

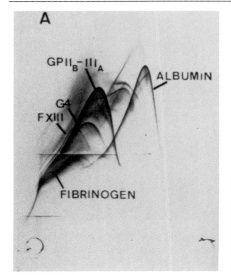
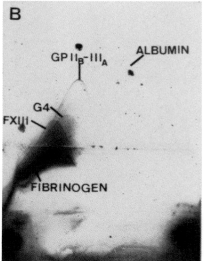
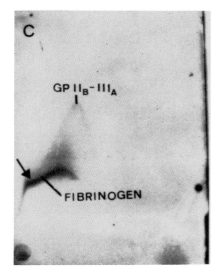

FIG. 10. Crossed immunoelectrophoresis of Triton X-100-solubilized platelets followed by incubation of the immunoplates with ^{125}I-labeled fibrinogen. (A) Coomassie Brilliant blue-stained immunoplate; (B) autoradiograph of immunoplate incubated with 0.1 mg ^{125}I-labeled fibrinogen/ml; (C) autoradiograph of immunoplate incubated with 0.01 mg ^{125}I-labeled fibrinogen/ml. Note the binding of fibrinogen to the GPIIb–GPIIIa complex, as well as to fibrinogen, factor XIII, and G4 (thrombospondin), proteins that bind to fibrinogen. At lower ^{125}I-labeled fibrinogen concentration, binding is specific for the GPIIb–GPIIIa complex and fibrinogen (fibrinogen binding to fibrinogen probably secondary to exchange with antibody in immunoprecipitate arc). (Gogstad et al.[7])

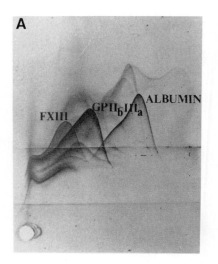 

FIG. 11. Factor XIII activity in immunoplate obtained after CIE of Triton X-100-solubilized platelets. (A) Coomassie Brilliant blue-stained immunoplate; (B) immunoplate incubated with factor XIII substrates casein and dansylcadaverine, followed by exposure to UV light. (Gogstad and Brosstad.[8])

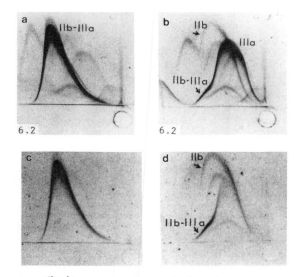

FIG. 12. Binding of $^{45}Ca^{2+}$ to the associated and EGTA-dissociated major antigen complex of the Triton-solubilized platelet membrane. (a) Fifty micrograms of control platelet membranes. (b) Membranes incubated with 0.5 mM EGTA for 5 min. The washed, blotted CIE slides were then incubated with $^{45}Ca^{2+}$ for 15 hr, followed by washing, blotting, and drying prior to autoradiography. (c and d) Autoradiograms that refer to (a) and (b), respectively, which have been stained with Coomassie Brilliant blue. Note absence of radioactivity for IIIa on autoradiogram (Karpatkin *et al.*).

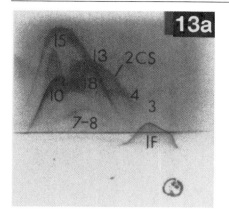 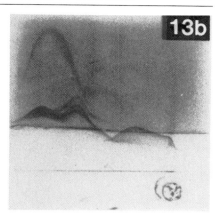

Fig. 13. Crossed immunoelectrophoresis of human platelet membranes employing immobilized concanavalin A in an intermediate spacer gel. (a) Spacer gel contains buffer; (b) spacer gel contains 250 μg/ml of concanavalin A bound to Sepharose. Note antigens 1F, 10, 13, and 18 have been retarded in their migration, indicating reactivity with the carbohydrate ligand, whereas 2CS, 3, 7–8, and 15 are not affected. (Shulman and Karpatkin.[2])

concentrations of intact washed platelets. Figure 15 demonstrates the graded disappearance of immunoprecipitate arcs representing the major antigen complex, fibrinogen, and other unidentified antigens with increasing absorption of antibody, indicating their surface location. Surface location of platelet membrane antigens (i.e., inside vs outside location) can be determined by first labeling the intact washed platelet surface by the ^{125}I-labeled lactoperoxidase technique[23] and then preparing platelet membranes for CIE. Figure 16 demonstrates that unlike the major antigen complex, platelet membrane fibrinogen is not located on the surface of the intact platelet.[12] Presumably, platelet "membrane" fibrinogen was released from α granules during platelet lysis and membrane preparation. Similar surface labeling studies using ^{125}I were helpful in revealing the presence of the GPIIb–GPIIIa complex in intracellular α granules of platelets.[24] Autoradiograms of CIE gels of α granules revealed the absence of ^{125}I labeling.

Recognition of Amphiphilic Proteins. Proteins with membrane insertions contain hydrophobic domains that can bind to nonionic detergents by hydrophobic interaction. In contrast, hydrophilic proteins do not bind to nonionic detergents. The ability of proteins to bind to ionic detergent

[23] D. R. Phillips, *Biochemistry* **11**, 4582 (1972).
[24] G. O. Gogstad, I. Hagen, R. Korsmo, and N. O. Solum, *Biochim. Biophys. Acta* **670**, 150 (1981).

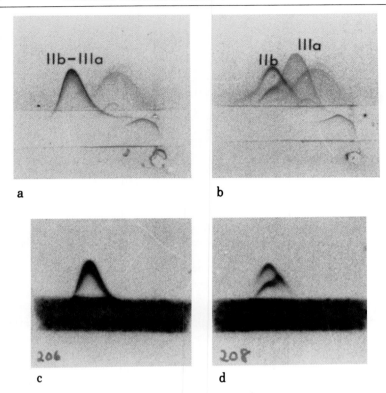

FIG. 14. Crossed immunoelectrophoresis of human platelet membranes, demonstrating the reactivity of monoclonal antibody 3B2 with the platelet membrane major antigen GPIIb–GPIIIa complex as well as with GPIIb. (a) Triton X-100-solubilized membranes processed in the presence of 2 mM CaCl$_2$; (b) same membranes incubated with 5 mM EGTA for 15 min at room temperature. An intermediate spacer gel was poured containing ^{125}I-labeled monoclonal antibody 3B2; (c) autoradiogram of (a); (d) autoradiogram of (b). Note that monoclonal antibody coprecipitates with the major antigen complex GPIIb–GPIIIa, as in (c), as well as with GPIIb and the undissociated complex, as in (d). (Varon and Karpatkin.[14])

micelles has been utilized in electrophoretic systems in which charge shifts are encountered. The positively charged ionic detergent cetyltrimethylammonium bromide and the negatively charged detergent deoxycholate will, when bound to amphiphilic proteins, result in a cathodal or anodal shift, respectively, in the first dimension when compared to the migration of the proteins in the absence of the ionic detergent but in the presence of the nonionic detergent alone (Triton X-100). This property has been utilized to identify platelet membrane amphiphilic proteins.[1]

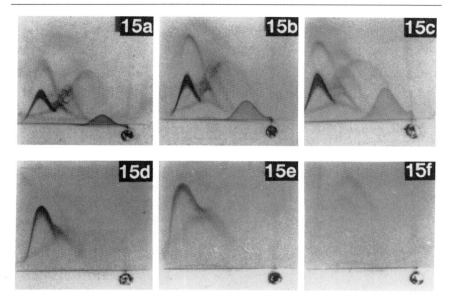

FIG. 15. Crossed immunoelectrophoresis of human platelet membrane preparation following absorption of rabbit anti-platelet membrane antibody with increasing concentrations of washed platelets. (a) 0 platelets; (b) 1.25×10^8 platelets; (c) 2.5×10^8 platelets; (d) 6×10^8 platelets; (e) 8×10^8 platelets; (f) 1×10^9 platelets. (Shulman and Karpatkin.[2])

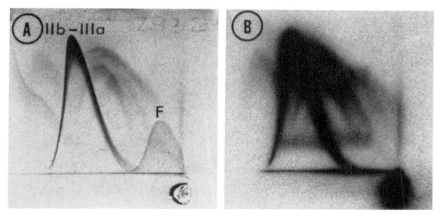

FIG. 16. Effect of surface labeling of intact washed platelets with ^{125}I on association of fibrinogen with Triton X-100-solubilized platelet membrane after CIE. (A) Coomassie Brilliant blue stain; (B) autoradiogram. Note absence of isotopic labeling of fibrinogen peak. (Karpatkin et al.[12])

Crossed hydrophobic interaction immunoelectrophoresis in the first dimension of CIE has also been employed to identify amphiphilic proteins. Amphiphilic proteins with hydrophobic domains are bound to a hydrophobic matrix, phenyl-Sepharose, whereas hydrophilic proteins are unaffected.[1]

Methodology

Preparation of Platelet Membranes. Platelet-rich plasma is obtained from ACD-A anticoagulated blood bank blood by centrifugation at 150 g for 15 min at 4°. Ethylenediaminetetraacetic acid (EDTA) (Vacutainer test tubes; Becton-Dickinson, Rutherford, NJ) as well as sodium citrate [0.38% (w/v) final concentration] have also been used as anticoagulant. The platelet-rich plasma is centrifuged at 2800 g for 15 min at 4° and then washed in a human Ringer's solution containing 127 mM NaCl, 4 mM KCl, 25 mM NaHCO$_3$, 2 mM Na$_2$HPO$_4$, 0.2 mM Na$_2$SO$_4$, 2 mM EDTA, 1% (w/v) ammonium oxalate to lyse red blood cells and also containing the protease inhibitors 10 mM benzamidine, 100 μg/ml soybean trypsin inhibitor, and 0.1 mM phenylmethylsulfonyl fluoride (PMSF). To obtain a pH of 7.1, 10% CO$_2$ plus 90% air is bubbled through the solution at 0°. The platelets are washed twice in the same buffer mixture.

Membranes are prepared by four cycles of freezing (acetone–dry ice) and thawing (37°) in a small volume of the above buffer mixture without EDTA, followed by sonication for 15 sec at 60 W at 4° in a Branson sonic power sonicator (W185; Branson Instruments Co., Plainview, NY). Cellular debris is removed by centrifugation at 150 g for 10 min at 4°, and the supernatant applied to a 30% sucrose cushion. The sucrose is dissolved in 0.01 M Tris buffer, pH 7.4 containing 0.15 M NaCl and the above protease inhibitors. The mixture is centrifuged for 1 hr at 60,000 g at 4° in a swinging bucket rotor and the membrane layer separated from the interface. The membranes are washed in their original buffer and recentrifuged at 100,000 g for 1 hr. The membranes are then solubilized in 1% (v/v) Triton X-100 containing 10 mM benzamidine, 100 μg/ml soybean trypsin inhibitor, 1 mM PMSF (with or without EDTA or EGTA). This preparation can be stored at 4° for at least 3 weeks.

Preparation of Anti-Platelet Membrane Antibody. Antiserum to purified platelet membrane is raised in rabbits. The platelet membranes from 1 unit of blood are suspended in 0.5 ml of saline (0.9% NaCl solution) and emulsified 1 : 1 (v/v) with Freund's complete adjuvant. If membranes are solubilized in 1% Triton containing 5 mM EGTA, an antibody is obtained with greater precipitating activity for separated GPIIb and GPIIIa. This is injected into the hind foot pads of a rabbit. Subsequent booster injections, obtained from 1 unit of platelets from multiple donors, are administered with Freund's incomplete adjuvant subcutaneously in divided doses every 2 to 3 weeks. Sera from five consecutive bleedings are pooled, and immunoglobulin is partially purified by precipitation with 50% (w/v) saturated $(NH_4)_2SO_4$. The precipitate is dissolved in one-fourth of its original serum volume and dialyzed against water, followed by sodium acetate buffer, pH 5.0. The solution is then dialyzed against 0.1 M NaCl containing 15 mM NaN$_3$.

Crossed Immunoelectrophoresis. Crossed immunoelectrophoresis is performed by a modification[2] of the Laurell technique[19] employing 0.02 M barbital, 0.07 M Tris buffer, pH 8.6, containing 1% (v/v) Triton X-100. Agarose gels (1%, w/v) are prepared in the above buffer by boiling. Melted 56°-heated gels are cast onto glass plates (50 × 50 × 0.6 mm, precoated with 56° heated 1% agarose in water with a paint brush) to give a volume-to-surface area ratio of 0.132 ml/cm^2. Wells, 4 to 7 mm in diameter, are prepared by applying suction to a Pasteur pipette. Samples (50 μg in 20 μl) are applied to the well and electrophoresed at 150 V for 2.5 to 3 hr in an immunoelectrophoresis cell. An agarose strip (12 × 50 mm) containing the antigens that had been subjected to electrophoresis is retained on the plate after removal of the rest of the gel, which is replaced with an adjacent gel (36 × 50 mm) containing anti-platelet membrane antibody (about 100–200 μl/ml; 2 mg/ml). Electrophoresis in the second dimension is performed at 55 V for 18 hr. Gels are then pressed with filter paper, washed six to eight times in 0.1 M NaCl, once in water, and then oven dried at 50°. Gels are stained with 0.25% (w/v) Coomassie Brilliant blue (dissolved in 45% ethanol, 9% acetic acid) for 15 min at 22°. The gels are destained in 45% ethanol, 9% acetic acid.

Affinity Crossed Immunoelectrophoresis. The initial step is identical to that described above. However, after electrophoresis in the first dimension, an agarose strip (30 × 50 mm rather than 40 × 50 mm, as above) is removed and replaced by an agarose gel (30 × 50 mm) containing anti-platelet antibody. An intermediate strip of agarose (10 × 50 mm) located between the antibody-containing gel and the first-dimension agarose strip (10 × 50 mm) is then removed and replaced with agarose containing a second specific antibody or an immobilized lectin (such as concanavalin A bound to Sepharose beads). In control plates, the spacer gel (10 × 50 mm) contains buffer alone.

[39] Use of Correlative Microscopy with Colloidal Gold Labeling to Demonstrate Platelet Receptor Distribution and Movement

By Ralph M. Albrecht, Olufunke E. Olorundare, Scott R. Simmons, Joseph C. Loftus, and Deane F. Mosher

Introduction

The study of platelet receptors has been approached from a number of directions. Many of the techniques utilized are presented elsewhere in this and a previous volume.[1] Considerable information with respect to kinds and numbers of receptors, binding constants, and association with cytoskeletal proteins has appeared based on a variety of predominantly biochemical procedures.[2-4]

What additional and unique information regarding receptors on platelet surfaces and receptor association with structural elements can be obtained by employing a microscopic approach? Features such as the overall distribution, that is, location on the platelet surface, of different receptor types, as well as their direction and rate of movement on living platelets relative to the activation process, can be followed microscopically.[5-7] Changes in platelet ultrastructure, particularly reorientation of the cytoskeleton and changes in receptor distribution, or lack of it, in response to the cytoskeletal changes can also be determined microscopically.[5,8] The position of surface receptor glycoproteins and/or antigens relative to one another and to surface and subsurface ultrastructural features can also be visualized using double- or triple-labeling procedures. It is also important to note that because the platelet morphology and overall distribution of receptors or surface antigens can be examined on individual platelets, differences within a population can be determined. Nonmorphological studies often

---

[1] S. Timmons and J. Hawiger, this series, Vol. 169, p. 11.
[2] J.E. Fox and D. R. Phillips, *Semin. Hematol.* **20,** 243 (1983).
[3] D. R. Phillips, in "Platelet Membrane Glycoproteins" (J. N. George, A. T. Nurden, and D. R. Phillips, eds.), p. 145. Plenum, New York, 1985.
[4] J. S. Bennett, in "Platelet Membrane Glycoproteins" (J. N. George, A. T. Nurden, and D. R. Phillips, eds.), p. 193. Plenum, New York, 1985.
[5] J. C. Loftus and R. M. Albrecht, *J. Cell Biol.* **99,** 822 (1984).
[6] R. M. Albrecht, S. L. Goodman, and S. R. Simmons, *Am. J. Anat.* **149,** 1989 (1985).
[7] R. M. Albrecht, O. E. Olorundare, H. W. Bielich, and S. R. Simmons, *Proc.—Annu. Meet., Electron Microsc. Soc. Am.* **45,** 556 (1987).
[8] J. C. Loftus, J. Choate, and R. M. Albrecht, *J. Cell Biol.* **98,** 2019 (1984).

rely on averages of many platelets that may not all be at the same stage of activation or may consist of several subtypes. Although certain data are unique to the microscopic approach, it has been our experience that microscopic and biochemical findings tend to be mutually confirmatory and an analysis of one in light of the other often leads to new insights regarding a particular mechanism of action or the validity of an existing hypothesis.

In this chapter we will present procedures that permit identification of receptor sites in the context of the external surface and internal structure of whole platelets. Thus far, conventional light microscopy coupled with immunofluorescent labeling techniques have provided much of the information on localization of surface molecules. However, the small size of the platelet and the limited resolving power of conventional light-based imaging procedures permit only a relatively generalized localization of molecules of interest. Due to limitations on spatial resolution and to some extent the intensity of the labeling, receptor localization is therefore difficult. More recent light microscopic instrumentation that employs rectified interference contrast procedures is especially useful for following changes in receptor distribution and platelet structure in living cells.[9,10]

Most of the work investigating the ultrastructure of platelets has thus far been generated using conventional transmission electron microscopy (TEM). Extensive investigations delineating the ultrastructure of resting and activated platelets have appeared in the literature.[8,11-13] Labeling studies have also been carried out effectively on specimens subsequently prepared for TEM.[14] The primary drawback to conventional TEM labeling studies is the need to plastic-embed and cut extremely thin sections. In the case of receptor distribution, many serial sections are required to develop a picture of an overall distribution, even on a single platelet. Using thin sections, it is also difficult to determine accurately the overall three-dimensional organization of the filamentous elements composing the platelet cytoskeleton.

Our approach has been to utilize both scanning electron microscopy (SEM) to define whole-cell surface structure, and stereo pair 1-MeV high-voltage transmission electron microscopy (HVEM) to visualize the internal structure of whole platelets. The exact same platelets can be viewed sequentially, first by HVEM and then by SEM. Low-voltage, high-resolu-

[9] S. Inoue, *J. Cell Biol.* **89,** 346 (1981).
[10] S. L. Goodman, K. Park, and R. M. Albrecht, *in* "Colloidal Gold: Principle, Methods, and Applications" (M. A. Hayat, ed.), Vol. 3, p. 369, Academic Press, San Diego, 1991.
[11] J. Boyles, J. E. Fox, D. R. Phillips, and P. E. Stenberg, *J. Cell Biol.* **101,** 1463 (1985).
[12] V. Nachmias, *J. Cell Biol.* **86,** 795 (1980).
[13] J. G. White, *Scanning Microsc.* **1,** 1677 (1987).
[14] O. Behnke, *J. Cell Sci.* **87,** 465 (1987).

tion field emission SEM (LV-HR-SEM) permits resolution in the SEM close to that attainable in the TEM. At present, using appropriate preparative methology, it is possible with this instrumentation[15,16] to visualize unlabeled individual molecular species such as fibrinogen on artificial substrates or cell surfaces.

Labeling is accomplished by direct conjugation of ligand or monoclonal antibody directed against receptor glycoprotein to colloidal gold beads of 3, 5, 18, or 30 nm. These labels are readily identified when viewed by LV-HR-SEM or HVEM and the smaller labels can provide the spatial resolution required to identify individual receptors. As in conventional indirect immunolabeling procedures, gold can also be conjugated to second antibody or to anti-ligand antibody. This permits addition of soluble unconjugated antibody or ligand as a first step. Direct conjugation of 3-nm gold to Fab fragments or to purified active sites of ligands forms very small probes with high specificity that provide excellent spatial resolution.[17] Protein A–gold conjugates can also be prepared and can be used as a secondary label to identify antibody bound to receptor glycoprotein on the platelet surface or antibody bound to attached ligand. However, spatial resolution is less precise with these indirect procedures.

In this chapter we will deal first with the chemistry and preparation of colloidal gold; second, with the attachment of ligand or antibody to the gold particles; and third, with the actual labeling of platelets and their subsequent preparation for electron microscopy. Because it is not possible to deal specifically with all the ligands or antibodies that may be of potential use, we present a description of the principles of the procedure generally applicable to all ligands or antibodies. This is followed by a specific example of a ligand, in this case fibrinogen, and an example of two antibodies, anti-fibrinogen and monoclonal 10E5 specific for the glycoprotein IIb–IIIa complex,[18] which serves as the platelet fibrinogen receptor. The monoclonal antibody was provided by Dr. Barry Coller (Department of Medicine, State University of New York at the Stony Brook Health Sciences Center).

---

[15] R. M. Albrecht, S. R. Simmons, and D. F. Mosher, in "Fibrinogen, Current Basic and Clinical Aspects" (M. Matsuda, S. Iwanaga, A. Takada, and A. Henschen, eds.), Vol. 4, p. 87. Elsevier, Amsterdam, 1990.

[16] S. R. Simmons and R. M. Albrecht, Scanning Microsc., Suppl. **3**, 27 (1989).

[17] S. R. Simmons, J. B. Pawley, and R. M. Albrecht, J. Histochem. Cytochem. **38**, 1781 (1990).

[18] B. S. Coller, E. I. Peerschke, L. E. Scudder, and C. A. Sullivan, J. Clin. Invest. **72**, 325 (1983).

Colloidal Gold

Colloidal gold is an inorganic metal colloid that can be produced through the reduction of its corresponding salt, hydrogen tetrachloroauric(III). The size of the colloidal particles, 1 nm up to 500 nm, can be regulated by changing either the amount or the type of the reducing agent. Some of the most commonly used reducing agents for the production of colloidal gold include phosphorus,[19] ascorbic acid,[20] sodium citrate,[21] and sodium citrate/tannic acid.[22] Colloidal gold suspensions have also been prepared using ultrasonics.[23] When sodium citrate is the reducing agent, particle size is varied by exposing a constant amount of tetrachloroauric(III) acid, 0.0002 M, to varying amounts of sodium citrate. The mean particle size increases as the amount of citrate decreases, making possible the production of gold particles in sizes ranging from 12- to 150-nm average diameter. Alternatively, the amount of sodium citrate can be held constant and varying amounts of tannic acid can be added. Again, as the amount of tannic acid decreases, particle size increases, resulting in particles ranging from 3 to 16 nm in diameter with less variation in particle size than is seen in gold colloid produced by reduction with sodium citrate alone. Because all of the gold is reduced in all of the preparations,[24] the resulting size is determined by the number of nuclei formed in each preparation. At higher citrate concentrations (0.0008 M), the number of nuclei formed is high and gold will crystallize on many centers. Because the total amount of gold ions is limited in such a dilute solution, the nuclei cannot grow very large and the resulting particles have an average diameter of 18 nm. With decreasing citrate concentration, the number of nuclei formed decreases. Because fewer nuclei are formed, the gold ions will collect on fewer points and the resulting particles will have a larger diameter. The formation of colloidal gold particles by sodium citrate has two components: the oxidation of citrate ion and the reduction of tetrachloroauric(III) acid. A major intermediate in the oxidation of citrate is acetone dicarboxylic acid. Acetone dicarboxylic acid is oxidized further to formaldehyde and formic acid, both of which are powerful reducing agents. Either may then act on the potassium gold oxide complex to form the resulting gold nuclei.

---

[19] H. B. Weiser, *in* "Inorganic Colloidal Chemistry" (H. B. Weiser, ed.), Vol. 1, p. 21. Wiley, New York, 1933.
[20] L. I. Larsson, *Nature (London)* **282**, 743 (1979).
[21] G. Frens, *Nature (London), Phys. Sci.* **241**, 20 (1973).
[22] J. W. Slot and H. J. Geuze, *Eur. J. Cell Biol.* **38**, 87 (1985).
[23] C. L. Baigent and G. Mueller, *Experientia* **36**, 472 (1980).
[24] J. Turkevich, P. C. Stevenson, and J. Hillier, *Discuss. Faraday Soc.* **11**, 55 (1951).

Preparation of Colloidal Gold

The method of Frens[21] as modified by Horisberger[25] is perhaps the most straightforward method for the preparation of colloidal gold. A 4% (w/v) solution of tetrachloroauric(III) acid trihydrate is prepared by dissolving 0.08 g in 2 ml distilled water. A 0.5-ml aliquot is added to 200 ml of distilled deionized water that has been filtered through a microporous membrane filter (0.22-μm pore size filter; Millipore Corp., Bedford, MA). The boiling flask used for the preparation of the colloid is washed in hot Linbro 7X cleaning solution (Flow Laboratories, Dublin, VA) and rinsed well with distilled water before use because the presence of small contaminants will affect the formation of nuclei and hence adversely affect the desired size and range of the gold particles. The solution is heated to boiling. The desired amount of freshly prepared and prefiltered 1% (w/v) trisodium citrate is added to the boiling solution with agitation. The solution quickly turns faintly blue, indicating nucleation, then changes to an orange-red or violet color, depending on the size of the particle, indicating the endpoint of the reaction. The mixture is refluxed for 30 min, cooled on ice, filtered through a prerinsed 0.22-μm pore size Millipore filter, and stored at 4°. If a disposable syringe is used to force the gold colloid through the filter unit, the syringe should first be rinsed with distilled water because certain manufacturers employ lubricating agents that will cause aggregation of the gold. The solution has an indefinite shelf life, provided it is maintained under sterile conditions and not allowed to freeze, which can cause aggregation.

For preparation of 3-nm gold particles by the sodium citrate/tannic acid method,[22] a 0.25-ml aliquot of the 4% $HAuCl_4$ solution is mixed with 80 ml of distilled water in the cleaned boiling flask. Four milliliters of 1% sodium citrate is filtered and mixed with 5 ml of 1% tannic acid (Mallinkrodt, St. Louis, MO) and neutralized with a volume of 25 mM K_2CO_3 equal to that of the tannic acid. Distilled water is added to bring this solution to a total volume of 20 ml. For larger gold particles, smaller quantities of tannic acid and K_2CO_3 are used in the 20-ml solution. Both solutions are warmed to approximately 60° and are rapidly mixed. The mixture is then brought to boiling to complete the reaction.

The average particle size and size distribution of each preparation can be determined from electron micrographs. Ten microliters of the freshly prepared gold stock solution is deposited on a Formvar-filmed nickel maxtaform grid and allowed to air dry. The grid is carbon coated and then examined by TEM. Measurements of 100 particles are made directly off the

---

[25] M. Horisberger and J. Rosset, *J. Histochem. Cytochem.* **25**, 295 (1977).

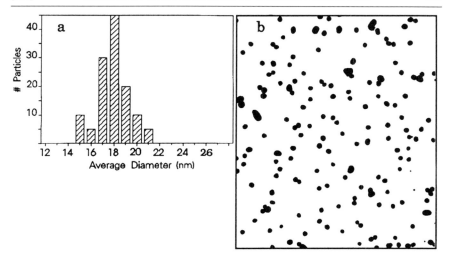

FIG. 1. (a) Typical size distribution for nominally 18-nm bead size. Smaller bead sizes have less absolute variation, although variation as a percentage of size is similar. (b) Transmission electron micrograph of an 18-nm colloidal gold preparation. The mean diameter is 18.38 nm with a standard deviation of 2.7 nm.

negative with a peak scale Lupe ×10 measuring magnifier; alternatively, particle numbers and size can be determined via computer analysis. A representative particle size distribution is shown in Fig. 1.

Colloidal gold granules possess a single peak of absorption in the visible spectrum in the range of 520–550 nm.[25–27] In general, peak absorption moves to longer wavelengths as particle diameter increases and there is a broadening of the absorption peak with increasing heterogeneity of particle shape. Monitoring these spectral properties helps ensure reproducibility of marker size, shape, and size distribution.

Stability of Colloidal Gold

Colloidal gold particles are maintained in solution by electrostatic repulsion. Individual particles carry a net negative charge resulting from the slow dissolution of the [$AuCl_2$] complex, which is theorized to be present on the surface of the particle.[28] Gold particles are also subject to electro-

[26] S. L. Goodman, G. M. Hodges, and D. C. Livingston, *Scanning Electron Microsc.* **2**, 133 (1980).
[27] M. Horisberger, in "Biotechnology and Bioapplications of Colloidal Gold" (R. M. Albrecht and G. M. Hodges, eds.), p. 19. Scanning Microscopy International, Chicago, 1988.
[28] K. J. Mysels, ed., "Introduction to Colloidal Chemistry," p. 171. Wiley (Interscience), New York, 1959.

magnetic effects that lead to London–van der Waals forces of attraction. The interactive energy of these attractive forces is inversely proportional to the sixth power of the distance between two particles (r^6). Thus, this interaction is strong only at very short distances. The charged ionic clouds that surround each particle exert their influence over longer distances and therefore are capable of overriding the influence of the attractive forces. The addition of electrolytes to the colloidal solution effectively compresses the ionic layer surrounding each particle, enabling two particles to approach each other more closely before their electrical layers overlap and they are repelled. As the concentrations of electrolytes in the solution increases, the particles will continue to approach each other more closely until they reach a critical distance where the attractive forces are equal to the repulsive forces. At less than this critical distance, the particles will agglutinate and fall out of suspension.[29,30]

Behavior of Adsorbed Macromolecules

Colloidal gold particles can be protected from the agglutinating effects of electrolytes by coating them with a stabilizing layer of macromolecules such as proteins, thus preventing the cohesion of one particle to another by keeping them sufficiently far apart. The adsorption of protein molecules onto metal surfaces in water is a complex phenomenon that is not completely understood. It depends on a number of factors, including the tertiary structure of the protein, pH, ionic strength, and temperature.[31] Polymers do not appear to be adsorbed over their entire length at any interface.[32] Protein molecules in particular are thought to remain globular when adsorbed onto nonwettable surfaces.[33,34] The flexibility and size of the protein are therefore important considerations. The larger the protein molecule, the greater the number of possible contacts with the surface.[35] Proteins such as bovine serum albumin and ovalbumin are adsorbed to colloidal gold as a monolayer approximately 2–3 nm thick.[36] Studies of the adsorption of immunoglobulins onto gold films has demonstrated that

---

[29] A. Sheludko, ed., "Colloidal Chemistry," p. 208. Elsevier, London, 1966.
[30] J. Overbeek, *in* "Colloidal Dispersions" (J. W. Goodwin, ed.), p. 1. Henry Ling Ltd., Dorchester, United Kingdom, 1982.
[31] F. R. Eirich, *J. Colloid Interface Sci.* **58**, 423 (1977).
[32] G. Steinberg, *J. Phys. Chem.* **71**, 292 (1967).
[33] A. Stilberg, *J. Phys. Chem.* **66**, 1884 (1962).
[34] J. L. Brash and D. J. Lyman, *J. Biomed. Mater. Res.* **3**, 175 (1969).
[35] M. Horisberger, *Scanning Electron Microsc.* **2**, 9 (1981).
[36] A. Rothen, *Rev. Sci. Instrum.* **28**, 283 (1957).

immunoglobulins are adsorbed to a thickness of 7–8 nm.[37,38] Once adsorbed on metal surfaces, proteins remain firmly attached and retain the ability to be recognized by homologous antibodies.[39,40] This would seem to confirm that proteins are not adsorbed over their entire length, which would significantly alter antigenicity. Rather, proteins appear to be somewhat flexible, with protruding chains that can be recognized by antibodies. In addition, proteins appear to retain their specific biological activity following adsorption to colloidal gold. Adsorption of enzymes has been used as a sensitive indicator of retention of bioactivity.[26,41] With only a few known exceptions, such as catalase,[42] nearly all adsorbed enzymes retain their ability to interact specifically with substrate.

In a solution where the concentration of electrolytes is below that causing agglutination of the colloidal particles, the adsorption of proteins is pH dependent. In solution, proteins behave as polyions. The net charge, due to ionization of constituent amino acids, is a function of pH and is neutral at the isoelectric point.[43] Although changes in pH may affect the net charge on protein molecules, the charge on colloidal gold is pH independent. Addition of proteins to colloidal gold at a pH value lower than the pI of the protein will result in rapid flocculation of the gold particles. At pH values below the pI of the protein, the protein will carry a net positive charge and the forces of the attraction to the negatively charged gold particles will be very strong. In this case, one protein molecule may be bound by several gold particles, resulting in the formation of large aggregates that cause flocculation. Because the zwitterion is the dominant species at pH values close to the pI, the protein has no net charge. Regions of maximal surface tension occur at or close to the pI and correlate with the point of least solubility for most proteins.[44] At this point, the attraction of the protein for the gold has decreased due to the lesser number of positive charges on the protein, and because the protein is only weakly hydrated, conditions are highly favorable for adsorption to the hydrophobic surface of the gold granule. As the pH of the colloidal solution increases above the pI of the protein, the amount of protein adsorbed will decrease due to the increase in the number of negative charges present on

---

[37] C. Mathot and A. Rothen, *J. Colloid Interface Sci.* **31**, 51 (1969).
[38] I. Giaver, *J. Immunol.* **110**, 1424 (1973).
[39] M. Horisberger and M. Vauthey, *Histochemistry* **80**, 13 (1984).
[40] A. Rothen and C. Mathot, *Helv. Chim. Acta* **54**, 1208 (1971).
[41] M. Bendayan, *J. Histochem. Cytochem.* **29**, 531 (1981).
[42] M. Horisberger, *Experientia* **34**, 721 (1978).
[43] C. Tanford, *in* "Physical Chemistry of Macromolecules" (C. Tanford, ed.), p. 1. Wiley, New York, 1966.
[44] J. R. Pappenheimer, M. P. Lepie, and J. W. Syman, *J. Am. Chem. Soc.* **58**, 1851 (1936).

the protein. Repulsion between the negatively charged protein and the negatively charged gold particle, combined with the increased solubility of the protein, favors solution and significantly hinders adsorption of protein.

Conjugation of Proteins to Colloidal Gold

The pH-Variable Adsorption Isotherm. The optimal pH for the adsorption of each protein is determined from a pH-variable adsorption isotherm.[45] All proteins to be adsorbed should be as salt free as possible or dialyzed against deionized, distilled water to remove electrolytes that could cause agglutination. Protein stock solutions are filtered through a 0.2-μm Millipore filter or are centrifuged to remove large aggregates, and aliquots are made up to 100 μg in 0.5 ml of distilled water. For molecules that tend to aggregate in distilled water, extremely low salt solutions, 5 mM or less, may be necessary. A series of colloidal gold solutions (2 ml each) are made up in a range of pH values from pH 5.0 to 9.0, as measured by a gel-filled combination electrode (#9115; Orion Research, Cambridge, MA), which has the advantage of having a low electrolyte flow rate.[46] Conventional electrodes are not used because they are quickly plugged by colloidal gold, which is flocculated by the KCl in the electrode. Gold solutions are made more acidic by the addition of 1 M acetic acid or more basic by adding 0.2 N potassium carbonate. The colloidal gold solutions are added to the protein solutions and mixed rapidly. After 2 min at room temperature, 0.5 ml of a 10% (w/v) NaCl solution is added, mixed rapidly, and the solution is allowed to stand for 5 min. Inadequate stabilization of the colloid results in flocculation, which can be judged visually or spectrophotometrically. If measured by spectrophotometer, the absorbance is read at 580 nm using 2 ml of gold, diluted to 3.0 ml with distilled water, as the blank. Inadequate stabilization of the colloid is indicated by an increase in absorbance and a color change from red to blue. In most cases, the optimal pH is found to be at, or slightly basic to, the reported pI of the protein. With smaller particles, for which the London–van der Waals forces are less, higher concentrations of ionic species can be tolerated. Should solubility be a problem, addition of certain molecular species such as polyethylene glycol can also improve protein solubility without the need to add ionic species.

Concentration–Variable Adsorption Isotherm. In addition to pH, the minimal concentration of protein required to stabilize the colloid is deter-

[45] W. D. Geoghegan and G. A. Ackerman, *J. Histochem. Cytochem.* **25**, 1187 (1977).
[46] W. D. Geoghegan, S. Ambegaonkar, and N. J. Calvanico, *J. Immunol. Methods* **34**, 11 (1980).

mined by a concentration–variable adsorption isotherm.[25,47] An insufficient amount of protein will produce an incomplete coating of the particles so that cohesion of the particles will still be possible on the addition of electrolytes. A series of protein dilutions, centrifuged or prefiltered on 0.2-μm Millipore filters, are made up to 0.5 ml in distilled, deionized water, giving a final concentration range of 1–100 μg. A 2-ml solution of colloidal gold, the pH of which was adjusted to the optimum pH determined by the pH variable adsorption isotherm, is added to the protein solution and mixed rapidly. After 2 min, 0.5 ml of a 10% NaCl solution is added, mixed, and allowed to stand for 5 min. Insufficient amount of protein results in flocculation, which is determined as described for the pH-variable adsorption isotherm. Considerable information regarding the production of colloidal gold and its conjugation is available in the literature.[48–50] A number of colloidal gold preparations, principally conjugated to second antibodies, are available from commercial sources. Labels consisting of an 0.8-nm, 11-atom gold core surrounded by an organic matrix (total diameter 2.0 nm), termed undecagold, and a similar 1.4-nm core, 2.7-nm total diameter have also been developed.[51–53] The particles are extremely uniform in size and the organic matrix permits covalent linkage to a variety of molecular species, including ligands, antibodies, and tRNA.[52–54]

Materials and Methods for Fibrinogen Receptor Identification

Preparation of Fibrinogen

Fibrinogen is purified from a fibronectin and fibrinogen-containing fraction prepared by precipitation of fresh-frozen human citrated plasma with 25% saturated ammonium sulfate, as previously described.[55] Fibrinogen free of fibronectin is isolated by ion-exchange chromatography on a

---

[47] W. P. Faulk and G. M. Taylor, *Immunocytochemistry* **8**, 1081 (1971).

[48] A. J. Verklejj and J. L. Leunissen, eds., "Immuno-Gold Labeling in Cell Biology," p. 1. CRC Press, Boca Raton, Florida, 1989.

[49] M. A. Hayat, ed., "Colloidal Gold: Principles, Methods and Applications," Vols. 1–3. Academic Press, San Diego, 1989.

[50] L. I. Larsson, ed., "Immunocytochemistry: Theory and Practice," p. 1. CRC Press, Boca Raton, Florida, 1988.

[51] P. A. Bartlett, B. Bauer, and S. J. Singer, *J. Am. Chem. Soc.* **100**, 5085 (1978).

[52] J. F. Hainfeld, *Science* **236**, 450 (1987).

[53] J. F. Hainfeld, F. R. Furuya, and R. D. Powell, *Proc.—Annu. Meet., Electron Microsc. Soc. Am.* **49**, 284 (1991).

[54] J. F. Hainfield, M. Sprinzel, V. Mandyan, S. J. Tomminia, and M. Boublik, *J. Struct. Biol.* **107**, 1 (1991).

[55] D. F. Mosher, *J. Biol. Chem.* **250**, 6614 (1975).

DEAE-cellulose column equilibrated with 0.01 M Tris, 0.05 M NaCl, pH 7.4, and run with a salt gradient of 0.05 to 0.3 M NaCl. Fibrinogen elutes as two peaks in this procedure.[56] The peak 1 product is dialyzed against 0.01 M Tris, 0.14 M NaCl buffer, pH 7.4. Aliquots of 10–15 mg/ml are snap frozen in an ethanol–dry ice bath and stored at $-70°$.

Conjugation of Fibrinogen to Colloidal Gold

pH-Variable Adsorption Isotherm. The optimal pH for the adsorption of fibrinogen to 18-nm colloidal gold is determined from a pH variable adsorption isotherm.[45] A series of 18-nm colloidal gold solutions of 2 ml each are made up in a range of pH values from pH 5.07 to 7.08 with 0.2 N K_2CO_3, as measured by a gel-filled combination electrode (#9115; Orion Research). Frozen aliquots of fibrinogen (1 mg/ml) are thawed in a water bath at 37° and dialyzed against 5.0 mM NaCl. Immediately before use, all fibrinogen solutions are centrifuged at 15,000 g for 10 min at 4° to remove any aggregates that may have formed during storage or dialysis. If filtration is necessary, a Nuclepore (microporous polycarbonate film) filter can be used with little fibrinogen loss. Final protein concentration is checked by a Bio-Rad (Richmond, CA) or similar protein assay. Fifty micrograms of fibrinogen or the ligand is made up to 0.5 ml in distilled, deionized water. The colloidal gold solutions are added to the protein solution and mixed rapidly. After 2 min at room temperature, 0.5 ml of a 10% NaCl solution is added and the solution allowed to stand for 5 min. Stabilization is measured spectrophotometrically at 580 nm, the wavelength of the color of the colloidal gold if it aggregates, using 2 ml of gold diluted to 3 ml with distilled water as a blank (Table I). Stabilization is indicated by an absorbance reading approaching zero, with the optimum pH chosen as the first point at which the curve parallels the x axis (Fig. 2b). From this graph, pH 6.3 is the optimal pH for the adsorption of fibrinogen.

Concentration-Variable Adsorption Isotherm. A series of fibrinogen solutions of increasing concentration are made up to 0.5 ml in deionized, distilled water. To each of the fibrinogen solutions, 2 ml of colloidal gold, the pH of which was adjusted to 6.3, is added and mixed rapidly. After 2 min at room temperature, 0.5 ml of a 10% NaCl solution is added and the solutions allowed to stand for 5 min. Stabilization is measured spectrophotometrically at 580 nm as for the pH isotherm, using 2 ml of gold diluted to 3 ml with distilled water as a blank. Stabilization is indicated by an absorbance approaching zero with the optimum concentration chosen

---

[56] J. S. Finlayson and D. W. Mosesson, *Biochemistry* **2**, 42 (1963).

TABLE I
FIBRINOGEN pH-VARIABLE ADSORPTION
ISOTHERM[a]

pH	Visible color change	$OD_{580\,nm}$
5.32	+ +	0.22
5.78	+ +	0.18
5.96	+ −	0.15
6.21	−	0.025
6.54	−	0.02
6.84	−	0.02
7.08	−	0.018

[a] Visible color change and optical density at 580 nm are given. A stable conjugate occurs at about pH 6.3, slightly alkaline to the isoelectric point of fibrinogen. Fibrinogen concentration made up to 0.5 ml with double-distilled H_2O; final concentration, 50 µg/ml. Colloidal gold added (2 ml), final volume, 2.5 ml.

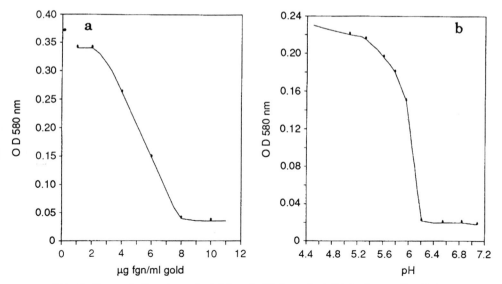

FIG. 2. Concentration-dependent (a) and pH-dependent (b) isotherms for fibrinogen (fgn)–gold conjugates.

as the point where the curve first parallels the x axis (Fig. 2a). A concentration of 8 μg/ml of colloidal gold appears as the optimal concentration.

Preparation of Fibrinogen–Gold Conjugate. From a stock solution of 18-nm colloidal gold, 10 ml (pH 6.3) is added to a 10% excess of fibrinogen and mixed by gentle inversion. After 5 min at room temperature, 0.5 ml of freshly prepared and prefiltered (Millipore 0.45-μm pore size filter) 1% polyethylene glycol (M_r 20,000) is added as a secondary stabilizing agent. The fibrinogen-labeled colloidal gold (FGN–Au) is concentrated and excess fibrinogen removed by centrifugation in polycarbonate tubes at 10,000 rpm in an angle rotor for 30 min at 4°. The supernatant is discarded and the concentrated red pool is resuspended to 1 ml with sterile filtered (Millipore 0.2-μm filter) protein-free Tyrode's buffer supplemented with 1 mM calcium. The absence of any unconjugated fibrinogen is determined by the presence of a single peak when the centrifuged solution is run over a Sepharose 6B column. The presence of fibrinogen on the gold particles is demonstrated by agglutination of the gold–fibrinogen conjugate by serial dilution with anti-human fibrinogen antibodies (#F-2506; Sigman Chemical Company, St. Louis, MO).

Immunolabeling of Receptors (Anti-IIb/IIIa–Gold Procedure)

Conjugation of Heterologous Immunoglobulin to Colloidal Gold

The immunoglobulin fraction of serum is obtained by sodium sulfate fractionation of normal, pooled human serum.[57] The resulting fraction is further purified by ion-exchange chromatography on DEAE-cellulose eluted with a gradient from 0.04 M Tris–phosphate, pH 8.4, to 0.5 M Tris–phosphate, pH 3.4. The resulting fraction is dialyzed against 0.1 M phosphate buffer, pH 7.4, and stored at $-70°$. The optimum pH and concentration of immunoglobulin is determined from pH-variable and concentration-variable adsorption isotherms in the same manner as described for fibrinogen. The optimum pH is determined as pH 7.5 and the optimum concentration for this antibody is 25 μg of immunoglobulin/ml of 18-nm gold stock solution. The immunoglobulin fraction of serum is conjugated to colloidal gold, concentrated by ultracentrifugation, and resuspended to 2 ml with sterile, filtered 0.1 M N-2-hydroxyethylpiperazine-N'-2-ethanesulfonic acid (HEPES) buffer, pH 7.4. The presence of immunoglobulin on the gold particles is demonstrated by agglutination of the gold–immu-

---

[57] K. Heide and H. G. Schwick, *in* "Handbook of Experimental Immunology" (D. M. Weir, ed.), Vol. 1, p. 7.1. Blackwell, Oxford, 1978.

noglobulin conjugate by serial dilution with anti-human immunoglobulin (#517305; Calbiochem-Behring, La Jolla, CA).

Conjugation of Bovine Serum Albumin to Colloidal Gold

Bovine serum albumin (BSA), radioimmunoassay (RIA) grade (#A-7888; Sigma) is dissolved in deionized, distilled water to produce a 2-mg/ml stock solution. The optimum pH and concentration for adsorption to 18-nm colloidal gold is determined from pH-variable and concentration-variable adsorption isotherms as described for fibrinogen. The optimum pH for this albumin preparation is pH 5.5 and the optimum concentration is 50 μg/ml 18-nm gold stock solution. Bovine serum albumin is conjugated to colloidal gold, concentrated by ultracentrifugation, and resuspended to 2 ml with sterile filtered 0.1 M HEPES buffer, pH 7.4. The presence of BSA on the gold particles is demonstrated by agglutination of the gold–albumin conjugate by serial dilution with anti-bovine serum albumin antibodies (#65-111-1; Miles Biochemicals, Elkhart, IN).

Conjugation of 10E5 Monoclonal Antibody to Colloidal Gold

Antibody Purification. The monoclonal antibody, 10E5, which is directed against the glycoprotein IIb–IIIa receptor complex, was provided by Dr. Barry Coller (Department of Medicine, State University of New York at Stony Brook Health Sciences Center). The specificity of 10E5 has been characterized fully in a previous report.[18] 10E5 antibody is purified from hybridoma culture supernatants by precipitation with 50% (w/v) saturated ammonium sulfate. The precipitate is resuspended to one-tenth of the original volume with 0.1 M sodium phosphate buffer, pH 8.0. The sample is dialyzed against the same buffer, then applied to a protein A-Sepharose 4B column equilibrated with phosphate buffer. The column is eluted with phosphate buffer until the optical density of the eluate returns to baseline, after which stepwise elution is accomplished with 0.1 M citrate buffers of pH 6.0, 4.5, 3.5, and 3.0, as described by Ey *et al.*[58] Protein elution is monitored by optical density at 280 nm and the appropriate fractions pooled and dialyzed against 0.15 M NaCl, 0.01 M Tris-HCl, pH 7.4., containing 0.05% (w/v) NaN_3. The antibody is of the IgG_{2a} subclass and has an isoelectric point at approximately pH 5.5. Protein concentration is estimated by absorption at 280 nm, assuming an extinction coefficient of 15. Aliquots of 1.5 mg/ml are stored at 4°.

Preparation of 10E5–Gold Conjugate. The optimum pH and concentration for adsorption of 10E5 to 18-nm colloidal gold, as determined by

[58] P. L. Ey, S. J. Prowse, and C. R. Jenkins, *Immunocytochemistry* **15**, 429 (1978).

adsorption isotherms, are pH 6.0 and 25 μg/ml 18-nm gold stock solution, respectively. 10E5 monoclonal antibody is conjugated to colloidal gold, concentrated by ultracentrifugation, and resuspended to 2 ml with sterile filtered 0.1 M HEPES buffer, pH 7.4. The presence of 10E5 antibody on the gold particles is demonstrated by agglutination of the conjugate by serial dilution with goat anti-mouse immunoglobulin (#M-8642; Sigma).

Preparation of Fab Fragments of Anti-Fibrinogen. Polyclonal sheep anti-human fibrinogen (ICN Immunobiologicals, Costa Mesa, CA) is digested for 5 hr at 37° with papain in 20 mM phosphate buffer containing 10 mM EDTA and 20 mM cysteine hydrochloride, pH 7.0. The immobilized papain (Pierce, Rockford, IL) is removed by centrifugation and the Fab fragments separated from the Fc fragments by passage over an immobilized protein G column (Pierce). The Fab fragments are dialyzed into Tris-buffered saline and stored at $-80°$.[59,60]

Conjugation of Fab Anti-Fibrinogen to Colloidal Gold. A 3-nm gold solution is brought to pH 6.5 with 0.2 N K_2CO_3. Because the optimal pH range for conjugation of this Fab fragment is very narrow, the protein solution is also buffered to pH 6.5 with 0.01 M Tris buffer. The minimum amount of protein necessary to stabilize the gold was determined to be 30 μg/ml of gold solution. Secondary stabilization of the gold with polyethylene glycol is not necessary with this protein–gold combination. The conjugated Fab gold is concentrated by centrifugation at 60,000 g for 45 min at 4° and resuspended in Tyrode's buffer.

Platelet Preparation and Gold Labeling

Adherent Platelets

Platelets are obtained from normal, healthy adult volunteers who have not taken aspirin during the previous week. Blood samples (10 ml) are collected in polypropylene tubes containing 10 mM EGTA and mixed by gentle inversion. Platelet-rich plasma (PRP) is prepared by centrifugation of whole blood at 180 g for 10 min at room temperature. Platelets are separated from plasma proteins by passage through a Sepharose CL-2B column having a 40-ml bed volume.[61] The column is equilibrated at room temperature with a calcium-free Tyrode's buffer, pH 7.3 (136 mM NaCl, 2.7 mM KCl, 0.42 mM $NaH_2PO_4 \cdot H_2O$, 12 mM $NaHCO_3$, 2 mM $MgCl_2$, 1 g/liter dextrose, and 1 g/liter albumin). Platelets are collected in the void volume and deposited on Formvar-filmed nickel maxtiform grids (other

[59] A. J. Barrett, this series, Vol. 80, p. 771.
[60] A. J. Barrett, *Trends Biochem. Sci.* **12,** 193 (1987).
[61] D. Tangen, H. J. Berman, and P. Marfey, *Thromb. Diath. Haemorrh.* **25,** 268 (1971).

polymers can be used but polymer type may influence spreading)[62] and allowed to settle and adhere at 37° in a moist chamber. The extent of spreading is monitored by differential interference contrast microscopy (DIC).[9,10,63] At various stages of spreading, grids containing adherent platelets can be removed and washed with protein-free buffer. Observation of platelet attachment and spreading by DIC is valuable for following the course of the deposition and spreading process, and to ensure that the platelets have adhered. From the time of contact with the Formvar film, the entire spreading process requires approximately 15–20 min. Maximum incubation time is 25 min.

Labeling of Live Platelets

Live platelets are labeled by incubating individual grids in a 20-μl drop of the FGN-Au or 10E5-Au suspension placed on a piece of Parafilm (American Can, Greenwich, CT) for 5 min at 37°. In control experiments, four grids of live platelets are incubated with albumin-labeled colloidal gold, colloidal gold conjugated to the IgG fraction of normal serum, a large excess of unlabeled soluble fibrinogen (2 mg/ml), or a large excess of unlabeled 10E5 antibody, respectively, prior to incubation in the FGN-Au suspension or the 10E5-Au suspension. Grids are then washed thoroughly in buffer to remove unbound label and processed for electron microscopy.

Video-Enhanced Light Microscopy

Platelets can also be labeled while under observation in the DIC light microscopy.[7,9,10,63,64] Platelets are allowed to adhere and spread on Formvar-coated grids that are contained in a flow cell on the microscope stage. The flow chamber permits the introduction of various solutions for labeling, rinsing, and fixing of the platelets. The modulation transfer for various interference-based light microscopy (asymmetric interference contrast and DIC microscopy) is such that small processes and other changes in structure not readily seen by phase-contrast microscopy are readily visible in the interference-based system.[9,65,66] It is also possible using the interference-based system coupled with available computer processing/analysis instrumentation to follow the movement of groups of label and

---

[62] S. L. Goodman, T. G. Grasel, S. L. Cooper, and R. M. Albrecht, *J. Biomater. Mater. Res.* **23**, 105 (1989).
[63] S. L. Goodman and R. M. Albrecht, *Scanning Microsc.* **1**, 727 (1987).
[64] R. M. Albrecht, O. E. Olorundare, and S. R. Simmons, in "Fibrinogen 3 Biochemistry, Biological Functions, Gene Regulation and Expression" (M. W. Mosesson, D. L. Amrani, K. R. Siebenlist, and J. P. DiOrio, eds.), p. 211. Elsevier, Amsterdam, 1988.
[65] B. Kachar, *Science* **227**, 766 (1985).
[66] R. D. Allen, N. S. Allen, and J. L. Travis, *Cell Motil.* **1**, 291 (1981).

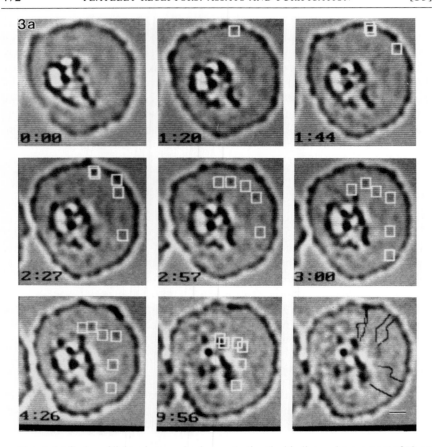

FIG. 3. (a) Series of light micrographs demonstrating the binding and movement of 18-nm fibrinogen–gold labels on the surface of an adherent, fully spread platelet. Boxes enclose individual gold particles that bind near the platelet margin and then move in the plane of the membrane across the platelet surface. Lines on the final frame indicate the paths taken by several of the receptor–fibrinogen–gold 18-nm complexes as they move from the initial binding site toward the center of the spread platelet. Bar: 1 μm. (b–e) Low-voltage, high-resolution scanning electron micrographs of the same platelet seen in 3(a). Images taken at 1.5 kV (b and d) in the secondary electron imaging mode demonstrate the minimal beam penetration seen at this accelerating voltage. High-resolution images of the platelet surface and of the protein (fibrinogen) coating the gold bead surfaces demonstrate the surface ultrastructure and the location of the gold–fibrinogen on the surface. The gold particles tracked in (a) (arrows) are readily visible but not distinguishable from other surface structures of similar size and shape (arrowheads). Images taken at 4.5 kV accelerating voltage in the back-scattered electron imaging mode (c and e) show less detail of the surface structure but positive identification of the dense gold particles is now possible (arrows). Bar (b and c): 1 μm; bar (d and e): 0.25 μm. (f) High-voltage (1000 kV) TEM stereo pair of the same platelet seen in (a–e). The gold particles tracked in (a) and seen relative to the platelet surface in

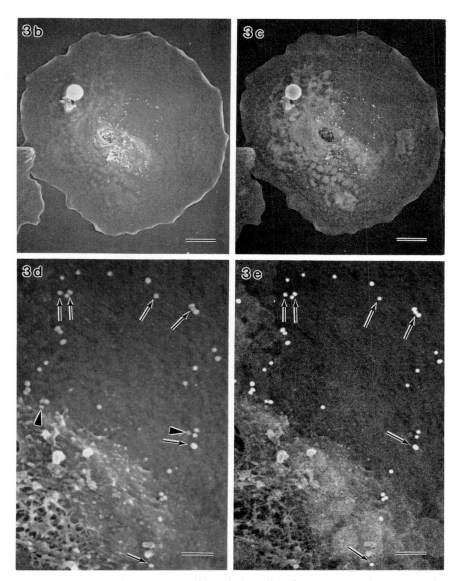

(b–e) appear as dark dots (arrows). Although the cell surface membrane is not seen, the underlying cytoplasmic structure is now visualized. The granulomere (g) is seen in the platelet center and is surrounded by three distinct cytoskeletal zones. The outermost peripheral web (p) is a dense meshwork of short interconnecting actin filaments; inside the peripheral web is the outer filamentous zone (o), which is a more open network of interconnecting actin filaments and some microtubules. The receptor–fibrinogen–gold complexes travel in the membrane over the subjacent outer filamentous zone and come to rest in the membrane over the inner filamentous zone (i), which is a dense meshwork of actin filaments, filament bundles, and some microtubules arranged in a curvilinear pattern around the granulomere. Bar: 1 μm.

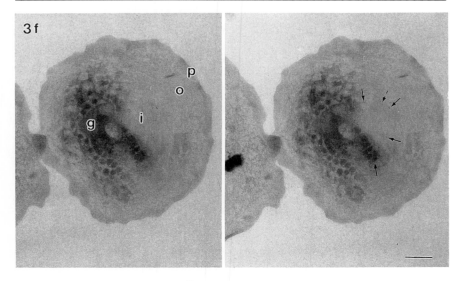

FIG. 3. (continued)

even individual gold–ligand or gold–antibody label, as small as 10 nm (100 Å), on living platelets.[10] This is possible due to the ability to detect an inflated diffraction image of the gold label, even though the size of the particles is well below the limits of resolution of the light microscope (Fig. 3a). The use of finder grids facilitates the subsequent correlative observation of the same platelets observed in the light microscope (Fig. 3b–f) by scanning or transmission electron microscopy. Thus, the identity of gold labels seen by DIC microscopy can be confirmed and correlated with ultrastructural details of the platelet surface and cytoskeleton.[6,10]

This microscope is based on that described by S. Inoue[9,67] and built on a Nikon (Garden City, NY) Diaphot modified for increased light transmittance to the TV camera. The microscope is equipped with a polarization rectified condenser[68] with DIC optics and a 1.25 NA × 100 plan apochromatic objective, A 100-W mercury lamp with a 546-nm interference filter is used for illumination. When DIC is used, a small degree of bias retardation can be helpful in imaging gold colloids, with the Wollaston prism adjusted for high transmittance. Images are projected to the television camera to give an optical and electronic magnification of ×6000.[10]

The video system utilizes an instrumentation-grade Dage (Michigan

[67] S. Inoue, "Video Microscopy." Plenum, New York, 1986.
[68] S. Inoué and H. L. Hyde, *J. Biophys. Biochem. Cytol.* **3**, 831 (1957).

City, IN) MTI Newvicon video camera modified for manual control of gain and black level. Manual control of camera gain and black level are essential for analysis because automatic cameras will continually adjust gain and black level to accommodate changes in the image. The automatic adjustments can make it impossible to compare brightness changes between video frames that occur, for instance, as a consequence of binding and movement of colloidal gold labels on the cell surface. To improve visualization further, a video processor and attendant synchronization stripper (Colorado Video 604 and 302-z; Boulder, CO) permit nonlinear expansion of contrast and brightness. Experiments are recorded to videotapes for subsequent playback and analysis using a high-resolution 3/4-in. U-matic or time-lapse video recorder. Micrographs are recorded on Plus-X film directly from the high-resolution analog monitor (Panasonic WV-5410; Secaucus, NJ) from high-resolution videotape playback, or as digitally processed images from the computer monitor. For computer analysis and morphometry, either real-time or playback images are routed to a 33-mHz MS-DOS 80486-based computer that is equipped with video capture and processing hardware and custom software (Image 1; Universal Imaging Corp., Media, PA). A second light source permits epiillumination for simultaneous fluorescence imaging for conventional immunolabeling procedures or for the identification of pH or Ca^{2+} levels. To capture fluorescent images of low intensity, a second video camera of either the silicon-intensified type or a charge-coupled device is necessary.[10,69]

Labeling of Prefixed Platelets

Platelets that are fixed prior to labeling are prepared as follows. Gel-filtered platelets are allowed to spread on Formvar-coated grids for 25 min at 37°. Grids are washed with 0.1 M HEPES buffer and fixed in 0.1 M HEPES-buffered 0.1% (v/v) glutaraldehyde, pH 7.3, for 20 min at room temperature. Fixed platelets are thoroughly washed with buffer and incubated for 30 min at room temperature in HEPES buffer containing 0.05 M glycine to block free aldehyde groups. Specimens are washed and labeled with 10E5–gold for 20 min at room temperature. Grids are washed repeatedly to remove unbound label and processed for electron microscopy. When platelets are labeled following fixation, it is necessary to demonstrate that prefixation does not result in nonspecific binding of gold-conjugated mouse immunoglobulin. To accomplish this, filtered platelets are fixed for 30 min at room temperature in 0.1 M HEPES-buffered 0.1%

---

[69] L. M. Waples, O. E. Olorundare, S. L. Goodman, S. L. Cooper, and R. M. Albrecht, *Proc. AAMI Cardiovasc. Sci. Tech. Conf.* Bethesda, MD, Dec. 2–4. p. 5 (1991).

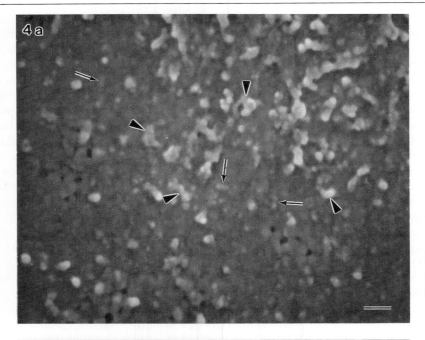
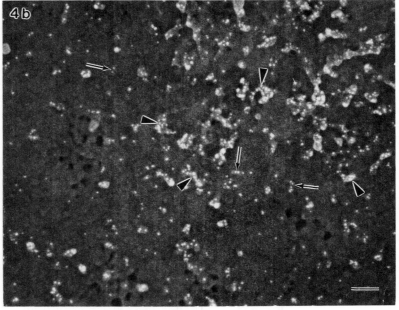

glutaraldehyde containing 0.05% (w/v) saponin and 0.2% (w/v) tannic acid, pH 7.3. Grids are washed with buffer and incubated in 0.05 M glycine for 20 min. Grids are then washed and labeled with 18-nm colloidal gold conjugated to mouse IgG (mouse immunoglobulin G, #I-5381; Sigma) for 20 min. In addition, normal human red blood cells and cultured Chang epithelial cells, both of which lack the GPIIb–IIIa fibrinogen receptor, are also fixed in 0.1 M HEPES-buffered 0.1% glutaraldehyde containing 0.05% saponin and 0.2% tannic acid, pH 7.3, for 30 min. Grids are washed, incubated with 0.05 M glycine, and labeled with 10E5-gold for 20 min at room temperature. All preparations are washed thoroughly to remove any unbound label and processed for electron microscopy. No labeling should occur in any of the above.

Control Preparations for Fibrinogen–Gold and Monoclonal Antibody–Gold Labeling

To ensure that nonspecific labeling is not occurring, the platelets are incubated with an excess of soluble fibrinogen. The soluble fibrinogen effectively blocks all fibrinogen–gold binding. In addition, platelets can also be incubated as previously described, but with the albumin–gold beads. Due to the absence of specific surface albumin receptors, little or no binding should be seen. In the case of the monoclonal antibody–gold beads, incubation with beads conjugated to monoclonal antibody of the same isotype as the specific antibody, but directed at an antigen not present on the platelet surface, serves as a control. Also, incubation with an excess of the unlabeled specific monoclonal antibody should block all binding of the specific monoclonal antibody–gold. Prior incubation of platelets with unlabeled mouse immunoglobulin, lacking antibody to the platelet fibrinogen receptor, should show no blocking of the binding of monoclonal antibody–gold. Normal human serum IgG–gold also should not show significant binding.

FIG. 4. Gold particles (3 nm) coupled to anti-fibrinogen Fab fragments are used to label soluble fibrinogen molecules bound to platelet surfaces. (a) Single fibrinogen molecules (arrows) and aggregates of fibrinogen (arrowheads) formed during the centralization process can be identified on the platelet surface using secondary electron imaging at 1.5 kV accelerating voltage. (b) Identical area as (a), but using the back-scattered electron mode at 5.0 kV accelerating voltage. The 3-nm gold is now clearly visible. Thus the fibrinogen can be positively identified via the anti-fibrinogen Fab–gold 3-nm complexes. Arrows and arrowheads indicate the same molecules and aggregates of fibrinogen as in (a). Bar (a and b): 0.1 μm.

Platelets in Suspension

For studies of nonadherent cells, activated or unactivated platelets are labeled and fixed in suspension or, in the case of prefixation studies, fixed and then labeled in suspension. These platelets can then be collected on one of several substrates, depending on the particular information desired. It should be noted that labeling after fixing and collection on the substrate is also possible. For subsequent scanning electron microscopic analysis, the most efficient means of collection is filtration, with gentle pressure, onto 22-mm diameter, 0.2-μm mean pore diameter, polycarbonate microporous filters (Nuclepore, Pleasanton, CA) or onto microporous filters of sintered silver oxide (Osmonics, Minnetonka, MN) or aluminum oxide (Alltech Associates, Inc., Deerfield, IL). Regardless of filter type, platelets must be fixed prior to deposition to avoid changes in structure that may result from the platelet response to the filter surface or from general effects of the surface on cell membranes. For high-voltage electron microscopy, suspension-fixed cells must be collected on filmed grids. Usually Formvar (polyvinyl formal) films are employed. These may be uncoated or, if necessary, coated with any of a variety of "adhesive-improving" molecular species. A 0.1% (w/v) solution of the M_r 90,000 form of poly(L-lysine) is perhaps the most widely used material for providing a more adhesive surface. Usually a 5-min exposure of the surface to the suspension, followed by a thorough washing with distilled water, is sufficient. The effects of the poly(L-lysine) on cell structure are controversial. It may directly or indirectly affect the membrane or the cytoskeleton of living cells; the response of fixed cells is less of a problem, although there is some evidence to suggest that poly(L-lysine) can influence cytoskeletal structure even after aldehyde fixation. Other adhesive proteins are available. A drop of the platelet suspension is placed on the surface of the filmed grid and the platelets allowed to sediment onto the surface. Subsequent preparation is as for adherent platelets. Light microscopic monitoring is helpful to document platelet adherence to the film.

Electron Microscopy

All specimens that are labeled live are then fixed in 0.1 M HEPES-buffered 1% (v/v) glutaraldehyde (#1050A; Tousimis Research Corp., Rockville, MD) containing 0.2% (w/v) tannic acid, 0.05% (w/v) saponin, pH 7.3, for 30 min at room temperature.[70] Specimens, including those prefixed prior to labeling, are postfixed in 0.1 M HEPES-buffered 0.05%

---

[70] P. Maupin and T. D. Pollard, *J. Cell Biol.* **6**, 51 (1983).

(w/v) osmium for 20 min at room temperature and stained in 1% (w/v) aqueous UMgAc for 10 min. Specimens then can be dehydrated through a graded series of ethanol (20, 30, 50, 70, 80, 85, 90, and 95%, 3 min each) to absolute ethanol, itself dried by storage over molecular sieve (molecule sieve type 3A; Linde Division, Union Carbide Corp., Chicago, IL). Samples are dried by the critical point procedure in a critical point dryer equipped with an in-line molecular sieve filter (#8782; Tousimis Research Corp., Rockville, MD) and a hydrophobic, water-excluding filter (type Ap25; Millipore) to remove particles, oils, and trace water from the liquid CO_2.[71] Some samples are then coated with a thin layer of carbon by evaporation at a vacuum of 5×10^{-6} Torr. All samples are stored in a sealed container over a molecular sieve to maintain an atmosphere of 0% relative humidity until examined with the 1-MeV electron microscope. Stereo pair micrographs are taken at tilt angles appropriate for specimen thickness and magnification (Fig. 3f).[72] Samples are then coated when necessary with approximately 1–2 nm of platinum using an ion beam sputter-coating apparatus. Specimens are examined on a Hitachi S-900 high-resolution, field emission scanning electron microscope at 1–25 kV accelerating voltage (Fig. 3b–e). (A tilt angle of 7° can be used for SEM stereo pairs.) Observing the specimens at low accelerating voltages, at which beam penetration into the specimen is on the order of a few nanometers, allows individual molecules to be resolved on the platelet surface at 1–2 kV (Fig. 4a).[15] Increasing the voltage to 4–5 kV and above progressively decreases the amount of surface detail in the image, but makes possible the positive identification of gold labels on or below the platelet surface.[17] Collection of back-scattered electrons further enhances detection of the gold particles (Fig. 4b).[73] Excellent resolution can be obtained at 5 kV if a YAG-Autrata type back-scattered electron detector is employed.[74,75] The number of FGN–Au labels bound per platelet at saturation is determined by direct counting of the labels on individual platelets. Micrographs of randomly selected, fully spread platelets are enlarged to a final print magnification of $\times 100,000$. Individual labels can be counted and marked on a transparent overlay. Various additional forms of image analysis are also possible. The image can be digitized either directly from the microscope or from a micrograph. Thresholding the image provides a black and white image showing only the gold particles, and the numbers of gold particles as well as their location relative to cell features can easily be determined by computer analysis.[10]

[71] H. Ris, *J. Cell Biol.* **100**, 1474 (1985).
[72] B. Hudson and M. J. Maken, *J. Phys. Educ. Sci. Instrum.* **3**, 311 (1970).
[73] R. M. Albrecht and B. Wetzel, *Scanning Microsc. Suppl.* **3**, 1 (1989).
[74] R. Autrata, *Scanning Microsc.* **3**, 739 (1989).
[75] P. Walther, R. Autrata, Y. Chen, and J. B. Pawley, *Scanning Microsc.* **5**, 301 (1991).

Author Index

Numbers in parentheses are footnote reference numbers and indicate that an author's work is referred to although the name is not cited in the text.

A

Abrams, C. S., 427
Abrams, D. I., 429
Achard, C., 109
Ackerman, G. A., 464, 466(45)
Adelman, B., 174, 175(85), 291, 420, 438
Adelstein, R. S., 59, 60, 78, 80, 82, 86, 87(20)
Adler, J. R., 143
Affolter, H., 215
Aggerbeck, L. P., 277
Aghajanian, G. K., 217
Agin, P. P., 156, 157, 158(24, 25, 26), 165(26), 174(26), 176, 245, 413, 414(4), 417, 418(4), 431
Aitken, J. W., 398
Akerboom, T. P. M., 37
Akkerman, J. W. N., 433
Albrecht, R. M., 122, 126, 456, 457(8), 458, 469(18), 471, 475, 479
Alexander, R. J., 156
Alexander, R. W., 183, 184(12, 24)
Allain, J. P., 421
Allen, N. S., 471
Allen, R. D., 471
Allen, R. W., 448
Allison, W. S., 144
Ambegaonkar, S., 464
Amrani, D., 311
Anderson, L. K., 42, 49(2), 50(2), 62
Anderson, T. F., 122
Andrews, H. E., 398
Andrews, R. K., 403
Anjaneyulu, P. S. R., 403, 404(4), 405
Annamalai, A. E., 144
Anstee, D. J., 290
Anstee, K. J., 278
Antoniades, H. N., 31
Aoke, A., 137, 139(8)

Aoki, M., 118
Appelmans, F., 30
Appling, D. R., 404, 408(19)
Archer, J. A., 400
Ardlie, N. G., 145, 354, 383, 399
Arro, E., 122, 123(36)
Asano, T., 429
Asch, A., 76, 121
Ashmun, R. A., 420, 427(3)
Aster, R. H., 215, 289, 430, 431(16), 438(21)
Auerswald, W., 398
Ault, K. A., 174, 175(85), 420
Austen, K. F., 420, 427(9)
Authi, K. S., 19, 20
Autrata, R., 479
Aviram, M., 391, 392(11), 397, 398(18)
Axelrod, J., 30
Aynaud, M., 109

B

Babb, L., 402
Bader, R., 232, 240, 263, 275
Baenziger, J. U., 246, 261(18)
Baenziger, N. L., 156, 158(23), 176(23)
Baigent, C. L., 459
Bainton, D. F., 30, 31
Bajaj, S. P., 331, 353
Baker, R. K., 399, 401(1), 402
Bamburg, J. R., 43
Banga, H. S., 132
Barber, A. J., 32, 33, 152, 187, 217, 292
Barber, A. L., 172
Barlow, G. H., 233
Barnard, M. R., 174
Barnett, D. B., 187
Barr, R. D., 430
Barrett, A. J., 470

Bartlett, P. A., 465
Basford, R. E., 38
Bauer, B., 465
Bauer, J. S., 38, 39(7), 41
Baumgartner, H. R., 3, 232
Beardsley, D. J. S., 430, 431(22), 433(22), 436(22), 438(22), 439(22), 440
Bebus, E., 126
Beckerle, M. C., 3, 100
Bednar, R. A., 144
Bednarek, M. A., 230
Beffer, S., 430
Behnke, O., 457
Benayan, M., 463
Bencuya, R., 214
Bengur, A. B., 81
Bengur, A. R., 86, 87(20)
Bennett, E. M., 253, 417, 418(12), 443, 448(17)
Bennett, J. P., 208
Bennett, J. P., Jr., 373
Bennett, J. S., 132, 143, 144, 152(18), 161, 253, 456
Bennett, W. F., 156, 157(16)
Bennett, W. P., 291
Bensadoun, A., 101
Berchtold, P., 439
Beretta, G., 361
Bergstrand, A., 29
Berliner, S. A., 230
Berman, C. L., 31, 427
Berman, H. J., 372, 383, 470
Berndt, M. C., 42, 43, 49(2), 50(2), 62, 174, 175(82), 176, 177, 179(9), 180(9), 277, 278, 285, 287, 403, 419
Bertocci, A., 133
Bessis, M., 121, 122(31)
Bettex-Galland, M., 59, 109, 110(7, 8)
Beutler, L., 285
Bevilacqua, M. P., 311
Beyer, E. C., 440
Bianco, B. S., 400, 402(9)
Bianco, C., 311
Bielich, H. W., 456, 471
Bienz, D., 174, 177, 287, 288
Bilheimer, D. W., 392
Billett, H. H., 126
Bina, M., 438
Binder, B. R., 393, 397(14), 398
Bishopric, N. M., 183, 184(9)

Bizzi, B., 291
Bjerrum, O. J., 370, 433, 440, 441, 442(1, 13), 443(2, 13), 444(13), 446(2), 448(1), 451(2), 452(1), 453(12), 454(1), 455(2)
Blair, I. A., 132, 182
Blanchard, D., 429, 440(11)
Blanchette, V. S., 430
Blankstein, K. B., 369
Blikstad, I., 43, 51(4), 54, 74, 77(42)
Block, L. H., 398
Bloom, J. W., 317, 318(8), 324(8), 345, 338(33), 360
Bloomfield, V. A., 318, 327(12)
Blumenfield, O. O., 170
Bodensteiner, D. C., 420
Boizard, B., 430, 431(27), 438
Bokoch, G. M., 181
Bolhuis, P. A., 243
Bolton, A. E., 331, 362
Bonate, P. L., 133
Bonner, W. A., 419
Bonner, W. M., 410
Booth, W. J., 403
Bormens, M., 429, 440(12)
Born, G. V. R., 4, 143, 201, 243, 373
Bottini, E., 264, 265(8)
Boublik, M., 465
Bouma, B. N., 361, 370, 382
Boven, E., 307
Bowie, E. J. W., 331, 353(17), 335(17)
Boyer, J. L., 70, 77(38)
Boyles, J. K., 42, 43(1), 49(2), 50(2), 53(1), 62, 77, 116, 414
Boyles, J., 50, 51(28), 457
Bradford, H., 361, 369
Bradford, M. M., 91, 235
Brandon, S., 187, 199, 200(45)
Brash, J. L., 462
Brasier, R. S., 183, 185(25), 188(25)
Braslawsky, G. R., 307
Bray, P. F., 245
Brenneman, T., 186
Bressler, N. M., 22
Brewer, H. B., Jr., 384
Bridgman, P. C., 123
Brinkhous, K. M., 232, 264, 421
Broche, G. N., 156, 158(23), 176(23)
Brochier, J., 174
Brodeur, B. R., 301

Broekman, M. J., 21, 22, 27(1), 28(1), 29(2), 30(1, 2), 31(1), 32(1, 2, 3, 4, 5, 6)
Bromberg, M. E., 78
Brook, J. G., 391, 392(11), 397, 398(18)
Brooks, D., 296, 298(5), 301(5), 312(5)
Brosstad, F., 277, 441, 443(7), 444(7), 446(9), 449(7), 450(8)
Brouet, J. C., 429, 440(12)
Brown, E., 403
Bruckdorfer, K. R., 396, 397, 398
Bryan, J., 60, 69, 88, 91, 92(1, 7), 94(4, 7, 12, 13), 95, 96, 97(24), 99(24)
Bryon, P. A., 9
Buck, C., 100
Buckley, S., 398
Buczynski, A. Z., 146, 427
Budzynski, A., 132
Buell, D. N., 400
Bull, H. A., 174, 175(82)
Bunn, P. A., Jr., 307
Burdash, N. M., 420
Burden, R. L., 324
Burke, T. A., 404
Burns, T. W., 183, 184(13)
Burnstock, G., 134
Burridge, K., 3, 71, 72(39), 73, 74(39), 91, 94(15), 100, 103
Burstein, M., 121, 122(31)
Byers-Ward, V., 240
Bykowska, K., 277
Bylund, D. B., 183, 184(13), 185(26)
Bylund, D., 199

C

Caasonato, A., 275
Caen, J. P., 37, 40(4), 177, 245, 276, 289, 419, 431, 432(32), 433
Caen, J., 419, 438
Cahmberg, C. G., 431, 432(34)
Calafat, J., 403
Calvanico, N. J., 464
Calvete, J. J., 245
Canfield, W., 358, 359(45)
Cannon, D. C., 375
Cantor, C. R., 385
Capitanio, A., 418
Carey, F., 19, 20
Carlsson, L., 43, 51(4), 54, 60, 74, 77(42)
Carnahan, G. E., 292
Caron, M. G., 132, 183, 185(27), 193
Carrell, N. A., 244, 245(5)
Carretero, O. A., 371
Carter, R. D., 403
Cartron, J. P., 429, 440(11)
Carty, S. E., 40, 214, 222(5)
Caruso, R., 429
Casella, J. F., 43
Castagnola, M., 291
Castaldi, P. A., 174, 175(82), 278, 285, 403, 419
Catcher, C. H., 360
Cazenave, J. P., 157
Cesura, A. M., 133
Champeix, P., 429, 440(11)
Chao, F. C., 119, 158, 159(43), 167(43), 248
Chaponnier, C., 96, 97(25), 99(25)
Charo, I. F., 245
Chediak, J., 311
Chelbowski, J. F., 371
Chen, J., 154
Chen, Y., 479
Cheng, C. M., 437
Cheng, Y., 203
Cheresh, D. A., 230
Chernoff, A., 21, 32(4)
Cheung, Y.-D., 187
Chinkers, M., 136
Choate, J. J., 156, 158(15), 159(15), 167(15), 176, 431, 432(34)
Choate, J., 122, 456, 457(8)
Chong, B. H., 174, 175(82)
Chowdhry, V., 159, 165(45)
Chung, D. W., 276, 278(6), 284(6, 8), 285(6, 8), 290, 292(12)
Chung, S. Y., 134, 154
Cimo, P. L., 427
Cines, D. B., 161, 253, 438
Cintron, J., 136
Clamp, J. R., 290
Clark, C., 372
Clement-Jones, V. V., 384
Clemetson, K. J., 46, 47(17), 174, 175(86), 177, 276, 277, 278(17), 286, 287, 288, 289, 290, 293(17), 417, 418, 419
Cochrane, C. G., 369, 370, 382(1, 3)
Cohen, C. M., 248
Cohen, C., 60
Cohen, I., 60, 119

Cohen, P., 21, 27(1), 28(1), 29(2), 30(1, 2), 31(1), 32(1, 2, 3)
Coleman, J. G., 371
Coles, E., 225
Colette, C., 403
Colibretti, M. L., 275
Coller, B. S., 175, 229, 264, 284, 292, 294, 295, 296(1), 297(1), 298(1), 301, 310, 312, 458, 469(18)
Coller, H. A., 301
Collier, E. I., 428, 430(1), 431(1)
Collier, N. C., 94, 99, 100, 101(3)
Colman, P. M., 371
Colman, R. F., 134, 143, 144, 145(9), 146, 152(18, 19), 154
Colman, R. W., 134, 143, 144, 146, 148, 152(18, 19), 154, 362, 369, 382
Colombani, J., 438
Colten, H. R., 96
Coluccio, L. M., 95
Conahey, J. J., 400
Cone, T. E., Jr., 430
Connolly, T. M., 181
Conti, M. A., 78
Cooper, B., 183, 184(12)
Cooper, D. M. F., 187
Cooper, H. A., 276, 291, 421
Cooper, S. L., 471
Corash, L., 429
Corbin, J. D., 83
Corcoran, M., 448
Corino, G. L., 403
Costa, J. L., 41
Coughlin, S. R., 131, 175
Courvalin, J.-C., 429, 440(12)
Cowen, P. J., 213
Cragoe, E. J., 132
Cragoe, E. J., Jr., 182, 199, 200(45)
Crastes De Paulet, P., 403
Crastes DePaulet, J., 403
Crawford, N., 10, 13(3), 15, 17, 19, 20, 59
Crissman, R. S., 298
Croall, D., 100
Croset, M., 20
Cross, A. J., 208, 212(13)
Cummingham, M., 427
Cunningham, D., 286
Cunningham, L. W., 292, 403, 404(7), 407, 408(7), 409(7), 410(7), 411(7)

Curtiss, L. K., 397
Cusack, N. J., 134, 137, 141(6)
Cutler, L., 4
Cuttitta, F., 307

D

D'Souza, S. E., 404
Da Parda, M., 37, 133
Dabrowska, R., 78
Dahlbäck, B., 329
Daijiyi, M., 183, 184(20)
Daneels, G., 435
Dangelmaier, C. A., 41, 156
Dangott, L. J., 136
Daniel, J. L., 60, 78, 84, 85(6, 16)
Daniel, J., 154
Danishesky, K. J., 161, 162(48)
Danon, F., 429, 440(12)
Daughaday, W. H., 399, 401(1), 402(1)
Davidson, M. M. L., 181
Davie, E. W., 276, 284(8), 285(8), 331, 353(20)
Davies, G. E., 403, 404(5)
Davies, P. J. A., 427
Davies, T. A., 157, 158(38, 39), 161(38, 39), 162(38, 39), 165(38, 39), 174, 175(77, 78)
Davies, T., 181
Davis, K. S., 122
Davis, T. M., 403
Davis-Ferreira, J. F., 109, 110(5), 121(5)
Dawson, G., 413, 415(3)
Dawson, R. M. C., 76
Day, H. J., 28
de Duve, C., 30
de Groot, P. G., 382
De Marco, L., 263, 264, 265, 268(4), 275
de Petris, S., 306
De Roia, D., 275
de St. Groth, S. F., 296, 298(4), 301(4)
Dean, G. E., 221, 222
Dean, J., 162, 170(53)
Dean, P. N., 421
Dechavanne, M., 9, 174, 288
DeCristofaro, R., 291
DeFeo, P., 154

Degos, L., 438
Delbart, C., 397
DeLean, A., 186
DeMarco, L., 232
DeMarinis, R. M., 183, 185(27)
DeMey, J., 435
DeMeyts, P., 402
Derksen, A., 21, 32(3)
Desai, K., 396, 397
DeSisto, M., 42
Detwiler, T. C., 33, 59, 155, 156, 158(17), 159(17), 161, 162(48), 177, 290
Devine, D. V., 438
Didry, D., 403, 438
Ding, A., 177
Dingus, J. D., 91, 94(12, 13)
Dingus, J., 88, 92(1), 95(1)
Dixon, F. J., 400
Dixon, F., 371
Dobryszycka, W. M., 13
Dobson, C. M., 41
Dockter, M. E., 55, 58(30), 420, 427(3)
Doleschel, W., 385, 391(8), 392(8), 394(8), 396(8), 398
Donovan, J. W., 407
Doolittle, R. F., 230, 233
Dorfman, D., 441, 446(11)
Dorfmeijer, H., 430, 431(20)
Driezen, P., 84
Drillings, M., 21, 32(4)
Drotts, D., 174, 175(77, 78)
Drummond, A. H., 145
Du, X., 285
Dubay, A., 291
Dubourque, D., 371
Duggan, K., 100
Dumont, E., 136
Dunstan, R. A., 420, 421(8), 427(8)
Dupuis, D., 177
Durack, B. E., 426
Duthilleul, P., 397

E

Earp, S., 100
Ebashi, F., 59
Ebashi, S., 59
Eckhardt, A. E., 178

Edgard, W., 145
Edgington, T. S., 375, 381
Edwards, H. H., 43, 58(5), 74, 77(43), 277
Eid, S., 317, 324(6)
Eide, L. L., 330, 331, 353(17), 355(16, 17), 356(16), 358(16)
Eirich, F. R., 462
Eisenberg, S., 384, 392
Eisman, R., 229, 244, 245(7)
El-Baz, Z., 402
El-Garf, T. A., 402
Elliott, J. M., 201, 206, 213
Ellison, N., 427
Endres, G. F., 169
Engelfriet, C. P., 430, 431(26), 438(26), 439
Epstein, L. B., 226
Erickson, H. P., 244, 245(5)
Eriksson, L., 60
Ernster, L., 29
Ersda, E., 382
Esch, F. S., 144
Escolar, G., 116, 119(17)
Esmon, C. T., 358, 359(44)
Ey, P. L., 301, 469

F

Faig, D., 429
Fain, J. N., 183, 184(13, 22)
Faires, J. D., 324
Falcoff, E., 403
Fallon, J. R., 76, 99, 101(4)
Farbiszewski, R., 398
Farr, A. L., 27, 168, 209, 362
Fasco, M. J., 282, 284(24)
Faulk, W. P., 465
Faust, K., 81
Feagler, J. R., 156, 157
Federici, A. B., 275
Fedorko, J., 307
Feinberg, H., 373
Feinberg, M., 143
Feinman, R. D., 156
Feinstein, M. B., 4
Feng, P. H., 158, 159(43), 167(43)
Fenton, J. W., 156

Fenton, J. W., II, 156, 158(17), 159(17), 282, 284(24)
Feramisco, J. R., 73
Fereacin, F., 398
Ferris, B. J., 143
Ferris, B., 244
Ferziger, R., 441, 446(11)
Figures, W. R., 134, 144, 146, 148, 152(19), 154
Filat, D., 162, 170(53)
Filion-Myklebust, C., 276, 290, 292(15), 308, 310(17)
Finch, C. A., 3
Finch, C. N., 286
Finlayson, J. S., 466
Fish, W. W., 384
Fishkes, H., 42, 133, 213, 214, 216, 217(9), 218, 221, 222
Fitzgerald, L. A., 47, 178, 229, 244, 245, 246, 248(19), 249(19), 250(19), 251(19), 253, 255(24)
Flanagan, M. D., 43
Fletcher, A. P., 33
Folch, J., 384
Follee, G., 174
Foote, L. J., 307
Forsyth, J., 311
Foster, W. B., 332, 337, 358
Fournier, D., 278, 419
Fox, J. E. B., 3, 42, 43, 49, 50, 51(28), 53(1), 55, 58(5, 30), 62, 74, 77, 99, 100, 101(5), 116, 174, 277, 287, 288, 414
Fox, J. E., 456, 457
Fraker, D. J., 269
Fraker, P. J., 362
Francke, U., 132
Fraser, C. M., 183, 185(28)
Frederiksson, B. E., 122, 123(36)
Frens, G., 459, 460(21)
Frieden, C., 88, 95(10), 144
Fries, E., 258
Fruchart, J. C., 397
Fujii, S., 370, 371
Fujikawa, K., 276, 278(6), 284(6, 8), 285(6, 8), 290, 292(12)
Fujimoto, T., 131, 232, 288, 375
Fujimura, K., 177, 253, 288
Fujimura, Y., 232
Fukami, M. H., 38, 39, 41

Furfine, C., 170
Furie, B. C., 4, 31, 427, 429
Furie, B., 4, 31, 427, 429
Furuya, F. R., 465

G

Gahmbery, L. G., 156, 158(15), 159(15), 165, 167(15), 176(15)
Galfre, G., 296, 298(3), 301(3)
Galgoci, B., 183, 184(15)
Gallagher, M., 61
Gallop, P. M., 170
Gancel, A., 403
Ganguly, P., 156, 157, 158(11), 289
Gantzos, R. D., 183, 185(25), 188(25)
Garcia-Sainz, A., 183, 184(13)
Garcia-Sevilla, J. A., 183, 184(19)
Gatti, L., 232, 263, 275(2)
Gautheron, P., 136
Gavin, J. R., 400, 402
Geaney, D. P., 201, 213
Gear, A. R. L., 174
Genco, P. V., 420
Geoghegan, W. D., 464, 466(45)
George, D., 233
George, J. N., 256, 413, 427, 443, 445(15)
Gerrard, J. M., 31, 60, 111, 113, 119
Geuze, H. J., 459, 460(22)
Giancotti, F. G., 244
Gianetto, R., 30
Giaver, I., 463
Gibbons, S., 428, 429(2)
Gilman, A. G., 181
Gilman, H., 183, 184(22)
Ginsberg, M. H., 31, 161, 232, 245, 311, 312, 314, 375, 404, 427
Girolami, A., 264, 265(4), 268(4), 275
Glen, K. C., 156, 157(16)
Glover, J. S., 251
Goding, J. W., 296, 298(2), 301(2)
Goeddel, D. V., 136
Goetze, A., 291
Goetzel, E. J., 225
Gogstad, G. O., 277, 370, 441, 443(7), 444(7), 446(10), 447, 448(20), 449(7), 450(8)
Gogstad, G., 441

Goldstein, I. J., 178
Goll, D. E., 49, 100, 101(5)
Gonzalez-Rodriguez, J., 245
Gonze, J., 38
Goodman, S. L., 456, 457, 461, 463(26), 471, 475
Göran, U., 43
Gorden, P., 400
Gordon, D. J., 70, 77(38)
Gordon, E. L., 136
Gordon, J. L., 134
Gordon, J., 436
Gorman, J. J., 403
Gough, G., 137
Gould, N. L., 157, 289
Govers-Riemslag, J., 317, 319(5), 324(5), 328(5)
Graber, S. E., 45
Graber, S., 229
Grabers, D. L., 136
Graf, M., 213
Grahame-Smith, D. G., 201, 206, 213
Gralnick, H. R., 174, 175(83), 403
Grant, J. A., 181
Grasel, T. G., 471
Greco, N. J., 157, 174(35)
Greenawalt, J. W., 30
Greenberg, J. H., 10, 33
Greenberg, J., 291
Greenberg-Sepersky, S. M., 156, 157, 158(38, 39), 161, 162(38, 39, 47), 165(38, 39), 166(39, 47), 173(28)
Greengard, J. S., 361, 373, 374(31), 375(31), 376(31), 377(31), 378(31), 380(31), 382
Greenwood, F. C., 251, 436
Gregory, C., 278, 419
Griffin, J. H., 361, 369, 370, 371, 372(6), 373, 374(31), 375(31), 376(31), 377(31), 378(31), 380(31), 382(1, 3)
Griffin, J. M., 382
Griffith, O. M., 25
Grinnel, F., 311
Grumet, F. G., 431
Guccione, M. A., 47, 143
Guichardant, M., 9, 15
Guillin, M.-C., 403
Guilly, M.-N., 429, 440(12)
Guimaraes, J. A., 371
Gurewich, V., 233

Gustafson, E. J., 382
Guyer, C. A., 187

H

Hack, N., 17, 20
Hageman, T. C., 169
Hagen, F. S., 276, 278(6), 284(6, 8), 285(6, 8), 290, 292(12)
Hagen, I., 31, 276, 277, 290, 292(15), 308, 310(17), 370, 433, 440, 441, 442(1, 13), 443(2, 7, 13), 444(7, 13), 446(2, 9), 448(1), 449(7), 451(2), 452(1), 454(1)
Hainfeld, J. F., 465
Hajek, A. S., 399, 401(1), 402(1)
Hall, S. A., 420
Hallam, R. J., 173
Han, Y. N., 370, 371
Hancock, K., 435
Handa, M., 276, 284(7), 285(7)
Handin, R. I., 21, 29(2), 30(2), 32(2, 3), 143, 174, 175(85), 183, 184(12, 24), 291, 292, 294(27), 420, 430, 431(22), 433(22), 436(22), 438(22), 439(22)
Handwerger, B. S., 400, 402(9)
Hannig, K., 20
Harker, L. A., 3, 31
Harmon, J. T., 33, 134, 156, 157, 174(33)
Harms, C. S., 372
Harpel, P. C., 31
Harrington, W. J., 429
Harris, D. A., 94
Harris, H. E., 43, 60, 70(18), 94, 95(19)
Harrison, J. H., 168
Hartree, E. F., 101, 220
Hartree, F. R. S., 220
Hartshorne, H., 78
Hartsuck, J. A., 371
Hartwig, J. H., 42, 59, 88, 92(5), 95(5)
Harvey, E. V., 81
Hasegunos, S., 433
Hashimoto, N., 370
Hasitz, M., 289, 291
Haslam, R. J., 137, 181
Hassall, D. G., 396, 397(16), 398
Hassanein, A. A., 402
Hathaway, D. R., 78
Hauser, H., 76

Hawiger, J., 44, 45, 47(14), 131, 134, 136(20), 169, 228, 229, 230, 232, 233, 234(32), 237(34), 238(1), 240, 241, 372, 375, 407, 437, 443, 446(18), 456
Hayat, M. A., 465
Hayes, C. E., 178
Hechter, O., 382
Heeb, M. J., 382
Heide, K., 468
Heidenreich, R., 229, 244, 245(7)
Heidrich, H.-G., 20
Helenius, A., 258
Helman, T., 301
Hemker, H., 317, 319(5), 324(5), 328(5)
Henschen, A., 245
Henson, E., 170
Hermann, F. H., 447, 448(20)
Hermann, J. H., 430, 438(28)
Hermodson, M. A., 331, 353(20)
Herzenberg, L. A., 421, 426(10)
Hess, D., 287, 289
Hibbard, L. S., 345, 338(33), 353
Hibbard, L., 330
Hickey, M. J., 276, 284(9)
Higgins, D. L., 318, 333, 338(25), 360(25)
Hilgers, J., 403
Hiller, M. C., 428, 430, 431(1, 16)
Hillier, J., 459
Hirsch, J., 157
Hixon, C. S., 144
Ho, J. S., 440
Hoard, D. E., 139
Hodges, G. M., 461, 463(26)
Hodson, E., 240
Hoffman, B. B., 183, 184(13, 22), 186, 187, 193
Hofmann, S. L., 354
Hogervost, F., 403
Hoglund, A. S., 122, 123(36)
Hollingsworth, P. J., 183, 184(19)
Hollingworth, J. W., 429
Holmer, G. R., 60
Holmsen, H., 4, 28, 39, 41, 44, 78, 84, 85(16), 156
Honda, Z., 133
Hoogenraad, J., 301
Hoogenraad, N., 301
Horisberger, M., 460, 461, 462, 463
Horne, M. K., 174, 175(83)
Horne, M. K., III, 403

Horne, W. C., 156, 168, 169(27), 171(27, 62), 173(27, 62), 174(29), 183, 184(24)
Horstman, D. A., 187
Horwitz, A., 100
Hou, D. C., 91, 94(12, 13)
Houdijk, W. P. M., 433
Houghten, R. A., 232, 278
Hourani, S. M. O., 137, 141(6)
Hovig, T., 28, 109, 110(6)
Howard, L., 441, 443(4), 446(12), 447(4), 448(4), 450(12), 451(12), 453(12)
Howe, S. E., 430
Hoxie, J. A., 253, 427
Hrinda, M. E., 264, 265(8)
Huang, C. K., 162, 168(52)
Huang, P.-K., 86
Hudson, B., 479
Hudson, E. N., 384
Hummel, B. C., 168
Hung, D. T., 131, 175
Hunter, M. W., 436
Hunter, W. M., 251, 331, 362
Huntson, D. L., 335
Hutton, R. A., 397
Hwang, S.-B., 226
Hwo, S., 96, 97(24), 99(24)
Hyde, H. L., 474
Hynes, R. O., 161, 162(49), 244, 311, 312, 403, 410

I

Ichihara-Tanaka, K., 288
Iida, K., 96
Ikari, N., 370, 371
Ikeda, Y., 119
Ingham, K. C., 65
Innerarity, T. L., 397
Inoue, S., 457, 471(9), 474
Inoué, S., 474
Insel, P. A., 183, 184(10, 17, 23), 187
Ito, K., 133
Iwanaga, S., 370, 371

J

Jackson, C. M., 316, 329, 334(4, 5), 351(4, 5), 370, 375

Jacobs, M. M., 430, 431(22), 433(22), 436(22), 438(22), 439(22)
Jacques, Y. V., 31
Jakabova, J., 431, 432(33)
Jakábová, M., 33, 49, 63, 64(32), 94, 277, 290, 347, 421
Jakobs, K.-H., 183, 184(14)
James, E., 174, 177, 288
Jamieson, G. A., 10, 32, 33, 134, 152, 156, 157, 172, 174(33–35), 187, 217, 232, 276, 286(5), 289, 290, 291, 292, 413, 418(5)
Jandl, J. H., 215
Jandrot-Perrus, M., 403
Janmey, P. A., 96
Jansonius, J. N., 371
Janssen, H., 403
Jarett, L., 399, 401(1), 402(1)
Jefferson, J. R., 134
Jenkins, C. R., 301, 469
Jenkins, C. S. P., 126, 276, 277, 278(17), 289, 419
Jennings, L. K., 43, 58(5), 74, 77(43), 244, 246(4), 261(4), 277, 410, 415, 420, 427(3)
Ji, T. H., 403
Johnson, A. J., 265
Johnson, M. M., 289
Johnson, M., 307
Johnson, R. G., 40, 42, 214, 222(5)
Johnson, S. L., 182
Joist, J. H., 399, 401(1), 402
Jones, D. B., 403
Jones, G. D., 174
Judland, P. S., 278
Judson, P. A., 290
Juengiaroen, K., 201
Julius, D., 133
Jung, S. M., 156, 174, 278, 286, 403, 419

K

Kabral, A., 278, 419
Kachar, B., 471
Kaczanowska, J., 277
Kadish, J. L., 298
Kahn, C. R., 400, 402(9)
Kahn, R. A., 246, 448
Kalomiris, E. L., 310
Kalomiris, E., 284
Kambayashi, J., 347
Kamerling, J. P., 286
Kane, N. H., 316
Kane, R. E., 65, 69
Kane, W. H., 329, 351(7), 354, 375
Kao, K.-J., 264
Kaplan, C., 429, 440(11)
Kaplan, K. L., 21, 32(4, 5, 6)
Karlson, R., 122, 123(36)
Karpatkin, M. H., 265
Karpatkin, M., 441, 446(12), 450(12), 451(12), 453(12)
Karpatkin, S., 402, 428, 429, 440, 441, 442(2), 443(2, 4), 446(2, 11, 12), 447(4), 448(4, 14), 450(12), 451(2, 12), 452(14), 453(12), 455(2)
Karpowicz, M., 277
Kartenbeck, J., 258
Katada, T., 181
Kato, H., 370, 371
Kato, T., 139
Katzmann, J. A., 332, 343(23), 353, 358
Kaufman, N., 382
Kawamoto, S., 86, 87(20)
Kaywin, P., 438
Kearney, J. F., 298
Keilin, D., 220
Kelton, J. G., 428, 429(2), 430
Kennedy, M. E., 200
Kennel, S. J., 307
Kenney, D. M., 119
Kerberiou, D. M., 361, 371
Kerry, R., 131
Keyes, S., 217
Khoranagand, J., 183, 184(15)
Kickler, T. S., 430, 438(28)
Kieffer, N., 412, 427, 438
Kikugawa, K., 137, 139(8)
Kindon, H. S., 168
Kingdon, H. S., 331
Kinlough-Rathbone, R. L., 44, 45(13), 47, 143, 157, 407
Kinoshita, T., 246, 261(17)
Kirk, K. L., 41
Kisiel, W., 331, 353(20), 358, 359(45)
Kitagawa, H., 174, 175
Klee, C. B., 82

Kloczewiak, M., 229, 230, 233(32), 234(32), 443, 446(18)
Klotz, I. M., 239, 305, 335, 379
Knight, L. C., 361, 362(3), 364(3), 365(3), 366(3), 367(3), 368(3), 413
Knight, L., 362, 382
Knorr, M., 398
Knupp, C. L., 158
Kobilka, B. K., 132
Kobilka, T. S., 132
Koller, E., 385, 391(8), 392(8), 393, 394(8), 396(8, 15), 397(14), 398
Koller, F., 385, 391(8), 392(8), 393, 394(8), 396(8), 397(14), 398
Komiya, M., 370
Kono, T., 399, 402(4)
Kopec, M., 277
Koppe, J. G., 430, 431(26), 438(26)
Korn, E. D., 70, 77(38), 81
Kornecki, E., 148, 413
Korrel, S. A. M., 286
Korsmo, K., 441, 446(9)
Korsmo, R., 370, 441
Kosaki, G., 175, 347
Koshy, A., 361, 362(3), 364(3), 365(3), 366(3), 367(3), 368(3), 369
Kostel, P. J., 232
Kotite, N. J., 403, 404(7), 407(7), 408(7), 409(7), 410(7), 411(7)
Koziol, J. A., 240
Kozloff, L. M., 436
Krebs, E. G., 144
Kreger, A., 291
Krishnaswamy, S., 316
Krumwiede, M., 116, 119(17), 121
Krutnes, M.-B., 370, 441, 443(7), 444(7), 446(10), 449(7)
Kuang, W. J., 136
Kucich, U., 382
Kuehl, W. M., 80
Kueppers, F., 369
Kuhar, M. J., 217
Kunicki, T. J., 177, 244, 253, 277, 289, 370, 427, 430, 431, 432(32), 433(32), 438(21), 440, 443(3), 448
Kuramoto, A., 288
Kurth, M. C., 88, 92(1), 94(4), 95(1, 3)
Kuylenstierna, B., 29
Kwiatkowski, D. J., 96

L

Lachman, M. F., 298
LaDue, J. S., 28
Laemmli, U. K., 50, 90, 101, 254, 281, 361, 433
Laerum, O. D., 421
Lagarde, M., 9, 15, 19
Lages, B., 42, 44
Lahav, J., 161, 162(49), 403
Lam, M.-H., 226
Lam, S. C.-T., 404
Lamb, M. A., 264
Landolfi, R., 291
Lane, R. D., 298
Langley, P. E., 183, 184(13)
Langone, J. J., 400
Lankford, P. K., 307
Laposata, M., 287
Larose, Y., 301
Larrieu, M.-J., 276, 289
Larsen, N. E., 156, 159(19), 161(19), 162(19), 165(19), 166(19), 168, 171(19, 62), 173(62), 176(19), 403
Larsson, L. I., 459, 465
Laskey, R. A., 410, 419
Laubscher, A., 201, 208, 209(6), 212(6)
Laurell, C. B., 447
Lawing, W. J., Jr., Lawler, J. W., 158, 159(43), 167(43), 293
Lawler, J., 248
Lawrence, J., 61, 62, 63(28), 64(28), 70(31), 75(28)
Lazarchick, J., 420
Lebret, M., 37, 40(4)
Lees, A. M., 391, 392(11)
Lees, M., 384
Lees, R. S., 391, 392(11)
Lefkovits, H., 398
Lefkowitz, R. J., 132, 183, 184(9, 11, 13, 22), 185(27), 186, 187, 188(36), 193, 204
Lehrer, S. S., 118, 121(21)
Leitman, S. F., 253
Leitner, A., 430, 431(16)
Lennette, E. T., 429
Lepie, M. P., 463
Lerner, W., 429
Lesniak, M. A., 402

Lesznik, G. R., 21, 32(4)
Lethielleux, P., 438
Leung, B., 246, 248(19), 249(19), 250(19), 251(19)
Leung, L. L. K., 31, 229, 443, 446(18)
Leung, L. L., 246, 261(17)
Leunissen, J. L., 465
Levine, S. N., 60
Levy, R. I., 392
Levy-Toledano, S., 438
Lewis, D. C., 413
Lewis, J. C., 99, 122
Leytin, V. L., 391, 392(12), 394(12)
Liesgang, B., 298
Lightsey, A. L., 428, 429(4)
Lightsey, A., 311
Lim, T. K., 318, 327(12)
Limbird, L. E., 132, 181, 182, 183, 184(21), 187, 188(21), 190(21), 193(21), 195(38), 199, 200
Lin, S., 43
Lind, S. E., 49, 60
Lindberg, U., 43, 51(4), 54, 60, 74, 77(42), 88, 91, 92(16)
Lindhout, M. J., 316, 375
Lindmo, T., 307
Lingappa, V. R., 245
Lipinska, I. B., 233
Lipinski, B., 233
Lipps, J. P. M., 144
Lipscomb, W. N., 371
Liss, A. R., 63, 64, 75
Livingston, D. C., 461, 463(26)
Lloyd, D. H., 76
Lo, S. S., 229
Locher, R., 398
Loftus, J. C., 122, 126, 456, 457(8)
Loftus, L. S., 233
Lombart, C., 232, 276, 286(5), 290
Lopez, J. A., 276, 278(6), 284(6, 8), 285(6, 8), 290, 292(12)
Lorand, L., 145, 169, 400
Lorenz, P. E., 162, 170(53)
Loscalzo, J., 233, 291, 292, 294(27)
Low, P. S., 76
Lowe, D. G., 136
Lowry, O. H., 27, 168, 209, 362
Lowry, P. J., 384
Lucas, R. C., 59, 61, 62, 63(29), 64(29), 66(29), 67(29), 68(29), 69(29), 70(31), 73(29), 76(29), 91, 94(14), 103
Lukas, T., 230
Lundberg, U., 122, 123(36)
Lundblad, R. L., 155, 156, 157, 168, 176(13), 331
Luscher, E. F., 109, 110(7, 8), 276, 289
Lüscher, E. F., 59, 277, 278(17), 287, 288, 417, 418, 419
Lux, S. E., 430, 431(22), 433(22), 436(22), 438(22), 439(22), 440
Lyerly, D., 291
Lyle, V. A., 286
Lyman, D. J., 462
Lynch, C. J., 183, 184(16), 185(16)
Lynch, D. M., 430
Lynnam, J. A., 181

M

Macfarlane, D. E., 137, 140(7), 142, 183, 184(18)
Maehama, S., 177, 288
Magruder, L., 174
Maguerie, G. A., 260, 312
Maguire, M. H., 137
Mahley, R. W., 248, 397
Maimon, J., 126
Majerus, D. W., 329, 334(9), 354(9)
Majerus, P. W., 156, 157, 158(12, 23), 176(23), 246, 261(18), 316, 329, 334(4, 5, 9), 351(4, 5, 7), 354, 370, 375
Maken, M. J., 479
Malhotra, O., 328
Mandyan, V., 465
Mann, J. I., 403
Mann, K. G., 168, 316, 317, 318, 319(3), 321(7), 324(6, 7, 8), 328, 329, 330, 331, 332, 333, 334, 335, 336(13, 16), 337, 338(6, 15, 31, 25), 339(6), 341(13), 342, 343(6, 23, 31), 344(14, 31), 345, 346(14), 347(14), 348(26), 349(15), 350(15), 353, 355(16, 17), 356(16), 358, 359(14, 45), 360, 375, 384
Mannucci, P. M., 275
Mansour, T. E., 144
March, J. P., 420, 427(9)

Marcus, A. J., 21, 22, 32(7)
Marder, V. J., 233, 428, 430, 431(1)
Marfey, P., 372, 383, 470
Margossian, S., 293
Marguerie, G. A., 161, 232, 311, 375, 381
Marguerie, G., 312
Mark, D., 165, 166(54, 55)
Markey, F., 43, 51(4), 54, 60, 74, 77(42), 88, 91, 92(16)
Marti, G. E., 174
Martin, B. M., 156
Martin, M. N., 4, 253, 417, 418, 443, 448(17)

Martin, S. E., 233
Martin, V., 420
Martin-Nizard, F., 397
Maruyama, K., 59, 88, 92(5), 95(5)
Mason, D. Y., 419
Mason, D., 430, 438(23), 439
Mathot, C., 463
Matsui, H., 132
Matthews, B. W., 371
Mattson, J. C., 122
Maupain-Szamier, P., 116
Maupin, P., 478
Mazurov, A. V., 391, 392(12), 394(12)
Mazzucato, M., 275
McClenaghan, M. D., 181
McConahey, P., 371
McDonough, M., 183, 184(23)
McDuffie, F. C., 330, 336(13), 337(13), 341(13), 375
McDuffie, F. W., 317
McEver, R. P., 4, 31, 246, 253, 261(18), 312, 417, 418, 427, 443, 448(17)
McFarlane, A. S., 146, 236
McGowan, E. B., 177, 290
McGregor, J. L., 174, 177, 288
McGregor, L., 174
McIntyre, T. M., 133
McKee, P. A., 264, 265
McLean, C. H., 384
McMichael, A. J., 419
McMillan, R., 429, 430, 438(23), 439
Meier, H. L., 369, 382(2)
Mejbaum-Katzenellenbogen, W., 13
Melamed, M. R., 421
Melnick, B., 292, 294(27)
Meloni, F. J., 382
Meltze, H. Y., 183, 184(20)

Menashi, S., 10, 13(3), 15, 19, 20
Mendelsohn, M. E., 233
Meyer, D., 276, 289, 312
Meyer, M., 286, 447, 448(20)
Meyers, K. M., 4
Meza, I., 91, 94(13)
Michael, H., 373
Michal, F., 201
Michel, T., 183, 184(22), 186, 187, 193
Michelson, A. D., 174, 175(85), 291, 292, 294(27), 420
Miki, I., 133
Miles, L. A., 361
Miletich, J. P., 329, 334(4, 5, 9), 351(4, 5), 354, 370
Milinas, F. C., 403
Millard, F., 439
Miller, G. L., 27
Miller, J. L., 286
Mills, D. C. B., 137, 140(7), 141(7), 142(3, 7), 146, 271
Milstein, C., 296, 298(3), 301(3)
Minami, M., 133
Minnich, V., 429
Mitchell, J. R. A., 201
Mitchell, T., 307
Miyamoto, T., 133
Moake, J. L., 232, 427
Moeremans, N., 435
Mohan Pai, K. R., 430
Mohri, H., 232
Mole, J. E., 96
Molinoff, P. B., 227
Molish, I. R., 78, 84, 85(6, 16)
Molnar, J., 145, 169, 400
Monnier, L. H., 403
Montgomery, R. R., 232, 253, 263, 275(2), 277, 448
Mooney, J. J., 183, 184(24)
Moore, A., 289
Moore, C. V., 429
Morgan, D. G., 404, 408(19)
Morinelli, T. A., 134, 144, 146, 148, 152(19), 154
Morita, T., 371
Moroi, M., 174, 278, 286, 291, 403, 419
Morrison, M., 400, 414
Morrissey, J. H., 50, 281
Morton, J., 419
Mosesson, M. W., 21, 32(5), 311, 313, 466

Mosesson, M., 313
Mosher, D. F., 156, 158(15), 159(15), 167(15), 176, 431, 432(34), 458, 465
Motulsky, H. J., 183, 184(10), 187
Motulsky, H., 183, 184(17)
Moulinier, J., 430, 431(24)
Mueller, G., 459
Mugli, R., 3
Mulac-Jerićevic, B., 96
Muller, J. Y., 429, 440(11)
Mullikin-Kilpatrick, D., 183, 184(22)
Munson, P. J., 228, 275, 315, 377, 379, 380(39, 42)
Mustard, J. F., 44, 45(13), 47, 143, 145, 157, 354, 383, 399, 407
Muszbek, L., 287
Myrmel, K. H., 330
Myrmel, K. W., 345
Mysels, K. J., 461

N

Nachman, R. L., 31, 143, 229, 244, 246, 261(17), 289, 443, 446(18)
Nachman, R., 419
Nachmias, V. T., 76, 99, 101(4), 122, 155
Nachmias, V., 121, 457
Nagen, I., 453(12), 455(2)
Nahorski, S. R., 187
Nakabayashi, H., 86
Nakamura, M., 133
Narang, H. K., 426
Nardi, M. A., 429
Needham, L., 134
Nelsestuen, G. L., 264, 318, 327(12)
Nelson, P. J., 42, 133, 213, 214, 216, 217(9, 10, 12), 218, 221, 222
Nesheim, M. E., 316, 317, 318(8), 319(3), 321(7), 324(6, 7, 8), 328, 329, 330, 332, 336(13), 337, 338(6, 31, 33), 339(6), 341(13), 342, 343(6, 23, 31), 344(14, 31), 345, 346(14), 347(14), 353, 358, 359(14, 45), 360, 375
Ness, P. M., 430, 438(28)
Neubig, R. R., 183, 185(25), 188(25)
Neville, D. M., Jr., 345, 402
Newhouse, P., 372
Newman, J., 265

Newman, K. D., 183, 184(9)
Newman, P. J., 246, 277, 427, 448
Nichols, B. A., 30
Nicholson, G. L., 76
Nicholson-Weller, A., 420, 427(9)
Niederman, R., 59
Nieuwenhuis, H. K., 433
Nievelstein, P. F. E. M., 433
Niewiarowski, S., 134, 144, 146, 148, 152(19)
Niija, K., 240
Nijenhuisle, L. E., 430, 431(26), 438(26)
Nina, M., 47
Noll, H. J., 208
Norberg, R., 43
Norman, N., 156, 174(29)
Northup, J. K., 181
Nouvelot, A., 397
Nunnari, J. M., 187, 195(38), 200(38)
Nurden, A. T., 37, 40(4), 177, 244, 245, 256, 276, 289, 370, 403, 419, 431, 432(32), 433, 438, 440, 443, 445(15)
Nurden, A., 441, 442(13), 443(13), 444(13)
Nyström, L. E., 60

O

O'Farrell, P. H., 180
O'Halloran, T., 3, 100
O'Neill, S., 233
O'Rourke, L. S., 360
Oda, K., 288
Odegaard, B. H., 316
Oh, E. H., 439
Ohara, S., 131, 232, 375
Ohno, H., 347
Okado, H., 133
Okamura, T., 232, 276, 286(5)
Okita, J. R., 277
Okumura, T., 289, 290, 292(1), 413, 418(5)
Olorundare, O. E., 456, 471, 475
Olsen, T. M., 277
Olson, J. D., 232, 427
Ordinas, A., 156
Ores-Carton, C., 126
Orkin, S. H., 96
Osawa, T., 286, 290
Osborn, M., 126, 152, 172
Osborne, J. C., Jr., 384

Ott, D. C., 139
Overbeek, J., 462
Owen, J. S., 396, 397, 398
Oyama, R., 288

P

Packem, M. L., 407
Packham, M. A., 44, 45(13), 47, 143, 145, 354, 383, 399
Pagani, S., 275
Pal, P. K., 144
Palek, J., 403, 404(5)
Pantazis, P., 31
Papayannopoulou, T., 276, 278(6), 284(6), 285(6), 290, 292(12)
Pappanheimer, J. R., 463
Parise, L. V., 178, 245, 257, 258(32), 259(16, 32), 260(16)
Parise, L., 229
Park, K., 457, 471(10), 475(10)
Parkinson, S., 229
Parks, D. R., 421, 426(10)
Pato, M. D., 82
Pawley, J. B., 458, 479
Paz, M. S., 170
Pearson, H. A., 430
Pearson, J. D., 134, 136
Pedbrazhensky, S. N., 391, 392(12), 394(12)
Peerschke, E. I. B., 76, 132, 175, 229, 294, 295, 296(1), 297(1), 298(1), 310(1), 458
Peet, K. M. S., 213
Penglis, F., 137
Pepper, D. S., 276, 292
Peroutka, S. J., 208, 211(12)
Perrie, D. W., 354
Perrie, W. T., 84
Perry, D. W., 143, 145, 383, 399
Perry, S. V., 84
Person, H. A., 430
Persson, T., 43, 51(4), 54, 60, 74, 77(42), 88, 91, 92(16)
Peters, J. R., 201, 206
Peters, K., 159, 403
Peterson, G. L., 260
Peterson, J. M., 330, 336(13), 337(13), 341(13), 375
Peterson, J., 317

Pethica, B. A., 4
Pettigrew, D. W., 144
Pexton, T., 135
Pfueller, S. L., 277, 278(17), 419
Philips, J. E. B., 120
Phillips, D. R., 31, 33, 35, 42, 43, 46, 47(18), 49, 50, 51(28), 53(1), 55, 58(5, 30), 60, 63, 64(32), 74, 77, 94, 100, 101(5), 116, 120, 156, 158(24, 25, 26), 165(26), 174(26), 176, 177, 178, 179(9), 180(9), 229, 244, 245, 246, 248(19), 249(19), 250(19), 251(19), 253, 255(24), 256, 257, 258(32), 259(16, 32), 260(16), 261(4), 276, 277, 289, 290, 347, 410, 412, 414(2, 4), 415, 417, 418(4), 421, 431, 432(33), 443, 445(15), 456, 457
Pickett, E. B., 427
Pidard, D., 244, 253, 277, 370, 440, 443(3), 448
Pierce, J. V., 369, 382(2)
Pierschbacher, M. D., 229, 230, 232, 311, 312
Pilch, J. R., 419
Piperno, J. R., 361, 362, 369(6)
Piperno, M., 403
Pizzo, S. V., 264
Pletscher, A., 37, 133, 201, 208, 209(6), 212(6), 213, 215, 398
Plow, E. F., 31, 232, 240, 245, 260, 311, 312, 375, 381, 397, 404
Podlubnaya, Z. A., 60
Podskalny, J. M., 399
Poh Agin, P., 289
Pollard, T. D., 59, 80, 81, 478
Pollard, T., 116
Poncz, M., 229, 244, 245(7)
Pong, S.-S., 226 Ponzio, A. D., 420
Pool, J. G., 265
Porter, K. R., 122
Portzehl, H., 59
Poste, G., 76
Potterf, R. D., 413
Poulsen, F. M., 41
Powell, R. D., 465
Powers, P., 430
Pozzan, T., 174
Prasanna, H. R., 415
Prater, T., 122
Prendergast, F. G., 332

Prentice, C. R. M., 145
Prescott, R. J., 403
Prescott, S. M., 133
Pressman, B. C., 30
Prodanos, C., 353
Proter, T., 122
Prowse, S. J., 301, 469
Prusoff, W. H., 203
Puszkin, E. G., 60, 126
Puszkin, S., 60, 265
Pytela, R., 312

R

Rabellino, E. M., 31
Radbruch, A., 298
Radley, J. M., 3
Rajewsky, K., 298
Rall, S. C., 244
Rall, S. C., Jr., 229
Randall, F. J., 362
Randall, R. J., 27, 168, 209
Rao, G. H. R., 60, 31, 119
Rapaport, S. I., 353
Rashidbaigi, A., 159
Raushek, R., 183, 184(14)
Ray-Prenger, C., 183, 185(26)
Read, M. S., 232, 264
Rechler, M. M., 399
Reeman, E. M., 83
Reese, T. S., 123
Regan, J. W., 132, 183, 185(27)
Reimers, H. J., 157
Reinherz, E. L., 420
Reinhold, V. R., 225
Rejkind, M., 170
Renaud, S., 136
Rendu, F., 37, 40(4)
Renshaw, L., 439
Repaske, M. G., 187, 195(38), 200(38)
Repin, V. S., 391, 392(12), 394(12)
Reynolds, A. C., 324
Reynolds, C. C., 42, 43(1), 49, 53(1), 77, 100, 101(5), 116, 287, 414
Ricca, G., 264, 265(8)
Richard, B., 397
Richards, F. M., 159, 162, 168(52), 405, 403, 408

Richmond, T. J., 405
Rickli, E. E., 287, 289
Rigmaiden, M., 78, 85(6)
Rink, T. J., 173, 174
Ris, H., 479
Robb-Smith, A. H. T., 109
Rodan, G., 4
Rodbard, D., 156, 157, 228, 275, 315, 377, 379, 380(39, 42)
Rodbell, M., 187
Rodier, M., 403
Rodriguez, H., 136
Roeder, P. E., 159
Rogentine, G., 431
Rogers, J. A., III, 76
Rohrbach, M. S., 358
Rongved, S., 156
Rosa, J.-P., 244, 245, 370, 440, 443(3)
Rosebrough, H. J., 27, 362
Rosebrough, N. J., 168, 209
Rosen, C. E., 420, 427(9)
Rosenberg, S., 61, 62, 63(28, 29), 64(28, 29), 66(29), 67(29), 68(29), 69(29), 70(31), 71, 72(39), 73, 74(39), 75(28), 76(29), 91, 94(14, 15), 103
Rosenfeld, C. S., 420
Rosenfeld, G. C., 91, 94(12, 13)
Rosing, J., 317, 319(5), 324(5), 328(5)
Roskam, J., 109
Ross, G. D., 289
Ross, R., 31
Rosse, W. F., 438
Rosset, J., 460, 461
Roth, G. J., 276, 278(6), 284(8, 6, 9), 285(6, 8), 290, 292
Roth, J., 399, 400, 402
Roth, R. H., 217
Rothen, A., 462, 463
Rouslahti, E., 312
Roy, S., 144
Ruan, C., 285, 419
Rudnick, G., 36, 42, 133, 213, 214, 216, 217, 218, 219, 221, 222
Ruggeri, Z. M., 230, 232, 240, 263, 264, 265(4), 265(8), 268(4), 275, 276, 278, 284(7), 285(7), 286
Ruis, N. M., 226
Ruoho, A. E., 159
Ruoslahti, E., 229, 230, 232, 244, 311, 312

Ruska, H., 109, 110(4)
Russell, S., 264, 265(4), 268(4), 286
Rust, N. A., 419
Rutterford, M. G., 213

S

Sadanandan, S., 441, 443(4), 447(4), 448(4)
Saganicoff, L., 38
Sage, S. O., 174
Sakariassen, K. S., 3, 243, 433
Sakon, M., 230, 347
Salacinski, P. R. P., 384
Salganicoff, L., 38, 39(7), 42, 78
Salzman, E. W., 134, 136(20)
Santoro, S. A., 135, 404, 407
Saradambal, K. V., 144
Saucerman, S., 427
Sauk, J. J., 122
Saunders, P. W., 426
Savona, S., 429
Scarpa, A., 40, 42, 214, 222(5)
Scatchard, G., 227, 239, 240(42), 368, 378, 399
Scearce, L. M., 146, 154
Schächter, M., 201, 213
Schaller, J., 287, 289
Schechter, A. N., 162, 170(53)
Scheidegger, D., 296, 298(4), 301(4)
Schelenberg, I., 286
Scheraga, H. A., 169
Schildkraut, J. J., 183, 184(24)
Schimme, P. R., 385
Schiphorst, M. E., 144
Schlossman, S. F., 420
Schmaier, A. M., 382
Schmidt, G. M. S., 430, 438(23)
Schnaitman, C., 30
Schnippering, W., 174, 177, 288
Schoenborn, B. P., 371
Schollmeyer, J. V., 60
Schreiber, A. B., 264, 265(8)
Schreiber, A. D., 438
Schubert, D., 233
Schuette, W. E., 174
Schuldiner, S., 42, 133
Schutsky, D., 382
Schwartz, E., 229
Schwartz, J. H., 94
Schwartz, M. A., 161, 162(49), 403
Schwartz, R. A., 400, 402(9)
Schwartz, S. A., 233
Schwick, H. G., 468
Scicli, A. G., 371
Scicli, G., 371
Scott, C. F., 362
Scottocasa, G. L., 29
Scrutton, M. C., 44, 131, 132, 181
Scudder, L. E., 175, 229, 284, 294, 295, 296(1), 297(1), 298(1), 310, 458
Scurfield, G., 3
Seachord, C. L., 4
Seaman, F. S., 361, 362, 364(3), 365(3), 366(3), 367(3), 368(3), 369
Searfoss, G. H., 264, 265(8)
Sears, D. A., 413
Seegers, W. H., 155, 156(2)
Seibert, K., 199, 200(45)
Selby, S. M., 325
Sellers, J. R., 81, 86, 87(20)
Senogles, S. E., 264
Senyi, A. F., 157
Setzer, P. Y., 30
Sevy, R. W., 78
Sexton, P. W., 358, 359(44)
Seyama, Y., 133
Shafiq, S., 62, 70(31)
Shaklai, M., 3
Shanon, A. E., 265
Shapiro, H. M., 421
Shapiro, S. S., 265
Sharp, A. A., 201
Shattil, S. J., 132, 183, 184(17, 23), 427
Sheludko, A., 462
Shima, M., 232
Shimizu, T., 133
Shimomura, T., 177, 288
Shimoyama, R., 399, 402(2)
Shreeve, S. M., 183, 185(28)
Shuldiner, S., 213, 222(1)
Shulman, N. R., 428, 430, 431(1, 16)
Shulman, R. G., 41
Shulman, S., 440, 441, 442(2), 443(2, 4), 446(2), 447(4), 448(4), 451(2), 453(12)
Shuman, M. A., 31, 429, 436(14), 438(14)
Siegl, A. M., 373, 379(29)
Siekevitz, P., 29
Siemankowski, R. F., 84
Silk, S. T., 21, 32(7)

Silver, S. M., 229
Simmons, S. R., 456, 458, 471
Simons, E. R., 132, 156, 157, 158(38, 39), 159(19), 161, 162(19, 38, 39, 47), 165, 166(19, 39, 47, 54), 168, 169(27), 171(19, 27, 62), 173(27, 28, 62), 174, 175(77, 78), 176(19), 403
Simpson, M. B., 420, 421(8), 427(8)
Sims, P. J., 136
Singer, S. J., 465
Singh, S., 136
Sinha, D., 361, 362, 364(3), 365(3), 366(3), 367(3), 368(3), 369
Sixma, J. J., 3, 4(1), 144, 243, 286, 433
Skinner, V. O., 398
Skrzydlewski, Z., 398
Slakey, L. L., 134, 136
Slayter, H. S., 31, 293
Slot, J. W., 459, 460(22)
Small, J. V., 123
Smirnov, V. N., 391, 392(12), 394(12)
Smith, A. L., 38
Smith, C. B., 183, 184(19)
Smith, J. R., 76
Smith, S. K., 183, 184(21), 187, 188(21), 190(21), 193(21)
Smith, S. V., 264
Smith, S. W., 174
Sneddon, J. M., 213
Snyder, S. H., 208, 211(12)
Soboeiro, M. S., 81
Sochyinsky, R., 419
Solum, N. O., 276, 277, 290, 292(15), 308, 310(17), 370, 433, 440, 441, 442(1, 13), 443(2, 7, 13), 444(7, 13), 446(2, 9, 10), 447, 448(1, 20), 449(7), 451(2), 452(1), 453(12), 454(1), 455(2)
Sonnenberg, 403
Sonnichsen, W. J., 156, 158(11)
Spaet, T. H., 126, 136
Speck, J. C., 269, 362
Speck, J. L., 187
Spiegel, J. E., 430, 431(22), 433(22), 436(22), 438(22), 439(22)
Spier, D. S., 420, 427(9)
Sprinzel, M., 465
Spudich, J. A., 68, 81
Srivastava, P. C., 137, 140(7), 141(7), 142(7)
Srivastava, S., 413
Srouji, A. H., 312

Staatz, W. D., 135
Stabaek, T., 276, 290, 292(15), 308, 310(17)
Stacher, A., 347
Stackrow, A. B., 282, 284(24)
Stacy, R. S., 201
Stadel, J. M., 186
Staehelin, T., 436
Stamford, N., 354
Stanley, G. H. S., 384
Staros, J. V., 403, 404, 405, 406(20), 408(19, 20)
Steck, T. L., 413, 415(3)
Steele, R. J., 122
Steer, M. L., 134, 136(20), 183, 184(15, 16), 185(16)
Stefanenko, G. A., 60
Steffen, P. K., 42, 49(2), 50(2), 62
Stein, T. M., 76
Steinberg, G., 462
Steinberg, M., 284, 310
Steiner, A. N., 60
Steiner, B., 178, 229, 244, 245(5), 417
Steiner, M., 119
Stenberg, P. E., 31, 50, 51(28), 457
Stendahl, O. I., 95
Stenflo, J., 329
Stevenson, P. C., 459
Stewart, G. J., 38, 39(7), 146
Stewart, G., 78, 85(6), 154
Stilberg, A., 462
Stossel, T. P., 49, 59, 60, 88, 92(5), 95, 96, 97(25), 99(25)
Stracher, A., 59, 61, 62, 63, 64(28, 29), 66(29), 67(29), 68(29), 69(29), 70(31), 71, 72(39), 73, 74(39), 75(28), 76(29), 91, 94(14, 15), 103
Strachurska, J., 277
Stricker, R. B., 429, 436(14), 438(14)
Strong, D. D., 230
Stump, D. C., 183, 184(18)
Suehiro, H., 137, 139(8)
Sugimoto, M., 264, 265(8)
Sugo, T., 370, 371
Sullender, J. S., 76, 99, 101(4), 121
Sullivan, C. A., 175, 229, 294, 295, 296(1), 297(1), 298(1), 310(1), 458
Sundkvist, I., 60
Surrey, S., 229
Suttie, J. W., 329
Suzuki, A., 60

Suzuki, M., 288
Suzuki, T., 370, 371
Swartz, D., 156, 174(29)
Sweatt, J. D., 132, 182
Switzer, M. E., 265
Sykes, J. E. C., 384
Syman, J. W., 463
Szewczyk, B., 436

T

Taatjes, H., 440
Tack, B. F., 162, 170(53)
Tager, J. M., 37
Tait, A. T., 217, 218(14)
Takamatsu, J., 174, 175(83), 403
Takemoto, M., 288
Takenishi, T., 139
Takio, K., 276, 284(7), 285(7)
Talvenheimo, J., 216, 217(9), 221
Tam, S. W., 33, 156, 158(17), 159(17)
Tandon, N. N., 157, 174(35)
Tandon, N., 156, 157
Tanford, C., 463
Tang, S. S., 427
Tangen, D., 470
Tangen, O., 372, 383
Tani, P., 430, 438(23), 439
Tanoue, K., 174, 175
Tans, G., 317, 319(5), 324(5), 328(5)
Taswell, J. B., 316, 317(3), 319(3), 342, 343(31), 344(31), 338(31)
Tavanheimo, J., 218, 219
Tavassoli, M., 3, 118
Taves, C., 448
Taylor, G. M., 465
Taylor, L. G., 122
Taylor, S. S., 144
Tellam, R., 88, 95(10)
ten Cate, J. W., 430, 431(26), 438(26)
Tenner, T. E., Jr, 157, 174(35)
Tenoue, K., 433
Terry, B. E., 183, 184(13)
Tharp, M. D., 183, 184(13)
Thomas-Maison, N., 161
Thorpe, D. S., 136
Thorstensson, R., 43
Timmons, S., 44, 47(14), 169, 229, 230, 233, 234(32), 237(34), 240, 241, 372, 407, 456
Tischler, M. E., 37

Titani, K., 276, 284(7), 285(7), 288
Tobelem, G., 419, 438
Todrick, A., 217, 218(14)
Toh, H., 133
Tolbert, M. E. M., 183, 184(22)
Tollefsen, D. M., 156, 157, 158(12)
Tomaski, A., 438
Tomminia, S. J., 465
Towbin, H., 436
Tracy, P. B., 316, 321(7), 324(7), 329, 330, 331, 332, 334, 335, 336(13), 337(13, 28), 338(15, 6), 339(6), 341(13), 342(6), 343(6, 23), 344(14), 345, 346(14), 347(14), 348(26), 349(15), 350(15), 353(14, 17, 23), 355(16, 17), 356(16), 358, 359(14), 360, 375
Tracy, R. P., 317
Tranzer, J. P., 37
Travis, J. L., 471
Treuting, J., 375
Triantaphyllopoulos, D. C., 33
Triplett, D. A., 372
Troll, J. H., 427
Tropier, G., 397
Trucco, M., 306
Truglia, J. A., 63, 347
Trzeciak, M.-C., 174
Tsai, B.-S., 183, 184(11), 187, 188(36)
Tsang, P., 301
Tsang, V. C. W., 435
Tshovrebova, L. A., 60
Tsien, R. Y., 174
Tsuji, T., 286, 290
Tsunehisa, S., 290
Tsuzynski, G. P., 154
Tucker, M. M., 317, 332, 358, 360
Turitto, V. T., 232
Turkevich, J., 459
Turnbull, J., 183, 184(23)
Turner, J. T., 183, 185(26)
Tuszynski, G. P., 78, 134, 361, 362, 369(6), 413
Tyler, D. D., 38

U

U'Prichard, D. C., 183, 184(20)
Ugurbil, K., 39, 41
Uhteg, L. C., 168
Uhteg, R. C., 331

Ui, M., 181
Umfleet, C. A., 313

V

Vaheri, A., 156, 158(15), 159(15), 167(15), 176, 431, 432(34)
Valeri, C. R., 41
Valone, F. H., 225, 226
van der Hart, M., 430, 431(20)
Van der Meer, R., 37
van der Ven, J. T. M., 439
van der Weerdt, C. M., 430, 431(25)
Van Dilla, M. A., 421
van Halbeek, H., 286
Van Iwaarden, F., 382
van Leeuwen, E. F., 439
van Loghem, J. J., 430, 431(20)
Varon, D., 441, 448(14), 452(14)
Vauthey, M., 463
Vecchione, J. L., 41
Venter, J. C., 183, 185(28)
Verheught, F. W. A., 430, 431(26), 438(26)
Verklejj, A. J., 465
Vetter, W., 398
Vicente, V., 278
Vilaire, G., 132, 143, 144(4), 161, 229, 253
Vincente, V., 230, 232
Vogel, C. N., 168, 331
Vogt, E., 398
von dem Borne, A. E. G. Kr., 430, 431(26), 438(26), 439
von Riesz, E., 430, 431(26), 438(26)
Vu, T. K. H., 175
Vu, T. K., 131
Vukovich, Th., 398

W

Wachtfogel, Y. T., 382
Wachtfogel, Y., 148
Wagner, R. H., 291, 421
Wajerus, P. W., 156
Waldmann, G. R., 371
Walker, F. J., 358, 359(44)
Wallace, W., 287
Walsh, C. M., 429
Walsh, J. J., 135

Walsh, P. N., 271, 329, 361, 362, 364(3), 365(3), 366(3), 367(3), 368(3), 369, 370, 372(6), 382
Walther, P., 479
Walz, D. A., 31
Wang, K., 94, 99, 100, 101(3), 408
Wang, L.-L., 60, 88, 92(1, 7), 94(7), 95(1, 7)
Wang, W. C., 420, 427(3)
Waples, L. M., 475
Wardell, M. R., 287
Ware, J., 286
Wasiewski, W. W., 156
Watanabe, T., 133
Watanabe, Y., 290
Watt, S., 68, 81
Wattiaux, R., 30
Wautier, J. L., 430, 431(27), 438(27)
Way, M., 96
Weber, G., 384
Weber, K., 126, 152, 172
Wechter, W. J., 144
Weeds, A. G., 43, 60, 70(18), 94, 95(19), 96
Weil, G. J., 174, 175(77, 78)
Weiland, G. A., 227
Weiland, T., 118
Weinstein, D., 101
Weintroub, H., 10, 13(3), 15(3)
Weiser, H. B., 459
Weisgraber, K. H., 248
Weiss, H. J., 232
Weksler, B., 419
Wenc, K. M., 298
Wencel-Drake, J. D., 31, 427
Wester, J., 3, 4(1)
Westheimer, F. H., 159, 165(45)
Westmoreland, N. P., 21, 27(1), 28(1), 30(1), 31(1), 32(1)
Wetzel, B., 479
Whatley, R. E., 133
Wheaton, V. I., 131, 175
Whigham, K. A. E., 145
White, G. C., 60
White, G. C., II, 155, 156, 157, 158, 168, 176(13)
White, J. G., 4, 31, 43, 59, 60, 109, 110, 111, 113, 114(12), 116, 117, 119, 120(11), 121, 122, 271, 402, 457
White, M. S., 122
Wicki, A. N., 174, 175(86), 288
Widmann, F. K., 33
Wiedmer, T., 136

Wietzerbin, J., 403
Wiggins, R. C., 370
Wild, F., 174
Williams, L. T., 183, 184(9), 204
Williams, S. A., 276, 284(9)
Williamson, J. R., 37
Wilson, A. L., 187, 199, 200
Wilson, W. E., 430
Wold, F., 403
Wolpers, C., 109, 110(4)
Wong, K. T. H., 21, 32(7)
Woo, J., 375
Woods, V. L., 439
Woods, V. L., Jr., 312, 439
Workman, E. F., 157
Workman, E. F., Jr., 155, 156, 168, 176(13)
Worowski, K., 398
Wright, B. L., 183, 184(18)
Wrobleski, F., 28
Wu, C., 291
Wu, Q. X., 134, 154
Wurtman, R. J., 30
Wyatt, J. L., 144, 145(9)
Wyatt, J., 144
Wyler, B., 287

Y

Yamada, K. M., 311
Yamaguchi, A., 433
Yamamoto, K., 174, 175, 290, 433
Yamamoto, N., 175, 433
Yamazaki, H., 174, 175, 433
Yanase, R., 137, 139(8)
Yang-Fang, T. L., 132
Yankee, R. A., 431
Yeo, E. L., 4, 31, 427
Yeo, K.-T., 290
Yin, H. L., 49, 60, 88, 92(5), 95(5, 8), 96, 97(25), 99(25)
Yoshida, N., 286, 419
Yoshikawa, M., 139
Yoshioka, A., 232

Z

Zaaishvilia, M. M., 60
Zaner, K. S., 60, 88, 95(8)
Zhou, F., 154
Zimmerman, G. A., 133
Zimmerman, T. S., 232, 263
Zito, R. A., Jr., 214
Zola, H., 174, 175(82), 278, 296, 298(5), 301(5), 312(5), 419
Zoller, M. J., 144
Zucker, M. B., 21, 32(5, 6), 126, 155
Zucker, M., 407
Zuichers, C. A., 122
Zurendock, P. F., 37
Zwaal, R., 317, 319(5), 324(5), 328(5)

Subject Index

A

N-Acetyl-β-glucosaminidase, as marker for acid hydrolase-containing organelles, 28, 30, 39–40
Acid hydrolase-containing organelles, localization of, enzymes used for, 28, 30
Acid hydrolase vesicles, marker enzyme, 39–40
Acquired immunodeficiency syndrome, 429
Actin, 284
  bundling of, 69
  depolymerization, effect of actin-binding protein and myosin on, 74–76
  globular state. *See* G-actin
  immunofluorescence microscopy of, 126
  of platelet membrane, 17–18, 143
  polymerization, in platelet activation, 42–43, 56
  purification of, 66–70, 76
    by bundling, 68
    effect of platelet activation on, 64
    isolation of platelet cytoskeleton for, 62–64
    preparation of platelets for, 60–61
  recombination of actin-binding protein and α-actinin with, 73–74
α-Actin, immunofluorescence microscopy of, 126
Actin-binding protein, 42–43, 59–60, 284, 288
  column purification of, 66–67
  effect on actin depolymerization, 74–76
  of platelet surface membranes, 18
  purification of, 64–66, 76
    effect of ionic strength on, 62–63
    effect of platelet activation on, 64
    isolation of platelet cytoskeleton for, 62–64
    preparation of platelets for, 60–61
  recombination with actin, 73–74
    competition with α-actinin, 73–74
    effect of actin state on, 73
    effect of Ca^{2+} on, 73
    effect of ionic strength on, 73
Actin dilution effect, 77
Actin filaments
  depolymerization of, 52
  in platelet lysates, assays of, 54–58
  in platelets, 42, 112, 117
  effect of centrifugation on, 77
  proteins associated with, identification of, 50–54
Actin–gelsolin interactions, assays of, 94–95
α-Actinin, 43, 59–60
  column purification of, 70, –72
  platelet, recombination with actin, 73–74
    competition with actin-binding protein, 73–74
  of platelet surface membranes, 17–18
  purification of, 70–72, 76
    isolation of platelet cytoskeleton, 62–64
    preparation of platelets, 60–61
  recombination with actin
    competition with actin-binding protein, 73–74
    effect of actin state on, 73
    effect of Ca^{2+} on, 73
    effect of ionic strength on, 73
  skeletal muscle, recombination with actin, 73–74
    competition with actin-binding protein, 73–74
Actin microfilaments, 6–7, 112–113
  isolation of, 59
Actin shell
  of platelet, 116–117
  of thrombin-activated platelet, 117–118
Actomyosin, platelet, 59
Adenosine diphosphate, 131, 236
  binding sites for, on platelets, 143–144, 155
  carbon-14, high-affinity sites for, on platelets, 143

in dense granules, 41
effect on inositol phosphate production, 182
effect on platelets, 137, 143
platelet receptor, 132, 134, 144
  FSBA interaction with, 143–155
platelet shape change mediated by, 142
  effect of FSBA on, 147–151
  effect of shape change inhibitors on, 151
released by thrombin, 41
role of, 229
in whole platelets, 41
Adenosine triphosphatase
  Ca^{2+}, Mg^{2+}-, of platelet membranes, 16, 19–20
  K^+-EDTA, of platelet myosin, 81–82
  Mg-, actin-activated, of platelet myosin, 81–82
Adenosine triphosphate
  in dense granules, 41
  released by thrombin, 41
  in whole platelets, 41
Adenylate cyclase, in human platelet surface and intracellular membranes, 16
α_{2A}-Adrenergic receptor, 199–200
α_{2B}-Adrenergic receptor, 199
  yohimbine-agarose chromatography of, 199
α_{2C}-Adrenergic receptor, 199
α_2-Adrenergic receptor agonist, interactions in detergent-solubilized preparations, altered properties of, 193–194
Adrenergic receptors, 133
α_2-Adrenergic receptors, 131
  digitonin solubilization of, 189–191, 193–194
    isolation of platelets and preparation of washed lysates for, 187–189
  human platelet, 181–200
    affinity for agonist and antagonist agents, effectors that modulate, 186–187
    digitonin-solubilized
      characteristics of, 192–193
      [^{3}H]yohimbine-binding assays of, 192
      identification of, 194
    high-affinity state, 186
    identification of, 183–187
    commercially available radioligands for, 184–185
    in digitonin-solubilized preparations, 191–192
    low-affinity state, 186
    yohimbine–agarose affinity chromatography of, 197–198
      synthesis of yohimbine–agarose matrix for, 194–197
    from metabolically labeled cells, micropurification of, by yohimbine-agarose chromatography, 199–200
  opossum kidney, 199
  porcine brain, yohimbine–agarose affinity chromatography of, 197–198
  role in activation of GPIIb–IIIa, 131–132
  subtypes, yohimbine–agarose chromatography of, 199
Affinity chromatography, of glycocalicin and/or GPIb, 310–311
Affinity crossed immunoelectrophoresis
  with antigen ligands, 447–448
  of human platelet membranes, methodology, 455
Aggregin, 134, 154–155
Albumin
  crossed immunoelectrophoresis of, 445
  human, with platelet membranes, crossed immunoelectrophoresis of coelectrophoresis of, 445
Aldolase, rabbit muscle, cross-linking of, 406–407
  assay, 408–409
Alloantibodies, platelet destruction by, syndromes of, 430–431
Alloantigens, platelet-specific, 431, 437–438
Amidolytic assay, of factor XI, 362
p-Amino[^{3}H]clonidine, 184
Amitriptyline, inhibition of binding of [^{3}H]LSD to human platelet membranes by, 211–212
ANPH-thrombin complex, derivatization and cross-linking of, 165
ANPH-TLCK-thrombin complex, approximate molecular size and reduced components of, 160
Antibody. *See also* Monoclonal antibody
  associated with quinidine purpura, 438

to platelets, 428
polyclonal
  to GPIb, 175
  to platelet antigens, screening of, 420
quinidine, 429
quinidine-dependent, 429
Antigen 10, 443
  relationship with antigens 13 and 18, on CIE of normal human platelet membranes, 446
Antigen 13, relationship with antigen 10, on CIE of normal human platelet membranes, 446
Antigen 16, 443
Antigen 18, relationship with antigen 10, on CIE of normal human platelet membranes, 446
Antigens
  platelet
    fluorescence flow cytometry analysis of, 420–427
    immunoblotting of, 428–440
    monoclonal and polyclonal antibodies against, screening of, 420
    platelet surface, relative surface location of, 448–451
    public, 429
Apolipoprotein A-I, specificity, 261
Apparent dissociation constant, 344
Aprotinin, in protease inhibitor cocktail, for platelet suspensions, 10
Apyrase
  effect on ADP- and azo-PGH_2–mediated platelet shape change, 151
  effect on platelets, 154
Arachidonic acid, 112
Ascites, production of, 301
Asialo-vWF, 264
ATPase. See Adenosine triphosphatase
Autoantibodies
  in autoimmune thrombocytopenia, 438–440
  to GPIIb–IIIa, 439
  to GPIIIa, 438–439
  HIV-associated, 429
  platelet destruction by, syndromes of, 428–430
Autoimmune thrombocytopenia. See Idiopathic thrombocytopenia purpura
2-Azidoadenosine diphosphate, 137

p-Azido[^{3}H]clonidine, 185
Azo-prostaglandin H_2, platelet shape change mediated by, effect of shape change inhibitors on, 151

B

Bacteriorhodopsin, specificity, 261
BASIC (computer program), subroutine for estimation of B (fraction of bound substrate) by bisection algorithm, 325
Bernard–Soulier syndrome, 3–4, 276, 288–289. See also Giant platelet syndrome
N,N'-Bis(2-nitro-4-azidophenyl)cystamine S,S-dioxide. See DNCO
Bis (sulfosuccinimidyl) suberate
  effect on collagen-induced platelet aggregation, 410
  effect on platelet aggregation, 408–409
  effect on SDS-polyacrylamide gel profile of platelet surface glycoproteins, 409–410
  structure of, 404
Blood collection, for platelet isolation, 8–9
Blood group antigens, identification of, 420
Bolton–Hunter assay, of LDL binding to platelets, 392
Botrocetin, 264
Bovine serum albumin
  adsorbed to colloidal gold, behavior of, 462
  conjugation of, to colloidal gold, 469
Bromo-LSD, inhibition of binding of [^{3}H]LSD to human platelet membranes by, 211
BS^3. See Bis (sulfosuccinimidyl) suberate
Buffy coat packs, processing, for platelet isolation, 9
(+)-Butaclamol, inhibition of binding of [^{3}H]LSD to human platelet membranes by, 210–212
(−)-Butaclamol, inhibition of binding of [^{3}H]LSD to human platelet membranes by, 210–212

C

Calcium
  in dense granules, 36, 41
  effect on CIE pattern from platelet membranes, 447

released by thrombin, 41
  in whole platelets, 41
Calcium ionophore A23187, 136
Calcium mobilization, in platelets, 20
Calmodulin, 60
Calpain, 277
Carbon-14 labeling, 162–164
Catalase, adsorbed to colloidal gold, behavior of, 463
CBP, of platelet intracellular membranes, 19
Cell surface labeling, 165
Cell suspensions, preparation of, 362
Centrifugation
  effect on actin filaments in platelets, 77
  effect on platelet cytoskeleton, 77
  of platelets, 45–47
Chlorimipramine, inhibition of [^{3}H]5-HT receptor binding by, 205
Chlorpromazine, inhibition of [^{3}H]5-HT receptor binding by, 205, 212
Cholesterol, in human platelet surface and intracellular membranes, 16
Cholesterol–phospholipid molar ratio, in human platelet surface and intracellular membranes, 16
Clones, freezing, 301
Clonidine, 181
  ^{3}H-labeled, 184
Clot retraction, 119–120
Clotspeed (computer program), 316–328
Coagulation proteins, assays of, 362
Colchicine-binding assay, of platelets, 119
Collagen, 131, 136
  platelet receptor for, 135–136, 245
Colloidal gold, 459–465
  conjugation of bovine serum albumin to, 469
  conjugation of Fab anti-fibrinogen to, 470
  conjugation of heterologous immunoglobulin to, 468–469
  conjugation of MAb 10E5 to, 469–470
  conjugation of proteins to, 464–465
  macromolecules adsorbed to, behavior of, 462–464
  preparation of, 460–461
  stability of, 461–462
Colloidal gold labeling
  with fibrinogen, 476–477
    control preparations, 477
  of live platelets, 471
  with monoclonal antibody, control preparations, 477
  of platelet receptors, 456–479
    anti-IIb/IIIa–gold procedure, 468–470
  of platelets, 470–478
    electron microscopy after, 478–479
    video-enhanced light microscopy of, 471–475
  of platelets in suspension, 478
  of prefixed platelets, 475–577
Colloidal gold particles, size distribution of, 461
Concanavalin A-Sepharose affinity chromatography, of lysed platelets, 247–248
Contractile apparatus. See Platelet contractile apparatus
CP/CPK
  effect on ADP- and azo-PGH$_2$–mediated platelet shape change, 151
  effect on platelets, 154
Crossed immunoelectrophoresis
  affinity
    with antigen ligands, 447–448
    of human platelet membranes, methodology, 455
  of coelectrophoresis of human albumin with platelet membranes, 445
  crossed hydrophobic interactive immunoelectrophoresis in, 454
  of human platelet membranes, 440–455
    after absorption of rabbit anti-platelet membrane antibody, 453
    assay for intrinsic biologic activity, 443–446
    with concanavalin A, 451
    effect of EDTA, EGTA, sodium citrate, and Ca^{2+} on, 447
    immunoprecipitate arcs
      antigen–antibody, sensitivity of, 446
      quantitation of, 446–447
    methodology, 454–455
  of platelets, 442, 444–445
    incubation of immunoplates with ^{125}I-labeled fibrinogen after, 449
  of rabbit anti-human platelet membrane antibody, quantitation of purified fibrinogen by, 450
Cross-linkers, membrane-impermeant, 404

Cross-linking
  of platelet membrane glycoproteins, 403–412
    experimental procedures, 405–408
    results, 408–410
  of platelet surfaces, 411–412
  complications of, 404
  of rabbit muscle aldolase, 406–407
    assay, 408–409
    spin, 405
  of thrombin, experimental procedure, 162–165
CTA_2, effect on ADP- and azo-PGH_2-mediated platelet shape change, 151
Cyclic AMP, synthesis, α_2-adrenergic receptor-mediated inhibition of, platelet model of, 181
Cyclooxygenase, of platelet membranes, 16, 19
Cyproheptadine, inhibition of binding of [^{3}H]LSD to human platelet membranes by, 211
Cytochalasin B, effect on platelets, 119–121
Cytocontractile apparatus. *See* Platelet contractile apparatus
Cytoskeletal proteins, platelet, 3
Cytoskeletal structures, platelet, 6–7
Cytoskeleton
  electrophoresis of, 53
  immunocytochemical labeling of, 50–51
  myosin-containing, 62
  platelet, 3, 6–7, 59
    effect of centrifugation on, 77
    effect of cold on, 77
    effect of local anesthetics on, 76
    effect of thrombin on, 64
    isolation of, 61–64, 76–77
      different solutions for, 76–77
    effect of platelet activation on, 64
    protein composition of, under various conditions, 63
    in Triton X-100 lysates, 42–58
  platelet contractile. *See* Platelet contractile apparatus
  polypeptides isolated with, identification of, 50
  Triton X-100-insoluble, isolation of, 48–49
Cytosol, platelet, localization, 28

D

6D1 affinity column, preparation of, 309
DAPA, prothrombin activation experiments monitored with, 342–344
DAPA–thrombin complex, 342
DEAE-cellulose chromatography, of P235, 102–105
DEAE-Sephacel chromatography, of gelsolin–actin complexes, 89, 92
DEAE-urea chromatography, 90
  of gelsolin, 93–94
Decay-accelerating factor, identification of, 420
Dendritic platelets, whole-mounted, 125
Dense bodies
  labeling of, 31
  separation of, by ultracentrifugation, after nitrogen cavitation, 27
Dense granule membrane vesicles, 221–224
  purified, 222–223
  serotonin transport into, assay for, 222–224
Dense granules, 6–7
  constituents of, 36, 41–42
  isolated, purity of, 40
  isolation of, 36–42
    yield, 40
  marker enzyme, 39–40
  nucleotide–divalent cation–amine interactions in, 41
  properties of, 41–42
  serotonin uptake and storage, 41–42
  subfractionation, on sucrose density gradient, 39–40
  sucrose gradient-enriched, marker enzymes in, 40
Dense tubular membrane system, 18–19
  platelet, 6–7
Detergent extraction, 122
Diacylglycerol lipase, of platelet membranes, 19
Diazepam, 212
Diazotized diiodosulfanylic acid labeling, of platelet surface proteins, 413
Diglyceride lipase, in human platelet surface and intracellular membranes, 16
Dihydroergonine, ^{3}H-labeled, 184
Dihydroeryocryptine, ^{3}H-labeled, 184
  as radioligand, 191–192
Diisopropyl fluorophosphate, 156–157
  as protease inhibitor, 23

3,3'-Dithiobis(sulfosuccinimidyl propionate)
  effect on collagen-induced platelet aggregation, 410
  effect on platelet aggregation, 408–409
  structure of, 404
  treatment of platelets with, 407
3,3'-Dithiobis(sulfosuccinimidyl propionate) disodium salt, synthesis of, 406
DNase I–agarose chromatography, of gelsolin–actin complexes, 89, 91–92
DNase I inhibition assay, of actin filaments, 54–56, 58
DNCO
  identification of thrombin–receptor complexes with, 166
  structure of, 166
  synthesis of, 162, 168–169
DNCO-thrombin, derivatization and cross-linking of, 165
DNCO–α-thrombin
  approximate molecular size and reduced components of, 160
  derivatization and cross-linking of, 165
  identification of thrombin–receptor complexes with, 165–166
DNCO-TLCK-thrombin
  approximate molecular size and reduced components of, 160
  derivatization and cross-linking of, 165
  identification of thrombin–receptor complexes with, 165–166
Dodecapeptide (γ400–411)
  effect on ^{125}I-labeled vWF binding as compared with ^{125}I-labeled fibrinogen binding, 241
  effect on ristocetin-induced binding, 241
Dopamine, inhibition of binding of [^{3}H]LSD to human platelet membranes by, 210–211
Dopamine D_2 receptors, 133
DTSSP. See 3,3'-Dithiobis(sulfosuccinimidyl propionate)

E

EDTA. See Ethylenediamine tetraacetic acid
EGTA
  effect on CIE pattern from platelet membranes, 447
  effect on platelet cytoskeleton, during isolation, 63–64
  effect on platelet suspension, in isolation technique, 23
Electron-dense bodies, platelet, 112
Electron microscope, 110
Electron microscopy. See also Scanning electron microscopy; Transmission electron microscopy
  characterization of glycocalicin by, 292–293
  of gold labeled platelets, 478–479
Endoplasmic reticulum, platelet, 19
  markers for, 28–30
Enzyme-linked immunosorbent assay, of glycocalicin, 293–294
Enzymes, adsorbed to colloidal gold, behavior of, 463
Epinephrine, 131, 236
  effect on inositol phosphate production, 182
  ^{3}H-labeled, as radioligand, 184, 186, 194
  platelet stimulation by, 181–182
  role of, 229
  in activation of GPIIb–IIIa, 132
Ergotamine, inhibition of binding of [^{3}H]LSD to human platelet membranes by, 211
Ethylenediamine tetraacetic acid
  effect on CIE pattern from platelet membranes, 447
  effect on platelet cytoskeleton, during isolation, 62
  effect on platelet membrane sorbitol density gradient fractionation, 13–15
  effect on platelets, during isolation, 23, 47
N-Ethylmaleimide, 277

F

4F8, 97
Fab fragments, of anti-fibrinogen
  conjugation to colloidal gold, 470
  gold particles coupled to, labeling of fibrinogen molecules bound to platelet surfaces with, 476–477
  preparation of, 470
F-actin, 117

… SUBJECT INDEX … 507

Factor V, 136
  binding interactions, 317
    with bovine platelets, 330
  binding parameters of, 336
  bovine, radiolabeling of, 332–333
  human, purification of, 353
  ^{125}I-labeled, binding to bovine platelets, measurement of, 334–336
Factor Va, 329–330
  binding interactions, 317
    with bovine platelets, 330, 336
    with factor V-deficient platelets, kinetic determination of, 355–358
    with platelets
      autoradiography of, 350
      as determined by direct binding measurements, 349
      kinetic determination of, 341–344
    with platelets and factor Xa, stoichiometry of, kinetic determination of, 355–357
  binding sites on platelets, functional, kinetic determination of, 343
  binding to platelets, mediated through component E, 348–351
  binding to platelet surface, 359–360
  bound to platelets
    activated protein C and factor Xa-induced cleavages in, 347
    electrophoresis and autoradiography of, 349
    proteolysis of, induced by complex formation with activated protein C and factor Xa, 345–347
    visualization of, 344–351
    autoradiographic, 346
  bound to unstimulated platelets, bound factor Xa and, ratio of, as function of factor Xa concentration, 340
  bovine
    binding interactions, with bovine platelets, 331–341
    radiolabeling of, 333
  component D, 358
    cleavage of, 358–359
  component D', 348
  component E, 348–349, 358
    binding of factor Va to platelets mediated through, 348–351
    factor Xa-induced cleavages of, autoradiography of, 352

    as part of receptor for factor Xa at bovine platelet surface, 351
  EDTA-inactivated, binding interaction with platelets, autoradiography of, 350
  human, binding parameters, 330–331
  ^{125}I-labeled
    binding to bovine platelets, measurement of, 334–336
    binding to unstimulated bovine platelets, coordinate with factor Xa binding, 338–340
  bound to platelets
    concentration-dependent binding of ^{125}I-labeled factor Xa to, 339
    electrophoresis and autoradiography of, 344–345
  platelet receptor for, at bovine platelet surface, component E as part of, 351
Factor X
  binding interactions, 317
  human, purification of, 353
Factor Xa, 330
  binding interactions, 317
    with bovine platelets, 330
    with component E bound to platelets, 352
    with normal thrombin-activated human platelets, kinetic determination of, 354–356
    with platelets
      derivation of expression for, 340–341
      parameters of, calculation of, 342
      requirements for, 329
    with platelets and factor Va, stoichiometry of, kinetic determination of, 355–357
    with thrombin-activated factor V-deficient platelets, kinetic determination of, 355–357
  binding to platelets, requirements for, 370
  bound to unstimulated platelets, and bound factor Va, ratio of, as function of factor Xa concentration, 340
  bovine
    binding interactions, with bovine platelets, 331–341
    radiolabeling of, 333–334

cleavages in platelet-bound factor Va induced by, 347
complex formation with activated protein C and platelet-bound factor Va, proteolysis of platelet-bound factor Va by, 345–347
human, binding parameters, 330–331
^{125}I-labeled
  binding to platelet-associated ^{125}I-labeled factor Va, concentration-dependent, 339
  binding to unstimulated bovine platelets, 336–338
    coordinate with factor Va binding, 338–340
  platelet receptor for, 329–360
    human, 351–357
    visualization of, 344–351
Factor XI
  activation of, 370
    by platelets, 361
  binding to platelets
    time course for, 382
    zinc and calcium ionic concentration dependence of, 382
  ^{125}I-labeled, binding to platelets, 381
  radiolabeling of, 362
Factor XIa
  amidolytic assay, 362
  binding to human platelets, 361–369
    affinity of, determination of, 368
    assay of, validation of, 363–364
    characteristics of, 365–369
    measurement of, 363
    saturability and reversibility of, 367–368
    sites for, determination of number of, 368
    specificity of, 366–367
    time course and requirements for, 365–366
  ^{125}I-labeled
    autoradiography of, 365
    binding to platelets
      progress curves of, 366
      saturable, 367
    binding to thrombin-treated platelets, 364
    SDS gel electrophoresis of, 365
  platelet-bound, characterization of, 368–369

platelet receptor for, 361–369
radioimmunoassay, 362
Factor XII
  activation of, 370
    by platelets, 361
  competition for binding of ^{125}I-labeled high molecular weight kininogen with, 377
Factor XIII
  activity, after CIE of platelets, 450
  crossed immunoelectrophoresis of, 445
Fatty acyl-CoA transferase, in human platelet surface and intracellular membranes, 16
Fibrinogen, 132
  antibody against, Fab fragments of conjugation to colloidal gold, 470
    preparation of, 470
  association with solubilized platelet membrane, effect of ^{125}I surface labeling of washed platelets on, after CIE, 453
  binding domains, 230
  binding to GPIIb–IIIa complex, 228–243
  binding to platelet receptors, 232–233
    ADP-mediated, 154
    effect of FSBA on, 148–151
    measurement of, 148–150
  binding to platelets, 229, 245
  binding to platelet surface, study of, 420
  bound to platelet surface, gold labeling of, with anti-fibrinogen Fab fragments, 476–477
  conjugation to colloidal gold, 466–468
  crossed immunoelectrophoresis of, 445
  GPIIb–IIIa receptor, 135
  in α granules, 30
  heparin-Sepharose affinity chromatography of, 247–248
  human, purified, quantitation of, by CIE of rabbit anti-human platelet membrane antibody with plasma fibrinogen, 450
  ^{125}I-labeled
    binding to platelets, 146–147, 238–243
      concentration-dependent, 238–239
      correlation with platelet aggregation, 242
      inhibition by dodecapeptide, 241
      inhibition by vWF, 240
      measurement of, 238

SUBJECT INDEX

platelet-bound, dissociation by FSBA, 150–151
synthesis of, 146
inhibition of ^{125}I-labeled vWF binding to human platelets by, 240
interaction with human platelets, 228–230
model of, 231
labeling of, 419
plasma, purified, quantitation of, by CIE of rabbit anti-human platelet membrane antibody with plasma fibrinogen, 450
platelet membrane, relative surface location of, 451
platelet receptor for, 132, 245, 370
identification of, by colloidal gold labeling, materials and methods for, 465–468
platelet storage site for, 21
preparation of, 465–466
purification of, 233–234
purified
binding to human platelets
assay, 237
measurement of, 233
cautions, 234
materials and methods for, 233–237
iodination of, 235–237
Fibrinogen–gold conjugates
concentration-variable isotherms for, 466–468
pH-variable isotherms for, 466–467
preparation of, 468
Fibrinogen–gold labeled platelets
high-voltage transmission electron microscopy of, 472–474
light microscopy of, 472
low-voltage high-resolution scanning electron microscopy of, 472–473
Fibrinogen–gold labeling, control preparations for, 477
Fibrinogen molecules, model of, 231
Fibronectin
binding to platelets, 311–316
assays of, 314–316
data analysis, 315–316
binding to platelet surface, study of, 420
platelet receptor for, 135, 245
platelet storage site for, 21

preparation of, 313
radiolabeling of, 314
Filamin, 42, 100
degradation of, 99–100
immunofluorescence microscopy of, 126
of platelet surface membranes, 17
Filopodia, 5
Flow cytometry
applications of, 420
platelet analysis by, 423–426
data analysis and system calibration, 426
determination of appropriate sizing gate for, 424
fluorescent staining, 421–422
immunofluorescent staining, 422
preparation of platelets for, 421–422
without fixation and in whole blood, 426–427
Fluorescence flow cytometry. *See* Flow cytometry
Fluorescent actin nucleation assay, of gelsolin, 88
5'-p-Fluorosulfonylbenzoyladenosine, 144–145
effect on ADP-mediated platelet shape change and aggregation, 147–151
effect on azo-PGH$_2$–mediated platelet shape change, 151
effect on epinephrine-induced platelet aggregation and fibrinogen binding, 154
effect on fibrinogen binding to platelet receptors, 148–151
effect on platelets, 154
interaction with platelet ADP receptor, 143–155
methods and results, 145–154
labeling of platelets with, 146
synthesis of, 145–146
5'-p-Fluorosulfonylbenzoylguanidine, 154
Fluoxetine
inhibition of binding of [^{3}H]LSD to human platelet membranes by, 211
inhibition of [^{3}H]5-HT receptor binding by, 205
α-Flupenthixol
inhibition of binding of [^{3}H]LSD to human platelet membranes by, 210–211

inhibition of [³H]5-HT receptor binding
   by, 205–207
β-Flupenthixol
   inhibition of binding of [³H]LSD to
      human platelet membranes by, 210–
      211
   inhibition of [³H]5-HT receptor binding
      by, 205–207
Formaldehyde-treated platelets, GPIb on,
   fluorescence flow cytometry analysis
   of, 423
Free flow electrophoresis
   of mixed membrane fractions, 11–15
   of platelet membranes, 8
Freeze-thawing, purification of glycocalicin
   by, 290

G

8G5, 97
G-actin, 117
β-Galactosidase, as marker for acid hydrolase-containing organelles, 28, 30
GDP, 134
Gelatin-Sepharose column, preparation of,
   312–313
Gel filtration
   of glycocalicin, 291
   of human platelet protein P235,
      105–108
   of platelets, 44–45
   in sodium dodecyl sulfate, of GPIb–IX,
      282–283
   of thrombin–receptor complex,
      172–173
Gelsolin, 60
   plasma, purification of
      with immunoreagents, 98–99
      large-scale, 96–97
   platelet, 88–99
      activity assays of, 94–95
      assays of, 88–89
      isolation of, 89–94
         by DEAE-urea chromatography,
            93–94
         by ion-affinity chromatography, 98
         strategies for, 95–99
         yields, 94
      of platelet surface membranes, 17–18

purification of
   by immunoaffinity chromatography,
      95–99
   by ion-exchange chromatography, 95–
      96
Gelsolin–actin complexes
   activity assays of, 95
   isolation of, 89–94
      DNase I–agarose method, 89, 91–92
      by immunoaffinity chromatography, 99
      by ion-exchange gel-filtration and
         hydroxylapatite chromatography,
         92–93
Giant platelet syndrome, 230–232. See also
   Bernard–Soulier syndrome
Glanzmann's thrombasthenia, 232, 245,
   417, 443–444
   platelet membranes of patient with,
      crossed immunoelectrophoresis of,
      446
Glucose-6-phosphatase, Triton-inhibited,
   as marker for endoplasmic reticulum,
   28, 30
β-Glucuronidase, as marker for acid hydrolase-containing organelles, 28, 30
Glutamate dehydrogenase, assay, for
   localization of mitochondria, 28, 30
Glutaraldehyde, 118
Glycerol lysis, of human platelet plasma
   membranes, 32–36
   buffers for, 35–36
   procedure, 33–35
   yield, 36
β-Glycerophosphate, as marker for acid
   hydrolase-containing organelles, 28, 30
α-Glycerophosphate oxidase, as marker for
   mitochondria, 39–40
Glycocalicin, 174, 276, 289–294
   carbohydrate portion, 292
   characterization of, 285–286, 292
      by electron microscopy, 292–293
   compositional analysis of, 293
   formation of, 277
   isolation of, by ELISA, 293–294
   purification of, 290–292
      by affinity chromatography, 310–311
   relationship with GPIb, 289–290
   structure of, 290–291
Glycocalyx, 111–112
Glycogen, platelet, 112
Glycophorin, specificity, 261

SUBJECT INDEX

Glycoprotein I, 156
Glycoprotein Ia
  of platelet surface membranes, 17–18
  proteins of, 432
Glycoprotein Ib, 53, 135, 174–176, 276–289, 438. *See also* glycoprotein Ib–IX complex
  affinity purification of, 310–311
  crossed immunoelectrophoresis of, 445
  on formaldehyde-treated platelets, fluorescence flow cytometry analysis of, 423
  identification of, with anti-GPIb antibody, 420
  immunofluorescent staining of, 422
  labeled, identification of, 418–419
  monoclonal antibodies against, 175, 295–311
    binding studies, 304–307
    fusion, 300–301
    immunization of mice for, 295–297
    iodination of, 303–304
    myeloma cells, 299
    preparation of, 298–301
    purification of, 301–303
      from ascites, 302
      chromatographic, 302–303
      from culture supernatant, 302
    radioimmunoelectrophoresis of, 308
    screening assay, 297
    spleen cells, 299–300
  polyclonal antibodies against, 175
  proteins of, 432
  purification of, by SDS-PAGE electrophoresis, 281
  structure of, 290
Glycoprotein Ib$_\alpha$, 276, 285
  antibodies against, preparation and characterization of, 287–288
  domains, 285, 287–288
  isolation of, 283–284
Glycoprotein Ib$_\beta$, 276, 285
  antibodies against, preparation and characterization of, 287–288
  isolation of, 283–284
Glycoprotein Ibα, of platelet surface membranes, 17–18
Glycoprotein Ibβ, of platelet surface membranes, 17–18
Glycoprotein Iba, molecular variants, 285
Glycoprotein Ic, proteins of, 432

Glycoprotein Icα, of platelet surface membranes, 18
Glycoprotein Icβ, of platelet surface membranes, 18
Glycoprotein IIa
  of platelet surface membranes, 17–18
  proteins of, 432
Glycoprotein IIb, 244–245. *See also* glycoprotein IIb–IIIa complex
  antibodies against, effect on EGTA-dissociated complex, 448
  association state, characterization of, methods for, 253–254
  functionally active, isolation of, 261–263
  identification of, 397
  labeled, identification of, 417–418
  properties of, 252
  proteins of, 432
  reactivity of MAb 3B2 with, demonstration of, by crossed immunoelectrophoresis, 452
  sedimentation of, 255
Glycoprotein IIb$_\alpha$, 244
Glycoprotein IIb$_\beta$, 244
Glycoprotein IIbα, of platelet surface membranes, 17–18
Glycoprotein IIbβ, of platelet surface membranes, 17–18
Glycoprotein IIIa, 244–245. *See also* glycoprotein IIb–IIIa complex
  antibodies against, effect on EGTA-dissociated complex, 448
  association state, characterization of, methods for, 253–254
  autoantibodies against, 438–439
  functionally active, isolation of, 261–263
  identification of, 397
  labeled, identification of, 417–418
  of platelet surface membranes, 17–18
  properties of, 252
  proteins of, 432
  sedimentation of, 255
Glycoprotein IIIb, 278
  labeled, identification of, 418
  of platelet surface membranes, 17–18
Glycoprotein IV, 278
  labeled, identification of, 418
  of platelet surface membranes, 17
  proteins of, 432
Glycoprotein V, 176–180, 278, 288, 438
  detection of, 177–178

proteins of, 432
purification of, 177
  procedure, 178–180
purified
  isoelectric forms, 180
  properties of, 180
  SDS-PAGE of, 179
Glycoprotein V_{fl}, 176
  purified, SDS-PAGE of, 179
Glycoprotein IX, 276, 284–285
  antibodies against, preparation and characterization of, 287–288
  labeled, identification of, 419
  proteins of, 432
Glycoprotein 17, of platelet surface membranes, 17
Glycoprotein Ib–IX complex, 3–4, 135, 230
  binding domain, 232
  binding of vWF via, 232
  characterization and properties of, 284–288
  purification of, 278, 284
    determination of yield, 284
    experimental procedures, 278–284
    by gel filtration in sodium dodecyl sulfate, 282–283
    by Q-Sepharose ion-exchange chromatography, 278–281
    by SDS-polyacrylamide gel electrophoresis, 283
    by thrombin-Sepharose affinity chromatography, 278, 282
    Triton X-114 phase separation method, 278–279
    by wheat germ agglutinin affinity chromatography, 277–278
    by wheat germ agglutinin-Sepharose 4B affinity chromatography, 279–280
  role in platelet adhesion and activation, 276
  solubilization, 276–279
Glycoprotein IIb–IIIa complex, 135, 244–263, 312, 443–444
  activation of, 132
  α subunits, 244
  autoantibodies against, 439
  β subunits, 244
  binding of vWF via, 232

characterization of, 252–257
crossed immunoelectrophoresis of, 445
demonstration of, 443
dissociated, HPLC separation of GPIIb from GPIIIa, 261–263
dissociation of, 261
fibrinogen and vWF binding to, 228–243
fibrinogen-binding properties of, 245, 260–261
function of, 245
  in platelet aggregation, 245
identification of, 420
incorporation into large vesicles, 259
incorporation into small sided vesicles, 258–259
in intact platelets, quantitation of, 255–257
isolated, sucrose density centrifugation of, 254–255
isolation of, 246–251
  by concanavalin A affinity chromatography, 247–248
  by heparin-Sepharose affinity chromatography, 247–248
  methods for, 246
  by Sephacryl S-300 gel chromatography, 249–250
  by wheat germ agglutinin-Sepharose chromatography, 250–251
purification of, steps, 251
purified, 260
radiolabeling of, 251–252
reactivity of MAb 3B2 with, demonstration of, by crossed immunoelectrophoresis, 452
reassociation of, 263
receptor function of, 233
reconstitution into phospholipid vesicles, 257–261
  methods for, 257–258
role of
  in platelet activation, 398
  in platelet function, 245
sedimentation of, 255
Glycoprotein IIb-IIIa–containing vesicles, ligand binding to, 259–261
Glycoprotein IIb-IIIa–dependent antibody, binding of, measurement of, 257
Glycoproteins
  platelet membrane, 17

cross-linking of, 403–412
   experimental procedures, 405–408
   results, 408–410
   labeled
      autoradiography of, 417
      identification of, 417–419
      SDS–polyacrylamide gel electrophoresis of, 417–419
   lactoperoxidase-catalyzed iodination of, 413–415, 418
   nomenclature for, 17
   periodate/sodium boro[^{3}H]hydride labeling of, 413, 415–418
   radiolabeled, SDS gel profile of, effect of BS^3 on, 411
   radiolabeling of, 413
   SDS-polyacrylamide gel profile of, effect of BS^3 on, 409–410
   surface labeling of, 412–419
   proteins of, 432
   Triton X-114-extracted, high-performance liquid chromatography of, 246
Glycoprotein Ib–vWF interaction, 263–275
GMP-140. *See* Granule membrane protein 140
Gradient centrifugation, of platelet homogenates, 24–26
Gramicidin, effect on platelet membranes, 214
Granule membrane protein 140, 4
α granules, 6–7
   assays of, 30–31
   marker enzyme, 39–40
   platelet, 112
   proteins located in, 21
   separation of, by ultracentrifugation, after nitrogen cavitation, 27
GTP, 134
Guanine nucleotides, affinity of α_2-adrenergic receptor for, 186–187

H

H$^+$, affinity of α_2-adrenergic receptor for, 186–187
Haloperidol, 210, 212
Heavy meromyosin, 121

Heparin-Sepharose affinity chromatography, of thrombospondin, 247–248
High-density lipoprotein$_2$, binding to human blood platelets
   characteristics of, 394
   in presence of inhibitory lipoprotein subclasses, 395
High-density lipoprotein$_3$, binding to human blood platelets
   characteristics of, 394
   in presence of inhibitory lipoprotein subclasses, 395
High-density lipoprotein-E, binding to human blood platelets, characteristics of, 394
High-density lipoprotein$_2$ + high-density lipoprotein$_3$, binding to human blood platelets, characteristics of, 394
High-performance liquid chromatography
   of glycocalicin, 291
   separation of GPIIb from GPIIIa by, 261–263
   of Triton X-114-extracted glycoproteins, 246
High-voltage transmission electron microscopy, 457–458
   of fibrinogen–gold labeled platelets, 472–474
Hirudin, 158, 165
Histamine, inhibition of binding of [^{3}H]LSD to human platelet membranes by, 210–211
Histidine-rich glycoprotein, in α granules, 31
HLA alloantigens, 430–431
Human immunodeficiency virus infection, 429
Hybridoma
   IgA–secreting
      growth in serum-free medium, 97
      purification of, 97–98
      purification of gelsolin with, 98–99
      purified, coupling to cyanogen bromide-activated agarose, 98
   IgG–secreting
      growth in serum-free medium, 97
      purification of, 97–98
      purification of gelsolin with, 98–99
      purified, coupling to cyanogen bromide-activated agarose, 98

Hydroxyapatite chromatography
  of gelsolin–actin complexes, 90, 93
  of glycocalicin, 291
N-Hydroxysuccinimidyl-6-(4'-azido-2'-nitrophenylamino)hexanoate, 162, 166
  identification of thrombin–receptor complexes with, 166
N-Hydroxysuccinimidyl-6-(4'-azido-2'-nitrophenylamino)hexanoate thrombin, derivatization, experimental procedure, 162, 164
N-Hydroxysulfosuccinimide sodium salt, synthesis of, 405–406
5-Hydroxytryptamine
  ^{3}H-labeled
    platelet aggregation induced by, inhibition of, IC_{50} values for, 206
    platelet binding, 201–207
      application of, 206
      assay of, 202–203
      components of, calculation of, 203–204
      high-affinity, 207
      kinetics of, 204
      nonspecific, 204–205
    receptor binding, inhibition of, 205–206
    receptor binding sites
      assay of, 206
      inhibition of, 205–206
    uptake of, 204–205
      inhibition of, 205
  inhibition of binding of [^{3}H]LSD to human platelet membranes by, 211–212
  inhibition of [^{3}H]5-HT receptor binding by, 205–206
  platelet receptors, 201–213
  shape change in resuspended human platelets induced by, inhibition of, 212
  tritiated. See 5-Hydroxytryptamine, ^{3}H-labeled
  uptake of, inhibitors of, 210
5-Hydroxytryptamine-2 receptor, 133

I

Idiopathic thrombocytopenia purpura, 428–429

AIDS-related, 429
  autoantibodies in, 438–440
Imipramine
  binding, effect of serotonin on, 218–219
  binding to detergent-solubilized membranes, calcium phosphate assay of, 220–221
  binding to digitonin-solubilized membranes, gel-filtration assay of, 221
  binding to membrane vesicles, filtration assay of, 220
  inhibition of binding of [^{3}H]LSD to human platelet membranes by, 211–212
Imipramine-binding assays, 219–221
Immune staining, of platelet proteins, 436
Immunoaffinity chromatography
  of gelsolin, 95–99
  of gelsolin–actin complexes, 99
Immunoblotting, 430
  of platelet antigens, 428–440
  of platelet surface proteins, technique for, 431–432, 434–435
Immunofluorescence microscopy, 125–126
Immunofluorescent staining, 422
  of GPIb, 422
Immunoglobulin
  adsorbed to colloidal gold, behavior of, 462–463
  heterologous, conjugation to colloidal gold, 468–469
  nonspecific binding to platelets, 437
Immunoglobulin A–secreting hybridoma growth in serum-free medium, 97
  purification of, 97–98
  purification of gelsolin with, 98–99
  purified, coupling to cyanogen bromide-activated agarose, 98
Immunoglobulin complement, platelet-associated, identification of, 420
Immunoglobulin G–secreting hybridoma growth in serum-free medium, 97
  purification of, 97–98
  purification of gelsolin with, 98–99
  purified, coupling to cyanogen bromide-activated agarose, 98
Immunolabeling, of platelet receptors, anti-IIb/IIIa–gold procedure, 468–470
Immunoprecipitation, 430
Indomethacin, 182
Inositol trisphosphate, 112

SUBJECT INDEX

Insulin
  binding to platelets, 399
    assay of, 400–401
      methods for, 399–401
      specificity of, 401
  catfish, binding to platelets, 401
  FITC-labeled, 399
  iodination of, 400
  platelet receptor for, 398–403
    significance of, 402–403
    site number and binding affinity of, 402
  porcine, binding to platelets, 401
Integrin receptors, 3, 135
Iodination
  of insulin, 400
  lactoperoxidase-catalyzed, of platelet membrane glycoproteins, 418
  of platelets, during thrombin-induced aggregation, 416
Iodogen labeling, of platelet surface proteins, 413
Ion-affinity chromatography, of gelsolin, 98
Ion-exchange chromatography
  of gelsolin, 95–96
  of gelsolin–actin complexes, 92–93
Ionophore A23187, 237
Isoprenaline, inhibition of binding of [^{3}H]LSD to human platelet membranes by, 211
ITP. See Idiopathic thrombocytopenia purpura

K

Ketanserin, inhibition of binding of [^{3}H]LSD to human platelet membranes by, 211
Kininogen, high molecular weight
  binding to platelets
    assays of
      controls for, 373–374
      methodology for, 372–373
      data, 374–381
      saturation data, analysis of, 376–381
      site specificity, 376
    binding to thrombin-stimulated platelets, 381–382
      LIGAND-derived binding constants for, 380

    histidine-rich region, 370–371
  ^{125}I-labeled, binding to platelets, 374–375
    competition with proteins for, 376–377
    as function of zinc and calcium ionic concentrations, 375–376
    high molecular weight kininogen concentration dependence of, 378–379
  platelet receptor for, 369–382
  purification of, 370

L

Lactate dehydrogenase, as marker enzyme for cytosol, 28
Lactoperoxidase-catalyzed iodination, of platelet membrane glycoproteins, 413–415, 418
Laminin receptor, 135
Laurell immunoelectrophoresis, 446
Lectin, iodinated, labeling of platelets with, 10
Lectin affinity chromatography, 246
Leucine aminopeptidase, in human platelet surface and intracellular membranes, 16
Leupeptin
  as protease inhibitor, 23
  in protease inhibitor cocktail, for platelet suspensions, 10
LIGAND-derived binding constants, for high molecular weight kininogen binding to thrombin-stimulated platelets, 380
Light microscopy
  of fibrinogen–gold labeled platelets, 472
  video-enhanced, of gold labeling of platelets, 471–475
Lipoprotein-binding proteins, identification of, 397
Lipoproteins. See also Low-density lipoprotein
  binding to human platelets
    assay of
      fluorescence method, 385–389
      methods for, 383–398
      radioactive method, 385
    characteristics of, 383–398
    number and affinity of sites, 393–395
    specificity of, 393
    temperature dependence of, 395

blood platelet receptor for, physiological function of, 396–398
fluorescence labeling of, 384–385
^{125}I-labeling of, 384
plasma, binding to human blood platelets, characteristics of, 394
platelet-bound, quantitation of, 389–391
platelet receptor for, 383–398
isolation of, 396
preparation of, 383–384
receptor binding studies involving, general considerations and problems in, 389–393
uptake of, 396
Lipoxygenase, in human platelet surface and intracellular membranes, 16
Local anesthetics, effect on platelet cytoskeleton, 76
Low-density lipoprotein
binding to human platelets
characteristics of, 394
in presence of inhibitory lipoprotein subclasses, 395
temperature dependence of, 395
binding to platelets, 392
assays of, 392–393
fluorescence assay of, 387
binding to proteins, relationship with platelet reactivity, 398
FITC-labeled, fluorescence assay of, 388–389
^{125}I-labeled
binding to blood platelets, time dependence of, 390–391
platelet-bound, dissociation of, kinetics of, 390–391
platelet receptor for, isolation of, 396
Lysergic acid diethylamide, 212
^{3}H-labeled
binding in human frontal cortex, inhibition of, 212
binding to human platelet membranes, 201, 208–213
application of, 210–213
assay of, 208–209
characteristics of, 209–210
inhibition of, 210–212
δ-Lysergic acid diethylamide, inhibition of binding of [^{3}H]LSD to human platelet membranes by, 211–212

Lysosomes, platelet, 6–7, 112
separation of, by ultracentrifugation, after nitrogen cavitation, 27

M

Macroglycopeptide, 276, 289
Macromolecules, adsorbed to colloidal gold, behavior of, 462–464
Magnesium
affinity of α_2-adrenergic receptor for, 186–187
in dense granules, 41
released by thrombin, 41
in whole platelets, 41
Masses, platelet, 112
Megakaryocyte, 3
Mescaline, inhibition of binding of [^{3}H]LSD to human platelet membranes by, 211
Metergoline, inhibition of binding of [^{3}H]LSD to human platelet membranes by, 211–212
Methiothepin, inhibition of binding of [^{3}H]LSD to human platelet membranes by, 211–212
5-Methoxydimethyltryptamine, inhibition of binding of [^{3}H]LSD to human platelet membranes by, 210–212
2-Methylthioadenosine diphosphate. See 2-Methylthioadenosine [β-^{32}P]diphosphate
2-Methylthioadenosine [β-^{32}P]diphosphate, 137–142
assay of, 140
bioassay of, 140
platelet binding sites, 141
platelet binding studies, 140–141
synthesis of, 137–140
precautions, 140
Methysergide
inhibition of binding of [^{3}H]LSD to human platelet membranes by, 211–212
inhibition of [^{3}H]5-HT receptor binding by, 205–206
MgATP, effect on platelet cytoskeleton, 62
Mianserin, 212
inhibition of [^{3}H]5-HT receptor binding by, 205–206

Microscopy. *See also* Electron microscopy
  immunofluorescence, 125–126
  light
    of fibrinogen–gold labeled platelets, 472
    video-enhanced, of gold labeling of platelets, 471–475
Microtubule-associated polypeptides, of platelet surface membranes, 17
Microtubules, platelet, 112
Microviscosity, of human platelet surface and intracellular membranes, 16
Mitochondria, platelet, 6–7, 112
  enzyme markers, 28, 30, 39–40
  separation of, by ultracentrifugation, after nitrogen cavitation, 27
Molecular sensors, 134
Monoamine oxidase assay, for localization of mitochondria, 28, 30
Monoclonal antibody
  A_2A_9, 257
  AP2, 257
  AP3, 257
  3B2, reactivity with GPIIb–GPIIIa complex and GPIIb, crossed immunoelectrophoresis demonstration of, 452
  7E3, 257
  10E5, 257
    conjugation to colloidal gold, 469–470
    purification of, 469
  to GPIb, 175, 295–311
    binding studies, 304–307
    fusion, 300–301
    immunization of mice for, 295–297
    iodination of, 303–304
    myeloma cells, 299
    preparation, 298–301
    purification of, 301–303
      from ascites, 302
      chromatographic, 302–303
      from culture supernatant, 302
    radioimmunoelectrophoresis of, 308
    screening assay, 297
    spleen cells, 299–300
  murine, purification of gelsolin–actin complexes with, 97, 99
  PAC-1, 257
  to platelet antigens, screening of, 420
  TAB, 257

Monoclonal antibody affinity chromatography, 246
Monoclonal antibody 10E5–gold conjugate, preparation of, 469–470
Monoclonal antibody–gold labeling, control preparations for, 477
Muscarinic cholinergic receptor, 133
Myoglobin, 405
Myosin, 43, 60
  effect on actin depolymerization, 74–76
  immunofluorescence microscopy of, 126
  platelet, 78–88
    assay of ATPase activities of, 81–82
    dephosphorylation of, 82
    phosphate assay of, 82
    phosphorylation of, 78, 82–83
    purification of, 79–80
  of platelet membrane, 18, 143
  purification of, effect of platelet activation on, 64
Myosin light chain
  phosphorylation of, measurement of, 83–85
  phosphorylation sites, one-dimensional phosphopeptide mapping of, 85–88
Myosin light chain kinase, 60

N

Na^+, affinity of α_2-adrenergic receptor for, 186–187
NADH–cytochrome c reductase
  in human platelet surface and intracellular membranes, 16
  rotenone-insensitive, as marker for endoplasmic reticulum, 28–30
Nagarse, treatment of platelets with, 38
NATP. *See* Neonatal alloimmune thrombocytopenia
Neonatal alloimmune thrombocytopenia, 428, 430
Neuraminidase, treatment of whole platelets with, 10–12
Neuraminidase/galactose oxidase/sodium boro[^{3}H]hydride labeling, of platelet surface proteins, 413
Nitrogen cavitation, platelet homogenization by, 21–32
  advantages of, 31–32

p-Nitrophenylphosphatase, in platelet plasma membrane localization, 28
Nomarski optics, 110
Noradrenaline, inhibition of binding of [³H]LSD to human platelet membranes by, 211
Norepinephrine, ³H-labeled, 185–186
Nortriptyline, inhibition of [³H]5-HT receptor binding by, 205
5'-Nucleotidase, in human platelet surface and intracellular membranes, 16
Nucleotides, in dense granules, 36

O

Open canalicular membrane system, 4, 6–7
Opsins, 133
Orosomucoid, specificity, 261
Ovalbumin
 adsorbed to colloidal gold, behavior of, 462
 competition for binding of ¹²⁵I-labeled high molecular weight kininogen with, 377
Ovomucoid trypsin inhibitor, competition for binding of ¹²⁵I-labeled high molecular weight kininogen with, 377

P

PAGEM, 4
Peanut agglutinin chromatography, 278
Pepstatin, in protease inhibitor cocktail, for platelet suspensions, 10
Periodate/sodium boro[³H]hydride labeling
 of cell surface glycoproteins, 418
 of platelet membrane glycoproteins, 413, 415–417
Peroxisomes, platelet, 112
Phase-contrast systems, 110
Phenoxybenzamine, ³H-labeled, 185
Phentolamine, ³H-labeled, 184
 D-Phenylalanyl-L-prolyl-L-arginine chloromethyl ketone–thrombin, 157
Phorbol myristate acetate, 237
Phosphatidylcholine, in human platelet surface and intracellular membranes, 16

Phosphatidylethanolamine, in human platelet surface and intracellular membranes, 16
Phosphatidylinositol
 in human platelet surface and intracellular membranes, 16
 metabolism of, in platelets, 20
Phosphatidylserine, in human platelet surface and intracellular membranes, 16
Phospholipase A_2
 activation, epinephrine or ADP stimulation of, 182
 in platelet membranes, 16, 19
Phospholipids, in human platelet membranes, 16
Phospholipid vesicles, binding interactions, 317
P_i
 in dense granules, 41
 released by thrombin, 41
 in whole platelets, 41
Pirenperone, inhibition of binding of [³H]LSD to human platelet membranes by, 211
Pizotifen
 inhibition of binding of [³H]LSD to human platelet membranes by, 211
 inhibition of [³H]5-HT receptor binding by, 205–207
Plasma
 platelet-rich
  platelet recovery in, 9
  preparation of, 9, 188
 removal from platelets, 145
Plasma fibrinogen, purified, quantitation of, by CIE of rabbit anti-human platelet membrane antibody with plasma fibrinogen, 450
Plasma lipoproteins, binding to human blood platelets, characteristics of, 394
Plasma membranes
 isolated, characterization of, 32–33
 isolation of, by glycerol lysis, 32–36
 platelet, localization, 28–29
Plasma membrane vesicles, 216–221
 imipramine-binding assays, 219–221
 serotonin transport assay, 216–218
Plasma proteins, separation of platelets from, 237

SUBJECT INDEX

Platelet-activating factor, 131, 133
  binding to intact platelets, assay, 225–226
  binding to platelet membranes, assay, 226–227
  binding to platelets, 224–228
    assay method, 224–228
    quantification and characterization of, 227–228
  platelet receptor for, 131, 133–134
Platelet adhesion, 3
Platelet agglutination, 430
Platelet aggregates, forward-scatter analysis of, 425
Platelet aggregation, 5
  collagen-induced
    effects of DTSSP, BS^3, and SSP on, 410
    inhibition of, 412
  effect of DTSSP or BS^3 on, 408–409
  5-HT-induced, inhibition of, IC$_{50}$ values for, 206
  thrombin-induced, iodination during, 416
Platelet agonists, 131, 236
Platelet antigens
  fluorescence flow cytometry analysis of, 420–427
  immunoblotting of, 428–440
  monoclonal and polyclonal antibodies against, screening of, 420
Platelet concentrates, 100
Platelet contractile apparatus, 3, 62
  ultrastructural analysis of, 109–127
    by immunocytochemistry, 125–127
    methods for, 114–127
    overview, 110
    thin-section techniques for, 114–121
    whole-mount method for, 121–125
Platelet-derived growth factor
  in α granules, 31
  platelet storage site for, 21
Platelet extracts, isolation of, 89–91
Platelet factor 3, 135
Platelet factor 4
  crossed immunoelectrophoresis of, 445
  in α granules, 30, 39–40
  platelet storage site for, 21
Platelet homogenates, gradient centrifugation of, 24–26
Platelet lysates
  preparation of, 246–247
  Triton X-100, actin filament content, methods of measuring, 54–58
  washes, 189
Platelet membrane glycoproteins, surface labeling of, 412–419
Platelet membrane polypeptides, 17
  that incorporate [^{3}H]SBA, SDS-PAGE of, 152–153
Platelet membranes, 3–4
  anatomy and structural physiology of, 112
  antibody against, preparation of, for crossed immunoelectrophoresis, 454
  digitonin solubilization of, 189–191, 193–194
  isolation of platelets and preparation of washed lysates for, 187–189
  digitonin-solubilized, 219–220
  exchange into 0.1% digitonin, 191–192
  fractions
    isolation of, 4
    protein analyses, 13, 59
  human
    assay for intrinsic biologic activity, 443–446
    crossed immunoelectrophoresis of, 440–455
      after absorption of rabbit antiplatelet membrane antibody, 453
      with concanavalin A, 451
  human albumin with, crossed immunoelectrophoresis of, coelectrophoresis of, 445
  intracellular, 6–7
    analytical and enzymatic characterization of, 15–20
    components of, 17–18
  isolated, by glycerol lysis
    purity of, 36
    yield of, 36
  isolation of, 5–20, 31
    by glycerol lysis, 32–36
    for studies of Ca^{2+} transport, 13
    for studies of transport systems, 13
  mixed fraction of
    free flow electrophoresis of, 11–15
    preparation of, 8–12
  of patient with Glanzmann's thrombasthenia, crossed immunoelectrophoresis of, 446

plasma, localization of, 28–29
potential change, fluorescence measurements of, 173–174
preparation of, 208
  for crossed immunoelectrophoresis, 454
subfractionation of, by free flow electrophoresis, 8
surface
  analytical and enzymatic characterization of, 15–20
  subfractionation of, on free flow electrophoresis, 13–15
  two-dimensional SDS–polyacrylamide/isoelectric focusing gel preparations of, 15–19
Platelet membrane skeleton, 42
Platelet particulates, synthesis of, 187–189
Platelet pool, circulating, analytical and functional heterogeneity in, 6–8
Platelet receptors, 131–136
  action of, 142
  for adenosine diphosphate, 132, 134
    FSBA interaction with, 143–155
    binding of fibrinogen to, 232–233
      measurement of, 148–150
    binding of von Willebrand factor to, 232–233
  for collagen, 135–136, 245
  colloidal gold labeling of, 456–479
  for factor Va, at bovine platelet surface, component E as part of, 351
  for factor Xa, 329–360
  for factor XIa, 361–369
  for fibrinogen, 132, 245, 370
    identification of, by colloidal gold labeling, materials and methods for, 465–468
  for fibronectin, 135, 245
  for high molecular weight kininogen, 369–382
  human, α_2-adrenergic, 181–200
  for 5-hydroxytryptamine, 201–213
  immunolabeling of, anti-IIb/IIIa–gold procedure, 468–470
  for insulin, 398–403
  for integrin, 3, 135
  for lipoproteins, 383–398
  nonintegrin, 135

  for platelet-activating factor, 131, 133–134
  reconstituted system, 261
  for serotonin, 131, 133
  for thrombin, 155–176
    activation by, 131
  for vitronectin, 135, 245
Platelet receptor–thrombin complexes, approximate molecular size and reduced components of, 160
Platelet-rich plasma
  platelet recovery in, 9
  preparation of, 9, 188
Platelets
  activated, 4–5
    anatomy and structural physiology of, 112–113
    suspensions of, preparation of, 47–48
  activation, 5, 112–114, 131
    events in, 5
    morphological changes in, 4–5
    by thrombin receptor, 131
  anatomy and structural physiology of, 110–114
  antibody against, 428
  bovine, binding interactions, 317
  centrifugation of, 45–47, 77
  colchicine-binding assay of, 119
  crossed immunoelectrophoresis of, 442, 444–445
  dendritic, whole-mounted, 125
  discoid, anatomy and structural physiology of, 110–111
  fixation of, 272–273
  fluorescence flow cytometry analysis of, 420–427
  fractionation procedure for, 38–39
  function of, 3
  granular organelles, 6–7
  homogenization
    with French pressure cell, 37–39
    by nitrogen cavitation, technique, 23–25
    problems with, 37
  immune destruction of, syndromes of, 428–431
  iodination, during thrombin-induced aggregation, 416
  isolation, 21–23, 43–48, 60–61, 145, 187–189, 278, 354, 399–400, 432–433

labeling of, 10
  with gold, 470–478
    electron microscopy of, 478–479
  with serotonin, 36–37
life span of, 3
lysed, concanavalin A affinity chromatography of, 247–248
masses, 112
membrane-mediated interactions of, 4
morphological changes, 58–59
nonactivated, 4–5
organelle zone, anatomy and structural physiology of, 112
peripheral zone, anatomy and structural physiology of, 111–112
plasmatic halo on, 8
preparation of, 169, 202, 225
  for flow cytometry analysis, 421–422
  for gold labeling, 470–478
production of, 3
protein composition of, 61
secretion of
  α_2-adrenergic receptor-induced, role of effector systems in, 182
  epinephrine-induced, 181–182
separation from plasma proteins, 237
shape of, 3–6
size of, 3–4
sol–gel zone, anatomy and structural physiology of, 112
sonication procedure for, 10–12
sorbitol density gradient fractionation, 11–12
spreading of, 3
structural analysis of, 5
structural and functional organization of, 3–5
subcellular fractionation
  difficulties of, 5–6
  nitrogen cavitation technique in, 21–32
subcellular organelles
  assays, 27–31
  markers for, 38–40
  separated by ultracentrifugation after nitrogen cavitation, 26–27
subfractionation, on sucrose density gradient, 39–40
surface labeling of, effect on association of fibrinogen with solubilized platelet membrane after CIE, 453

volume of, 3
washed, preparation of, 270–272, 383
washing of, 37, 278, 354, 371
whole, neuraminidase treatment of, 10–12
whole-mounted
  detergent extraction, 122
  drying techniques, 122–123
  negative stain, 124, 126
Platelet serotonin transporter, 213–224
  reversibility of, 217
Platelet surface antigens, relative surface location of, 448–451
Platelet surface proteins, 431–436
  amphiphilic, recognition of, 451–454
  autoradiography of, 437
  immunoblotting of
    electrophoresis for, 433–436
    immune staining, 436
    results, 437–440
    technique, 431–432, 434–435
    transfer to nitrocellulose, 436
  incorporation of [^{3}H]SBA into, determination of, 150–154
  transglutaminase labeling of, 413
Polyclonal antibody
  to GPIb, 175
  to platelet antigens, screening of, 420
Polypeptides, platelet membrane, 17
  that incorporate [^{3}H]SBA, SDS–PAGE of, 152–153
Posttransfusion purpura, 428, 430
PP$_i$
  in dense granules, 41
  released by thrombin, 41
  in whole platelets, 41
Prazosin, 181
Prekallikrein, competition for binding of ^{125}I-labeled high molecular weight kininogen with, 377
Procoagulant, 135
Profilin, 60
Prostacyclin, 427
Protease, 330
Protease inhibition
  in platelet isolation, 23
  in platelet membrane isolation, 33
Protease inhibitor cocktail, for platelet suspensions, 10
Protein C, activated, 329

cleavages in platelet-bound factor Va induced by, 347
complex formation with factor Xa and platelet-bound factor Va, proteolysis of platelet-bound factor Va by, 345–347
Protein–gold conjugates
concentration-variable isotherms for, 464–465
pH-variable isotherms for, 464
Protein kinase, of platelet membranes, 20
Protein P235, 3
human platelet, 99–109
assay of, 101
degradation of, 99–101
extraction of, 102–103
molecular properties of, 108
purification of, 100–109
by DEAE-cellulose chromatography, 102–105
by gel filtration followed by phosphocellulose, 107–108
by phosphocellulose followed by gel filtration, 105–107
by SDS–polyacrylamide gel electrophoresis, 101–104
Proteins
adsorbed to colloidal gold, behavior of, 462–464
amphiphilic, recognition of, 451–454
conjugation of, to colloidal gold, 464–465
lipoprotein-binding, identification of, 397
plasma, separation of platelets from, 237
platelet surface, 431–436
amphiphilic, recognition of, 451–454
autoradiography of, 437
immunoblotting of, results, 437–440
incorporation of [^{3}H]SBA into, determination of, 150–154
transglutaminase labeling of, 413
Prothrombin, 316
activation of
as catalyzed by prothrombinase complex at platelet surface, model of, 359
DAPA-monitored experiments, 342–344
initial rate of, calculation of, 323–326
prothrombinase-catalyzed, rationalization of, 328

binding distribution of, calculation of, 323–326
binding interactions, 317
binding kinetics of, equations for, 319–323
bulk and local concentrations of, calculation of, 323–326
human, purification of, 353
Prothrombinase, 136
assembly at platelet surface, mediation of, by component E, subunit of factor Va, 348–349
binding distribution of, calculation of, 323–326
binding kinetics of, equations for, 319–323
binding on vesicles
parameters of, 324
simulation of
kinetics required for, 324
parameters and solution to binding equations on, 326
bulk and local concentrations of, calculation of, 323–326
mediation of, 357–358
model of, 359
model of, 316–328
attributes, uses, and limitations of, 327–328
BASIC subroutine for estimation of B (fraction of bound substrate) by bisection algorithm for, 325
concepts of, 318–319
magnitude of δ, 326–327
properties of, rationalization of, 327–328
P-selectin, in α granules, 31
Pseudopods, 3–4
PTP. See Posttransfusion purpura
Purinergic P_2y receptor, 132

Q

Q-Sepharose ion-exchange chromatography, of GPIb-IX, 278–279
Quinidine antibodies, 429
Quinidine-dependent antibodies, 429
Quinidine purpura, antibodies associated with, 438
Quipazine, inhibition of binding of [^{3}H]LSD to human platelet membranes by, 210–211

R

Radioimmunoassay, of factor XI, 362
Radioimmunoelectrophoresis, of monoclonal antibody against GPIb, 308
Rauwolscine, ^{3}H-labeled, as radioligand, 184–185, 191
Ristocetin, 264

S

Scanning electron microscopy, 457
  low-voltage high-resolution field emission, 457–458
  of fibrinogen–gold labeled platelets, 472–473
Scatchard analysis, of saturation binding data of high molecular weight kininogen, 378–380
  least-squares fit, 379
  nonlinear curve fit, 380
SDS–agarose gel electrophoresis, of radiolabeled vWF, 270
SDS–polyacrylamide gel electrophoresis
  of GPIb, 281
  of GPIb-IX, 282–283
  of GPIIb-IIIa complex, 249–250
  of GPV and GPV$_{fl}$, 179
  of P235, 101–104
  of platelet membrane glycoproteins, 417–419
    effect of BS^3 on, 409–411
  of platelet membrane polypeptide that incorporates [^{3}H]SBA, 152–153
  of proteins, 433–436
  of thrombin–receptor complexes, 172
Sedimentation assay
  of actin filaments, 56–58
  of gelsolin, 88
Sensors, 134
Sephacryl S-300 chromatography, of GPIIb-IIIa complex, 249–250
Sephadex G-200 chromatography, of gelsolin–actin complexes, 89, 92–93
Sequence RGDF (α95–98), 230
Sequence RGDS (α572–575), 230
Serotonin
  bound to human platelets, uptake mechanisms, 133
  in dense granules, 36, 41
  effect on imipramine binding, 218–219
  as marker for dense granules, 39–40
  platelet labeling with, 36–37
  platelet receptor, 131, 133
  platelet transporter, 213–224
    reversibility of, 217
  released by thrombin, 41
  transport into dense granule membrane vesicles, assay of, 222–224
  transport into intact platelets, assay of, 214–216
  transport into plasma membrane vesicles
    assay of, 216–218
    electrogenicity and stoichiometry, 217
    ionic requirements, 216–217
  in whole platelets, 41
Sodium citrate, effect on CIE pattern from platelet membranes, 447
Sonication, purification of glycocalicin by, 290
Sorbitol density gradient fractionation, of platelets, 11–12
Soybean trypsin inhibitor, competition for binding of ^{125}I-labeled high molecular weight kininogen with, 377
Spectrin, 42
Sphingomyelin, in human platelet membranes, 16
Spiperone, inhibition of binding of [^{3}H]LSD to human platelet membranes by, 210–212
SQ 29,538, 182
SSP. *See* Sulfosuccinimidyl propionate
Storage granules, platelet, 4
  isolation of, 4
Storage pool deficiency, 42
Succinate–cytochrome c reductase, assay, for localization of mitochondria, 28, 30
Sucrose density centrifugation, of GPIIb-IIIa, 254–255
Sulfonylbenzoyladenosine, ^{3}H-labeled
  incorporation into platelet surface protein, determination of, 150–154
  platelet membrane polypeptide that incorporates, SDS–PAGE of, 152–153
Sulfosuccinimidyl active esters, reaction with amino groups to form amide bonds, with loss of N-hydroxysulfosuccinimide, 405
Sulfosuccinimidyl propionate, effect on collagen-induced platelet aggregation, 410

Surface-connected open canalicular system, 4, 6–7
Surface labeling
  of platelet membrane glycoproteins, 412–419
  of platelets, effect on association of fibrinogen with solubilized platelet membrane after CIE, 453
Systemic lupus erythematosus, 429

T

Tachykin receptors, 133
Talin, 3, 100
  of platelet surface membranes, 17–18
Taxol, effect on platelets, 119
Tenase complex, 136
Thiocarbohydrizide, 118
Thrombin, 131, 136, 236
  active-site blocking of, 168
  derivatization and cross-linking, experimental procedure, 162–165
  derivatization of, by reductive amination, 169–170
  effect of preexposure to TLCK-thrombin, 158
  effect on platelet cytoskeleton, 64
  effect on platelets, 113–114
  platelet membrane binding site, 174–175
  platelet receptors for. See Thrombin receptors
  purification of, 168
  role of, 229
α-Thrombin, 156
  derivatization and cross-linking, experimental procedure, 162–163
  platelet receptors for
    determination of number of thrombin molecules covalently bound to, 172
    high-affinity, nature of, 174–175
    identification of, 159, 165–166
    number and affinity of, 157
    radiolabeling of, by reductive methylation, 170–171
Thrombin–platelet coupling, 171–172
Thrombin–receptor complex
  gel filtration and isolation of, 172–173
  identification of, 165–166
  reduction and alkylation of, 173
  SDS–PAGE electrophoresis of, 172
Thrombin receptors, 155–176
  activation by, 131
  derivatization and cross-linking, methods, 167–175
  detection of, 158–159
    with photoreactive ligands, 159–160
  determination of number of thrombin molecules covalently bound to, 172
  experimental procedures, 162–165
  high-affinity, 156–157, 159, 161
    determination of, 159
    nature of, 174–175
    structure of, 166–167
  low-affinity, 156–157, 159, 161
    structure of, 167
  properties of, 175–176
  role in activation of GPIIb–IIIa, 132
  structure of, 166–167
Thrombin-Sepharose affinity chromatography, of GPIb-IX, 278, 282
Thrombin-Sepharose 4B, preparation of, 282
Thrombin-Sepharose 4B chromatography, of glycocalicin, 291
Thrombocytopenia
  autoimmune. See Idiopathic thrombocytopenia purpura
  drug-induced, 429, 438
  heparin-associated, 430, 438
  immune, 429
  immune-mediated, 428
  neonatal alloimmune, 428, 430
β-Thromboglobulin, platelet storage site for, 21
Thrombospondin
  crossed immunoelectrophoresis of, 445
  in α granules, 31
  heparin-Sepharose affinity chromatography of, 247–248
  labeling of, 419
Thromboxane A_2, 131
Thromboxane synthase, in platelet membranes, 16, 19
TLCK–thrombin. See Tosyllysyl chloromethyl ketone–thrombin
Tosyllysyl chloromethyl ketone–thrombin
  derivatization and cross-linking, experimental procedure, 162–163

effect of preexposure to, 158
identification of, 165–166
platelet receptor for
  determination of, 159
  determination of number of thrombin molecules covalently bound to, 172
  number and affinity of, 156–157
  radiolabeling of, by reductive methylation, 170–171
Transferrin, specificity, 261
Transglutaminase labeling, of platelet surface proteins, 413
Transmission electron microscopy, 457
  high-voltage, 457–458
    of fibrinogen–gold labeled platelets, 472–474
Tritium labeling, 162
Triton X-100, platelet lysis with, 43
Triton X-114-extracted glycoproteins, high-performance liquid chromatography of, 246
Triton X-100 platelet lysates, actin filament content, methods of measuring, 54–58
Triton X-114 separation, of GPIb-IX, 278–279
Tropomyosin, 60, 121
  bound to F-actin, 118
  immunofluorescence microscopy of, 126
  of platelet surface membranes, 17–18
Tubulin, of platelet intracellular membranes, 19

U

UK14,304, ^{3}H-labeled, 185–186
Ultracentrifugation, of platelet subcellular organelles, after nitrogen cavitation, 26–27
Unit membrane, 112
UTP, 134

V

Very low-density lipoprotein, binding to human blood platelets, in presence of inhibitory lipoprotein subclasses, 395
Video-enhanced light microscopy, of gold labeling of platelets, 471–475

Vinculin
  immunofluorescence microscopy of, 126
  of platelet surface membranes, 18
Vitronectin
  GPIIb–IIIa receptor, 135
  platelet receptor, 245
VLA-2, 135
VLA-6, 135
von Willebrand factor
  binding domains, 232
  binding to GPIb complex, 263–275
  binding to GPIIb–IIIa complex, 228–243
  binding to platelet receptors, 232–233
  binding to platelets, 245
    mechanisms of, 230–232
      ristocetin-dependent, 232
      thrombin-, ADP-dependent, 232
    ristocetin-induced, 273–275
    thrombin-induced, 232
  binding to platelet surface, study of, 420
  crossed immunoelectrophoresis of, 445
  function of, 263
    mechanisms underlying, 263
  GPIIb–IIIa receptor, 135
  ^{125}I-labeled, binding to platelets, 238–243
    concentration-dependent, 239
    correlation with platelet aggregation, 242
    inhibition by dodecapeptide, 241
    inhibition by fibrinogen, 240
    measurement of, 238
    ristocetin-induced, 274
  inhibition of ^{125}I-labeled fibrinogen binding to human platelets by, 240
  isolation of, 233–234
  labeling of, 269–270
  multimeric structure of, 267, 270
  platelet binding sites for, 263
    GPIb-related, 264
  platelet receptor, 245
  platelet storage site for, 21
  purification of, 234–235, 265–268
  purified
    binding to human platelets
      assay, 237
      measurement of, 233
      cautions, 234

materials and methods for, 233–237
  iodination of, 235–237
  radiolabeled, SDS-agarose gel electrophoresis of, 267, 270
  ristocetin cofactor activity of, 264
von Willebrand's disease, 286–287

W

Wheat germ agglutinin affinity chromatography, of GPIb-IX, 277–278
Wheat germ agglutinin-Sepharose 4B affinity chromatography, of GPIb-IX, 279–280
Wheat germ agglutinin-Sepharose chromatography, of GPIIb-IIIa complex, 250–251
Wheat germ lectin-Sepharose 6-MB chromatography, of glycocalicin, 291
White's saline, stock solutions for, 115
Whole-mounted platelets
  detergent extraction, 122
  drying techniques, 122–123
  negative stain, 124, 126

Y

Yeast pheromone receptors, 133
Yohimbine, 181
  ^{3}H-labeled, as radioligand, 184–185, 191, 194
Yohimbine-agarose affinity resin, 200
  synthesis of, 194–197
Yohimbine-agarose chromatography
  of α_2-adrenergic receptors, 197–199
  from metabolically labeled cells, effectiveness of, 199–200
  of α_2-adrenergic receptor subtypes, 199
[^{3}H]Yohimbine-binding assay, of digitonin-solubilized platelet α_2-adrenergic receptors, 192

Z

Zetterquist's veronal buffer, 115
Zinc ions, role in interaction of high molecular weight kininogen with platelet surface, 371
Zwitterion, adsorbed to colloidal gold, behavior of, 463